WIT Transactions

Transactions Editor

Carlos Brebbia
Wessex Institute of Technology
Ashurst Lodge, Ashurst
Southampton SO40 7AA, UK
Email: carlos@wessex.ac.uk

Editorial Board

Ecosystems and Sustainable Development VIII

Editors:

Y. Villacampa
Universidad de Alicante, Spain

C.A. Brebbia
Wessex Institute of Technology, UK

Ministerio de Ciencia e Innovación (CTM2010-10878-E)

Universitat d'Alacant
Universidad de Alicante

Ciudad de Alicante
SEDE UNIVERSITARIA
Universitat d'Alacant

WIT*PRESS* Southampton, Boston

Editors:

Y. Villacampa
Universidad de Alicante, Spain

C.A. Brebbia
Wessex Institute of Technology, UK

Published by

WIT Press
Ashurst Lodge, Ashurst, Southampton, SO40 7AA, UK
Tel: 44 (0) 238 029 3223; Fax: 44 (0) 238 029 2853
E-Mail: witpress@witpress.com
http://www.witpress.com

For USA, Canada and Mexico

Computational Mechanics Inc
25 Bridge Street, Billerica, MA 01821, USA
Tel: 978 667 5841; Fax: 978 667 7582
E-Mail: infousa@witpress.com
http://www.witpress.com

British Library Cataloguing-in-Publication Data

A Catalogue record for this book is available
from the British Library

ISBN: 978-1-84564-510-6
ISSN: 1746-448X (print)
ISSN: 1743-3541 (on-line)

The texts of the papers in this volume were set individually by the authors or under their supervision. Only minor corrections to the text may have been carried out by the publisher.

Printed in Great Britain by Martins the Printer.

This book is dedicated to honour the memory of our colleague and friend, Professor Enzo Tiezzi

Preface

This book is dedicated to honour the memory of our colleague and friend, Professor Enzo Tiezzi, who died on 25th June 2010, still at the peak of his career. The Volume contains most of the papers presented at the 8th International Conference on Ecosystems and Sustainable Development (ECOSUD), of which Enzo was to be Co-Chairman.

The ECOSUD Meeting has been held in many different locations around the world with the active participation of Enzo, who was a major personality in the ecological systems field, and recognised as one of Prigogine's disciples.

Enzo was Professor of Chemistry at the University of Siena but his interests were wide ranging, from poetry to chemistry, bringing new ideas and deeper insight to all fields of knowledge. His main scientific contributions were in the topics of evolutionary processes and ecosystems.

Enzo was closely associated to the Wessex Institute of Technology and its publishing arm, WIT Press, which has been fortunate to have published several scientific books authored by Enzo as well as one of his novels, which gives an idea of the breadth of his interests. He was co-editor of many WIT Press books and a main Editor of the International Journal of Design and Nature, also closely associated to the ECOSUD Conference.

The Prigogine Gold Medal Award, instituted by the University of Siena and the Wessex Institute of Technology, following the death of Professor Ilya Prigogine in 2003, has been given every year to the best researcher working in the field of ecosystems modeling. Enzo received the award in 2005 at the Aula Magna of the University of Cadiz. His life continues in the work of the many researchers he has guided over his long and prolific career, some of them authors of papers contained in this book.

The Prigogine Medal 2011 was awarded to Professor Larissa Brizhik from the Bogolyubov Institute for Theoretical Physics in the Ukraine where she is Head of the Department of Nonlinear Condensed Materials Physics. The full text of her special lecture "On the role and impact of electromagnetic files in ecosystems" can be seen in the Volume 6, No 3 of the International Journal of Design & Nature and Ecodynamics, 2011.

The ECOSUD Conference was organised by the University of Alicante, the Wessex Institute of Technology and the University of Siena with the support of the Ministerio de Ciencia e Innovacion in Spain (CTM 2010-10878-E) and the Caja del Mediterraneo of Alicante.

The Editors are grateful to all participants for their support and in particular to the authors who have contributed to the present book. They are also indebted to the members of the International Scientific Advisory Committee and other colleagues who helped to select the papers.

Yolanda Villacampa
Carlos A. Brebbia

Alicante, 2011

Obituary

The Wessex Institute of Technology regrets having to announce that our friend and colleague Professor Enzo Tiezzi had just died. He was a towering personality in the ecological systems field and through his pioneering work following Prigogine's ideas. Enzo's official title was Professor of Chemistry at the University of Siena, but his interests were wider, ranging from poetry to chemistry. He was a true Renaissance man, bringing new ideas and deeper insight to all fields of knowledge. He was able to discover the essence of different phenomena. His contributions to evolutionary processes and ecosystems in particular have been enormous.

WIT will sorely miss him because of his involvement in many of the Institute's activities. WIT Press is proud of having published scientific books co-authored by Enzo and his researchers as well as a novel. He was Co-Editor of many WIT Press books and a main Editor of the International Journal of Design & Nature and Ecodynamics, which has recently put together an outstanding issue on the Nature of Water, which Enzo expressed a special wish to see before his departure.

It will take too long to describe the major works of Enzo and his collaboration with WIT. It is important however to point out that he was the prime motivator behind the establishment of the Prigogine Gold Medal Award, instituted by the University of Siena and the Wessex Institute of Technology, following the death of Professor Ilya Prigogine in 2003. This important award has been given every year since 2004 to the best researcher working in the field of ecosystems modelling, particularly those following the work of Prigogine. Enzo received the award in 2005 at the Aula Magna of the University of Cadiz. He was also a Director and

Fellow of the Wessex Institute, who awarded him the Eminent Scientist medal in 2003. Enzo received many other awards and recognitions throughout his career.

Enzo was a man of integrity and died in accordance with his principle. He had what Spanish speakers call a 'good death' (*una Buena Muerte*) which is an idea that conveys not death itself but a sense of closure. He was aware of the terminal nature of his disease but with tremendous courage, decided not to try to prolong it as usual by treatment or surgery. He waited patiently to the end surrounded by a loving family and community.

His life continues in the work of the many researchers that he has guided over his long and prolific career. It also perdures in his books, papers, poems and other works; a legacy that will live for ever after his departure.

Finally, and equally important, he was for all the members of our community a good friend in which you could always rely in times of need.

Carlos A. Brebbia

Contents

Section 3: Environmental indicators

Section 4: Sustainable development studies

Section 5: Data Mining applications (Special session by A. Rabasa)

Section 6: Energy and the environment

Section 7: Soil contamination

Section 8: Waste management

Section 9: Ecotoxicity

Section 1
Ecosystems modelling

On the non-thermal effects of electromagnetic fields on charge transport processes in ecosystems

L. Brizhik
Bogolyubov Institute for Theoretical Physics, Kiev, Ukraine

Abstract

It is shown that periodic electromagnetic (EM) fields cause non-thermal effects on nonlinear charge transport processes in living organisms and ecosystems. In particular, such processes take place during respiration or photosynthesis and in hydrogen-bonded water chains present in interfacial water in cells, in coherence domains of water in the atmosphere and oceans. The nonlinear charge transport is mediated by solitons and bisolitons which are formed by single charge (electron, proton) or, respectively, two electrons in a singlet state, that are self-trapped in the self-induced local distortion of macromolecules or water chains. Due to oscillations of their velocity caused by discreteness of chains, (bi)solitons emit EM radiation of characteristic frequencies, determined by the intensity of the redox processes. In the presence of the periodic field (bi)solitons attain additional oscillations, and, therefore, the frequency of their EM radiation becomes modified, which affects redox processes in particular and the whole metabolism in general. It is shown that (bi)solitons have some characteristic frequencies, with respect to which periodic EM fields have non-thermal resonant effect on their dynamics and stability. Relatively strong enough unbiased periodic EM radiation can cause drift of (bi)solitons in molecular chains, and, therefore, it affects charge transport processes. These results show that changes in the EM solar activity can have effects on metabolism of living organisms and functioning of ecosystems and indicate a possible danger of the EM smog in ecosystems.

Keywords: ecosystems, soliton, bisoliton, charge transport, electromagnetic field, ratchet dynamics, water, electromagnetic smog.

WIT Transactions on Ecology and the Environment, Vol 144, © 2011 WIT Press
www.witpress.com, ISSN 1743-3541 (on-line)
doi:10.2495/ECO110011

1 Introduction

The existence of complex dynamical systems, such as ecosystems, implies exchange between the sub-units of the system not only with matter and energy, but what is even more important, also with information. While the first two exchange processes are intact, short-range, the exchange with information is a long-range process, which should occur very fast to provide the effective stability in the dynamical change of the conditions that are considered as an environment for sub-units. Here the question arises: what is the messenger for such fast long-range exchange with information? Our hypothesis (see Brizhik *et al.* [1, 2]) is that it is the electromagnetic field (EM), and here we develop further this theory. We show that such EM field not only is produced in the processes that occur in the system, but it also regulates these processes in the self-consistent way, providing thus the dynamical stability of the system and the emergence of the hierarchy of mutually interacting ecosystems. These interactions in and between sub-units are mediated by water which plays an essential role in all these processes due to its very peculiar and unique properties (see, e.g., Del Giudice and Tedeschi [3], Pollack [4], Voeikov [5]).

Electromagnetic field (EMF) is responsible for the matter formation and its properties, both in non-animated and alive, as well as in complex systems comprising them both, such as ecosystems. It is also the main tool of information storage and transfer in modern technologies. Till late 80s of the last century EMF was studied and technologically applied in linear systems whose optical properties almost do not depend on the field intensity. Moreover, in most cases EMF was assumed to be an external field for the problems studied. This is often the case, indeed. Nevertheless, there is a large class of systems which change the properties of the field itself, and systems with endogenous EMF, when the field itself is an intrinsic property of the system and is dynamically self-consistent with it. An important example of such systems is living organisms whose endogenous EMF is essential not only for their functioning, but provides their whole metabolism. No surprise that external EMF affects such systems, and that the caused bioeffects are very complicated and intensity and frequency dependent. In this paper we study the effect of EMF on soliton mediated charge transport in nonlinear systems, such as living organisms and water chains (Davydov [6], Antonchenko *et al.* [7]) or coherence domains of water (Del Giudice [8] and references therein). It is worth to recall here that term 'soliton' means a localized solitary wave solution of nonlinear equation(s), which is stable due to mutual compensation of nonlinearity and dispersion, and which is stable with respect to external perturbations and collisions with similar solitons. In the context of present paper solitons are charge carriers in macromolecules or hydrogen-bonded chains of water molecules, bound with the self-induced chain deformation (see Davydov [6]). In Section 2 we describe briefly main properties of solitons and bisolitons, and in Section 3 we study dynamics of solitons and bisolitons in periodic EMF.

2 Main characteristics of solitons and bisolitons

It is known that charge transport accompanies redox process in organisms with photosynthetic or respiratory system. In such organisms certain parts of the charge transport chain (so called Krebs cycle) in redox processes involve macromolecular enzymes which form alpha-helical conformation consisting of three almost parallel polypeptide chains, stabilized by hydrogen bonds along them. According to Davydov [6], charge transport along these enzymes is provided by soliton mechanism, when electrons released at the initial stage of the photosynthesis or respiration, cause deformation of the macromolecule and get self-trapped in soliton states. Similarly, solitons are formed in hydrogen bonded water molecules (see Davydov [6], Antonchenko *et al.* [7]). The Hamiltonian of a molecular chain with extra charge carriers consists of three terms, which are the Hamiltonians of free charge carriers, of the lattice and their interaction. In the site representation such Hamiltonian leads to the system of nonlinear coupled equations, which can be solved numerically or studied analytically in the continuum approximation, when discrete index n which numerates unit cells in a molecular chain is substituted by a continuous variable $x=na$. Here a is a lattice constant (distance between the nearest molecules). In a particular case of hydrogen bounded polypeptide chain in macromolecules with one or two extra charge carriers with opposite spins the resulting system of equations can be reduced to the nonlinear Schroedinger equation (see Davydov [6], Brizhik and Davydov [9]):

$$i\hbar\frac{\partial\Psi(x,t)}{\partial t}=-J\frac{\partial^2\Psi(x,t)}{\partial x^2}-2JG\,|\,\Psi(x,t)\,|^2\;\Psi(x,t)\,. \qquad (1)$$

Here $\Psi(x,t)$ is the wave function of a charge carrier which determines the amplitude of the probability of the charge presence at position x at time moment t, $\hbar$ is the Planck constant, J is the electron resonant interaction energy between the nearest sites, and G is the nonlinearity constant. The nonlinear term in eqn. (1) is the deformational potential which accounts the deformation of the chain induced by charges due to the electron-lattice interaction. The nonlinearity constant G in eqn. (1) for one ($G=G_s$) or two charge carriers in a singlet state ($G=G_{bs}$) takes the following values

$$G_s=g\,,\;G_{bs}=2g\,, \qquad (2)$$

respectively, where g is the dimensionless electron-lattice coupling constant:

$$g=\frac{\chi^2}{2Jw(1-s^2)}\,,\quad s^2=\frac{V^2}{V_s^2}\,. \qquad (3)$$

In eqn. (3) χ is the strength of electron coupling with the lattice, w is the elasticity of the chain, V is the velocity of the charge carrier, and V_s is sound velocity in the chain.

Eqn. (1) admits normalized to unity soliton-type solution

$$\Psi(x,t)=\frac{1}{2}\sqrt{G}\frac{\exp(ikx-iEt/\hbar)}{\cosh\{G[x-R(t)]/2a\}}, \tag{4}$$

where E is the energy, $R(t)=Vt$ is the c.m. coordinate of the charge carrier, k is its wave-vector. In the case of one extra charge carrier it is called soliton, and in the case of two charge carriers with opposite spins solution (4) describes bisoliton.

Thus, soliton is a bound state of a charge carrier (electron or proton) with self-induced local deformation of the chain. As any solitons in other nonlinear systems, solitons in molecular chains are exceptionally stable due to the compensation of wave dispersion and nonlinearity, their energy is lower than the energy of free charge carriers, they propagate along the molecular chain with the velocity, lower than the sound velocity in the system, and as a result, they can propagate on macroscopically large distances with very low energy dissipation. From eqns. (2)–(4) we conclude, that the width of the bisoliton is twice smaller than the width of a soliton:

$$l_s=\frac{2\pi}{g},\ l_{bs}=\frac{\pi}{g}. \tag{5}$$

The energy of the system, including the energy of the chain deformation, is lower than the energy of free charge carriers, $E_{0(s,bs)}$, and is given below:

$$E_s(V=0)=E_{0(s)}-J\frac{g^2}{12},\ E_{bs}(V=0)=E_{0(bs)}-\frac{2}{3}Jg^2. \tag{6}$$

Therefore, the binding energy of two charge carriers in the bisoliton state is

$$E_{bind}(V=0)=\frac{1}{2}Jg^2. \tag{7}$$

The effective mass of a (bi)soliton is bigger than the mass of free charge carrier(s), since it includes also the effective mass of the chain deformation, and is determined below for a soliton and bisolton states

$$M_s=m+M_{def(s)},\ M_{def(s)}=\frac{mM\chi^4}{6\hbar^2w^3}, \tag{8}$$

$$M_{bs}=m+M_{def(bs)},\ M_{def(bs)}=\frac{4mM\chi^4}{3\hbar^2w^3}, \tag{9}$$

respectively. Here m is the mass of a free charge carrier, and M is the mass of a molecule in a chain.

3 Effect of EMF on (bi)solitons

It has been shown by Brizhik *et al.* [10] that solitons have some characteristic frequency

$$\omega_s=g\frac{V_S}{\pi}, \tag{10}$$

which is determined as inverse time, necessary for the sound to pass the width of the soliton localization. Taking into account eqn. (5), we see that the characteristic frequency of a bisoliton is twice bigger than that of a soliton:

$$\omega_{bs} = 2g\frac{V_S}{\pi}. \tag{11}$$

According to Brizhik *et al.* [10], the external periodic EMF has a resonant effect on the dynamics of a soliton with respect to its characteristic frequency (10). Similar situation takes place for bisolitons. Namely, in the oscillating fields (bi)solitons start to oscillate with the frequency of the field ω as quasi-particles in the field of external periodic force. In low-frequency (at $\omega < \omega_{(s,bs)}$) and high frequency fields (at $\omega > \omega_{(s,bs)}$) (bi)olitons are stable, while at the resonant frequency $\omega = \omega_{(s,bs)}$ they intensively absorb energy from the field, and radiate it in the form of sound waves and EM radiation, rapidly decaying from a (bi)soliton state into the state of delocalized free charge carriers. Of course, such changes affect the efficiency of the redox processes and, therefore, the whole metabolism of the living organism or functioning of the corresponding ecosystem.

Another aspect of the effects of EMF on (bi)solitons is related to the so-called ratchet phenomenon, which consists in the ability of an excitation to attain a nonzero velocity (start to drift in the field of deterministic or stochastic unbiased (zero mean-value) periodic in time forces (see review by Reinmann [11]). Ratchet dynamics of solitons in nonlinear systems, described by nonlinear equations, such as sine-Gordon, was proven numerically (see, e.g., Zolotaryuk and Salerno [12], Gorbach *et al.* [13]). Also it has been shown by Brizhik *et al.* [14] that charged solitons can drift in molecular chains in the non-biased a.c. EMF. Appearance of such uni-directed motion is a critical phenomenon with respect to the intensity of the field and its frequency. This phenomenon can be studied by computer simulations of the coupled system of nonlinear discrete equations describing charge carriers in a deformable molecular chain. Some aspects of the ratchet dynamics can be described also within the approach of collective coordinates, taking into account the presence of the periodical in space lattice potential *U(R)* (so-called Peierls-Nabarro potential). In this latter case (bi)soliton dynamics in the external field *E(t)* with account of the energy dissipation has the form

$$M_{s(bs)}\frac{d^2R}{dt^2} = \gamma\frac{dR}{dt} - \frac{dU(R)}{dR} + qE(t), \tag{12}$$

where effective mass of a (bi)soliton is determined in eqns. (8)–(9), γ is the friction coefficient, and q is the charge of a particle ($q=e$ for a soliton, $q=2e$ for a bisoliton, e is the elementary charge). In the simplest case the lattice potential is represented as a double-well periodic potential

$$U(R) = U_1\cos\left(2\pi\frac{R}{a}\right) + U_2\cos\left(4\pi\frac{R}{a} + \varphi\right) \tag{13}$$

where U_1 and U_2 are the heights of the harmonics in a double-well potential, and φ accounts their asymmetry. An example of such soliton drift in a harmonic unbiased field $E(t) = E_0 \sin(2\pi t/T)$ for the intensity of the field E_0=8.7 and period T=2000 is shown in fig 1 (the intensity of the field is measured in units *ea/J* and time in units of $\hbar/J$).

As it has been shown in Brizhik *et al.* [14], such drift takes place only when the intensity of the field exceeds some critical value, which depends on the parameter values of the system, and in the fields whose frequency is not too high, namely, if it is less than the characteristic frequency of (bi)soliton (10) or (11), respectively. These conditions are necessary to allow (bi)soliton to overcome the potential barrier and the velocity (and frequency of oscillations) should be low enough so that the chain deformation could follow fast dynamics of electrons.

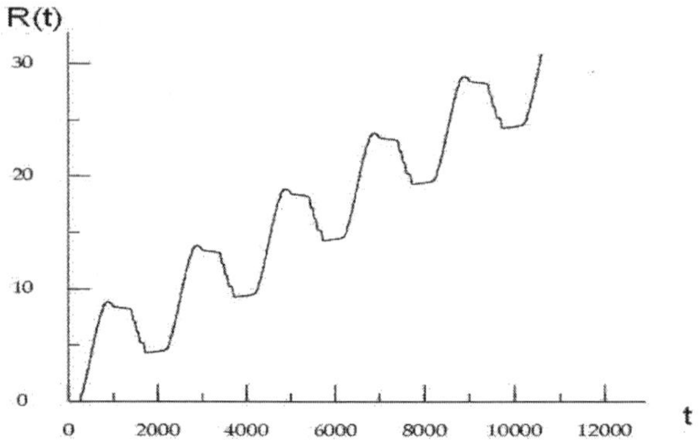

Figure 1: Position of the c.m.c. of a soliton, R(t), as function of time, t, in the harmonic field with the parameters E_0=8.7, T=2000 at γ=400. Parameters of the double-well potential in eqn. (13) are U_1=1, U_2=0.25, φ=-1.25.

4 Conclusions

Present study shows that external EMFs even of small intensities modify dynamics of charge carriers in living organisms and ecosystems, and that these effects depend on the frequency of the field. According to Davydov [6], charge transport in redox processes in living organisms and water systems is mediated by solitons and their bound states (bisolitons). In the presence of an oscillating EMF (bi)solitons oscillate with the frequency of the field, and the effective mass of (bi)soliton is a non-monotonous function of the frequency. At the resonant frequency (bi)solitons are unstable, they decay into unbound states of free electrons and lattice deformation.

The very fact, that (bi)solitons start to oscillate in the external periodic field, is very important, since during such oscillations (bi)solitons excite vibrational modes in a corresponding chain. Moreover, these oscillations, according to the Maxwell equations, will result in the emission by (bi)solitons of their own EMF. According to Brizhik *et al.* [15], frequency of the main harmonic of this radiation is determined by the average velocity of the soliton. In the absence of the field such radiation is caused by the discreteness of molecular chains and adds up to the endogenous EMF of the system.

Via exchange with the electromagnetic radiation, (bi)solitons get synchronized, and as a result, the system can attain coherence. This electromagnetic radiation is the fast messenger in the information exchange between the subunits of the complex system (cells, organs, individuals, ecosystems) which leads to the appearance of the long-range order in the system. The non-local self-consistent dynamics of matter and EMF is able to provide the development of dissipative structures, increasing the network of individuals, processes, and information and resulting in the appearance of the hierarchy of dissipative systems able to support fast and long-range exchange of the information and interconnectedness of the system (see Prigogine and Nicolis [16]).

In the non-biased EMF (bi)solitons can attain additional drift due to the ratchet effect. In this case their velocity is not just the velocity, determined by the rate of redox processes, but includes also the drift velocity. Therefore, the appearance of the drift modifies frequency of the (bi)soliton radiation, i.e., modifies intrinsic EMF of the system. This, of course, affects redox processes, and in such a way, metabolism of the living organism or functioning of an ecosystem. White noise can also result in the ratchet effect, as it has been shown by Luczka *et al.* [17], Bartussek *et al.* [18], Gomez-Marin and Sancho [19]. Therefore, we can expect that noise can also cause uni-directed current of (bi)solitons in molecular systems. This suggests a possible mechanism of the effects of the solar activity on biosystems via its influence on the nonlinear charge transport processes. It follows from these results technologically produced electromagnetic smog even of low intensities can be potentially dangerous, since, as we have shown above, the bioeffects of oscillating fields have resonant nature, and, therefore, are non-thermal. Finally, we remark that living systems and ecosystems, are not only complex open systems, they are also ontic, (see Tiezzi *et al.* [20]) so that although each sub-system is protected against excessive external influences, it remains open, which allows functioning of its complex internal organization in fast varying conditions. Such openness is not only because of the exchange with matter and energy, but also with the information, which occurs via self-consistent (endogenous) electromagnetic field.

Acknowledgements

I express my sincere gratitude to Prof. E. Del Giudice for stimulating discussions and co-operation. I acknowledge Prof. W. J. Zakrzewski and Drs. A. A. Eremko and B. M. A. G. Piette for our joint research, the results of which made this paper possible. I also acknowledge the Organizers of the ECOSUD Conference for the financial support, thanks to which I got the possibility to attend the event, which was so important to my friend and co-author Enzo Tiezzi, to whose memory I dedicate this contribution.

References

[1] Brizhik L., Del Giudice E., Jorgensen S.E., Marchettini N., and Tiezzi E. The role of electromagnetic potentials in the evolutionary dynamics of ecosystems. *Ecological Modelling*, **220**, pp. 1865-1869, 2009.

[2] Brizhik, L., Del Giudice, E., Tedeschi, A., Voeikov, V., The role of water in the information exchange between the components of an ecosystem, submitted to *Ecological Modelling*, 2011.

[3] Del Giudice, E.; Tedeschi, A., Water and the autocatalysis in living matter. *Electromagn. Biol. Med.*, **28**, pp 46-54, 2009.

[4] Pollack, G.H., Water, energy and life: Fresh views from the water's edge. *Int. J. Des. Nat. Ecodyn.* **5**, pp 27-29, 2010.

[5] Voeikov, VL. Fundamental Role of Water in Bioenergetics. In: *Biophotonics and Coherent Systems in Biology*. Beloussov, L.V., Voeikov, V.L., Martynyuk, V.S., editors. Springer, N.-Y., 2007, pp 89-104.

[6] Davydov, A.S., *Solitons in Molecular Systems* (2nd ed.), Reidel, Dordrecht, 1985.

[7] Antonchenko, V.Ya., Davydov, A.S., Zolotaryuk, A.V., Solitons and proton motion in ice-like structure, *Phys. Stat. Sol. (b)* **115** , pp 945-952, 1983.

[8] Del Giudice, E. Old and new views on the structure of matter and the special case of living matter, *J. Phys.: Conf. Series*; **67**, 012006, 2006.

[9] Brizhik, L.S., Davydov, A.S., The electrosoliton pairing in soft molecular chains, *Low Temp. Phys.*, **10**, pp 748-753, 1984.

[10] Brizhik, L., Cruzeiro-Hansson, L., Eremko A., Influence of electromagnetic radiation on molecular solitons, *J. Biol. Phys.*, **24**, pp 19-39, 1998.

[11] Reimann, P., Brownian motors: noisy transport far from equilibrium, *Phys. Rep.,* **361**, pp 57-265, 2002.

[12] Zolotaryuk, Y., Salerno, M., Discrete soliton ratchets driven by biharmonic fields, *Phys. Rev. E* **73**, 066621 (12 pages), 2006.

[13] Gorbach, A.V., Denisov, S., Flach, S., Optical ratchets with discrete cavity solitons, *Opt. Lett.*, **31**, pp 1702-1704, 2006.

[14] Brizhik, L., Eremko, A., Piette, B., Zakrzewski, W., Ratchet dynamics of large polarons in asymmetric diatomic molecular chains. *J. Phys.:Cond. Matt.*, **22**, 155105, 9 pp., 2010.

[15] Brizhik, L., Eremko, A., Nonlinear model of the origin of endogenous alternating electromagnetic fields and self regulation of metabolic processes in biosystems, *Electromagnetic Biol. and Med.*, **22**, pp 31-39, 2003.

[16] Prigogine, I., Nicolis, G., *Self-Organization in Non-Equilibrium Systems*. Wiley, New York, 1997.

[17] Luczka, J., Bartussek, R., Hänggi, P., White-noise-induced transport in periodic structures, *Europhys. Lett.,* **31**(8), pp 431-436, 1995.

[18] Bartussek R., Hänggi P., Lindner, B., Schimansky-Geier, L., Ratchets driven by harmonic and white noise, Physica D, **109**, pp 17-23, 1997.
[19] Gomez-Marin, A., and Sancho, J.M., Brownian pump powered by a *white-noise* flashing *ratchet. Phys. Rev. E,* **77**, 031108, 2008.
[20] Tiezzi, E., Cecconi, G., Marchettini, N., Confined ontic open systems. *Int. J. of Design and Nature and Ecodynamics,* **5**, 3-9, 2010.

The effect of human disturbance on site habitat diversity and avifauna community composition in suburban conservation areas

P. F. Mooney
School of Architecture and Landscape Architecture, University of British Columbia, Canada

Abstract

This study evaluated the causes and effects of habitat heterogeneity in suburban conservation areas of the metropolitan Vancouver area. Site disturbance, soil conditions and plant associations were evaluated for their relationship to habitat heterogeneity. Habitat heterogeneity was correlated with breeding bird populations and the type of birds found.

Three, urban conservation sites were surveyed for site conditions and avian populations. Data were spatially referenced in a geographic information system (GIS) and a habitat diversity index rating was calculated for each site. Site habitat diversity ratings were compared with the avian species richness and the birds recorded were classified as urban exploiters, adapters or avoiders.

It was found that human disturbance and biotic succession had produced a mosaic of seral stage plant communities that increased habitat heterogeneity. Site-level habitat heterogeneity was correlated with avian species richness. All categories of urban and non-urban birds were recorded.

Suburban conservation areas can support high avian species richness, including urban avoiders. Site-level habitat heterogeneity and avian species richness are strongly correlated. These findings can assist in selecting and managing urban conservation sites to maintain regional biodiversity.

Keywords: urban biodiversity, site-level habitat diversity, habitat diversity index, urban exploiters, urban adapters, urban avoiders, conservation.

1 Introduction

The extensive and intensive alteration of land cover associated with urbanization negatively impacts biodiversity in urban regions [1] with more biodiverse

WIT Transactions on Ecology and the Environment, Vol 144, © 2011 WIT Press
www.witpress.com, ISSN 1743-3541 (on-line)
doi:10.2495/ECO110021

regions being at greater risk for biodiversity loss [2]. In most urban regions, non-native species abound, habitats are fragmented, habitat quality is poor and species migrations are disrupted, leading to high rates of local extinctions [2, 3].

While these regional impacts of urbanization are well established, the causes of local land cover change, its impacts and the appropriate measures to support site-level biodiversity are less well understood. This study evaluated human disturbance as a cause of habitat diversity (or heterogeneity) and the effects of habitat heterogeneity on avian species richness and avifauna community composition, in suburban conservation areas, in order to better manage such sites to support regional biodiversity.

These findings can assist in selecting and managing urban conservation sites to maintain regional biodiversity.

Despite the negative effects of urbanization, many cities were founded in biologically diverse eco regions that still support significant species richness and remnant populations of uncommon species [4, 5]. Such regions may play crucial roles in supporting regional and even hemispheric biodiversity [2, 4, 6]. Preserving the remaining biodiversity in urban regions will require conservation actions that are informed by direct local knowledge [7–9].

2 Habitat heterogeneity and species richness

Habitat heterogeneity is a measure of the diversity and evenness of variation in land cover. It has been consistently correlated with avian diversity [10–16].

Studies of patch level avian habitat have typically used point source data of structural or floristic diversity as measures of both habitat diversity and quality. However, habitat diversity is dependent, in part, on the scale of observation and must be measured at the corresponding landscape scale [17]. This study used a site habitat diversity index (HDI) to measure the effect of habitat heterogeneity on avian species richness. The HDI is a more suitable measure of site level diversity than point source data, since it incorporates the variation among all habitats on a site and not just the diversity within individual vegetation patches.

Human disturbances, such as industrial activity, clearing and filling are intrinsic to cities. When sites are abandoned after disturbance, biotic succession follows. Grasses and herbaceous plants colonize the bare soil, followed by pioneer woody plants [18]. Small patches of incremental disturbance and subsequent plant succession within a site result in the development of a shifting mosaic of different seral stage plant communities.

Diverse plant associations may also develop in response to variation in environmental gradients such as soil moisture, fertility and pH [19–21]. Thus, the plant associations found on any site may result from biotic succession in response to site disturbance or the environmental gradients of the site, or a combination of the two.

3 Urban birds

Birds that occupy early seral stage landscapes increase in number in suburban environments by taking advantage of the supplemental food and nest sites provided by these anthropogenic habitats and are termed urban adapters [22–26]. Urban adapters include edge species, omnivores, ground foragers, corvids, seedeaters and aerial sweepers such as swifts [25, 27]. Swallows *Tachycineta spp.*, Bewick's wren *Trogolodytes bewickii,* red-winged blackbird *Agelaius phoeniceus,* American robin *Turdus migratorius,* Anna's hummingbird *Calypte anna,* American goldfinch *Carduelis tristus* and chestnut-backed chickadee *Parus rufescens* are representative species [28].

Other birds have been categorized as urban avoiders and urban exploiters [27]. Both urban adapters and urban exploiters are synanthropic species i.e. species associated with humans [24, 26] while urban avoiders are considered to be non-synanthropic. Urban exploiters include cliff dwellers, cavity nesters, ground foraging seedeaters and omnivores [25]. Peregrine falcons *Falco peregrinus,* house sparrows *Passer domesticus,* European starlings *Sturnus vulgaris* and European rock doves *Columba livia* are urban exploiters and may be found in the downtown cores of many cities [25, 27]. Urban avoiders include interior nesting birds, tree foraging insectivores, neotropical migrants, and ground nesting birds [25, 27]. Representative species include western wood pewee *Contopus sordidulus* and Hutton's vireo *Vireo huttoni* [27]. Research has shown that urban exploiters, adapters and avoiders are non-randomly assembled in urban environments [25].

The designation of bird species as urban exploiters, adapters and avoiders is incomplete. Not all birds have been categorized; some authorities differ as to the appropriate category for a species and some species may behave differently in different regions. For example, Blair [27] considered Steller's jay *Cyanocitta californica* to be non-synanthropic, while Donnelly and Marzluff [26] considered it to be a synanthropic species. Because it is frequently found at residential bird feeders in this region it was considered to be an urban adapter in this study.

These differences in classification may result from the fact that there is a continuum from true urban avoiders that never venture into urban realms to adaptive-avoiders that make minimal use of urban/suburban habitats [26, 29]. Since the study sites were suburban, it would be expected that urban adapters and exploiters would be numerous while urban avoiders would be absent or minimally present.

4 Methods

Three urban conservation areas were selected for study because their similarities maximized control of both site conditions and offsite factors. As well as similar size, location and vegetation, the three sites shared a history of industrial use, abandonment, biotic succession and subsequent protected status. Each site contained a combination of ruderal vegetation, remnant islands of natural vegetation and some managed vegetation. The sites were in coastal locations and had protected status. Two were municipal parks and the third was a designated wildlife management area. All three sites were surrounded by suburban residential communities, with second growth conifer forest on the inland side (see figure 1).

Figure 1: Location of research sites in the metropolitan Vancouver area.

5 Site inventories

Inventories of site disturbance and soil moisture and fertility were conducted to determine the causal factors of the mix of habitat types on each site. Birds were surveyed for species richness and the composition of the avian community.

5.1 Soil moisture and fertility

Native indicator plants of coastal British Columbia were used to infer soil nitrogen and moisture gradients. The method developed by Klinka et al. [30] uses native plants of coastal British Columbia to indicate site climate and a site's soil moisture and nitrogen gradients. In this classification system, soil nutrient regimes range from very poor to very rich and each plant species indicates a gradient of nutrient and moisture levels.

5.2 Vegetation surveys

Aerial photo analysis was used to identify vegetation association patches. This was followed by field surveys to establish patch boundaries and quadrat inventories that recorded the plant associations of each patch on each site. The plant associations of each vegetation patch were converted to habitat types, digitally mapped using GIS Arcview and the area of each habitat patch determined.

5.3 Bird surveys

Birds were surveyed using line transects of each habitat patch on each of the three sites. Surveys were conducted in the early mornings of four consecutive days between mid-May and mid-June. Only birds that were foraging or singing from perches were recorded. Territory mapping was used to transform multiple sightings into breeding territories [31]. All birds were classified as urban adapters, exploiters or avoiders. Those birds that fit the criteria of urban avoider but that nevertheless were recorded on the suburban conservation areas in this study were labeled adaptive-avoiders.

5.4 Time since disturbance

Since taking tree core samples was not an option on these public sites, an aerial photo analysis was used to determine the age of each vegetation patch on each site. Stereoscopic aerial photos of the study sites were obtained for the period 1926 to 1999. (After 1999, all the study sites were protected and no new disturbances occurred.) Aerial photography was conducted approximately every 7 years during the period 1926-1999.

The aerial photo analysis showed major disturbances, i.e. site clearing or filling that substantially altered the vegetative community of a patch (see Figure 2). The date of the survey showing the last major disturbance for each

Figure 2: Site 2: Maplewood Flats 1979, showing the extent of disturbance due to extensive filling.

polygon was subtracted from 1999 to calculate the approximate time since disturbance for each patch.

6 Habitat diversity index

To establish the relationship between habitat heterogeneity and avian species richness on the study sites, it was necessary to first measure the habitat heterogeneity of each site. Table 1 was constructed using the habitat maps that had been developed from the vegetative inventories.

Table 1: Habitat areas and percent of site area across all study sites.

Habitat Type	**Sites**					
	1) Cates Park		2) Maplewood Wildlife Area		3) Shoreline Park	
	Area (m^2)	% of Site	Area (m^2)	% of Site	Area (m^2)	% of Site
Clearing	11,170	5	3,859	1	5,572	2
Meadow	-	-	8,105	3		
Old Field	-	-	100,906	36		
Lawn & Trees	65,677	30	-	-	63,009	17
Hedgerow	-	-	4,338	2		
Shrub	-	-	13,349	5	50,379	13
Forest Deciduous	-	-	122,040	44	96,031	26
Forest Coniferous	23,380	11	-	-	85,084	23
Forest Mixed	120,153	54	-	-	35,248	19
Wetland Freshwater	-	-	4,896	2	538	0.1
Wetland Salt	-	-	17,845	7	36,667	10
Totals	220,380	100	275,338	100	372,528	100

A habitat diversity index (HDI) was calculated for each site using the Simpson's reciprocal diversity index. Simpson's diversity index is often used in ecology to quantify the diversity of a habitat [32]. It measures both the richness or number of species per sample and the evenness or relative abundance of different species in the sample.

Simpson's Index (D) is calculated as:

$$D = \frac{\sum n(n-1)}{N(N-1)}$$

where n= the total number of organisms of a particular species and N= total number of organisms of all species. In this study, N = total number of habitats within a site, n = percent area of each habitat type within a site. Used in this way an index of 0 equals infinite diversity and 1 = no diversity. Because this is counter-intuitive, the Simpson's reciprocal index (1/D) was used. For the Simpson's reciprocal index, (1/D), the higher the number the greater the diversity. The HDI rating for each site was then compared to the number of bird species using the site.

7 Results

7.1 Soils

All sites were found to be relatively fertile and different habitat types shared similar moisture and fertility levels. All soils had moisture levels of fresh to very moist, except wetland habitats that were wet to very wet. The soil analysis also revealed that soil fertility did not differ greatly between the study sites or the different habitat types. Of twelve habitat types, eight were rich to very rich and the rest were in the medium to rich categories in terms of soil nutrients. Neither soil fertility nor moisture was found to be a limiting factor to plant growth on the study sites. Moisture levels were high, even in summer, and fertility levels were medium to very rich.

7.2 Disturbance

The time since disturbance analysis showed that while the different habitat patches shared similar moisture and fertility levels they differed in age. Patches ranged from a maximum of 50 years to a minimum of 7 years since disturbance.

7.3 Birds

Sixty-five bird species recorded on the three study sites. The majority of the bird species found were urban adapters (39) or exploiters (6). This list of synanthropic species included such urban exploiters as European starling, rock dove, and such urban adapters as American goldfinch and northern flicker *Colaptes auratus.* The 45 synanthropic species comprised 69 percent of the species recorded. This included diving and dabbling waterfowl not usually associated with suburban areas but which are common on protected salt water habitats in this region. Twenty species of urban adaptive- avoiders were found on these suburban conservation sites.

7.4 Habitat diversity and avian species richness

The HDI ratings for the three sites ranged from 2.5 – 5.3. The number of bird species recorded on each site ranged from 28 – 46. Figure 3 shows that bird species richness increased with increasing habitat diversity for all three study sites.

Figure 3: Site habitat diversity index versus bird species richness.

8 Conclusions

The key findings may be summarized as follows:

1) Unplanned, incremental, human disturbance and subsequent biotic succession created the mosaic of multiple seral stages found on each of these sites.
2) The seral stages function as multiple habitats creating the HDI measurements recorded.
3) As HDI increases, avian species richness also increases.
4) Many bird species that had been classified as urban avoiders were found to use these suburban, disturbed conservation areas.

9 Discussion

9.1 Habitat heterogeneity and its interactions

In forest ecosystems in this region, multiple seral stages increase site level habitat heterogeneity and forests having a greater diversity of vegetative composition have greater avian species richness [33].

The number of bird species within a habitat patch has been found to increase as the patch size increases. For example, a broadly based study of species/area relations in the United Kingdom and Scandinavia found that bird species richness in 15 hectare woods generally approached a doubling of those found in 1 hectare woods [34]. However within fragments of equal size, those fragments that are more heterogeneous will tend to support a greater number and variety of species than those which are more homogeneous. This may be because heterogeneous habitat fragments are less susceptible to local extinction than more homogeneous habitats [35].

Most studies of urban adapters have looked at the matrix of the suburban landscape [26, 27, 36]. This study found that suburban parks and conservation areas are special patches within the suburban landscape matrix that have higher and different habitat values than the general matrix. Proportionately larger numbers of birds that are considered to be urban avoiders used these habitats (31% of all birds recorded). The urban avoiders recorded in this study were comprised of interior nesting birds such as pileated woodpeckers *Dryocopus pileatus,* neotropical migrants like Hutton's vireos, western tanagers *Piranga ludoviciana* and Wilson's warblers *Wilsonia pusilla* and shorebirds such as solitary and spotted sandpipers *Tringa solitari* and *Actitis macularia.* Habitat heterogeneity was found to support increased avian species richness for edge, synanthropic species and also for non-synanthropic urban avoiders. This suggests that in suburban parks and conservation areas, where there are often limited opportunities to increase the size of habitat, increasing the HDI may be a useful tool to increase site level avian diversity and support so called urban avoiders.

In assessing the relationship between avian species richness and landscape attributes, it is necessary to consider both the local and the regional landscape [35, 36]. The effect of local habitat heterogeneity is moderated by other ecological factors such as habitat area, quality, isolation, and the landscape matrix surrounding the habitat patch. These and other factors are simultaneously functioning to determine bird species distribution in the landscape mosaic [37–40].

In the Metro Vancouver region, significant areas of second growth forest are protected as watersheds that provide the regional drinking water. These large patches of intact conifer forest provide habitat for urban avoider species and likely contributed to their presence on the nearby suburban study sites.

9.2 Site trajectory

As the study sites are parks and conservation areas, they are now protected from the kind of small scale, intermittent disturbance that led to their present condition. In the absence of active management, these sites will, through succession, develop a more uniform plant community. This will result in a reduction of the site habitat diversity rating with attendant reductions in the numbers of vertebrate species.

9.3 Role of disturbance

Within the three study sites the same disturbance that created mixed seral stages has allowed the introduction and spread of invasive plant species. The most widespread of these is Himalayan blackberry, *Rubus discolor*. This plant has important habitat value; it provides thicket cover for various sparrows, especially fox and song sparrows, *passerella iliaca* and *melospiza melodia.* The berries are eaten copiously by many birds and by black bears, *ursus americanus* and the plant is a nectar source for several butterflies. Another invasive alien plant, Japanese knotweed, *Polyganum cuspidatum* is becoming increasingly widespread on the three study sites. These invasive alien plants alter the local biotic community directly through their presence and indirectly through such responses as niche displacement and species evolution in response to the invader [41]. Over time we may expect that the negative effects of invasive plants will increase unless effective control programs are introduced. We would expect blackberry to become a smaller component of the sites as it is shade intolerant and will die out under the forest canopy [42]. The knotweed will continue to spread and, if left unchecked, will in time form a monoculture on the forest floor.

Local land cover, invasive plant species, disturbance, the cessation of disturbance and other unseen forces such as climate change and habitat loss elsewhere will affect the levels of biodiversity found on such suburban sites. These influences are happening now and will continue to effect change. It is expected that climate change will contribute to the spread of invasive species and decline in the populations of native species.

9.4 Implications for conservation

These findings have implications for the selection and management of conservation sites in urban ecosystems. In selecting new conservation areas, sites that contained multiple seral stage plant communities and that had high HDI ratings would be preferred over sites having lower HDI ratings, if the conservation goal was to maximize general avian diversity in the urban region.

In managing suburban conservation sites like those in this study, human energy inputs in the form of controlled disturbance or ecological restoration could be used to increase their HDI rating and control invasive plants. This might mean periodically thinning the woody plants in an old field habitat, removing invasive plants or restoring a degraded wetland. The study has shown that past, unplanned human disturbance has inadvertently increased habitat diversity and species richness. It has also demonstrated that high avian diversity values can be found on such sites. Intentional interventions that increase habitat heterogeneity can now be used to increase site habitat diversity indices and enhance site level avian diversity in suburban conservation areas.

If we are to maintain global biodiversity, the global application of ecosystem management and restoration will be required [43]. Within regional conservation agencies, there exists a great need to develop and institutionalize adaptive ecosystem management based on the scientific understanding of local human ecosystems, rather than simply setting land aside in conservation/recreation.

References

[1] Lutz, W., Sanderson, W.C. & Scherbov, S., (eds).The End of World Population Growth in the 21st Century: New Challenges for Human Capital Formation and Sustainable Development, Earthscan: London and Sterling VA, pp. 341, 2004.

[2] Connor, E.F., Hafernik, J., Levy, J., Moore, V.L. & Rickman, J.K., Insect conservation in an urban biodiversity hotspot. *Journal of Insect Conservation,* 6 (4), pp. 247-259, 2002.

[3] Marzluff, J.M., Worldwide urbanization and its effects on birds (Chapter 2). *Avian Ecology in an Urbanizing World*, eds. J.M. Marzluff, R. Bowman & R. Donnelly, Kluwer Academic Publishers: Boston, pp. 19-48, 2001.

[4] Ricketts, T. & Imhoff, M., Biodiversity, urban areas, and agriculture: locating priority ecoregions for conservation. *Conservation Ecology,* 8(2), pp. 1, 2003.

[5] UK Biodiversity Action Plan (UK BAP). Online. http://www.ukbap.org.uk/.

[6] Butler, R.W. & Campbell, R.W., *Birds of the Fraser River Delta: Populations, Ecology, and International Significance*, Canada Wildlife Service Occasional Paper: Delta, BC, 65, 1987.

[7] Cook, E.A., *Ecological Networks in Urban Landscapes*, Wageningen University: Wageningen, The Netherlands, pp. 201, 2000.

[8] Cowling, R.M., Pressey, R.L., Roget, M. &Lombard, A.T., A conservation plan for a global biodiversity hotspot-the Cape Floristic Region South Africa. *Biological Conservation,* 112(1-2), pp. 191-216, 2003.

[9] Alberti, M., The effects of urban patterns on ecosystem function. *International Regional Science Review,*28(2), pp. 168-192, 2005.

[10] James, F.C., Ordinations of habitat relationships among breeding birds. *The Wilson Bulletin,* 83, pp. 215-236, 1971.

[11] Wiens, J.A. & Rotenberry, J.T., Habitat associations and community structure of birds in shrubsteppe environments. *Ecological Monographs,* 51(1), pp. 21-42, 1981.

[12] Forman, R.T.T. & Godron, M., *Landscape Ecology*, John Wiley and Sons: New York, pp. 619, 1986.

[13] Boecklen, W.J., Effects of habitat heterogeneity on the species-area relationships of forest birds. *Journal of Biogeography,* 13(1), pp. 59-68, 1986.

[14] Bohning-Gaese, K., Determinants of avian species richness at different spatial scales. *Journal of Biogeography,*24(1), pp. 49-60, 1997.

[15] Kruger, O. & Lindstrom, J., Habitat heterogeneity affects population growth in goshawk *Accipiter gentiles*. *The Journal of Animal Ecology,*70 (2), pp. 173-181, 2001.

[16] Tews, J., Brose, U., Grimm, V., Tielborger, K., Wichmann, M.C., Schwager, M. & Jeltsch, F., Animal species diversity driven by habitat heterogeneity/diversity: the importance of keystone structures. *Journal of Biogeography,* 31(1), pp. 79-92, 2004.

www.witpress.com, ISSN 1743-3541 (on-line)

[17] Bissonette, J.A. & Storch, I., Fragmentation is the message clear? *Conservation Ecology* 6(2), pp. 14, 2002.
[18] Kimmins, J., Forest Ecology: a Foundation for Sustainable Forest Management and Environmental Ethics in Forestry, Pearson Prentice Hall: New Jersey, pp. 531, 2004.
[19] Whittaker, R.H., Gradient analysis of vegetation. *Biological Reviews*, 42 (2), pp. 207-264, 1967.
[20] Austin, M.P., Searching for a model for use in vegetation analysis. *Plant Ecology*, 42, pp. 11-21, 1980.
[21] Partel, M., Local plant diversity patterns and evolutionary history at the regional scale. *Ecology*, 83(9), pp. 2361-2366, 2002.
[22] Blair, R.B. & Launer, A.E., Butterfly diversity and human land use: Species assemblages along an urban gradient. *Biological Conservation*, 80, pp. 113-125, 1997.
[23] Blair, R., Birds and butterflies along urban gradients in two ecoregions of the U.S. (Chapter 3). *Biotic Homogenization*, eds. J.L. Lockwood & M.L. McKinney, Kluwer Academic/Plenum Publishers: New York, pp. 33-56, 2001.
[24] Johnston, R.F., Synanthropic birds of North America (Chapter 3). *Avian Ecology in an Urbanizing World*, eds. J.M. Marzluff, R. Bowman & R. Donnelly, Kluwer Academic Publishers: Boston, pp. 49–67, 2001.
[25] McKinney, M.L., Urbanization biodiversity and conservation. *BioScience,* 52, pp. 883-890, 2002.
[26] Donnelly, R. & Marzluff, J.M., Relative importance of habitat quality, structure and spatial pattern to birds in urbanizing environments. *Urban Ecosystems,* 9, pp. 99 -117, 2006.
[27] Blair, R., Land use and avian species diversity along an urban gradient. *Ecological Applications,* 6(2), pp. 506-519, 1996.
[28] Marzluff, J.M., Island biogeography for an urbanizing world: how extinction and colonization may determine biological diversity in human-dominated landscapes. *Urban Ecosystems,* 8, pp. 157-177, 2008.
[29] Schaefer, V. Personal communication, 25 July 2008, Restoration of Natural Systems Program Director, University of Victoria, Victoria, BC
[30] Klinka, K., Krajina, V.J., Ceska, A. & Scagel, A.M., *Indicator Plants of Coastal British Columbia,* The University of British Columbia Press: Vancouver, pp. 288, 1989.
[31] Bibby, C.J., Burgess, N.D. & Hill, D.A., *Bird Census Techniques*, Academic Press: London, 1992.
[32] Offwell Woodland and Wildlife Trust. Simpson's Diversity Index. Online. http://www.countrysideinfo.co.uk/simpsons.htm
[33] Easton, W.E. & Martin, K., The effect of vegetation management on breeding bird communities in British Columbia. *Ecological Applications,* 8(4), pp. 1092-1103, 1998.
[34] Hinsley, S.A., Bellamy, P.E., Enoksson, B., Fry, B., Gabrielsen, L., McCollin, D. & Schotman, A., Geographical and land-use influences on

bird species richness in small woods in agricultural landscapes. *Global Ecology and Biogeography Letters,* 7(2), pp. 125-135, 1998.
[35] Collinge, S.K., Ecological consequences of habitat fragmentation: implications for landscape architecture and planning. *Landscape and Urban Planning,* 35, pp. 59-77, 1996.
[36] Melles, S., Glenn, S. & Martin, K., Urban bird diversity and landscape complexity: Species–environment associations along a multiscale habitat gradient. *Conservation Ecology,* 7(1), pp. 5, 2003.
[37] Law, B.S. & Dickman, C.R., The use of habitat mosaics by terrestrial vertebrate fauna; implications for conservation and management. *Biodiversity and Conservation,* 7, pp. 323-333, 1998.
[38] Mason, C.F., Woodland area, species turnover and the conservation of bird assemblages in lowland England. *Biodiversity and Conservation,* 10(4), pp. 495-501, 2001.
[39] Poulsen, R.O., Avian richness and abundance in temperate Danish forests: tree variables important to birds and their conservation. *Biodiversity and Conservation,* 11(9), pp. 1572-9710, 2002.
[40] Brotons, L., Herrando, S. &Martin, J.L., Bird assemblages in forest fragments within Mediterranean mosaics created by wild fires. *Landscape Ecology,* 19, pp. 663-675, 2004.
[41] Mooney, H.A. & Cleland, E.E., The evolutionary impact of invasive species. *Proceedings of the National Academy of Sciences of the United States of America,* 98 (10), pp. 5446-5451, 2001.
[42] Hoshovsky, M., *Element Stewardship Abstract for Rubusdiscolor (Rubusprocerus)* The Nature Conservancy: Virginia, pp. 1-11, 2000.imapinvasives.org/GIST/ESA/esapages/rubarme.html
[43] Nellemann, C. & Corcoran, E., (eds). Dead Planet Living Planet-Biodiversity and Ecosystem Restoration for Sustainable Development: A rapid Response Assessment, United Nations Environment Programme: United Nations, pp. 112, 2010.

Modelling the value of wetlands in the Kilombero Valley, Tanzania, using community preferences

F. Mombo[1,2], S. Speelman[1], D. Phillip[2] & G. van Huylenbroeck[1]
[1]*University of Ghent, Ghent, Belgium*
[2]*Sokoine University of Agriculture, Morogoro, Tanzania*

Abstract

Wetland degradation has recently received considerable attention in research. Although wetlands are valuable ecosystems, their actual value is difficult to measure because the services they provide do not have market values. The current study tries to define the value of wetlands in the Kilombero Valley, central Tanzania. Choice modelling is used to determine communities' preferences over wetland conditions associated with various management options. The results show that the inhabitants of the Kilombero Valley desire improvements in the wetland's conditions, suggesting that the ongoing degradation is not socially optimal. Local farmers are however highly dependent on the wetland for their livelihood. In order to reduce pressure on wetlands, it is therefore necessary and imperative to explore the options of alternative income generating activities and the application of proper technologies to improve efficiency and effectiveness in crop production.
Keywords: wetlands, choice modelling, valuation and willingness to pay.

1 Introduction

In recent years, degradation of wetlands has received considerable attention in research. While in the past, wetlands were mostly neglected because they were considered as marginal areas, authors like Fischer [1] and Turner et al. [2] have recognized their importance. Wetlands play a unique role as ecosystem and provide diverse services to communities such as climate regulation, flood control and provision of food and fiber. To preserve the functions played by wetlands a sustainable management is of paramount importance. One of the prerequisites for

WIT Transactions on Ecology and the Environment, Vol 144, © 2011 WIT Press
www.witpress.com, ISSN 1743-3541 (on-line)
doi:10.2495/ECO110031

sound management of natural resources is a correct valuation [3, 4]. It is necessary to know the value of products and services accrued from wetlands to be able to ensure sustainable management of these vital ecosystems throughout the world [5, 6]. Given this context the valuation of wetland services has recently been the subject of study in many countries both in developing and developed world [7–9]. Because the services provided by wetlands are often not traded in the market most research was based on non-market valuation techniques such as Contingent valuation, Travel Cost Methods, Hedonic Pricing and others [10]. Other issues which constrain proper management of wetlands are the lack of knowledge of different stakeholders (farmers, conservationists, recreationists, etc.) and their values and attitudes and acknowledgement of the conflicting multiple objectives of stakeholders. Community involvement in the formulation of wetland management plans could provide policy alternatives that are more acceptable to the community. It is therefore interesting to incorporate stakeholder preferences and multiple objectives to evaluate management options [11, 12].

This study considers the Kilombero valley wetlands in Tanzania. As in other areas in Africa wetlands play a significant role in the livelihoods of the rural communities of this region [13]. The ability of wetlands to store water during the wet season and release it during the dry season provides farmers living in semi-arid areas opportunities to grow crops all-year round. This improves their food security and incomes. Besides crop production, wetlands also provide other services that support human welfare such as livestock grazing and watering, water supply, fishing and natural products [13]. Degradation of wetlands can cause loss of these functions. Following several authors [5, 8, 12, 14] we use a Choice Experiment (CE) to expose public preferences over the range of possible future landscape configurations for the Kilombero wetlands. The method has the advantage to enable us to show through various models the unobserved and unconditional heterogeneity in preferences. As indicated by various authors [11, 14, 15] the conflicting multiple objectives of stakeholders, make it important to recognize the distinct groups showing different concerns. This information is important for policy makers during policy formulation [16, 17]. In our study the utilities of different individuals who are primary and secondary users of wetland services are estimated and used to see how environmental and socioeconomics attributes influence utilities and hence households willingness to pay for an improved status of the wetlands. The results of this study can be used by policy makers and other development practitioners in developing countries to see how wetlands can be sustainably managed under these typical conditions of diverse preferences of the users and varying wetlands attributes. Moreover in the establishment of management plans the CE can also be used as a conflict avoidance tool or conflict resolution tool [12].

2 The methodology

2.1 The study area

This study was conducted from August 2009 to March 2010 in the Kilombero Wetlands in Morogoro region, South Tanzania. The wetlands are located between the Udzungwa Mountains and the Mahenge escarpment. The valley is divided by the Kilombero River and falls within two districts of Kilombero and Ulanga. The wetland area covers 7,967 km^2 with a catchment's area of about 40,000 km^2. Many rivers, permanent and seasonal, feed the floodplain. The area is characterised by a sub-humid tropical climate with an annual rainfall of about 2000-3100mm. It has two seasons of rains where the long rains are from March to May and short ones from October to December. Temperatures normally vary from 20°C to 30°C.

2.2 The design of the choice experiment and data collection method

Choice experiments are based on two important theories: the traditional microeconomic theory of consumer behaviour and random utility theory [7, 18]. In this study the various wetland attributes, used to design wetland management options, were identified through consultations with experts at Sokoine university of Agriculture (SUA) and with officials of various government departments and NGOs. In order to ensure that these attributes are realistic various focus group discussions with communities around the wetlands were conducted. The government officials, experts and locals in this way identified the attributes which they consider to be most important for wetlands management. A similar identification process was found in several studies [8, 12, 14]. The definition and levels of these attributes are described in table 1.

The flood plain area attribute was considered important by the stakeholders because the areas cultivated by farmers are more and more in conflict with the wetlands. The main effect of crop cultivation is soil erosion, which has resulted into siltation of dams downstream. This attribute was also included in the study of Westerberg et al. [12]. Biodiversity was included as attribute because most communities depend on fishing as one of their sources of income and food. Moreover, tourism activities, a sector that contributes a lot to the Gross Domestic Product (GDP) of the country, is dependent on the occurrence of animal species. To continue having all these benefits it is important to conserve the biodiversity in this area. Biodiversity is a typical attribute to include in the evaluation of management plans. It is found in many studies [8, 11, 12, 14, 15]. The area under controlled livestock farming and the vegetation cover in the catchment areas were also identified as wetland attributes. This is because there is agreement on the negative effects of free range livestock farming and deforestation. It is argued by the officials that the actual number of livestock exceeds the carrying capacity of the wetland, resulting in negative externalities such a decrease in water retaining capacity of soils and drying of swamps and rivers. The vegetation cover is most affected by deforestation for timber extraction and fuel wood utilization. More than 90% of the households in the Kilombero district use fire wood for cooking. Another problem that causes deforestation is the expansion of farming area by the local people. A number of problems are related to deforestation: mainly loss of water retention of the soils and erosion. The payment attribute is included in the experiment to be able to measure the willingness to pay for changes in the other attributes. An addition to the water bill was chosen because

Table 1: The attributes and levels for Various Management option in Kilombero Valley wetlands.

Attributes	**Definition**	**Management levels**
Flood plain area	Flood plain area that remains unconverted into agriculture	-**Bad**: Increase of area under shifting cultivation and decrease in seasonal and permanently flooded streams, swamps and ponds in the flood plain -**Status quo** of the area under shifting cultivation and seasonal and permanent streams, swamps and ponds in the flood plain -**Good**: Decrease of area under shifting cultivation and increase in seasonal and permanently flooded streams, swamps and ponds in the flood plain.
Area under free range grazing	Wetland area that is used for free range grazing	-**Bad**: Continued increase in area under free range grazing -**Status quo** of area under free range grazing -**Good**: Protecting some parts of the wetland area from grazing
Water catchment area	Catchment area that remains covered by the vegetation	-**Bad**: Continued decline in vegetation cover -**Status quo**: current vegetation cover is maintained. -**Good**: Increase of area under vegetation cover from current state by 30%
Biodiversity	The number of different species of plants, wild animals and fish and their population levels.	-**Low**: Probability to catch/see plants and animal species decreases -**Status quo** -**High**: Probability to catch/see plants and animal species increases.
Payment	A one off payment to go to the wetland unit in the Wildlife department	Percent increase of the amount paid as water dues. 0% 5% 10% 15%

a link between the wetland management and water delivery is perceived. Moreover the local communities and experts agreed that water bills would constitute a convenient way to get most people to pay the costs of wetlands management. It was decided to use percentage increase in water bills so as to make each person pay depending on water use capacity as the payment of the bills is per cubic meter used.

Using SAS macros (see [19] for details) a generic D-optimal choice design was developed. From the attributes and their levels 9 choice cards, each with 3 management options and 1 status quo option, were constructed. To reduce the burden upon the respondents the 9 sets were blocked into 3. In this way each individual has to complete 3 choice tasks.

Data were collected face to face which gave us the opportunity to clearly explain the choice sets. In order to enhance this, cheap talk was devised following Nam and Bennet [9] who argue that in developing countries' situation, cheap talk makes it simple for people to grasp the management options and their levels. During the cheap talk various photos were portrayed to demonstrate the three status levels of the wetlands. The photos were taken from places where the destruction was evident to show what it meant by bad and those from conserved areas to show what it meant by good and to some places where the destruction was average to show the status quo as an average condition country wide. In the cards to be chosen the words were shortened to make an individual choice easy and meaningful. The design of the card is as portrayed in table 2.

Table 2: Choice experiment card for various management options.

<table>
<tr><td></td><td>Option A</td><td>Option B</td><td>Option C</td><td rowspan="6">Neither management option A, option B nor management option C</td></tr>
<tr><td>Biodiversity</td><td>High</td><td>Low</td><td>Low</td></tr>
<tr><td>Water catchment area</td><td>Good</td><td>Bad</td><td>No change</td></tr>
<tr><td>Flood plain area</td><td>Bad</td><td>No change</td><td>Bad</td></tr>
<tr><td>Area under free range grazing</td><td>Bad</td><td>No change</td><td>Good</td></tr>
<tr><td>Payment</td><td>10%</td><td>15%</td><td>5%</td></tr>
<tr><td colspan="5">I will prefer: Option A Option B Option C Neither….
(Please tick as appropriate)</td></tr>
</table>

A total of 408 households were randomly selected for the interview. The sampling frame consisted of all rural communities in and around the Kilombero wetlands, who are the primary users of the wetlands. As the study meant to advice the policy makers on how the wetlands can be sustainably managed in the whole country of Tanzania it was seen important to extend the sample into secondary users of the wetland and hence households in Morogoro municipal were randomly selected to represent the secondary users of the wetlands.

2.3 Data analysis

The responses from the CE were analysed using a Random Parameter Multinomial Logit Model. In this type of model which, unlike the conditional logit model, is not based on the Independent from Irrelevant Alternative (IIA) assumption, unobserved heterogeneity in preferences across respondents can be accounted for [18, 20]. This is important because an understanding of who will be affected by a policy change in addition to understanding the aggregate economic value associated with such changes is necessary [8, 16]. Furthermore the model can explicitly account for the repeated nature of the choices made [21]. In practice a person faces a choice among the alternatives in choice set j on each of the occasions they make a choice. The utility that respondent n obtains from alternative j in choice situation t consists of a systematic and a stochastic part (Ref. equation 1)

$$U_{njt} = \alpha_{nj} + \gamma_j S_n + \beta_n\, x_{njt} + \varepsilon_{njt} \tag{1}$$

where x_{njt} is a vector of attributes and S_n is a vector of socio-economic characteristics. The alternative specific intercept α_{nj} captures an intrinsic preference for the alternative and $\gamma_j S_n$ captures systematic preference heterogeneity as a function of individual characteristics. The coefficient vector β_n varies among the population with density f $(\beta|\theta)$, where θ is a vector of the true parameters of the taste distribution.

If the ε_{njt} are IID type I extreme value, we have a random parameter logit model [22]. The conditional probability of alternative j for individual n in choice situation t is then (Ref. equation 2).

$$P_n(jt\beta_n) = \frac{\exp\,(\alpha_{jn}+\gamma_j S_n+\beta_n\, x_{njt})}{\sum_{i\in A_k} exp\,(\alpha_{in}+\gamma_i S_n+\beta_n\, x_{nit})} \tag{2}$$

where Ak={A1,. . ., AN} is the choice set.

In this paper we will use a simulated maximum likelihood estimator, using Halton draws, to estimate the models [20]. It is furthermore also necessary to make an assumption regarding the distribution of each of the random coefficients. In principle any distribution could be applied. However, the choice is often limited by difficulty of model estimation and availability of econometric software. Following [8] we use a normal distribution for the random coefficients. Even though unobserved heterogeneity can be accounted for in the RPL model, the model fails to explain the sources of heterogeneity. Therefore we include interactions of respondent-specific socio-economic and attitudinal characteristics with choice specific attributes. In this way our model is able to pick up preference variation in terms of both unconditional taste heterogeneity (random heterogeneity) and individual characteristics (conditional heterogeneity), and hence the model fit is improved [8, 23]. The characteristics selected were education level, income, ownership of land for farming, ownership of livestock, and gender.

3 Results and discussion

3.1 Socio-economic characteristics of the respondents

The sample consisted of individuals with various socio-economic characteristics. Most (60%) of the respondents followed at least primary education, only 4% completed adult education and 20% completed secondary education. The remaining 26% had a higher education degree. The majority of the households

(52%) had an income between 100 000 and 1 million TAS (*1$ is approximately equivalent to 1450 Tanzanian Shillings, TAS*). It is interesting to note that the persons with the highest educational level and highest household income lived in Morogoro Municipality. Gender distribution was nearly even, with male respondents representing 58% of the respondents. The respondent's age ranged from 17 to 81 years old.

3.2 Communities preferences over management option

The results of the random parameter model without interaction terms reveal that in general people from Morogoro and Kilombero are in favour of better management strategies for the wetlands. This is in line with other studies [12, 15], who also found a positive public attitude towards wetland conservation. Here this can be deduced from the positive and significant coefficients for all attributes. For each attribute one has an estimated coefficient for the mean of the distribution, and one for the variance of the distribution. Associated with each of these is an estimate of the standard error, so one can draw standard inferences about the significance of the coefficient. If the estimate of the variance is not different from zero, then one can infer that the preference parameter is constant across the population. If the mean coefficient is zero, but the variance estimate is significant one cannot infer that the attribute does not affect choice, but rather that there is a diversity of preferences, both positive and negative. For an attribute to be declared as having no impact on choices, both the estimate of the mean and the variance has to be insignificantly different from zero. In contrast to the other attributes the coefficient for the payment vehicle is negative, meaning that a higher price reduces the probability of a choice option being chosen. From the four attributes biodiversity has the highest coefficient. This might be due to the communities' dependency on wild game, tourism and fishing (Table 3). Tanzania National Park Authority (TANAPA) which manages Udzungwa National Park has highly invested in conservation of the area through awareness creation and legal enforcement. Furthermore one has to note that 25% of Kilombero council income comes from fishing activities, which for sure also has implication on household income for most of the communities around Kilombero Valley. Finally fish and game meat still have an important role in the diets in the region which also leads to a high preference for a management option that would increase the biodiversity of the Kilombero Valley ecosystems.

The following attribute after biodiversity is the increase in vegetation cover in the catchment areas. This is an interesting finding, since it is an indication that the national campaign on tree planting and vegetation conservation has helped to create awareness in Tanzania's communities on the value of forests and trees. The vegetation cover in catchment areas and other forested land in Tanzania is in an alarming situation. Data from UN [24] suggest that the deforestation rate is currently at *412,000/ha/yr* and the forested land is 34 million hectares. It is therefore encouraging to have this general preference from the communities which indicates that if there would be proper programming in forest management there is a willingness among communities to cooperate into improving the

Table 3: The Basic Random Parameter Logit Model and with interaction estimates for wetlands management attributes.

	The basic model		The model with interaction	
Attribute	Coefficient	Coef. std(s.e)	Coefficient	Coef std(s.e)
Biodiv	2.07***	13.15 (0.15)	2.15***	12.40(0.16)
Catchment	0.80***	15.02 (0.11)	2.21***	15.76(0.24)
Flood plain	0.71***	8.37 (0.10)	0.69***	9.10(0.11)
Free range	0.37***	8.30 (0.10)	0.23**	9.41(0.10)
Payment	-0.004***	-	-0.004***	-
Flood Plain*Landown	-	-	-0.004**	-
Flood Plain*High inc	-	-	4.39***	-
Catchm*Gender	-	-	- 0.33	-
Biodiv*Landown	-	-	- 0.03**	-
Biodiv*High inc	-	-	0.03	-
Catchm*Low edu	-	-	-0.62***	-
Biodiv*Livest Own	-	-	-0.004***	-
Observations	*4884*		*4884*	
Log likelihood	*-3204.770*		*-3124.054*	
Pseudo R Square	*0.036*		*0.035*	
Chi-Square	*241.470*		*227.794*	
Degrees of freedom	*4*		*4*	

** Significant at 5%, ***significant at 1%

situation. Other studies also show that it is not the intention of the communities to cause deforestation, but that they have no alternatives for firewood as energy source for cooking [25–27]. It is up to the government to come up with policies that provide alternatives to people.

The smallest coefficient is found for the free range grazing attribute. Probably the fact that a decrease in free ranging area is perceived as a call for a reduction in the stock size; which would directly touch upon the livelihood of many of the respondents could explain this. Policy makers should always consider such impacts on livelihoods when coming up with policies. One should consider that livelihoods might be affected when certain strategies are sought for the improvement of the wetlands status. Much of this will be noted in the model where some socio-economic characteristics were interacted with some attributes.

After having the general results from the basic model we extended it by including interaction of some of the attributes with certain socio-economic characteristics. After extensive testing of various interactions of the four wetland management attributes with the respondents' social, economic and attitudinal characteristics collected in the survey, the model including interactions with education level, income level, ownership of land for farming, ownership of livestock, and gender was found to best fit the data .

The results of the interactions confirm that livelihoods affect preferences of respondents. The probability that respondents who owned crop farms in the valley choose for a decrease in shifting cultivation is for example smaller than that of the average respondents (negative coefficient of flood plain-landownership interaction). This observation shows that policy makers should carefully think about the policies they formulate and how they affect different groups. As a matter of advice as was noted in the scholarly documentation that in order to bring cooperation then the issue of who benefit and who incur cost into such undertakings should guide the conservation strategies [2, 28, 29]. The research revealed that while the crop farmers are less in favour of conservation practices, households with high incomes which are mostly from Morogoro municipal are clearly opting for an increase in flooded area throughout the year. This is not surprising because they depend on the flood plains for their water supply and have other alternatives for food and income sources, which the farmers who live in the floodplains do not have. A similar reasoning can be made for the biodiversity-land ownership interaction. For many of the attributes our study thus confirms the findings of [15], who found that livelihood strategies and location influence the preferences for wetland conservation. They therefore decided that the spatial distribution of stakeholder preferences for conservation and development is very important in shaping the wetland related policy in unprotected areas. The interaction between the vegetation cover attribute and education offers another important finding. Those who are lowly educated are far less in favour of decreases in tree extraction. This is thought to be due to the fact that people in the rural areas make a high proportion of illiterate and uninformed folks about the values of vegetation cover in wetlands. Also, they are primarily dependent on these resources for sustaining their livings in terms of cash, energy and habitation. This puts them into a dilemma of whether to conserve the vegetation or harvest it to sustain their living. As it has been discussed in the general findings it is therefore advised to come up with strategies that would help to reduce pressure on these areas. Although the results were marginally significant at 10%, gender also shows a distinct pattern, where males are opting for low status wetlands management in terms of vegetation cover in the catchment area; this is partly contributed by how conservation programmes are carried on in Tanzania. It is reported by Songorwa and Sikira [30] that more women are involved in these programs. Another reason might be that men as main bread earners have more insight in the potential livelihood effects. Studies by Ellis et al. [31] and URT [32] support this argument.

3.3 Determination of willingness to pay

The willingness to pay and implied ranking from the general basic model for each attribute was calculated following [33] (see table 4). In terms of ranking, biodiversity goes first followed by increase in catchment area vegetation cover. The details are as we have earlier discussed when we were determining communities' preferences. The overall marginal willingness to pay per household for improving all attributes is 0.683 $USDHh^{-1}month^{-1}$. This is a very encouraging amount for a community in a developing country like Tanzania. Moreover as of to date no budget goes specifically to the conservation activities in our wetlands ecosystems, institutionalisation of this payment could already make a big difference.

It is however important to come up with proper incentive mechanisms that would bring all stakeholders together into conserving the wetlands. The possible way to do this is through application of economic instruments, an avenue which is available since the country has a specific regulation in the newly enacted Environmental Management Act of 2005 [35].

Table 4: Willingness to pay and implied rankings in Management options of Kilombero Valley Wetlands.

Attribute	Coeff	Implied Ranking	WTP (USD)$Hh^{-1}Month^{-1}$
Biodiversity	2.071	1	0.357
Catchment area	0.804	2	0.139
Flood plain area	0.714	3	0.123
Free range grazing	0.373	4	0.064

1 USD is equivalent to 1450 TAS

4 Conclusion

In general there is a preference among communities to improve management of the wetlands, which reveals that the inhabitants of the Kilombero Valley and Morogoro Municipality desire improvement of the wetland's conditions.This suggests that the ongoing degradation is socially not optimal. Secondly, looking at the attributes associated with the wetlands, the study shows that land ownership and cultivation type are very important factors. Farmers are more reluctant to shift their practice and to give up some of their land in order to enhance the conditions of the wetland because this would drastically affect their household income. This being the case it is therefore important for the policy makers to have enough information of what are the really causes of the wetlands degradation in specific areas before they develop conservation strategies. Communities' involvement in finding such strategies is important to ensure effectiveness and sustainability of the strategies to be developed. Besides of ensuring sustainability, such approach would also reduce the cost of management of the wetlands as takes into account the best interest of all stakeholders. The study therefore recommends that in order to bring participation in wetlands conservation, the policies should consider compensation of the negatively affected group such as small farmers, whose income is entirely dependent on shifting cultivation. It is further recommended that other income generating activities which would reduce the pressure on the wetlands should be explored. But even more, investing in the development of proper technologies that would improve efficiency and effectiveness in crop production should be pursued.

References

[1] Fischer, G., *Land and Water Use of Wetlands in Africa: Economic Values of African Wetlands.* Interim Report No. IR- 02-063. Luxemburg, Austria, 2002.

[2] Turner, R.K., van den Bergh, J.C.M., Soderqvist, T, Barendregt, A., van der Straaten ,J; Maltby, E & van Ierland, E.C. Ecological-economic analysis of wetlands: scientific integration for management and policy. *Ecol Econ*, **35,** pp. 7– 23, 2000.

[3] Pearce, W.D. & Tunner, R.K., *Economics of Natural Resources and the Environment*. BPCC Wheatons Ltd: Exeter- Britain, pp. 378, 1990.

[4] Turpie, J. K., The existence value of biodiversity in South Africa: how interest, experience, knowledge, income and perceived level of threat influence local willingness to pay. *Ecological Economics,* **46 (2)**, pp. 199-216, 2003.

[5] Birol, E., & Cox, V., Using choice experiments to design wetland management programmes: The case of Severn Estuary wetland, UK. *Journal of Environmental Planning and Management,* **50**, pp. 363-380, 2007.

[6] Jenkins W.A., Murray B.C., Kramer R.A. & Faulkner, S. P. Valuing ecosystem services from wetlands restoration in the Mississippi Alluvial Valley. *Ecological Economics,* **69**, pp. 1051–1061, 2010

[7] Alpizar, F., Carlsson, F. & Martinsson, P., (eds). *Using Choice Experiments for Non-Market Valuation,* Working Papers in Economics, No. 52, Department of Economics, Gothenburg University, 2001.

[8] Birol E., Karousakis K., & Koundouri P., Using a choice experiment to account for preference heterogeneity in wetland attributes: The case of Cheimaditida wetland in Greece. *Ecological economics,* **60**, pp. 145 – 156, 2006.

[9] Nam, D and Bennet, J., Estimating wetland biodiversity values: a choice modelling application in Vietnam's Mekong River Delta. *Environment and Development Economics,* **14**, pp. 163–186, 2008.

[10] Adamowicz, W., Louviere, J., & Williams, M., Combining revealed and stated preference methods for valuing environmental amenities. *Journal of Environmental Economics and Management,* **26**, pp. 271–292, 1994.

[11] Herath G., Incorporating community objectives in improved wetland management: the use of the analytic hierarchy process. *Journal of Environmental Management,* **70**, pp. 263–273, 2004.

[12] Westerberg, V.H., Lifran, R., & Olsen, S.D., To restore or not? A valuation of social and ecological functions of the Marais des Baux wetland in Southern France. *Ecological Economics,* **69**, pp. 2383–2393, 2010.

[13] Jogo W., Hassan R., Balancing the use of wetlands for economic well-being and ecological security: The case of the Limpopo wetland in southern Africa. *Ecological Economics,* **69**, pp. 1569–1579, 2010.

[14] Birol E., Hanley N., Koundouri P., & Kountouris Y., Optimal management of wetlands: Quantifying trade-offs between flood risks, recreation, and biodiversity conservation. *Water Resources Research,* **45 (11426),** 2009.

[15] Wattage, P., & Mardle S., Stakeholder preferences towards conservation versus development for a wetland in Sri Lanka. *Journal of Environmental Management,* 77, pp. 122–132, 2005.

[16] Boxall, P.C., & Adamowicz, W.L., Understanding heterogeneous preferences in random utility models: a latent class approach. *Environmental and Resource Economics*, 23, pp. 421–446, 2002.
[17] Milon, W., & Scrogin, D., Latent preferences and valuation of wetland ecosystem restoration. *Ecological Economics*, **56 (2),** pp. 162–175, 2006.
[18] Louviere, J.J., Hensher, D.A. & Swait, J.D., *Stated choice methods: analysis and application.* Cambridge University Press: Cambridge, U.K., 2000.
[19] Kuhfeld, W.F. "Marketing Research Methods in SAS.", Online:www.support.sas.com/resources/papers/tnote/tnote_marketresearch.html
[20] Carlsson, F., Frykblom, P., & Liljenstolpe, C., Valuing wetland attributes: an application of choice experiments. *Ecological Economics*, **47**, pp. 95–103, 2003.
[21] Rigby, D & Burton, M., Capturing preference heterogeneity in stated choice models: a random parameter logit model of the demand for GM food. Discussion Paper Series No. 0319. School of Economic Studies: The University of Manchester, 2003.
[22] Train, K.E., Recreation demand models with taste differences over people. *Land Economics*, **74**, pp. 230–239, 1998.
[23] Revelt, D., & Train, K., Mixed logit with repeated choices: households' choice of appliance efficiency level. *Review of Economics and Statistics,* **53**, pp. 647–657, 1998.
[24] UN, The Millennium Development Goals Report. New York, 2007.
[25] Holmes, C.M., Cultural variation, decision making, and local institutions: An examination of fuelwood use in western Tanzania. *Society & Natural Resources*, **18**, pp. 61-73, 2005.
[26] Naughton-Treves, L; Kammen, D.M & Chapman, C., Burning biodiversity: Woody biomass use by commercial and subsistence groups in western Uganda's forests. *Biological Conservation*, **134**, pp. 232-241, 2007.
[27] Lusambo, L. P., Economics of Household Energy in Miombo Woodlands of Eastern and Southern Tanzania. Ph. D. dissertation. University of Wales: Bangor, UK., pp 493, 2009.
[28] Costanza, R., Thinking Broadly About Costs and Benefits in Ecological Management; Integrated Environmental Assessment and Management. TBACAB, pp. 166–173, 2005.
[29] Birdyshaw, E. & C .Ellis, Privatizing an open-access resource and environmental degradation. *Ecological Economics,* **61(2-3),** 2007.
[30] Songorwa, A. and Sikira, A., Women participation and empowerment in community-based wildlife management: The case of serengeti regional conservation project, Tanzania. Annual meeting of the International Congress for Conservation Biology, Convention Center, Chattanooga, Online. www.allacademic.com/meta/p243744_index.html
[31] Ellis, A. Blackden M., Cutura, J.; MacCulloch, F. & Seebens, H., (eds). *Gender and Economic Growth in Tanzania; Creating Opportunities for Women.* The World Bank: Washington, D.C., 2007.

[32] United Republic Of Tanzania (URT), Draft Report on Annotated Outline For National Strategy For Growth And Reduction Of Poverty. Dar es Salaam, Tanzania, 2010.
[33] Hanley, N. & Adamowicz, W., Using Choice Experiments to Value the Environment. *Environmental and Resource Economics.* **11(3-4),** pp. 413-428, 1998.
[34] United Republic Of Tanzania (URT), *The Environmental Management Act.* Dar es Salaam, Tanzania, 2005.

Section 2
Natural resources management

A participatory approach for modeling alternative future land use scenarios around Nairobi National Park using Bayesian Belief Networks

J. T. McCloskey[1], R. J. Lilieholm[1], R. Boone[2], R. Reid[2], S. Sader[1], D. Nkedianye[3], M. Said[3] & J. Worden[4]
[1]*School of Forest Resources, University of Maine, USA*
[2]*Warner College of Natural Resources, Colorado State University, USA*
[3]*International Livestock Research Institute, Nairobi, Kenya*
[4]*African Conservation Fund, Nairobi, Kenya*

Abstract

We develop an adaptive and flexible framework for engaging experts and stakeholders at the household- and community-level that have different livelihoods and land use interests within Kenya's Athi-Kaputiei Plains (AKP). We use Bayesian Belief Networks linked to GIS data layers to integrate empirical data and elicited stakeholder knowledge. The process is designed to address problems with past conservation-development strategies by allowing participants to build relationships among people with different land use interests in order to clarify opportunities and constraints, examine assumptions at the design phase of a project, and determine future actions and potential development scenarios. We use an example of four different livelihood groups in the AKP to demonstrate how the process might work to identify suitable areas for alternative land uses (e.g., wildebeest and livestock grazing, crop cultivation, and urban development), and to identify future compatibilities and conflicts between these different land use interests. The modeling process provides a maximal coverage strategy that allows decision makers to target and prioritize areas for protection or development, and to set specific strategies in the face of changing ecological, social, or economic processes. The process is iterative so that revised models can be developed as new data and knowledge arise, thereby

WIT Transactions on Ecology and the Environment, Vol 144, © 2011 WIT Press
www.witpress.com, ISSN 1743-3541 (on-line)
doi:10.2495/ECO110041

helping communities and other interests learn from past successes and failures, and better evaluate the impacts of alternative land uses.

Keywords: Bayesian Belief Networks, collaborative management, community-based conservation, integrated conservation and development projects (ICDPs), landscape change, land use, livelihoods, Nairobi National Park, participatory modeling, pastoralism, wildlife.

1 Introduction

The challenges and limitations facing conservation strategies such as integrated conservation and development projects (ICDPs), community-based conservation (CBC), and traditional protected areas like national parks are well documented in the literature [1–4]. These include the need for thematic coherence at the design phase, an understanding of community heterogeneity and complexity, including local communities in all phases of a project, mechanisms for monitoring and evaluation with respect to project goals and objectives, collaboration with other relevant projects, and the adoption of adaptive and flexible management approaches [1, 4].

Here, we address the shortfalls learned from past conservation and development strategies by developing a participatory framework that demonstrates how a diverse set of stakeholders with different land use interests can be engaged to identify suitable areas for wildlife migration, livestock grazing, crop cultivation, and urban development in the Athi-Kaputiei Plains (AKP) – the region south of Kenya's Nairobi National Park (NNP). The method uses Bayesian Belief Networks (BBN) linked to spatial data in order to identify suitable areas for alternative land uses, and to identify future compatibilities and conflicts. The result is a stakeholder-driven land use planning and analysis tool that provides decision makers with multiple options for targeting and prioritizing areas for conservation protection and economic development.

2 Background

Sustainability science recognizes the need to integrate social, economic and ecological information. Adaptive management is the continuous integration of design, management and monitoring to systematically test assumptions in order to adapt and learn [5]. The ultimate goal is to design resilient institutions that have the flexibility to modify policy in response to changing ecological and socio-economic environments. To build truly adaptive institutions, however, scientists, stakeholders and planners must collaboratively interpret and integrate scientific knowledge, evaluate the outcomes of management decisions, and revise and improve future management policies and actions [6]. The proximate goal, then, is to identify the relationships between scientific research, past experience, and public policy, thereby forging a mechanism for effective policy change (e.g., [7–9]).

Sociologists suggest that for meaningful policy change to occur, three aspects of the social structure must interlock and reinforce each other: (1) the allocation of resources and power that governs our ability to take effective action; (2) the myths, paradigms, or ideologies that frame our activities; and (3) the rules and norms that organize human actions [10]. These factors affect not only the ability of an institution (or individual) to detect and understand new scientific knowledge, but also its capacity to alter its worldview, which is implicit for successful knowledge integration and adaptive management.

First, the allocation of resources and power (or lack thereof) is often cited as a primary factor limiting the success of conservation efforts such as ICDPs, CBC projects, and national parks and reserves [1–4]. Indeed, few conservation strategies succeed without long-term financial support and effective laws, policies and regulations [2]. Meeting these requirements necessitates the integration of factors across multiple scales – from international and national policy advocacy, to site-based participation from local stakeholders. Wells et al. [11] review lessons learned from past ICDPs and offer advice for financial sustainability. Here, an important first-step is the creation of a participatory process that brings together diverse interests during the conceptual phase of a project to ensure that local- and national-level policies and conservation-development linkages are better understood and represented [3, 12].

The second factor cited above – the myths, paradigms, or ideologies that frame human activities – has also been identified as a weakness of many conservation strategies [1–4]. Indeed, ICDPs and CBCs have emerged over the last few decades in response to the recognition that humans and protected areas must successfully coexist in order to ensure the long-term viability of conservation efforts. These models emerged in reaction to the "guns and fences" approach that dominated conservation strategies until the 1990s, and which too-often failed to account for the interactions and complexities of coupled socio-ecological systems (SES) [13–15].

While effectively operating within such complexity requires greater effort in planning and decision-making, if individuals working together share a common world-view (i.e., paradigm), complex decisions can be easier to address [10]. However, the stronger a paradigm, the less receptive an institution will likely be to new knowledge [10]. Thus, any successful conservation-development strategy will need to have adaptable and flexible institutions and decision-making processes. Furthermore, complicated ecological modeling methods may actually add to the uncertainty of scientific knowledge, thereby complicating the decision-making process for practitioners and laypeople [9, 16]. We agree with Salafsky and Margoluis [5], who suggest that adaptive management does not necessarily need complicated methods, but instead should concentrate on designing conceptual models of relationships capable of testing assumptions, monitoring natural and human SES features, and systematically testing alternative actions. Any adaptive process should also encourage participation and learning by a diverse set of institutions, organizations, stakeholders, and individuals.

Finally, the rules and norms that organize our activities – together with power structures and paradigms (i.e., the other two aspects of the social structure) – combine to shape our basic assumptions of the world. A synthesis of ICDP and

CBC case studies suggests that weak assumptions, not identified until long-after implementation, have constrained the success of many projects [2, 11, 12]. These same studies show, however, that many false assumptions could have been discovered at the design phase of the project through in-depth discussion with a diverse set of participants [2, 11, 12]. Case study evaluations have consistently called for improved collaborative decision-making between planners and communities [1, 3, 4, 12, 17, 18].

We use this view of social structure and its links to policy change as the foundation for a participatory process designed to accommodate the knowledge, values, and opinions of a diverse set of decision makers, scientists, and stakeholders. We feel that this suite of interests, often excluded from protection strategies in the past, provides the key to strengthening the sustainability of future conservation and development efforts.

3 Study area

Kenya's AKP cover over 2,590 km^2 of rolling plains that once supported the migration of wildlife populations second in size to only the Mara-Serengeti region [19], fig. 1. Nairobi National Park covers a small portion of the northern AKP system, but serves as a crucial reserve for wildlife during the dry seasons. The Park is fenced on three sides and bordered to the north by Nairobi – one of the largest and fastest-growing cities in Africa [20].

Nairobi's population has increased from 500,000 people in 1970 to over 3 million people today [20]. This growth has been characterized by residential and commercial expansion and intensified land use. With limited land use planning, growth has outpaced infrastructure and human services to create large slums and unplanned settlements in peripheral areas, fig. 2. Lack of planning, combined with physical constraints and mounting environmental impacts, threaten the sustainability of both human and natural systems, fig. 2. These include the viability of urban centers, Maasai pastoral livelihoods, and broader landscape processes such as globally significant wildlife migration patterns [20].

4 Socio-ecological (SES) setting: identifying participants

Nkedianye et al. [21] have shown four distinct livelihood strategies in the AKP – diversified agro-pastoralists, pastoralists with wildlife income, marginal pastoralists, and wage-earning agro-pastoralists. Pastoralism remains the most important source of income for many in the AKP. Nevertheless, wildlife-based tourism is still viewed as the basis for sustainable development across many of East Africa's rangelands [22]. The Maasailands of Kenya and Tanzania alone generate an estimated $1.5 billion in tourism revenues each year [22]. Past policies and many development projects are based on the assumption that community-based or private conservation initiatives will help reduce poverty while maintaining wildlife populations. In reality, the benefits from such initiatives are often limited, and there is gathering evidence that these assumptions are unrealistic – both in terms of increasing wildlife populations and

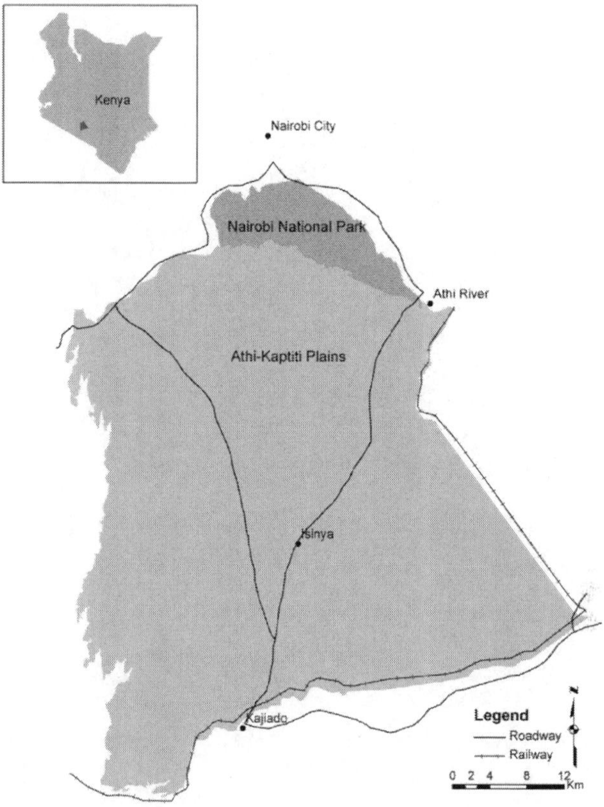

Figure 1: Nairobi National Park and the Athi-Kaputiei Plains of Kenya.

reducing poverty [22–25]. In fact, most wildlife populations across Kenya have declined considerably over the last 30 years despite the work of numerous governments, non-government organizations (NGOs), entrepreneurial initiatives, and scientific research efforts [26, 27].

In terms of poverty reduction, Homewood et al. [22] demonstrate that across Maasailand, wildlife earnings are poorly distributed among local communities and contribute little earnings when compared to households that rely on livestock, farming, and/or non-farm wage work. In the rare communities where wildlife revenues do generate a significant proportion of income, it is primarily distributed to households who own land near high-visitation conservation areas. Furthermore, households earning the most from wildlife often invest in land use practices (e.g., cultivation) that may undermine wildlife or pastoral livelihoods [22]. This is consistent with other studies that dispute the widespread assumption that by providing alternative income opportunities, ICDPs can reduce the use of natural resources by local people [1, 17].

Figure 2: A cement plant in the north-central portion of the top photo borders the eastern edge of Nairobi National Park, shown to the west. The town of Athi-River lies to the east. Such towns develop around industrial areas and transport corridors as a result of ad-hoc land use planning. The bottom photo shows Nairobi's Kibera slums, a densely-settled area located just 3km north of Nairobi National Park (courtesy of Google Earth).

Homewood et al. [22] offer three explanations for the apparent gap between the expectations and assumptions of policies and decisions based on conservation theory and the reality for individual households and wildlife populations across Maasailand. These include: (1) how ecotourism revenues are distributed; (2) how pastoral production systems are perceived and valued by different people and institutions; and (3) how project impacts are measured. These considerations are essentially the same as those of any social structure identified above by Westley [10] as crucial for linking knowledge to meaningful changes in policy (i.e., the allocation of power and resources, paradigms/world-views, and rules and norms). The first step, then, is to identify participants from the community and household-level who can represent the four livelihood strategies identified by Nkedianye et al. [21], as well as decision makers and experts who understand the policies, regulations, and institutional arrangements of the area (e.g., planners, developers, conservationists, agricultural interests, and scientists), fig. 3.

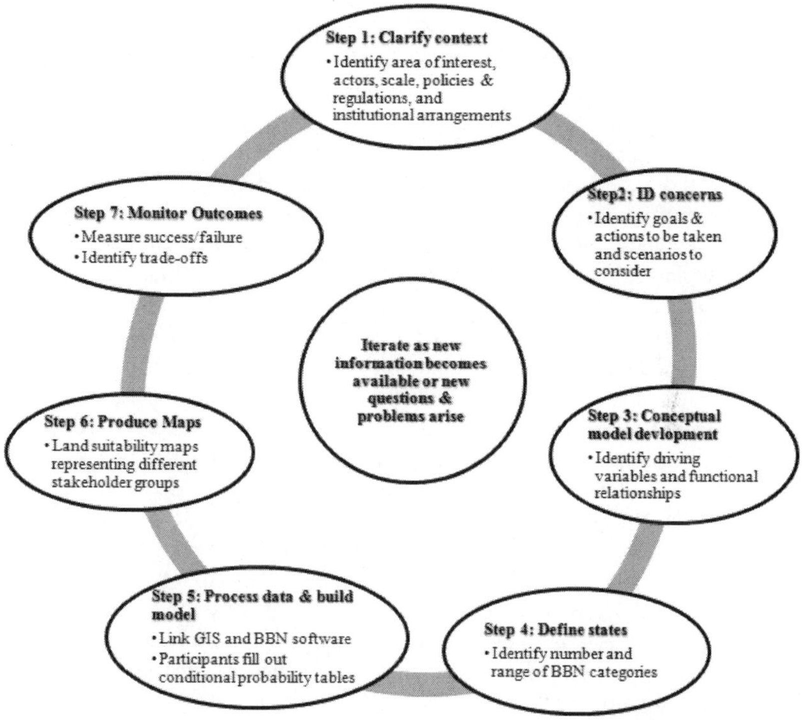

Figure 3: A framework for a participatory process to identify the suitability and potential conflict of different land uses.

5 Translating SES setting into spatial variables

A number of historic and geopolitical events have marginalized or excluded some interests and benefited others in the AKP region, thus contributing to the structural poverty of the area [22, 28]. In addition, impacts from biophysical events (e.g., drought) and the loss of mobility and access due to changes in land use (i.e., gazetting of protected areas, urbanization, and large-scale cultivation) have contributed to conjunctural poverty [22, 28]. Alleviating these two types of poverty requires different and complex strategies and tactics over various scales. In either case, those that have been well-positioned during past events (e.g., adjacent to wildlife reserves or perennial water sources) have an advantage over those less-well placed [22]. Indeed, like everywhere, different stakeholders in the

AKP have different types of power and resources. Consequently, they have different world views, understandings of local processes, and political and economic interests and influences [22]. These differences in the social structure among AKP participants influence the identification, threshold settings, and ranking of variables deemed important in any participatory modeling process.

5.1 Stakeholder modeling workshops

Using this knowledge, we designed a process similar to W.E. Deming's continual improvement model [29] that includes four separate workshops – one for each livelihood strategy – to begin to engage household and community-level participants and build relationships, fig. 3. During the workshops, participants first clarify the freedoms and constraints (e.g., policies and regulations) in order to identify the range of possibilities. Next, the actions that are needed and scenarios that might arise are considered. These first two steps begin to identify real and perceived assumptions. Next, we identify important environmental and economic variables that can be spatially associated with the suitability of land for wildebeest and livestock grazing, crop cultivation, and urban development. The variables are identified by conducting a review of the current literature, engaging community and household-level stakeholders in workshops and focus groups, and holding multiple meetings with scientists and planners with expertise in ecology, economics, cultural anthropology, and agriculture. During the workshops, participants identify important biophysical metrics to be used as spatial variables intended to identify areas that stakeholders deem to be suitable for wildebeest, livestock, crop cultivation, and urban development. These stakeholder- and expert-defined metrics are then used to create BBNs describing the functional relationships between variables. Participants then rank the importance of the *a priori* spatial conditions identified by the empirical data. If there is debate over the inclusion or exclusion of variables or definitions, differences can be explored to determine consistency of outcomes or to allow explicit weighting of alternative views. During the workshops, scenarios are identified for further analysis of alternative futures for wildebeest, livestock, and crop cultivation, in order to assess the likely consequences and tradeoffs between conserving these areas and allowing urban development.

Prior to the workshops, we use the expertise of scientists to identify potential variables used to develop "pilot" BBN influence diagrams for each land use, fig. 4. The first box (i.e., land availability) in each diagram acts as a filter that, under an idealized scenario, provides a considerable amount of land for development while protecting important areas (e.g., wetlands). The amount of land potentially available for each use will be determined from a land cover map based on high-resolution SPOT data acquired in October 2010. For example, the development diagram in fig. 4 contains six GIS and remotely sensed data layers (variables) thought to influence commercial and residential development in the AKP. The initial land cover map is then re-coded to represent the two states – available and unavailable for future development in the AKP. Because we are interested in future development, we assume existing urban areas to be unavailable. Likewise, because we are interested in finding common ground

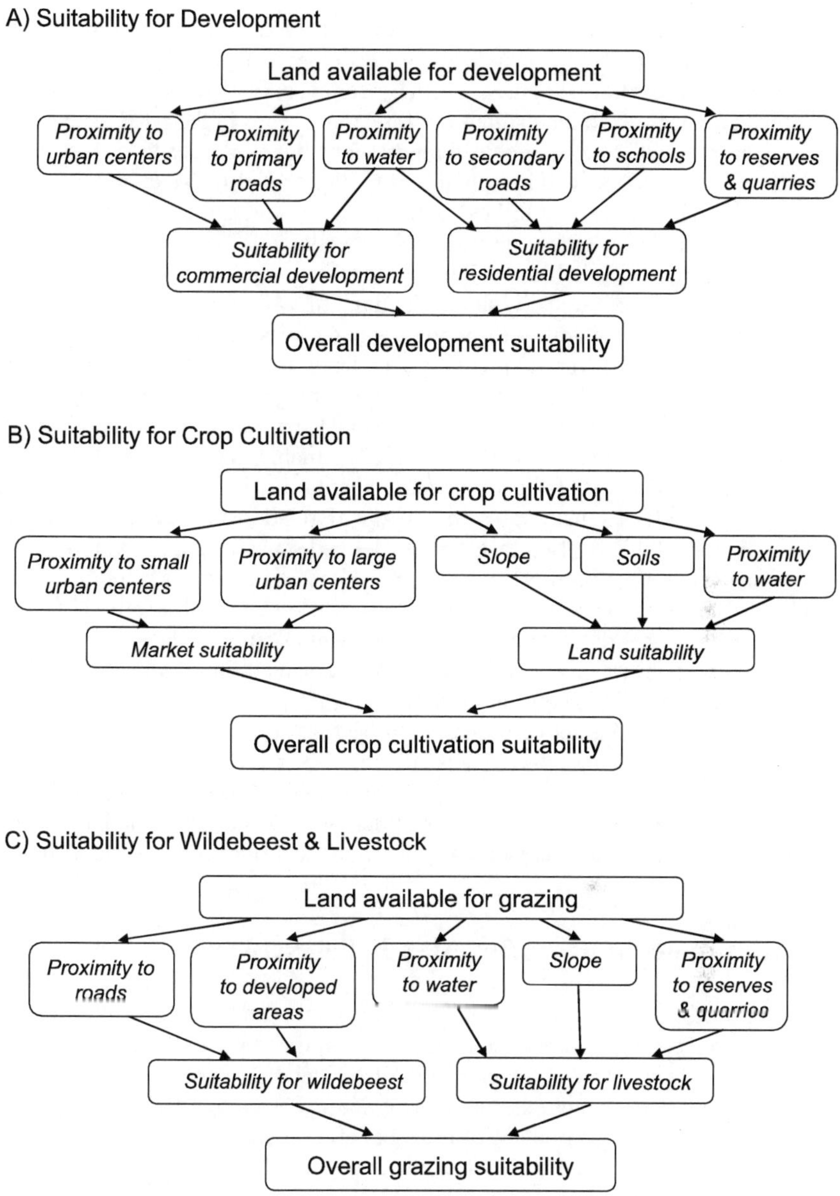

Figure 4: Influence diagrams showing key factors affecting development (A), crop cultivation (B), and wildlife and livestock grazing (C) in the AKP.

between development and conservation interests, riparian areas are assumed unavailable for development. Areas already classified as current conservation lands are also assumed to be unavailable for development.

The crop cultivation diagram, fig. 4B, contains five variables thought to influence the market and land suitability for certain crops in the AKP. The initial land cover map will be re-coded differently than above in order to represent the two states (i.e., available and unavailable) for all potential future crop cultivation lands in the AKP. Unavailable land includes urban areas, roads, water, current conservation lands, quarries, nurseries, and woodlands. All other land cover types are considered available as potential future crop land.

Finally, the wildlife and livestock diagram, fig. 4C, includes five variables thought to affect the suitability for wildebeest and livestock grazing in the AKP. Again the initial land cover map will be re-coded to represent the two states (i.e., available and unavailable) for all potential future grazing lands in the AKP. Unavailable land includes urban areas, roads, quarries, nurseries, and current crop lands. All other land types are considered available as grazing land.

The nodes and links of the influence diagrams in fig. 4 will eventually be developed into BBNs that represent functional relationships among variables [30]. Each node will have two-to-four user-defined states with a probability table that expresses the probability of each state either as prior distributions or as conditional on the probability of each state for the source nodes [30]. The prior probability tables are derived from empirical spatial data, whereas the conditional probability tables (CPT) are entered manually based on stakeholder and expert opinion elicited during the workshops [31].

The output from the above process will be four suitability maps for each of the four land uses. These four maps are then over-layed with each other to identify areas of potential conflict and compatibility [32]. A final workshop brings together the four groups of stakeholders in order to get an understanding of how the different land uses complement and conflict under various future development scenarios.

6 Advantages of using Bayesian Belief Networks

As simplifications of reality, all models have strengths and weaknesses, and none provide a panacea for understanding and managing complex natural and human systems [33]. This is true whether they be BBN models, expert system models, agent-based models, or neural networks. When it comes to BBNs, they can be limited by the quality and extent of prior beliefs or conditions, their inability to deal with unusual or unanticipated events, and the computational difficulty of exploring previously unknown networks [33].

In this case study, the first limitation is addressed by specifying the prior probability tables from case files of empirical spatial data [31, 32, 34]. In addition, the use of expert knowledge and diverse stakeholder groups to identify the appropriate variables, their structural and functional relationships, and explore alternative scenarios, allows us to consider unanticipated events and explore different network structures – features that we consider to be strengths, not weaknesses. For example, identifying suitable areas for wildlife, livestock, crop cultivation, and urban development will provide multiple potential locations for future development projects that do not interfere with the needs of pastoral people and ecotourism use and development. However, there are factors not considered in the current suitability models (e.g., ownership and land value). Thus, higher resolution spatial data will likely be needed for specific, smaller-scale (i.e., parcel-level) planning. This will require new BBN models, informed by the existing modeling framework, to allow planners and stakeholders to continue to build relationships and learn from past experience. Thus, we agree with others who have documented the advantages of BBNs over traditional modeling efforts and believe the advantages of BBNs outweigh potential limitations [14, 15, 33].

Another advantage of BBNs and the framework presented here is that we recognize the uncertainty of scientific knowledge [33]. Our own experience with BBNs [32] as well as others' [34], suggest that they are easy to calibrate, validate, and update as new information becomes available. Thus, BBNs fit well with the concepts of adaptive management [35] and can be a useful tool for organizing current thinking, generating testable hypotheses, comparing alternatives, and incorporating the uncertainty of scientific data. However, their use must be guided by observation, inference, and careful thinking, thereby underscoring the need for multiple working hypotheses [36].

7 Conclusions

The participatory modeling process proposed here provides an appropriate experimental design that can be implemented to link stakeholder and expert knowledge to actual policy change and land use planning decisions [16]. Bringing a diverse set of individuals together provides the capacity to build relationships, monitor responses, evaluate effectiveness, and consider economic opportunity and social equity for otherwise marginalized people, which we see as the keystone to sustaining natural and human systems [16, 37].

The community- and household-level stakeholder-led framework developed here directly links SES components. The process allows planners and stakeholders to learn how to use spatial and non-spatial data to explore real-world problems and scenarios, and identify areas suitable for development needs and various livelihood strategies (e.g., livestock, cropping, and wildlife uses). We acknowledge that there are no "one-size-fits all" solutions, and that no single modeling approach will be a panacea for natural resource and sustainability issues in Maasailand or anywhere else. Instead, the process described here encourages input from local people who rely on the AKP system for their livelihood and cultural identity in order to identify the impacts that are most important to them. Thus, we offer methods that can be easily understood and measured by local decision makers and stakeholders.

Finally, this process can be easily adapted as new information becomes available, or as new problems arise. The current model structure provides a maximal coverage strategy that allows diverse land use interests to be brought together to target and prioritize areas for protection or development, and to set specific strategies in the face of changing ecological, social, and economic processes [32]. Having multiple options can generate new hypotheses and decisions at the local scale, and may provide insight into solving more specific conservation needs not yet identified by stakeholders and decision makers. Subsequently, new models can be developed using the same process, but with higher-resolution data, thereby helping communities and other interests evaluate the impacts of alternative land uses between different prioritized areas at finer scales. As a result, future land use decisions can be made using the most up-to-date SES information [16].

Diverse and non-traditional stakeholder involvement ensures transparency and defensibility and should increase stakeholder capacity to develop, understand and react to alternative futures. By using this process to engage stakeholders, we expect to foster increased collaboration, expanded social capital, and better-targeted development and conservation proposals. To the extent that these outcomes are realized, we would expect to gain incremental improvements in quality-of-place, and more sustainable rural and urban economies across the AKP region.

Acknowledgements

This project was supported by NSF awards DEB-0919383 and EPS-0904155. Additional support was provided by the Maine Agricultural and Forest Experiment Station, and the Center for Research on Sustainable Forests.

References

[1] Browder, J.O., Conservation and development projects in the Brazilian Amazon: Lessons from the community initiative program in Rondonia. *Environmental Management*, **29(6)**, pp. 750-762, 2002.

[2] McShane, T.O. & Newby, S.A. Expecting the unattainable: The assumptions behind ICDPs (Chapter 4). *Getting Biodiversity Projects to Work: Towards More Effective Conservation and Development*, eds T.O. McShane & M.P. Wells, Columbia University Press: New York, pp. 49-74, 2004.

[3] Linkie, M., Smith, R., Zhu, Y., Martyr, D., Suedmeyer, B., Pramono, J. & Leader-Williams, N., Evaluating biodiversity conservation around a large Sumatran protected area. *Conservation Biology*, **22(3)**, pp. 683-690, 2008.

[4] Blom, B., Sunderland, T. & Murdiyarso, D., Getting REDD to work locally: Lessons learned from integrated conservation and development projects. *Environmental Science & Policy*, **13**, pp. 164-172, 2010.

[5] Salafsky, N. & Margoluis, R. Using adaptive management to improve ICDPs (Chapter 16). *Getting Biodiversity Projects to Work: Towards More Effective Conservation and Development*, eds T.O. McShane & M.P. Wells, Columbia University Press: New York, pp. 372-394, 2004.

[6] Gunderson, L.H., Holling, C.S. & Light, S.S., (eds). *Barriers and Bridges to the Renewal of Ecosystems and Institutions,* Columbia University Press, New York, 1995.

[7] Song, S.J. & M'Gonigle, M., Science, power, and system dynamics: The political economy of conservation biology. *Conservation Biology*, **15**, pp. 980-989, 2001.

[8] Endter-Wada, J., Blahna, D., Krannich, R. & Brunson, M., A framework for understanding social science contributions to ecosystem management. *Ecological Applications,* 8, pp. 891-904, 1998.

[9] Finlayson, A.C., *Fishing for Truth: A Sociological Analysis of Northern Cod Stock Assessments From 1977-1990*, Institute of Social and Economic Research, Memorial University of Newfoundland, St. John's, Newfoundland, Canada, 1994.

[10] Westley, F.R., Governing design: The management of social systems and ecosystem management. *Barriers and Bridges to Renewal of Ecosystems and Institutions*, eds. L. Gunderson, C.S. Holling, & S. Light, Columbia University Press, New York, pp. 391-427, 1995.

[11] Wells, M.P., McShane, T.O., Dublin, H.T., O'Connor, S. & Redford, K.H., The future of integrated conservation and development projects: Building on what works. *Getting Biodiversity Projects to Work: Towards More Effective Conservation and Development*, eds T.O. McShane & M.P. Wells, Columbia University Press: New York, pp. 397-421, 2004.

[12] Mistry, J., Berardi, A., Simpson, M., Davis, O. & Haynes, L., Using a systems viability approach to evaluate integrated conservation and development projects: Assessing the impact of the North Rupununi adaptive management process, Guyana. *The Geographical Journal*, **176(3)**, pp. 241-252, 2010.

[13] Anderson, D.R., Burnham, K.P. & Thompson, W.L., Null hypothesis testing: Problems, prevalence, and an alternative. *Journal of Wildlife Management,* **64**, pp. 912-923, 2000.

[14] Clarke, J.S. & Gelfand, A.E., A future for models and data in environmental science. *Trends in Ecology and Evolution,* **21**, pp. 375-380, 2006.

[15] Grunkemeier, G.L. & Payne, N., Bayesian analysis: A new statistical paradigm for new technology. *The Society of Thoracic Surgeons,* **74**, pp. 1901-1908, 2002.

[16] Pullin, A.S., Knight, T.M., Stone, D.A. & Charman, K., Do conservation managers use scientific evidence to support their decision-making? *Biological Conservation* **119**, pp. 245-252, 2004.

[17] Herrold-Menzies, M., Integrating conservation and development: What we can learn from Caohai, China. *The Journal of Environment & Development*, **15(4)**, pp. 382-406, 2006.

[18] Rinzen, C., Vermeulen, W.J.V., Wassen, M.J. & Glasbergen, P., Nature conservation and human well-being in Bhutan: An assessment of local community perceptions. *The Journal of Environment & Development*, **18(2)**, pp. 177-202, 2009.

[19] Gichohi, H., Gakahu, C. & Mwangi, E., Savannah ecosystems. *East African Ecosystems and their Conservation,* eds. T.R. McClanahan & T.P. Young, Oxford University Press: Oxford, UK, pp. 191-207, 1996.

[20] Mundia, C. & Aniya, M., Analysis of land use/cover changes and urban expansion of Nairobi city using remote sensing and GIS. *International Journal of Remote Sensing*, **26**, pp. 2831-2849, 2005.

[21] Nkedianye, D., Radeny, M., Kristjanson, P. & Herrero, M., Assessing returns to land and changing livelihood strategies in Kitengela. *Staying Maasia?*, eds. K. Homewood, P. Kristjanson & P.C. Trench, Springer: New York, pp. 115-149, 2009.

[22] Homewood, K., Kristjanson, P. & Trench, P.C., Changing land use, livelihoods, and wildlife conservation in Maasailand. *Staying Maasia?*, eds. K. Homewood, P. Kristjanson & P.C. Trench, Springer: New York, pp. 1-42, 2009.

[23] Murombedzi, J., Devolution and stewardship in Zimbabwe's CAMPFIRE program? *Journal of International Development*, **11(2)**, pp. 287-293, 1999.

[24] Anderson, D. & Berglund, E., *Ethnographies of Conservation: Environmentalism and the Distribution of Privilege*, Berghahn Books, New York, 2003.

[25] Mapedza, E. & Bond, I., Political deadlock and devolved wildlife management in Zimbabwe. *The Journal of Environment & Development*, **15(4)**, pp. 407-427, 2006.

[26] Ottichilo, W.K., de Leeuw, J., Skidmore, A.K., Prins, H.H.T. & Said, M.Y., Population trends of large non-migratory wild herbivores and livestock in the Maasai Mara ecosystem, Kenya, between 1977 and 1997. *African Journal of Ecology*, **38**, pp. 202-216, 2001.

[27] Homewood, K., Trench, P.C. & Kristjanson, P., Staying Maasai? Pastoral livelihoods, diversification and the role of wildlife in development. *Staying Maasia?*, eds. K. Homewood, P. Kristjanson & P.C. Trench, Springer: New York, pp. 369-408, 2009.

[28] Galaty, J. & Bonte, J., *Herders, Warriors and Traders. Pastoralism in Africa*, Westview Press: Boulder, CO, 1991.

[29] Heizer, J. & Render, B., Operations Management, 6th ed. Prentice Hall, New Jersey, 2001.

[30] Steventon, J.D., Sutherland, G. & Arcese, P., A population-viability-based risk assessment of Marbled Murrelet nesting habitat policy in British Columbia. *Canadian Journal of Forest Research,* **36**, pp. 3075-3086, 2006.

[31] Marcot, B., Steventon, J., Sutherland, G. & McCann, R., Guidelines for developing and updating Bayesian belief networks applied to ecological modeling and conservation. *Canadian Journal of Forest Research*, **36**, pp. 3063-3074, 2006.

[32] McCloskey, J.T., Lilieholm, R.J. & Cronan, C.S., Using Bayesian Belief Networks to identify potential compatibilities and conflicts between development and landscape conservation. *Landscape and Urban Planning*, **in press**, 2010.

[33] Pourret, O., Naim, P. & Marcot, B., *Bayesian Networks: A Practical Guide to Applications*, John Wiley & Sons, Ltd: West Sussex, England, 2008.

[34] Smith, C., Howes, A., Price, B. & McAlpine, C., Using a Bayesian belief network to predict suitable habitat of an endangered mammal – The Julia Creek dunnart (*Sminthopsis douglasi*). *Biological Conservation*, **139**, pp. 333-347, 2007.

[35] Prato, T., Bayesian adaptive management of ecosystems. *Ecological Modeling*, **183**, pp. 147-156, 2005.

[36] Chamberlain, T. C., The method of multiple working hypotheses. *Journal of Geology*, **5**, pp. 837-848, 1987. Reprinted in *Science* **148**, pp. 754-759, 1965.

[37] Leonard, H.J., *Environment and the Poor: Development Strategies for a Common Agenda*, Transaction Books, New Brunswick, New Jersey, 1989.

Characterisation of the spatial variability of waterlogging in the Blue River Basin (Argentina)

Y. Villacampa[1], P. Sastre[2], M. Gandini[2], A. Cañibano[2] & M. Cortés[1]
[1]*Departmento de Matemática Aplicada, Universidad de Alicante, España*
[2]*Facultad de Agronomía, Universidad Nacional del Centro de la Provincia de Buenos Aires, Argentina*

Abstract

Considering the soil as a complex, dynamic and heterogeneous system, chaos theory is a useful tool for gaining new knowledge into its variability. Fractal geometry has a close relationship with chaos theory, which describes dynamic and chaotic events with a development in time and space that can be predicted probabilistically. Furthermore, the concept of fractal dimension can also be applied to the study of spatial and temporal phenomena that possess and retain some kind of irregular geometric pattern within a range of scales. It is increasingly used to characterise the various properties of soil and to describe and model the characteristic functions of water retention and mobility in soil. This study is based on the need to estimate waterlogging of agro-livestock farms in the Blue River Basin (6,237 km^2), with a view to recommending the type of land use for each plot. By analysing Landsat satellite images, those with no cloud cover were adjusted spatially in order to build mosaics for eight points in time covering the entire study area. The images were then rated and processed, resulting in eight binary masks of non-flooded / waterlogged areas.

This paper presents a method for calculating the fractal dimension based on the box counting process, in order to test its use in differentiating between varying degrees and distributions of soil moisture. The results indicate eight situations of soil moisture by means of the fractal dimensions calculated using this method. Using the parallelism test, the slopes of the simple adjusted linear

WIT Transactions on Ecology and the Environment, Vol 144, © 2011 WIT Press
www.witpress.com, ISSN 1743-3541 (on-line)
doi:10.2495/ECO110051

regression models were compared statistically, and significant differences were observed.
Keywords: fractal, soil moisture, dynamic system.

1 Introduction

According to Sommer et al. [11], soil is a complex, dynamic and heterogeneous system, defined as a natural, four-dimensional body (length, width, depth and time) variable in space and time. Being a complex system, it has two generic properties: invariance to scaling or self similarity and universality. In addition, in order to study complex systems it is necessary to analyse their main attributes, these being: entropy, lacunarity, degree of occupation of the space of interest, roughness and multifractality (Bullock and Brennen [1], Cheng [2], Logsdon et al. [6]).

The variability of the structural pattern of soil in space and time depends on the environmental conditions. In addition, said variability is a consequence of the natural heterogeneity and genesis of the materials of geological and biological origin, as well as the temporal variations of the processes that act on them. Of particular importance is erosion, climate changes, soil degradation, aquifer recharge, evapotranspiration, pollution, etc. (Zhang [14]).

Characterisation of spatial variability is useful with any attempt to model processes. Indeed, the study of heterogeneity on different scales was the prime incentive behind the introduction, in soil science, of fractal geometry, which at least in principle allows us to analyse its geometric complexity and characterisation with just a few numbers, the fractal indexes.

An object is considered to be fractal when its measurement dimension, **D_b is greater than its topological dimension, D_t and less than its immersion dimension, D_i.** Therefore, D_b contains the idea of the occupation of space. So, as D_b increases, the degree of space filling, disorder or irregularity also increases, as something continuous, able to assume non-whole values. All of this differs from the common classic idea of geometric or Euclidean dimension. To sum up, by determining the fractal dimension of an object, we are obtaining information about its capacity for occupation of space. (Pachesky et al. [7]).

Analysis of the spatial structure of soil humidity may a priori provide information needed for other studies, such as: what the sample size needs to be, how far apart samples need to be taken, if the humidity is invariable with regard to changes in scale, or how spatial interpolation should be carried out. In this study, the aim was to estimate the probability of flooding of agricultural-stock raising farms in the Arroyo del Azul waterway (6,237 km^2), in order to recommend the appropriate land use for each plot.

2 Fractal dimension and box-counting dimension

The fractal dimension concept is used to study continuous temporal or spatial phenomena that show some kind of geometric pattern within a range of scales and which, in this case, is appropriate for describing the spatial distribution of the humidity retention function of the soil.

One of the interpretations of dimension, possibly the most natural, is associated with the capacity of objects to occupy the Euclidean space in which they are immersed. Thus, quantifying fractals means defining the proportion of the physical space in which they are inscribed, that is filled by them. The fractal dimension is an indicator of how the fractal object occupies the space in which it is immersed, or what kind of variation predominates when dealing with long-term time series (Gaytán-Ortiz [4]). A fractal dimension of a time series with a value of less than 1.6 indicates the predominance of long-term variations, while values of 1.5 to 2 indicate the predominance of short-term variations and that the phenomenon is unpredictable.

One of the different versions of fractal dimension numerical calculation is called box-counting, which is related to the concept of auto-similar structure. The methodology consists of placing a grid with a mesh of sizes, ε, over the structure to be analysed, forming a covering of boxes. The number of boxes that contain values of the structure to be characterised is Nε(k), which varies with the value of ε chosen.

The box-counting dimension concept, which allows an algorithm to be deduced for numerical estimation, was developed by Kolmogorov [5], but has a predecessor in that described by Pontrjagin and Schnirelmal [9]. Let $N(\varepsilon)$ be the least number of radius sets less than or equal to ε needed to cover the compact set $F \subset R^n$. The lower and upper box-counting dimensions of set F are defined as eqn (1) and eqn (2), respectively:

$$\underline{\dim}_D(F) = -\underline{\lim}_{\varepsilon\to 0}\frac{\log N(\varepsilon)}{\log \varepsilon} \tag{1}$$

$$\overline{\dim}_D(F) = -\overline{\lim}_{\varepsilon\to 0}\frac{\log N(\varepsilon)}{\log \varepsilon} \tag{2}$$

If the two amounts thus defined are the same, then we have the box-counting dimension of set F, eqn (3):

$$\dim_D(F) = -\lim_{\varepsilon\to 0}\frac{\log N(\varepsilon)}{\log \varepsilon} \tag{3}$$

Let us look at a simpler geometrical interpretation of the box-counting dimension. Let $N(\varepsilon)$ be the least number of balls of radius ε needed to cover compact set $F \subset R^n$. Then, the D-measure of F satisfies, approximately, $B^D(F) \propto N(\varepsilon)\varepsilon^D$ then for certain $c > 0$

$$N(\varepsilon) \approx \frac{c}{\varepsilon^D} \tag{4}$$

$$\begin{gathered}\log\big(N(\varepsilon)\big) = \log c - D\log\varepsilon \\ \log\big(N(\varepsilon)\big) - D\log\tfrac{1}{\varepsilon} = \log c \\ \frac{\log(N(\varepsilon))}{\log\frac{1}{\varepsilon}} - D = \frac{\log c}{\log\frac{1}{\varepsilon}}\end{gathered} \tag{5}$$

When $\varepsilon \to 0^+$ the term on the right decreases indefinitely, which suggests the following definition for box-counting dimension, eqn (6):

$$D_B(F) = \lim_{\varepsilon\to 0} \frac{\log(N(\varepsilon))}{\log\frac{1}{\varepsilon}} \tag{6}$$

3 Algorithm for estimating the box-counting dimension

In the process of determining the box-counting dimension, we replace the evolution of the variable ε with a succession $\{\varepsilon(k)\}$. Let $\{\varepsilon(k)\}$ be a succession of positive numbers that satisfy the conditions eqn (7):

$$\begin{gathered} \lim_{k\to\infty} \varepsilon(k) = 0 \\ \lim_{k\to\infty} \frac{\log \varepsilon(k)}{\log \varepsilon(k-1)} = 1 \end{gathered} \tag{7}$$

Thus,

$$D_B(F) = \lim_{k\to 0} \frac{\log N_{\varepsilon(k)}}{\log\left(\frac{1}{\varepsilon(k)}\right)} \tag{8}$$

With which we manage to make the infinite limit process discrete, eqn (8).

The balls can be replaced by another type of enclosure that better adapts to the construction of an algorithm. So, let $n = 2$ and diadic boxes, a succession of boxes with sides 2^{-k} configured as a succession of grids.

Let N'_k be the minimum number of boxes (square) with sides $1/k$ needed to cover F. We define the box-counting dimension as the limit (if it exists):

$$D_B(F) = \lim_{k\to\infty} \frac{\log N'_k}{\log 2^k} \tag{9}$$

However, for physical world fractals, the infinite limit process, eqn (9), makes no sense.

To calculate an estimated D_B, we overlay the structure to be analysed with a grid of sizes, ε, creating a layer of boxes. The number of boxes that contain values of the structure to be characterised is $N_{\varepsilon(k)}$, which varies with the value of ε chosen. To make the corresponding calculation of the fractal dimension, we count the number of boxes $N_{\varepsilon(k)}$ on different scales. The variation in scale from one step to the next depends on the size of the image and the multiples of that size. The value of the dimension corresponds to the slope of $\log(N_{\varepsilon(k)})$ Vs. $\log(1/\varepsilon)$.

The important advantage of this methodology is that it can be applied systematically and simply when creating the layer of boxes. In addition, the corresponding recount can be carried out quickly, even with high-resolution images (Vicsek [12]).

4 Processing digital information and methodology

We started with Landsat 5 TM and 7 ETM+ image analysis (scenes 225-85 and 225-86), and selected those of a high quality (zero cloud cover). The images were geometrically corrected to the Gauss Krüger projection system (Datum Campo Inchauspe 1969, Argentina Zone 5) to start constructing mosaics in order to cover the whole study area. Mixed classifications were then made of the satellite images, processing them using masking and obtaining eight binary images of non-flooded/flooded areas.

The importance of the study lies in the need to estimate the probability of the flooding of farmland, in order to be able to recommend the land use of each plot. The area studied is the Arroyo del Azul basin, covering 6,237 km^2. The Arroyo del Azul waterway has its source near the town of Chillar and stretches northwards, until it is lost before reaching the River Salado, known as the Arroyo Gualicho waterway in its lower reaches. The course is finally crossed and largely drained by Canal 11, which was adopted as the northern limit of the study area. The basin consists of a large plain (lower sub-basin) and a smaller area of low hills on the south (upper sub-basin), where the headwaters of the Arroyo del Azul are found. The upper area is connected to the plain by the foothills (middle sub-basin). The average slope of the land is 5% in the upper sub-basin and 0.2% in the lower sub-basin, while varying from 0.5 to 0.8% in the middle area. The overall behaviour of the region is that of a plains environment (Sala et al. [10]).

For each situation, we calculated the corresponding fractal dimension, using the box-counting method. We used the "parallelism" test (Zar 1984), to statistically compare the slopes of the eight adjusted simple linear regression models (one for each situation). The hypotheses of these tests are: $H_0: \beta_1 = \beta_2 = \cdots = \beta_k$ and $H_1: \beta_i \neq \beta_j$, using a conventional F-test with a specific F calculated (F_c) where β_i represents the slope of the adjusted regression model (fractal dimension), with $i = 1, 2, \ldots, k$. The statistic used is eqn (10):

$$F_c = \frac{(SC_c - SC_p)/(k-1)}{(SC_c/GL_p)}, \tag{10}$$

where:

- SC_c = Common residual sum of squares;
- SC_p = Grouped residual sum of squares (sum of the sums of squares for each regression);
- GL_p =Grouped degrees of freedom;
- k = Number of regressions.

To establish the differential effect of the degrees of humidity studied, represented by the fractal dimensions, we compared pairs of slope values by calculating a critical value, $q = (b_i - b_j) / SE$, where b_i and b_j are the slopes associated with the regression models of the degrees of soil humidity i and j, respectively, with eqn (11):

$$SE = \sqrt{\frac{\frac{SC_c+SC_j}{GL_i+GL_j}}{\sum x^2 - \frac{(\sum x)^2}{n}}} \quad (11)$$

$q_{\alpha,v,p}$

with:

α = level of significance
v = grouped degrees of residual freedom
p = number of simple linear regression models (degrees of humidity).

5 Results and discussion

The table shows the estimates found for the adjusted regression lines $\log(N(\varepsilon)) = \log c - D \log \varepsilon$, the Adjusted R Square and their errors for the 8 cases or periods. In all cases, the significance tests (ANOVA) for the regression adjustments, showed highly significant values. The t-student tests, to check whether the coefficients were non-null, showed highly significant values, which means that the estimated coefficient values were different to zero in all the regressions.

Table 1: Regression adjustments.

Year	Adjusted Regression	Adjusted R Square	Std. Error of the Estimate	State H. A	Fractal Dimension
April 2001	$\log N(\varepsilon) = 10{,}124 - 2{,}046 \log \varepsilon$	,955	,354	Normal	2,046
April 2002	$\log N(\varepsilon) = 9{,}588 - 2{,}070 \log \varepsilon$	,997	,090	Normal	2,070
August 1993	$\log N(\varepsilon) = 9{,}241 - 2{,}026 \log \varepsilon$	,999	,056	Normal	2,026
March 1998	$\log N(\varepsilon) = 9{,}042 - 2{,}028 \log \varepsilon$	1,000	,032	Normal	2,028
May 1986	$\log N(\varepsilon) = 8{,}858 - 2{,}056 \log \varepsilon$	,995	,112	Normal	2,056
November 1997	$\log N(\varepsilon) = 8{,}542 - 2{,}182 \log \varepsilon$	,991	,171	El Niño	2,182
September 2001	$\log N(\varepsilon) = 7{,}349 - 1{,}890 \log \varepsilon$	,999	,044	Dry	1,890
November 2001	$\log N(\varepsilon) = 6{,}186 - 1{,}745 \log \varepsilon$	,977	,214	Dry	1,745

The "parallelism" test (Zar [13]) was applied to statistically compare the slopes of the eight adjusted simple linear regression models (one for each situation). The hypotheses of this test are: $H_0: \beta_1 = \beta_2 = \cdots = \beta_k$ and $H_1: \beta_i \neq \beta_j$, and the values calculated are: $SC_c = C_1 - \frac{B_1^2}{A_1} = 167{,}48 - \frac{82{,}41^2}{41{,}10} = 2{,}21$,

$$SC_p = \sum_{i=1}^{K=8} SS_i = 1{,}59,, \qquad \text{with} \qquad Gl_p = \sum_{1=1}^{k=8} (n_i - 2) = 7*8 = 56,,$$

$$F_c = \frac{\dfrac{SC_C - SC_p}{k-1}}{\dfrac{SC_c}{GL_p}} = \frac{\dfrac{2{,}21-1{,}59}{8-1}}{\dfrac{2{,}21}{56}} = 2{,}244.$$

Giving the result: Probability F_c = 0.977763789 and Probability $F_{(2,244,7,56)}$= 0.04383, so the slopes are not equal. This means that the fractal dimensions estimated for each period significantly differ from each other.

In order to establish the differential effect of the degrees of humidity studied, represented by the fractal dimensions, we compared pairs of slope values by calculating a critical value, $= (b_i - b_i)/SE$, where b_i and b_j are the slopes associated with the regression models of the degrees of soil humidity i and j, respectively, within eqn (11).

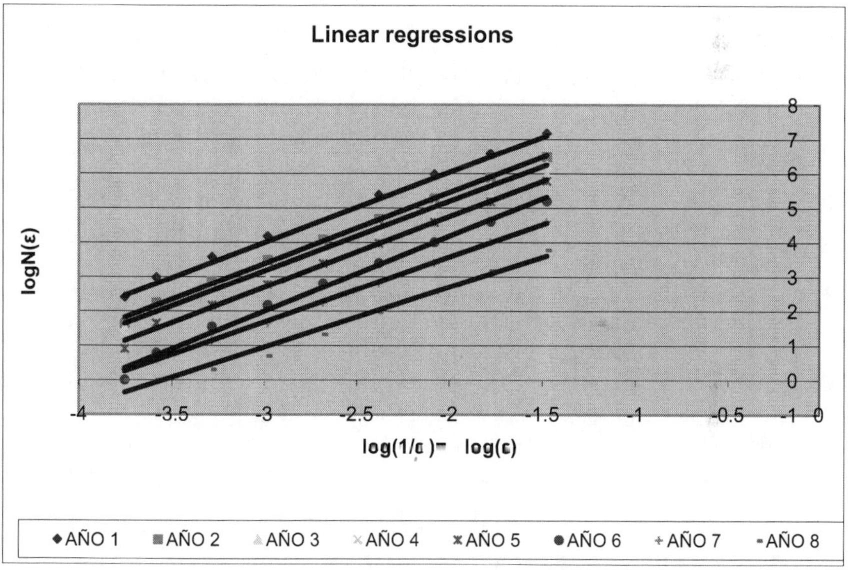

Figure 1: Slope values.

It can be seen that there are 3 groups of slopes that coincide with humidity statuses in the basin being studied and climatic phenomena such as "*El Niño*" There is a group of intermediate values that corresponds to "normal" water status states of the basin and a high value (2.182) that is related to the "*El Niño*" event that affected the basin, flooding a large area. In addition, the slope values for November and September 2001 are related to a low humidity water status, in which shallow bodies of water dried out completely.

6 Conclusion

The fractal dimensions of a plain basin, calculated and analysed using satellite images, vary according to the different humidity statuses of the soil. This would indicate that the fractal dimension estimated in this way could be used to measure the probability of events leading to cases of flooding. This would be of great use for farmers, as they could then recommend the type of land use for each plot.

It is necessary to further this study, increasing the type and amount of situations considered.

Acknowledgement

This paper has been supported by the project CTM2010-10878-E of Ministerio de Ciencia e Innovación, Spain. The authors wish to express their gratitude for such a grant.

References

[1] Bullock, S.S. & Brennen, G.K., Qudit surface codes and gauge theory with finite cyclic groups. *J. Phys. A: Math theor.* **40**, pp. 3481-3505, 2007.

[2] Cheng, S.S. 2008a. Spatial scaling analyses of soil physical properties: a review of spectral and wavelet methods. Vadose zone Journal. Soil science Society of America. Madison. USA.

[3] Cheng, Q. 2008b. Modeling local scaling properties for multiscale mapping, sate key laboratory of Geological Processes and mineral resourses. China Univ. of Geosciences.

[4] Gaytán-Ortiz, L.A. 2007. Auto afinidad y ruido f-β de temperaturas extremas mensuales. Tesis de Maestría en Matemática Educativa. Universidad Autónoma de Zacatecas.

[5] Kolmogorov, A. 1959. Entropy per unit time as a metric invariant of automorphisms. *Dokl. Akad. Nauk. SSSR*, 124:754.

[6] Logsdon S.D., E. Perfec and A. M. Tarquis. 2008. Multiscale soil investigations: physical concepts and mathematical techniques. Published in vadose zoen Journals. Madison, USA.

[7] Pachesky Y., E. Perfect and M.A. Martín. 2006. Fractal Geometry Applied to Soil and Related Hierarchical Systems. Geoderma. 134 : 237-239.

[8] Avances en Recursos Hidráulicos Número15, Mayo De 2007. Medellín

[9] Pontrjagin and Schnirelmal (1932).

[10] Sala, J.M., Kruse, E, Aguglino L. 1987 Investigación hidrológica de La cuenca Del Arroyo AZUL. Informe 37. CIC. La Plata Argentina.

[11] Sommer M., Gerke H. H., D. Deumlich. 2008. Modelling soil landscape genesis: a time split approach for hummocky agricultural landscapes.

[12] Vicsek, T. 1992 Fractal growth phenomena Publisher: World Scientific (Singapore and New Jersey).

[13] Zar, J.H., 1984. Bioestatistical analysis. Department of Biological Sciences, Northern Illinois University. EEUU. 842 P.
[14] Zhang X. 2006. Consolidation theories for saturated-unsaturated soils and numerical simulation of residential buildings on expansive soils. Texas A&M University. http://handle.tamu.edu/1969.1/2747

Central Baltic Sea herring: effect of environmental trends and fishery management

R. Aps[1], M. Fetissov[1], N. Holmgren[2], N. Norrström[2] & S. Kuikka[3]
[1]*Estonian Marine Institute, University of Tartu, Estonia*
[2]*University of Skövde, Skövde, Sweden*
[3]*University of Helsinki, Helsinki, Finland*

Abstract

Uncertainty is an endemic condition of the Baltic Sea herring (*Clupea harengus membras*, L) fishery management. It is a condition exacerbated by the fishing fleet overcapacity and consequent exploitation of the herring stock at a level believed to be unsustainable. Some sources of uncertainty are mainly related to biology and fishing technique: the unsolved problem of herring assessment and management units, the recruitment–environment relationship and the reduction in mean weights-at-age, uncertain ageing of fish, the problem of unaccounted fishing mortality caused by the fish selection through the trawl net. Fishing fleet overcapacity is believed to be behind of the regulatory overfishing when setting the Total Allowable Catches (TACs) higher than the scientific advice (decision overfishing) and tolerating the extensive underreporting of catches (implementation overfishing). Two scenarios for the Central Baltic Sea herring fishery management options are constructed and the Bayesian networks are used to represent and update uncertainties encountered in the process of the management related situation assessment. First scenario represents the current status of the fishery management resulting in fishing mortality (F) higher than F_{MSY} – the fishing mortality that corresponds to the Maximum Sustainable Yield (MSY). The second scenario demonstrates the assumed potential impact of economic incentives (e.g. zoning, individual transferable quotas (ITQs), territorial use rights etc.) on the reduction of excessive fishing capacity and bringing actual fishing mortality closer to F_{MSY}.

Keywords: Central Baltic Sea herring, fishery management, Bayesian belief network, management scenario simulation, uncertainty communication.

www.witpress.com, ISSN 1743-3541 (on-line)
doi:10.2495/ECO110061

1 Introduction

Uncertainty is an endemic condition of the Baltic Sea herring (*Clupea harengus membras*, L) fishery management. First of all, uncertainty is behind of the biological foundations of the management measures taken during the last decades. The scientific discussion on how best to establish the Baltic Sea herring assessment and management units in accordance with assumed different number of spawning components of this fish has been heated and many approaches have been taken over the years [1].

There is a number of studies published on the Central Baltic Sea herring recruitment–environment relationship [2, 3] and on substantial decrease in growth rate of the herring that has been observed since the beginning of the 1980s [4–6]. It is reported [7] that the decline in the Central Baltic Sea herring spawning stock biomass was partly caused by the reduction in mean weights-at-age. The reduction in herring mean weight-at-age was likely induced by a decline in the herring's main zooplankton prey, which in its turn is believed to depend on increased food competition with sprat or a decline in the salinity of the Baltic [6]. Herring mean weights-at-age have stabilized since the late 1990s, but remain at a low level.

Despite of in-depth studies into the structure of the Baltic Sea clupeid fish otoliths [8] the ageing of these fishes is still far too uncertain. Considerable problems associated with difficulties in ageing of the Baltic Sea herring are being transformed into the uncertainty of the calculated fishing mortality (F) and the year-class relative abundance estimates. Significant between-reader differences in ageing of the Baltic Sea herring are documented [9, 10].

The problem of the Baltic Sea herring's unaccounted fishing mortality caused by the fish selection through the trawl net has attracted attention of scientists and managers since 1990s [11, 12].

A review of recovery of the depleted Baltic Sea fish stocks, including the Central Baltic Sea herring, showed that for all internationally regulated fish stocks the Total Allowable Catches (TACs) have been set systematically more than the scientific advice based on sustainable exploitation [13]. Building on a language of Eagle and Thompson [14] it is interpreted as "decision overfishing" while the evidences of extensive underreporting of catches are interpreted as "implementation overfishing". Despite the measures that have been taken to combat the situation this still means that a management body was knowingly accepting a situation of overfishing.

It is stated [15] that the policy decisions on complex environmental risks often involve contested science and typically there are no 'facts' that entail a unique correct policy. Wardekker *et al.* [16] argue that communication of uncertainties aimed at policymakers, as well as other parties involved in policymaking, is important because uncertainties can influence the policy strategy that is selected while it is also a matter of good scientific practice, accountability and openness towards the general public. This is perfectly valid for the situation with fishery management.

This paper attempts to address the main sources of uncertainty related to the management of the Central Baltic Sea herring fishery, including the herring in the Gulf of Finland. The Bayesian Belief Networks (BBNs) are used to communicate the major sources of uncertainty related to the management of the Central Baltic Sea herring fishery. Two scenarios for the Central Baltic Sea herring fishery management options are constructed and the Bayesian networks are used to represent and update uncertainties encountered in the process of the management related situation assessment. First scenario represents the current status of the fishery management resulting in fishing mortality higher than F_{MSY} – the fishing mortality that corresponds to the Maximum Sustainable Yield (MSY). The second scenario demonstrates the assumed potential impact of economic incentives (e.g. zoning, individual transferable quotas (ITQs), territorial use rights etc.) on the reduction of excessive fishing capacity and bringing fishing mortality closer to F_{MSY}.

2 Material and methods

Conceptual model of the Central Baltic Sea herring stock management is developed using the Bayesian Belief Network (BBN) methodology [17] and the HUGIN RESEARCHER software. Converting a state of knowledge into a probability assignment is a problem that lies at the heart of Bayesian probability theory. The prior probabilities are obtained from the knowledge on the herring fishery management and the setting of conditional probabilities has been done externally based on the relevant published materials [18–20].

Interpretation of the probability statements expressed in this study is building on the notion of *formally singular probability statement* when it ascribes a probability to a single occurrence, or to a single element of a certain class of occurrences. Referring to objective interpretation of probability statements Popper defines the formally singular probability as follows [21]:

$$\alpha Pk(\beta) = \alpha F(\beta) \qquad (k \in \alpha)$$

This is expressed in words as: the formally singular probability that the event k has the property β – given that k is an element of the sequence α – is, by definition, equal to the probability of the property β within the reference sequence α. This means that only objective formally singular probability statement, when it ascribes a probability to a single occurrence, is *empirically testable*.

According to Popper [21] the subjective interpretation of probability statements about single events does not enable us to predict what the property of the event in question will be, but it enables us to express all we know about it by means of a formally singular probability statement – an *indefinite prediction* about the particular event in question. Author underlines: “I do not object [subjective interpretation of probability statements] … so long as we clearly recognize that the objective frequency statements are fundamental, since they alone are *empirically testable*. I reject, however, any interpretation of those [subjective] formally singular probability statements – these indefinite predictions – as statements about the objective state of affairs, other than the objective statistical state of affairs".

Given a scenario of fishery management option, a Bayesian network depicts graphically the cause and effect relationship between various elements of the system concerned. In doing so it also demonstrates conditional dependence i.e., which factors are more relevant and directly affect a given output and which factors are less relevant in the sense that knowledge regarding these factors become redundant once the direct causes are known.

3 Results and discussion

3.1 Uncertain science – uncertain management

According to EU policy paper on fisheries management [22] the ICES classifies the Central Baltic Sea herring stock under the category 3 because it is fished outside safe biological limits [7]. In addition, the assessment of the Central Baltic Sea herring stock and the corresponding forecast are regarded by ICES to be acceptable although the accompanying uncertainties in underlying assumptions and data are critically influencing the quality of the fishing mortality estimates, including the quality of F_{MSY} estimates.

3.1.1 Stock assessment and management units

Ojaveer and Kalejs [23] argue that the Baltic Sea different areas display marked differences in salinity, oxygen content and temperature of water layers, timing of the peak production cycle, and the other basic environmental conditions while the biota in the Baltic Sea has formed under wide variations in environmental conditions induced by the climate change since the last glaciations. It is suggested [24] that the Baltic Sea regions differ in essential environmental conditions and in the process of adaptation to their habitats, fishes like the Baltic Sea herring have developed local sub-species. Among marine exploitable fishes, spring spawning herring is the only species adapted to all regions of the Baltic Sea, forming 10 local populations. According to the latest studies by Jørgensen *et al.* [25] the Baltic herring can be divided into three genetically separated stocks: western, southern and northern stocks.

For the practical stock assessment reasons the Central Baltic Sea herring populations have been grouped by ICES for assessment purposes in different combinations over the period 1974–1990. Finally, the main reason for the creation of a large Central Baltic Sea herring assessment unit in 1990 (Subdivisions 25–29 and 32) was to try to eliminate the influence of migrations of herring in the Baltic Sea. Finally, the optimal solution was seen in seeking a reasonable compromise between biological knowledge and practical constraints set by the availability of data (lack of generally accepted routine method for separation of the herring sampled from the different local biological populations and lack of sufficient tuning data for assessment of the smaller stock units).

3.1.2 Effect of environmental trends

Latest studies into the Central Baltic Sea herring recruitment–environment relationship [2] revealed as important stock recruitment predictors the temperature and the stock weight-at-age of the spawners. Moreover, the food supply is also considered as a significant predictor, suggesting that changes in climate and/or food web structure may indirectly affect herring recruitment *via* prey availability for the recruits or spawners.

Substantial decrease in growth rate of the Central Baltic Sea herring, attributed mainly to the changes in fisheries, trophic interactions and the climate, has been observed since the beginning of the 1980s [4–6]. In relation to the Gulf of Finland it is shown [26] that the long-term changes in the herring growth are correlated positively with salinity and with the abundance of the marine zooplankton species *Pseudocalanus minutus elongatus*. According to that study the density-dependent mechanism was not likely to explain the changes in herring weight-at-age in the Gulf of Finland while the strong negative correlation between herring weight and sprat biomass may indicate competition for food between these species.

According to ICES [7] the regional differences in the Central Baltic Sea herring growth rate result in a high proportion of small individuals in the North (SD 28.2, 29 and 32) and large individuals in the South (SD 25 and 26). At the same time the growth rate tends to change due to salinity variations and fluctuations in the herring stock and competition with the Baltic Sea sprat. Furthermore, the Central Baltic Sea herring stock dynamics is highly affected by cod predation.

Rahikainen and Stephenson [27] argue that the growth rates of herring in the northern Baltic Sea differ among areas, and have changed substantially over time in some areas in response to environmental change. It is stated that the growth variation has implications for stock assessment and management: differences suggest a need for considering a smaller spatial structure, at least at the scale of the ICES subdivision, in the case of northern Baltic herring, and the herring in different areas of the northern Baltic Sea probably require different reference points and possibly different management strategies, as a consequence of differences and variability in growth characteristics.

3.1.3 Problems with ageing

Main features of the Baltic Sea sprat and hering otoliths used for ageing of these fishes are discussed in literature [8]. However, Eklund *et al.* [9] are reporting the result of comparative ageing of the Baltic Sea herring from ICES Sub-division 29 where none of the comparisons between readers reached 50% agreement and variations were considerable in old fish. It was further suggested that the variability of the age determinations should be taken into account as part of the data, and possibly incorporated as an error factor in, e.g., population assessments. It is stated [10] that "The actual importance of age variability in population modelling and fisheries management depends on the variability of other data used in the models, on the type of calculations, and also on the lines of management action that are considered. In Baltic herring, subject to the rapid changes in the ecosystem of the Baltic sea ... the imprecision of the age determinations constitutes a risk of misinterpreting biological signals both in fisheries management and in environmental monitoring". However, significant improvement was reported in 2008 when average agreement with the modal age of the Baltic Sea herring was on average 86.9% [28].

3.1.4 Effect of fishing

The problem of the Baltic Sea herring's unaccounted fishing mortality caused by the fish selection through the trawl net has been addressed in literature [5, 11, and 12]. The survival experiments indicated that survival is positively linked to the size of the escapee. However, the most of the herring that escape die shortly after. Furthermore, the results reported by Rahikainen *et al.* [29] showed that the herring of the age group 0 and 1 are discarded underwater in larger numbers than landed while the actual fishing mortality at age 1 is estimated to be more than twice higher than its estimates based on the unadjusted data.

Raid [30] has analysed how equal are the "equal" catches in tonnes? Author showed that the equal herring catches in tonnes are resulting in considerably higher herring catches in numbers in the Gulf of Finland than in the open part of the Northern Baltic Sea. Indicator – the mean catch in numbers per 1000 tons of landings was proposed to communicate the possible spatial variability in the Baltic Sea herring fishing mortality.

ICES [7] states that the pelagic fisheries take a mixture of herring and sprat and this causes uncertainties in catch levels because the extent to which species misreporting has occurred is not well known. The recent trend to move fishing effort northwards results in high catches of small individuals.

According to Aps *et al.* [31] the performance of fishery management was evaluated using two indicators proposed by FAO [32]: the ratio of the agreed TAC and the recommended catch (TAC/tac) as the response indicator, which could be seen as a measure of decision overfishing; and the ratio of the actual catch and the recommended catch (C/tac) as the pressure indicator and the measure of implementation overfishing. In essence, indicator values >1 represent regulatory overfishing.

It is stated [13] that incomplete or missing advice from ICES from 1984–1989 and from 1992–1999 proved to have been a major obstacle to setting TACs for herring in Subdivisions 25-29 and 32 (excluding Gulf of Riga herring). Under conditions of a continuous decline in SSB, TAC have been set largely in excess of the advice ($0.98 < TAC/tac < 2.55$). In addition, the officially recorded catches exceeded the recommended catches, with the exception of 1991, 2003, and 2004 ($0.73 < C/tac < 1.83$). Although the F values have mostly been in excess of the Fpa, they dropped to this level in 2005, followed by a moderate increase in SSB (Figure 1, a and b).

Because of unreliable catch statistics for the Baltic Sea sprat, fished as a mixture with herring, the ICES advice has been incomplete or missing in 1986, 1987, 1993, 1994, and 1997–1999 [13, 31]. In other years, the TAC largely exceeded recommended catches ($0.88 < TAC/tac < 2.44$). The official catch statistics never exceeded the TAC and thus exceeded the recommended catch by

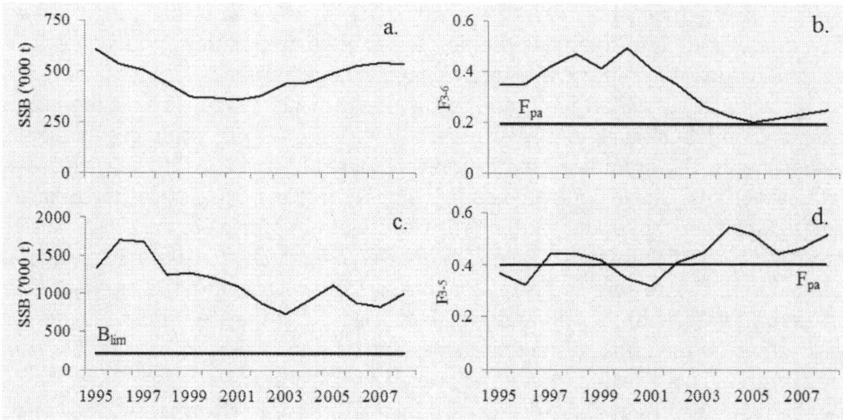

Figure 1: Annual estimates of SSB (left panels) and fishing mortality (F, right panels), 1995–2008, with indication of the limit (Blim) and precautionary (Fpa) reference points for (a and b) Baltic Sea herring (Subdivisions 25–29 and 32; excluding the Gulf of Riga; reference points agreed in 2002); and (c and d) Baltic Sea sprat (Subdivisions 25–32; Fpa operational since 2000) [13].

a smaller amount (0.51 < C/tac < 1.90). F has constantly exceeded Fpa since 2002, while the SSB have always been above Blim (Figure 1, c and d).

3.2 Central Baltic Sea herring fishery management: Bayesian uncertainty communication

3.2.1 Bayesian model of fishery management

BBN is constructed to assess a variety of situations that are likely to be generated by the different fishery management scenario options. Attempt is made to communicate the relationship between the fishing fleet capacity, the public pressure and the level of fishing mortality in relation to F_{MSY}. This BBN network (Figure 2) contains two information variables: 1) “Flet capacity” (fishing fleet capacity is balanced/not balanced with the fishery resources available), 2) “Public pressure” (reasoning in support of sustainable fishing – strong/weak), and one hypothesis variable - “Harvest” (fishing mortality is less or equal to the F_{MSY}/ fishing mortality exceeds the F_{MSY}) that is presenting the one yearly cycle of the fishery activities. The rest are intermediate nodes (“Control and enforcement” and “TAC decision” that help propagate evidence from the information variables to the hypothesis variable.

Let X be a discrete random variable taking the values $\{x_1, \ldots, x_n\}$ representing our current knowledge. If our current state of knowledge is *total ignorance* then we will not be able to distinguish between various alternatives and this leads to the uniform probability distribution $P(x_1) = 1/n$. Thus, we assign a probability distribution to our hypotesis variable such that all its possible values are equally probable.

After the network has reached an equilibrium with these initial conditions we instantiate both information variables to their corresponding evidence values according to scenario presenting the recent state of affairs.

In order to simulate the current status of the fishery management the “Fleet capacity” is instantiated as “Not balanced” with the fishery resources available. According to the opinion of the European Court of Auditors [20] the “[Recent] overcapacity is a factor in over-fishing. It influences the behavior of the industry and the legislator, by reducing the profitability of enterprises in the sector, which have to bear the costs associated with the overcapacity of their means of production. The fishing industry is exposed to the temptation of exploiting its over-sized vessels to the full and to fish beyond the authorised limits. There is a risk of underdeclaring of their catches or throwing the less profitable part overboard”. The variable “Public pressure” is instantiated as being weak based on the Baltic Sea Regional Advisory Council (BS RAC) performance [18, 19].

As a result, it is seen that the actual fishing mortality of the Central Baltic Sea herring is higher than F_{MSY} and the harvest is believed to be not sustainable. Furthermore, the value of the calculated F_{MSY} is quite uncertain e.g. just because of the insufficient quality of the catch data. European Court of Auditors [20] states that the incompleteness and unreliability of catch data prevent the TAC and quota system, which is a cornerstone in the management of Community fisheries resources, from functioning properly while the regulatory framework and the procedures in force guarantee neither the exhaustiveness of data collection, nor the detection of inconsistencies during validation.

The states of the intermediate nodes “Control and enforcement” and “TAC decision” agree with the published information. It is stated [20] that “... the effectiveness of inspection and control systems is impeded by the existence of a very high inherent risk associated with the existence of significant fishing overcapacity.”

State of the “TAC decision” node is also in accordance with the opinion of the European Court of Auditors [20]: “When the Council of Ministers adopts the TAC and quotas it acts as an arbiter between divergent short-term interests, environmental on the one hand, and socio-economic on the other. In doing so it may authorize catch quantities that are higher than those recommended by the scientists, in order to protect the immediate social and economic interests of those employed in the industry. Furthermore, the Council’s choices are made without any means of evaluating objectively how the catch levels will impact on any of those interests”.

The posterior probabilities get calculated when the network reaches equilibrium again and we use these to demonstrate this scenario related uncertainty (Figure 2).

For the first year the hypothesis variable “harvest_1” is showing that fishing mortality exceeds the F_{MSY} and this corresponds to the ICES latest assessment [7]. If no changes in economic incentives occur then the overfishing will most probably continue also during the second year of simulation.

Scenario that simulates the impact of strong economic incentives to reduce the fishing fleet capacity is presented in Figure 3.

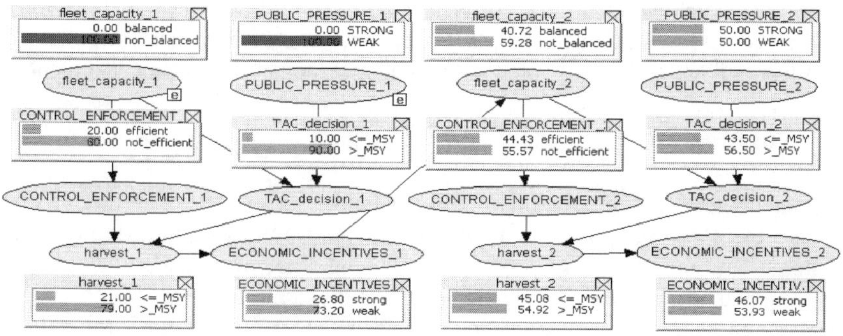

Figure 2: BBN representing the state of the knowledge for the current status of the Central Baltic Sea herring fishery management.

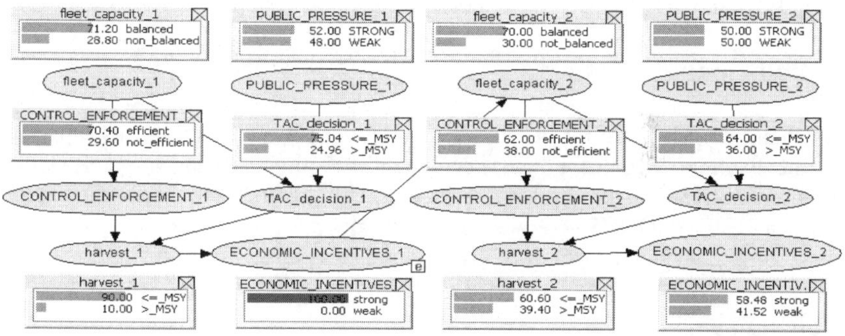

Figure 3: BBN representing the scenario with strong economic incentives to reduce the fishing fleet capacity in the Central Baltic Sea herring fishery.

The node "Economic incentives" is instantiated as being "Strong", and as a result the fishing fleet capacity is expected to be in a state "Balanced" with the fishery resources available. By virtue of that, it is expected that the control and enforcement, TAC decisions and public pressure will improve finally resulting in the actual fishing mortality value moving closer to the F_{MSY} value.

Traditional fishery management system is by its nature based on command and control with the regulator accepting most of the responsibility for management while rights based management rely much more on self-interest. Scott [33] states that "There are three general powers of ownership: to manage the asset, to transfer or sell it, and to take the income from it. The [Individual Transferable Quota] ITQ certainly gave the fisherman the third power. Instead of merely allowing him to go out and compete with others until the TAC was reached, it entitled him to a definite fixed percentage of the TAC - that is to the yield itself." The substantial body of published evidences shows that with open-access, non-exclusive fishing rights, too many resources are concentrated into fishing. It is believed that the ITQ system remarkably improves the allocation of resources because the fishing right-holders have a greater vested interest in the resource and are better motivated to take more responsibility for management.

4 Conclusions

Bayesian network simulation of the Central Baltic Sea herring fishery management is used to communicate the major sources of uncertainty in the system. This simulation shows that the fishing fleet overcapacity is the most important driver behind the decision and implementation overfishing and the related uncertainty. The environmental trends are influencing first of all the stock-recruitment relationship and the weight-at-age of the herring and they are contributing this way to the uncertainty of the calculated fishing mortality and the F_{MSY}. It is shown that assumed implementation of the strong economic incentives to reduce the fishing fleet capacity would bring the actual fishing mortality closer to the F_{MSY} value.

Acknowledgements

We thank Dr. Tiit Raid for his constructive comments. This study was supported by the Estonian Science Foundation grant No.7609, Estonian target financing program SF0180104s08 and by the Baltic Organizations Network for Funding Science EEIG (BONUS+) Project IBAM "Integrated Bayesian risk analysis of ecosystem management - Gulf of Finland as a case study".

References

[1] Aps, R. Management of shared Baltic fishery resources. *In: Management of Shared Fish Stocks. Ed. by A. I. L. Payne, C. M. O'Brien, and S. I. Rogers. Blackwell, Oxford*, pp. 190–201. 2004.

[2] Cardinale, M., Möllmann, C., Bartolino, V., Casini, M., Kornilovs, G., Raid, T., Margonski, P., Grzyb, A., Raitaniemi, J., Gröhsler, T., Flinkman, J. Effect of environmental variability and spawner characteristics on the recruitment of Baltic herring *Clupea harengus* populations. *Mar. Ecol. Prog. Ser. Vol. 388,* pp. 221–234. 2009.

[3] Margonski, P., Hansson, S., Tomczak, M.T., Grzebielec, R. Climate influence on Baltic cod, sprat, and herring stock–recruitment relationships. *Progress in Oceanography, 87*, pp. 277–288. 2010.

[4] Horbowy, J. Growth of the Baltic herring as a function of stock density and a food resources. *Acta Ichthyologica et Piscatoria, Vol. XXVII (1),* pp.27-39. 1997.

[5] Raid, T., Järvik, A. Baltic herring fisheries management in Estonia: a biological, technical, and socioeconomic approach. In: Herring: expectations for a New Millennium. Alaska Sea Grant College Program, AK-SG-01-04, pp. 703-720. 2001.

[6] Casini1, M., Bartolino, V., Molinero, J.C., Kornilovs, G. Linking fisheries, trophic interactions and climate: threshold dynamics drive herring Clupea harengus growth in the central Baltic Sea. *Mar. Ecol. Prog. Ser.* Vol. 413, pp. 241–252. 2010.

[7] ICES. ICES Advice 2010, Book 8, pp. 21-28. 2010.

[8] Aps, R., Ustinova, L., Gentzen, B., Grygiel, V., Paat, A., Uder, Y.-O., Fetter, M., Groth, B., Kestner, D., Wyshinski, M. Guide for the use of Baltic sprat and herring otoliths in fisheries studies. *Fischerei-Forschung, 29 (Sonderheft), Rostock,* 42 p. 1991.

[9] Eklund, J., Parmanne, R., and Aneer, G. Between-reader variation in herring otolith ages and effects on estimated population parameters. *Fisheries Research, 46,* pp. 147–154. 2000.

[10] Peltonen, H., Raitaniemi, J., Parmanne, R., Eklund, J., Nyberg, K., Halling, F. Age determination of Baltic herring from whole otoliths and from neutral red stained otolith cross sections. *ICES J. Mar. Sci.,* 59, pp. 323–332. 2002.

[11] Suuronen, P., Erickson, D.L., Orrensalo, A. Mortality of herring escaping from pelagic trawl codends. *Fish. Res. 25*, pp. 305–321. 1996.

[12] Suuronen, P., Perez-Comas, J.A., Lehtonen, E., Tschernij, V., Size-related mortality of herring (Clupea harengus L.) escaping through a rigid sorting grid and trawl codend meshes. *ICES J. Mar. Sci., 53*, pp. 691–700. 1996.

[13] Aps, R., Lassen, H. Recovery of depleted Baltic Sea fish stocks: a review. – *ICES J. Mar. Sci., 67*, pp. 1856–1860. 2010.

[14] Eagle, J., and Thompson, B. H. Answering Lord Perry's question: dissecting regulatory overfishing. *Ocean and Coastal Management, 46*, pp. 649–679. 2003.

[15] Van der Sluijs J.P., Petersen, A.C., Janssen, P.H.M., Risbey, J.S., Ravetz, J.R. Exploring the quality of evidence for complex and contested policy decisions. *Environ. Res. Lett., 3 (2)*, pp. 1-9. 2008.

[16] Wardekker, J.A., van der Sluijs J.P., Janssen, P.H.M., Kloprogge, P., Petersen, A.C. Uncertainty communication in environmental assessments: views from the Dutch science-policy interface. *Environmental Science & Policy, 11,* pp. 627-641. 2008.

[17] Jensen, F.V., Nielsen, T.D. Bayesian Networks and Decision Graphs. *Springer,* 448 p. 2007.

[18] http://www.bsrac.org/ooizzCMS/DA/statementsandrecommendations.

[19] Aps, R., Fetissov, M., Kell, L, Lassen, H. Baltic Sea Regional Advisory Council as a hybrid management framework for sustainable fisheries. *In: C.A. Brebbia, E. Tiezzi (Eds.). Ecosystems and Sustainable Development. UK: WIT Press*, pp. 163 - 172. 2009.

[20] ECA. The European Court of Auditors' Special Report No 7/2007 on the control, inspection and sanction systems relating to the rules on conservation of Community fisheries resources, 71 p. 2007.

[21] Popper, K. The logic of scientific discovery. *Routledge, London & New York*, 513 p. 2009.

[22] EC. Communication from the Commission: consultation on fishing opportunities for 2011. COM(2010) 241 final, 20 p. 2010.

[23] Ojaveer, E., Kalejs, M. The impact of climate change on the adaptation of marine fish in the Baltic Sea. *ICES J. Mar. Sci., 62,* pp. 1492-1500. 2005.

[24] Ojaveer, E., Kalejs, M. On ecosystem-based regions in the Baltic Sea. *Journal of Marine Systems*, 74, pp. 672-685. 2008.

[25] Jørgensen, H.B.H., Hansen, M.M., Bekkevold, D., Ruzzante, D.E., Loeschke, V. Marine landscapes and population genetic structure of herring (*Clupea harengus L.*) in the Baltic Sea. *Molecular Ecology* 14, pp. 3219-3234. 2005.

[26] Rönkkönen, S., Ojaveer, E., Raid, T., Viitasalo, M. Long-term changes in Baltic herring (Clupea harengus membras) growth in the Gulf of Finland. *Can. J. Fish. Aquat. Sci.* 61(2), pp. 219–229. 2004.

[27] Rahikainen M, Stephenson R.L. Consequences of growth variation in northern Baltic herring for assessment and management *ICES J. Mar. Sci.* 61 (3), pp. 338-350. 2004.

[28] ICES. Report of the Workshop on Age Reading of Baltic Herring (WKARBH). ICES CM 2008/ACOM: 36, 37 p. 2008.

[29] Rahikainen, M., Peltonen, H., Pönni, J. Unaccounted mortality in northern Baltic Sea herring fishery—magnitude and effects on estimates of stock dynamics. *Fisheries Research 67,* pp. 111–127. 2004.

[30] Raid, T. On the effect of fishery on herring stocks in the North-Eastern Baltic. *ICES, Baltic Fish Committee, C.M. 1996/J:27*. 16 p. 1996.

[31] Aps, R.; Kell, L.; Lassen, H.; Liiv, I. Negotiation framework for Baltic fisheries management: striking the balance of interest. *ICES J. Mar. Sci.,* 64(4), 858 - 861. 2007.

[32] FAO. FAO. Indicators for sustainable development of marine capture fisheries. *FAO Technical Guidelines for Responsible Fisheries: 8.* 68 pp. 1999.

[33] Scott, A. Introducing Property in Fishery Management, pp. 1-13. *In*: R. Shotton (ed.). *FAO Fisheries Technical Paper 404/1. Use of Property Rights in Fisheries Management. Proceedings of the FishRights99 Conference, Fremantle, Western Australia 11 - 19 November 1999, Rome.* 2000.

Unsustainable use of groundwater resources in agricultural and urban areas: a persistent scatterer study of land subsidence at the basin scale

G. Righini[1,2], F. Raspini[1], S. Moretti[1] & F. Cigna[1]
[1]*Earth Sciences Department, University of Florence, Italy*
[2]*ENEA, Bologna, Italy*

Abstract

Multi-temporal analysis of space-borne radar images through Persistent Scatterer Interferometry (PSI) is exploited for mapping subsidence at basin scale in the Gioia Tauro plain (Italy), a densely urbanized coastal area whose land deformation are controlled by aquifer overexploitation. Both historical (1992–2001; ERS1/2 images) and present (2002-2006; ENVISAT images) scenarios are analyzed to solve the spatial variability and temporal evolution of ground displacements affecting the plain. Average deformation rates as high as 10–15 mm/yr are observed from 1992 to 2006, with highest velocities (i.e. 22.8 mm/yr) occurred between 1992 and 2000 within the central part of the basin, in the area of Rizziconi (5 km ESE of Gioia Tauro). The outcomes of this PSI study will support the future improvement of groundwater management and the implementation of best strategies for risk mitigation, land use planning, sustainable use of groundwater resources, and consequent reduction of economic and social impacts.

Keywords: subsidence, groundwater overexploitation, Permanent Scatterers (PSInSAR), Persistent Scatterer Interferometry, SAR Interferometry, Gioia Tauro, Calabria.

1 Introduction

Densely urbanized and heavily industrialized areas are frequently affected by land subsidence, often caused by human activities and, in particular, subsurface

www.witpress.com, ISSN 1743-3541 (on-line)
doi:10.2495/ECO110071

water withdrawal. In many urban areas, subsidence induced by groundwater overexploitation and associated aquifer compaction has strong economic impacts and usually causes building damaging, road cracking, well-casing failure and deformation of water supply and sewage networks. Moreover, in low-lying coastal areas, decrease of ground elevation, combined with a global sea-level rise, may lead to severe flooding and inundations.

Advanced remote sensing techniques, such as multi-temporal InSAR (Synthetic Aperture Radar Interferometry), may contribute to the understanding of spatial and temporal patterns of ground subsidence, exploiting long time series of satellite images acquired at different times over the investigated areas, and reconstructing their deformation history.

We exploit the contribution of Persistent Scatterer Interferometry (PSI) through the multi-temporal analysis of space-borne radar images, for mapping land subsidence at basin scale in the Gioia Tauro plain (Italy), a densely urbanized coastal area built over unconsolidated fine-grained sediments. Available archives of satellite SAR images acquired since 1992 and currently operational SAR missions allow the study of both past and present ground displacements; the main objectives are to facilitate the analysis of hazard and risk scenarios, and to support the improvement of groundwater management in order to minimize future subsidence, and to help the implementation of best strategies for risk mitigation, land use planning and consequent reduction of economic and social impacts.

2 Persistent Scatterer Interferometry

Differential InSAR (DInSAR), a technique that exploits the phase difference of two SAR images gathered at different times on the same target area, has been widely used since late '80s [1, 2] to detect, map and quantify surface deformations with centimetre precision over large contiguous areas [3, 4]. In the last two decades, space-borne DInSAR techniques have demonstrated and improved their capability and their potential as ground deformation measurement tools in a wide range of application fields.

Conventional DInSAR is limited by temporal and geometrical decorrelation. Moreover, the impossibility of removing phase distortions introduced by atmospheric effects can degrade and limit the accuracy of the results.

Multi-interferometric approaches, such as PSI techniques, overcome the main limitations encountered in single-pair interferograms; the use of long temporal series of SAR data makes possible the removal of atmospheric artefacts or, at least, the reduction of their impact on the estimated displacement values.

PSI techniques represent powerful tools for identifying and mapping man-induced surface movements and for reconstructing the temporal evolution of the investigated areas, considerably improving the applicability of SAR Interferometry for the measurement of ground motions [5–9]. Extracting phase information from multiple SAR images, these techniques are able to identify ground resolution elements (pixels) dominated by single scatterers, the so-called PS (Persistent Scatterers). For this reason, PSI methods have successfully been exploited for the analysis of urban areas, where stable targets produce strong reflectors, typically corresponding to man-made structures (e.g. buildings, roads, dams, pipelines, pylons) or stable natural reflectors (e.g. rock outcrops). These targets allow the measurement of deformation velocities (both average ground deformation rate and displacement time series, acquisition by acquisition) along the satellite Line Of Sight (LOS), with millimetre precision [10, 11].

3 Gioia Tauro plain

3.1 Geological and geomorphologic setting

The Gioia Tauro plain, also known as “Rosarno Plain”, is an urbanized coastal area located in southern Calabria Region, Italy (Fig. 1), including 33 municipalities (entirely belonging to Reggio Calabria district), with a total extension of about 500 km^2. The area has nearly 150,000 inhabitants, with a very high population density (300 inhabitants/km^2).

From a geological point of view, the Gioia Tauro basin is a 25-30 km long, NNE–SSW oriented half-graben, spreading from the Valley of Mèsima to the Massif of the Aspromonte and elongated between the Calabrian Arc Mountains and the Tyrrhenian Sea [12]. The uplifted Aspromonte and Serre ridges form the structural backbone of southwest part of the Calabrian Arc.

The main geological feature of the Calabrian Arc is the 180 km long normal fault belt that strongly controls the morphology of western-facing mountain slopes, separating the main uplifted mountain range from a series of extensional sedimentary basins (i.e. Mesima, Gioia Tauro, Reggio Calabria and Barcellona basins), located on the Tyrrhenian side of the arc [13–18].

To the east, the Gioia Tauro basin is bounded by Cittanova fault, a 35 km normal fault juxtaposing the basement of the uplifted Aspromonte and Serre mountains with the Pleistocene deposits of the plain. S. Eufemia fault, an 18 km ENE-striking normal fault bounds to the south the basin. To the NE, the Coccorino and Nicotera normal faults abruptly truncate and separate the plain area from the uplifted promontory of Capo Vaticano.

While the uplifted blocks and the bedrock are characterized by crystalline basement rocks (gneiss and granites), the basin is filled by a 700 m thick succession of marine and continental sediments of Pliocene and Quaternary Age. The Late Pliocene-Early Pleistocene marine sequence consists of about 70 m of sands and calcarenites, overlaid by a layer of compacted clays and silty clays of great thickness (500 m or more), capped by 100 m of regressive sands and conglomerates.

The marine sequence is unconformably overlain by the most superficial layers of the plain (up to a depth of 50-70 m from the ground level), that is prevalently constituted by granular saturated soils consisting of Middle-Upper Pleistocene alluvial conglomerates and sands, truncated by a wide planation surface [13, 15].

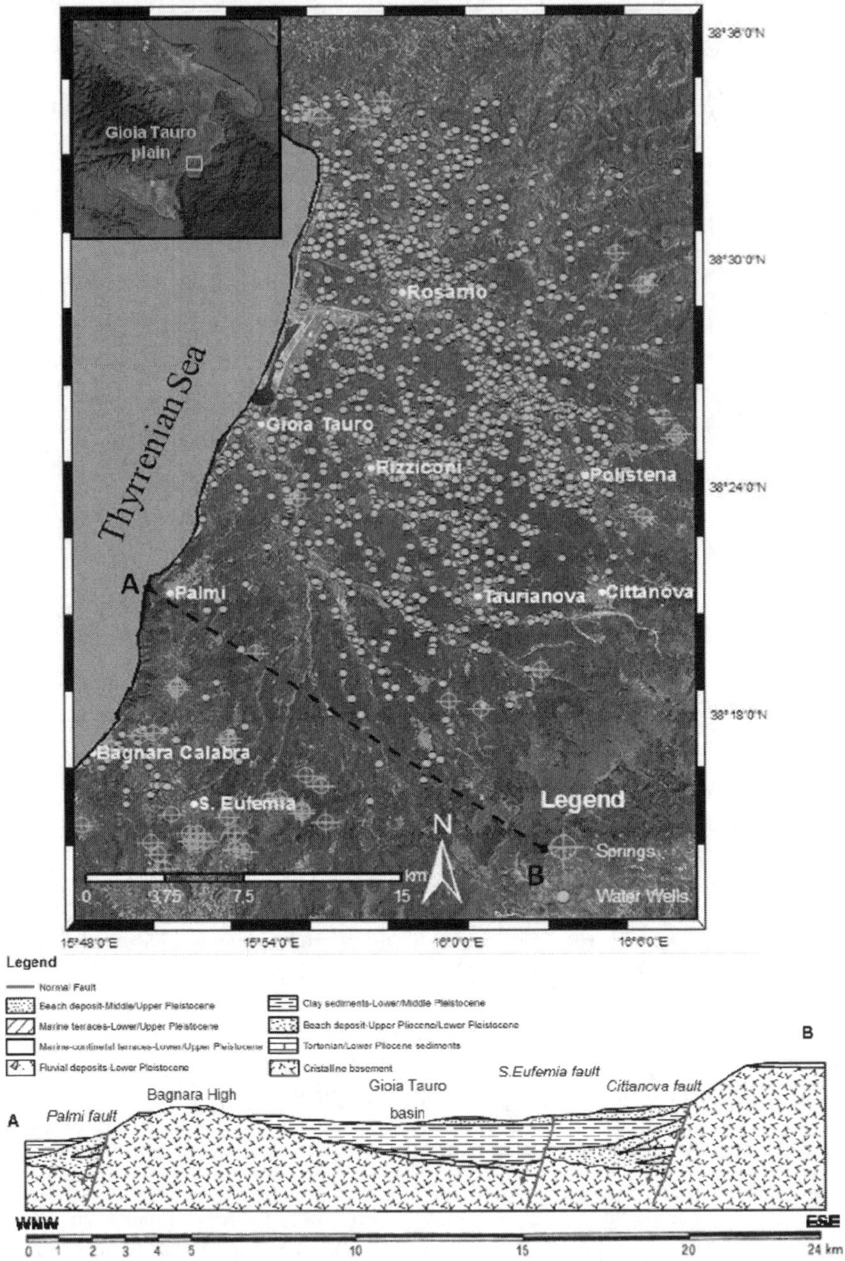

Figure 1: Geographical setting of the area of Gioia Tauro, Italy (above). Simplified geological cross-section of the basin (below), modified from Jacques et al. [15].

3.2 Agricultural activities and groundwater management

Agriculture is the most important economic activity in the area: olive trees and seed crops (mainly wheat) are cultivated extensively. The citrus cultivation represents a key factor for the economic livelihood of local population. In the whole plain, the yearly production of citrus (especially tangerines and oranges) amounts to 700,000 tons, i.e. 3% of the total product of Mediterranean countries or 0.7% of the world product [19].

The farming system in Calabria is based on a much fragmentized model, with hundreds of small plots, self-sufficient from the water supplying point of view. Farms and estates of less than 2 ha made up 69% of the total in 2000 [19].
In addition to the municipal network for water supply and distribution (Fig. 1), that has serious inefficiencies, many independent and usually uncontrolled water wells have been recently constructed and, in the whole plain, 45% of water needs are currently provided by private water wells [19]. Recently increased groundwater demands for irrigation purpose have caused the gradual lowering of the water table, with related compaction of the aquifer deposits, to progressively accelerate.

Different elements at risk are present in the plain, such as the Gioia Tauro Harbour, the largest container terminal of the Mediterranean Sea (about 440 ha of land and 1.8 million m^2 of stocking yards), the road infrastructure with the State Road n.18 and the motorway A3, and the railway network through the Rosarno Station. Many new works, aimed at improving the harbour and the industrial activity of the area have also been designed and are currently being built or under feasibility studies. In 2008, after the submission of an Environmental Impact Assessment, authorization for the construction of a huge liquefied natural gas terminal in the Gioia Tauro port area has also been obtained. The construction is planned to start in 2011 and the system is expected to be operative in 2014, with a capacity of 12 billion m^3 of gas per year, over 10% of the Italian gas demand.

3.3 Satellite data and PSI analysis

The satellite analysis of land subsidence in the Gioia Tauro basin was performed using the PSInSAR (Permanent Scatterers InSAR) technique, which belongs to the multi-interferometric approaches [5], and integrating several datasets of satellite SAR data acquired by different ESA (European Space Agency) missions.

Forty-eight SAR images acquired in 1992-2001 by ERS1/2 satellites along descending (when satellite passes over the target area approximately from north to south) and ascending (south to north) orbits were used to study past ground displacements. Eighteen SAR images acquired by ENVISAT satellite along descending orbits and twenty-six images along ascending orbits were then used to monitor recent land deformation.

The PSInSAR analysis provided estimates of yearly deformation velocity, referred to both historical (1992–2001; ERS images) and recent (2002–2007;

ENVISAT images) scenarios (Tab. 1), allowing the analysis of the spatial variability and temporal evolution of Gioia Tauro subsidence.

The density of PS turned out to be significantly higher in urbanized areas in comparison to the agricultural and woodland sectors: according to the 1st classification level of the land cover map (CORINE Land Cover, 2000), PS density is distributed as ca. 350 PS/km^2 in urban and artificial lands, 20 PS/km^2 in agricultural and pasture areas, and only 5 PS/km^2 in the woodland and semi-natural areas.

Table 1: Number of images, time period, number of identified PS and average PS density (differentiated for land cover categories) for ERS and ENVISAT data stacks used for the PSI analysis. Desc, descending; Asc, ascending.

Satellite and orbit	N° images	Time range	N° PS	PS density (PS/km^2)		
				Urban areas	Agricul. areas	Wood Land
ERS_desc	48	20/10/1992-05/01/2001	34,064	532	24	7
ERS_asc	48	08/09/1992-24/11/2000	31,935	473	25	7
ENVI_desc	18	26/12/2003 01/09/2006	8,161	129	6	1
ENVI_asc	26	29/11/2002-08/12/2006	22,192	325	18	6

3.4 Results and discussion

ERS1/2 and ENVISAT PSI data, both in ascending and descending geometry, allows the analysis of the spatial variability and, through the displacement time series, the temporal evolution of Gioia Tauro subsidence in 1992-2006.

PSInSAR results for ERS dataset (Fig. 2 and Fig. 3) show that subsidence was widespread throughout Gioia Tauro basin in the time period from 1992 to 2000, with a mean rate of 11 mm/yr. In particular, an ellipse-shaped area, SW-NE oriented, corresponding to the central part of the plain, is affected by intense subsidence, with mean deformation velocity of 15 mm/yr. Two areas of maximum deformation located at Rizziconi (within the central part of the basin, 5 km ESE of Gioia Tauro) and Rosarno (in the northern part of the plain) are observed in 1992-2000, whose higher measured deformation velocities are 22.8 mm/yr and 18.8 mm/yr, respectively.

PSInSAR results for ENVISAT dataset (Fig. 4 and Fig. 5) indicate, in the time period from 2002 to 2006, a general subsidence decrease, for both intensity and extension of the affected area. Compared with the previous period, average ground deformation rate highlights that the SW-NE oriented ellipse-shaped area corresponding to the central part of the Gioia Tauro basin, is affected by a subsidence with a mean value of 7 mm/yr. Maximum land deformation

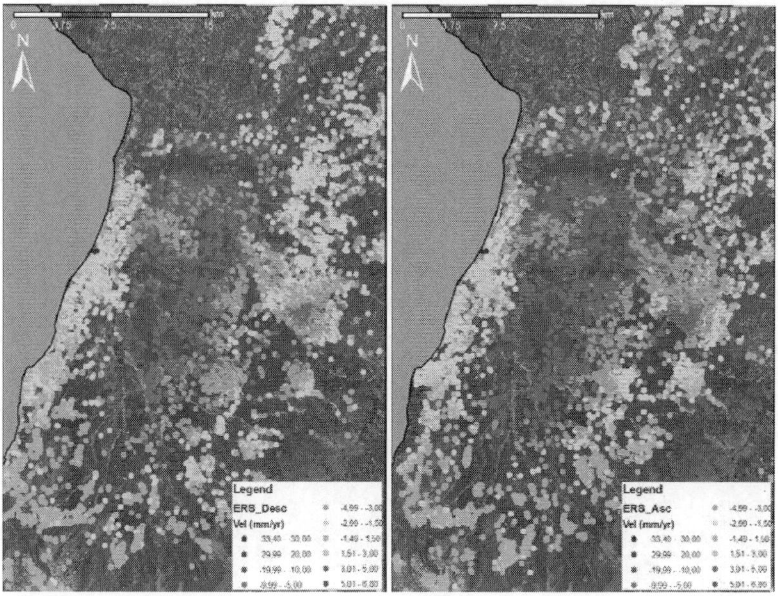

Figure 2: Average displacement rates along the LOS of ERS satellites in 1992-2000.

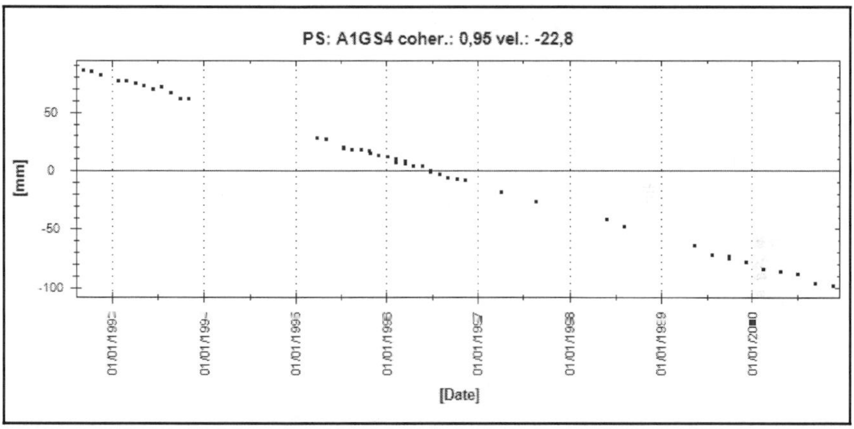

Figure 3: Example of displacement time series of a PS located in the area of highest deformation.

velocities, detected in 2002-2006 are 17.7 mm/yr and 11 mm/yr, measured at Rizziconi and north of Rosarno, respectively, which remain the areas of maximum subsidence rate.

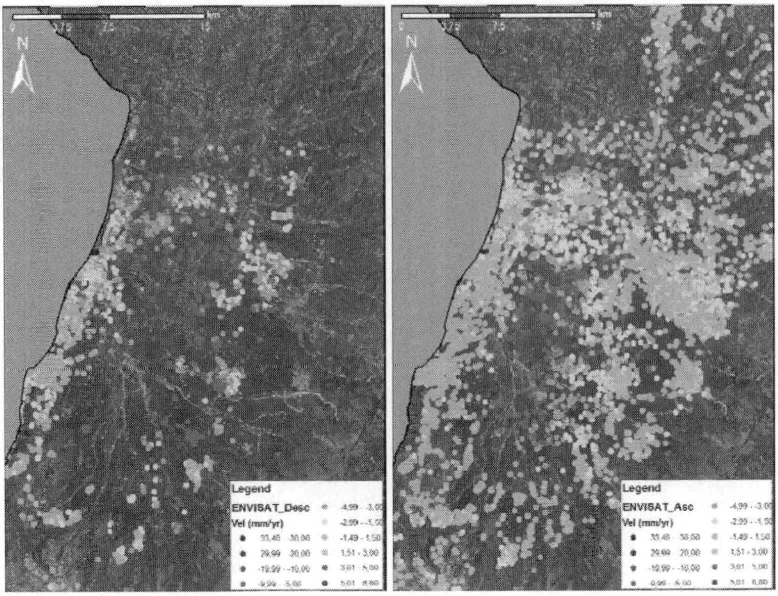

Figure 4: Average displacement rates along the LOS of ENVISAT satellite in 2003-2006.

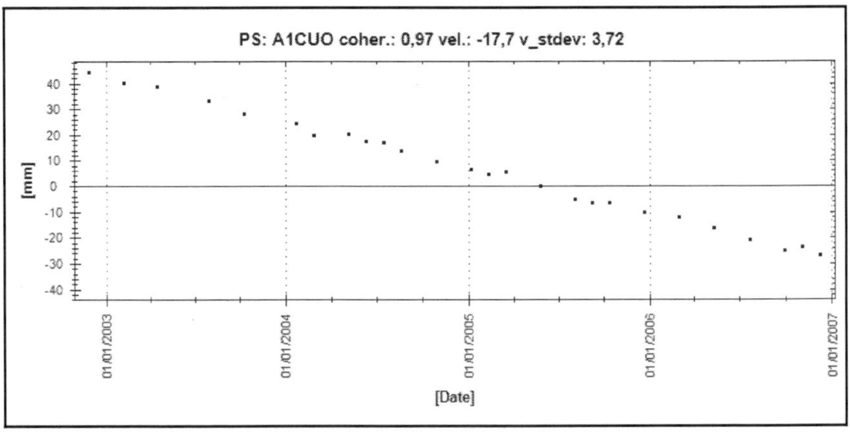

Figure 5: Example of displacement time series of a PS located in the area of highest deformation.

Analysis of PS displacement time series enable the analysis of the temporal variability of ground displacements, highlighting that, during both 1992-2001 (Fig. 3) and 2003-2006 (Fig. 5), subsidence is characterized by a constant rate of deformation.

Cumulating in time the displacement measured through the PS time series in 1992-2000 (ERS dataset), a maximum subsidence of 184 mm and 162 mm was observed at Rizziconi and Rosarno, respectively. On the other hand, for the 2003-2006 interval (ENVISAT dataset), a slightly lower deformation was measured, with maximum subsidence of 74 mm and 48 mm, observed at Rizziconi and Rosarno, respectively.

Fig. 6 and Fig. 7 show a 3D view of the cumulated LOS displacement field in 1992-2000 and 2003-2006 in the whole basin of Gioia Tauro. The ground deformation surface is interpolated with the IDW (Inverse Distance Weighted) method, assigning values to unmeasured locations by using the displacement values from scattered sets of surrounding PS.

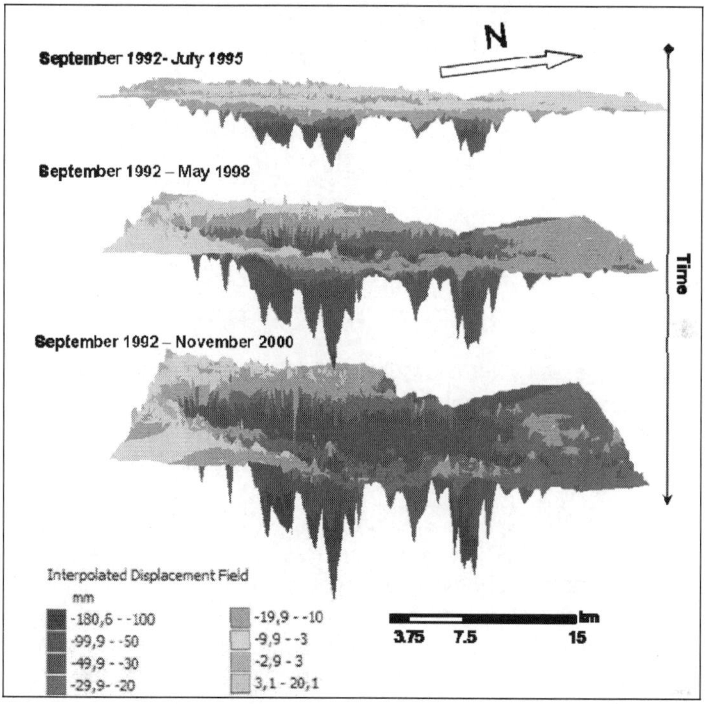

Figure 6: 3D view of cumulated LOS displacement in 1992-2000.

The three snapshots of Fig. 6 represent the interpolated ground deformation surface, corresponding to three different time intervals (whose length is about 950 days) during 1992–2000: September 1992–July 1995; September 1992–May 1998; September 1992–November 2000.

The two snapshots of Fig. 7 correspond to two different time intervals (whose length is about 750 days) during 2003–2006: November 2002–January 2005; November 2002–December 2006.

The observed subsidence in the Gioia Tauro plain is thought to be induced by groundwater overexploitation, with related water table decline and fine-grained aquifer compaction. Land subsidence is a displacement phenomena characterized by very high vertical component, hence ground deformation measurements have

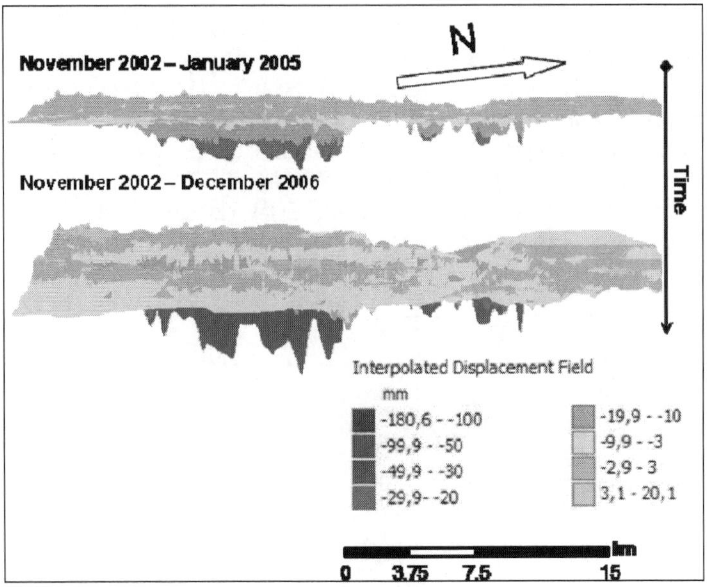

Figure 7: 3D view of cumulated LOS displacement in 2002-2006.

to be considered mainly vertical–subsidence (negative values, when PS move away from the sensor) and uplift (positive values, when PS move towards the sensor). Considering also the flat geometry of the plain, ascending and descending geometries provide very similar values of mean deformation rates of the same target areas.

Comparing the geographic location of water extraction sites (Fig. 1) and the deformation maps provided by the PSI analysis (Fig. 2 and Fig. 4), it is possible to highlight a slight correlation between the occurrence of intense subsidence and the distribution of water wells and springs. However, significant spatial variability of deformation velocities throughout the whole basin is also observed. Such variability may be induced by differences in both the geotechnical properties of the aquifer sediments and the compressible layer thickness and, secondarily, the water amounts pumped from the different water wells in the plain. Moreover, an incomplete census of water extraction sites cannot be ruled out for this area.

Identification of independent wells and respective water extraction rates, will be an urgent topic of discussion and analysis in the future, to better understand subsidence causes and patterns, and consequently improve water and land management strategies for a future sustainable use of groundwater resources.

4 Conclusions

The PSI analysis allowed the analysis of the spatial variability and temporal evolution of ground displacements affecting the Gioia Tauro plain in 1992-2006.

Both historical (1992-2001; ERS1/2 images) and present (2002-2006; ENVISAT images) scenarios reveal the occurrence of significant ground displacements affecting the unconsolidated fine-grained sediments of the plain. Average deformation rates as high as 10-15 mm/yr are observed in 1992-2006, with highest velocities (i.e. 22.8 mm/yr) occurred between 1992 and 2000 within the central part of the basin, in the area of Rizziconi (5 km ESE of Gioia Tauro).

Land deformation is probably controlled by groundwater overexploitation and consequent aquifer compaction. Slight correlation of intense subsidence with the distribution of water wells and springs is observed. Nevertheless, significant spatial variability of deformation patterns throughout the whole basin is also registered and justified by the differences in the geotechnical properties and thickness of the compressible layer, water extraction rates, and a probable incompleteness of the water well census.

The outcomes of this PSI study will support the future improvement of groundwater management and the implementation of best strategies for risk mitigation, land use planning, sustainable use of groundwater resources, and consequent reduction of economic and social impacts.

Acknowledgements

This analysis was carried out in the framework of the Terrafirma project, a pan-European ground motion hazard information service supported by the Global Monitoring for Environment and Security (GMES) initiative promoted by the European Space Agency (ESA).

References

[1] Zebker, H. A., & Goldstein, R. M., Topographic Mapping From Interferometric Synthetic Aperture Radar Observations. *Journal of Geophysical Research*, **91**, pp. 4993–4999, 1986.

[2] Gabriel, A. K., Goldstein, R. M. & Zebker, H. A., Mapping Small Elevation Changes Over Large Areas: Differential Radar Interferometry. *Journal of Geophysical Research*, **94**, pp. 9183–9191, 1989

[3] Canuti P., Casagli N., Ermini L., Fanti R. & Farina P., Landslide activity as a geoindicator in Italy: significance and new perspectives from remote sensing. *Environmental Geology*, **45 (7)**, pp. 907-919, 2004.

[4] Tarchi D., Casagli N., Moretti S., Leva D. & Sieber A.J., Monitoring landslide displacements by using ground-based synthetic aperture radar interferometry: application to the Ruinon landslide in the Italian Alps. *Journal of Geophysical Research*, **108 (8)**, pp. 2387-2400, 2004.

[5] Ferretti A., Prati C. & Rocca F., Non-linear Subsidence Rate Estimation Using Permanent Scatterers in Differential SAR Interferometry. *IEEE Transactions On Geoscience and Remote Sensing*, **38**, pp. 2202-2212, 2000.

[6] Colombo D., Farina P., Moretti S., Nico G. & Prati C., Land subsidence in the Firenze-Prato-Pistoia basin measured by means of spaceborne SAR

interferometry. *Proceedings of IGARSS 2003,* Toulouse: pp. 2927-2929, 2003.

[7] Farina P., Moretti S., Colombo D., Fumagalli A. & Manunta P., Landslide risk analysis by means of remote sensing techniques: results from the ESA/SLAM project. *International Geoscience and Remote Sensing Symposium (IGARSS)*: pp. 62-65, 2004.

[8] Crosetto M., Crippa B., Biescas E., Monserrat O., Agudo M. & Fernandez P., Land deformation monitoring using SAR interferometry: state of the art. *Photogrammettrie Fernekundung Geoinformation*, **6**: pp. 497-510, 2005.

[9] Tomàs R., Màrquez Y., Lopez-Sanchez J. M., Delgado J., Blanco P., Mallorquì J. J., Martìnez M., Herrera G. & Mulas J., Mapping ground subsidence induced by aquifer overexploitation using advanced Differential SAR Interferometry: Vega Media of the Segura River (SE Spain) case study. *Remote Sensing of Environment*, **98**, pp. 269-283, 2005.

[10] Ferretti A., Prati C. & Rocca F., Permanent Scatters in SAR Interferometry. *IEEE Transactions on Geoscience and Remote Sensing,* **39**, pp. 8-20, 2001.

[11] Colesanti C., Ferretti A., Novali F., Prati C. & Rocca F., SAR monitoring of progressive and seasonal ground deformation using the Permanent Scatterers technique. *IEEE Transactions on Geoscience and Remote Sensing*, **41**, pp. 1685-1701, 2003.

[12] Cucci L., D'Addezio G., Valensise G. & Burrato P., Investigating seismogenic faults in Central and Southern Apennines (Italy): modeling of fault-related landscape feature. *Annali di Geofisica*, **39 (3)**: pp. 603-618, 1996.

[13] Tortorici L., Monaco C., Tansi C. & Cocina O., Recent and active tectonics in the Calabrian arc (Southern Italy). *Tectonophysics*, **243**: pp. 37-55, 1995.

[14] Monaco C. & Tortorici L., Active faulting in the Calabrian arc and eastern Sicily. *Journal of Geodynamics*, **29**, pp. 407-424, 2000.

[15] Jacques E., Monaco C., Tapponnier P., Tortorici L. & Winter T., Faulting and earthquake triggering during the 1783 Calabria seismic sequence. *Geophysical Journal International*, **147 (3)**, pp. 499-516, 2001.

[16] Galli P. & Bosi V., Paleoseismology along the Cittanova fault: implications for seismotectonics and earthquake recurrence in Calabria (southern Italy). *Journal of Geophysical Research*, **107,** pp. 1-19, 2002.

[17] Tortorici L., Bianca M., de Guidi G., Monaco C., Tortorici L., Fault activity and marine terracing in the Capo Vaticano area (southern Calabria) during the Middle-Late Quaternary. *Quaternary International*, **101**, pp. 269-278, 2003.

[18] Catalano S., De Guidi G., Monaco C., Tortorici G. & Tortorici L., Active faulting and seismicity along the Siculo-Calabrian Rift Zone (Southern Italy). *Tectonophysics*, **453**, pp. 177-192, 2008.

[19] CIRCA, www.circa.europa.eu/irc/dsis/regportraits/info/data/en/itf6_eco.htm

Assessment of water resource consumption in building construction in India

S. Bardhan
Dept. of Architecture, Jadavpur University, India

Abstract

Building construction is a highly resource intensive process, concerning use of materials, land, energy and water. Since buildings are required to fulfil our primary need for shelter, consumption of these resources is simply unavoidable. However, with shrinking stock of natural resources and degrading eco-system services, the consumption process has to be wise, judicious and non-wasteful. Search for alternative and recycled materials, multiple floor spaces and use of renewable energy are some of the human responses against the perceived threat of materials, land and energy constraints respectively. Water resource management in building construction and operation, however, has still a long way to go, especially because the amount of water used per unit area of construction largely remains undocumented. There has been some pioneering research on embodied water measurement of several non-residential buildings in Australia. In this backdrop, the present paper seeks to understand and assess the quantity of fresh water used in contemporary urban multi-storeyed residential buildings and reports a study conducted for a real-life project in Calcutta (now, Kolkata), India. Since the use of construction water directly varies with the type of construction, a steel and glass building will have its embodied water-footprint mainly on account of that of its materials while on-site water use plays a major role in case of a cast-in-situ reinforced cement concrete and brick building. Thus, water efficiency at the production stage is required in the first case while the second category demands concerns and actions at the consumption stage. This paper examines some of the issues related to the subject like water demand at material production as well as construction stage and the resultant embodied water of typical urban constructions in India, which was found to be in the range of about 27 Kilolitres/Sq m of total built-up area.

Keywords: construction water, ground water, site water management.

WIT Transactions on Ecology and the Environment, Vol 144, © 2011 WIT Press
www.witpress.com, ISSN 1743-3541 (on-line)
doi:10.2495/ECO110081

1 Introduction

Fresh Water is a very precious natural resource and it is widely reported that the world reserve of fresh water is fast depleting. Life on earth is not possible without water. Climate change concerns across the globe has predicted severe water scarcity due to degrading eco-system services and World Business Council for Sustainable development (WBCSD) in its Vision 2050 statement has estimated an additional load of more than 500 million people under water-stress conditions in the BRIC (Brazil-Russia-India-China) countries by 2030 [1]. In India, with the National Action Plan for Climate Change (NAPCC) and its constituent National Water Mission, a target of 20% increase in water-use efficiency has been set to counter the challenge [2]. The National Water Policy of Government of India recognizes the need of planning the economic activities like agriculture, industries and urban development in conformity with availability of this life-sustaining resource and has recommended water-zoning [3]. From the point of view of urban development, building industry is known to be a major consumer of water resource, which in many cases, is the ground water extracted through bore-wells. There also had been reports of major subsidence threats due to over-exploitation of this important reserve in a particular segment of the city of Calcutta (now, Kolkata), India.

In a building's entire life-cycle, water and energy are considered to be the two major pillars that support human activities in it. While there are many studies on embodied energy and recurring energy consumption in buildings, such database appears scanty in the field of water. Similarly, embodied energy of building constituent materials is also available along-with their embodied CO_2 emission factor, which is again not the case when water resource is concerned. Even industry related documents, in most cases, barely touches upon water use in their raw materials' inventory. Thus, there seems to be a big void when similar baseline research data availability with respect to water is looked for. Out of one of the few studies, United Nations Environment Program [4] has indicated that over an entire life-cycle, the building industry consumes a global average of 30% of fresh water and generates 30% of world's effluents. In India, the manual on water supply and treatment [5] has indicated water consumed per unit production of steel and it is, thus, one of the few building materials having a recorded water foot-print. Other major building materials like bricks, cement, aluminium, glass etc. use considerable quantity of water in their manufacturing process through extraction and processing, though data on their water-foot print is not readily available. In this backdrop, the present paper seeks answers to the following research questions:

- Is it possible to know the capital water content or water foot-print of a building in India? This is precisely its embodied water.
- To know the above, it will also be important to know the water foot-print or embodied water of its constituent materials, at least the common ones. Is there any such data or reported research finding for Indian products in this area?

- It will also be important to understand the amount of water used during the actual construction process, both direct and indirect. How can this construction water be estimated for Indian buildings?

The above key issues have been deliberated in the course of the paper to the extent possible on the basis of available data, documentation and an actual case study.

2 Research so far

Internationally, one particular research team has been known to have done extensive research in this field. Researchers in Australia [6, 7] and their studies suggest that embodied water is many times greater than the volume of water used during a building's life. The team working at Deakin University, Victoria, looked at 17 non-residential case studies in Australia and found that embodied water could be as much as 20 kilolitres per m^2 of gross floor area, but that water required during the construction process was minimal. The buildings studied include examples from the health, education and commercial sectors; and the team looked at the embodied water of the structure, substructure, finishes, roof and services. The UK concrete industry [8] has agreed on a 'Concrete Industry Sustainable Construction Strategy' and has reported that *"though no studies have been done on homes in the UK, a 2004 study in Australia estimated that a typical Australian house represents about 15 years worth of operational water i.e. water used for cooking, cleaning, washing, drinking, toilet flushing and gardening. This study estimated that a kilo of concrete has about two litres of embedded water, a kilo of timber about 20 litres, a kilogram of steel about 40 litres, a kilo of aluminium about 88 litres, and that a kilogram of plastic has about 185 litres of embedded water."* This reporting shows that there has been sufficient work in finding out the embodied water in building materials and structures in Australia, though no such research has been reported from India. In contrary, operational water use has been under recent focus by one of the leading national organizations. Centre for Science and Environment, New Delhi, India [9] has taken up a study for developing water efficiency labeling and standards in Indian conditions. This study is aimed at rating fixtures like dual flush toilets, low water using/ sensor based/waterless urinals, sensor faucets/faucets with flow restrictors, water efficient showers etc. according to their efficiency, much like what has already been done for electrical appliances. However, studies on embodied water content of buildings being relatively complex involving multi-stage activities – raw material extraction, manufacture and construction – have not yet been conclusively reported by either industry or academic groups. It may be presumed that even if research is on-going in this area, results from more than one source will only substantiate the findings and will help to draw closer to accuracy by collating the output data.

3 Significance in national context

Reducing water consumption and improving water efficiency in buildings is a major step towards sustainable water management [8]. Unless the embodied water content of a building is known or measured, water management is not possible. Rampant misuse of water will continue and the objectives of National Water Policy of better water management practices and improvement in operational technology will remain unfulfilled. Further, this is integrally linked with sustainable building concepts and sustainable development issues. Life Cycle Analysis (LCA) of buildings and its corresponding eco-profile is considered to be a rational method of quantifying the 'environmental impact' or 'greenness' of a building [10]. However, this is dependent on real-life resource consumption amounts and other relevant data that are at present not available or are difficult to ascertain because of lack of base-line research in this area. Hence, LCA of buildings are currently carried out with assumed figures and therefore, seldom give the true picture. The present study intends to contribute in filling up these research gaps. With water shortages likely to grow in significance around the world due to climate change and Indian construction industry growing at an astounding rate of 10% per annum over the last ten years [11], this is an issue that needs immediate attention. The indigenous green building rating system titled Green rating for Integrated Habitat Assessment (GRIHA) has stipulated points for buildings which are resource efficient. The results of this study can be integrated into the ratings to determine a measurable and objective way of assessing their sustainability. This will give greater credence to the rating system and its scientific basis. It is expected that the architect community, real estate industry and general public will be more conscious and judicious in future water-use. Studies on multiple live case studies can create baseline data to help in framing appropriate policies for achieving water use reduction. This will support the goals of sustainable development, National Water Policy and National Water Mission under NAPCC.

4 Embodied water assessment in buildings

A group of multi-storied residential apartment building of steel and RC construction located in the southern part of the city of Calcutta in the Eastern Indian state of West Bengal was selected for studying its embodied water content. This project is a prototype of the developmental activities taking place in the housing sector in the entire country. It is meant for people belonging to high income group and is also designed for luxurious life-style. Thus, the choice of materials and the quality of construction are both superior. Understandably, for the type of construction mentioned above, steel and cement form the two most volumetrically significant as well as water-intensive materials, the latter requiring water at both production and construction stage. The total built-up area of the apartments together exceeds three million square feet and was constructed between the years 2002 and 2008. The site had a total of six bore wells to supply water for construction of the said building group.

4.1 Methodology

The methodology presented here involves two stages- one that assesses the embodied water of the materials used in the building and the second assesses the amount of water used during the construction process. The project office of the building was approached to access the materials' purchase records and the assessment was carried out on the basis of the figures provided by them. However, it is important to note that a large amount of construction was made from Ready Mix Concrete, whose embodied water content needs specific study and has not been covered here. The major building materials having the highest stake in the constructed volume such as bricks, cement and steel were considered in the first stage of assessment as presented in table 1. The second stage had two optional methods- the first one involving theoretical calculation of the water requirement for concrete mixes and curing of brick masonry as well as concrete castings, while the other depends on data collection on water consumption directly from site sources to have an idea on the various indirect and otherwise unrecorded heads for which water is used during construction like watering for sub-grade stabilization, dust control, water line testing and cleaning, use by on-site resident construction labours, so on and so forth. Research on concrete mixes in Indian conditions have estimated a savings of 7,500,000 tonnes of cement as well as 3,750,000 Kilolitre of potable water by the use of water-reducers for concrete mixes with 50 million tonnes of cement, indicating 2 Kilolitres of water requirement per ton of cement used in concrete mixes [12]. To account for the indirect component of water usage, this paper follows the second line of investigation and presents the water consumption assessment during construction stage from data released by the project office on the electrical energy expended on account of water pump operation at site. These figures were converted to equivalent water quantity on the basis of known yield of the bore wells as presented in table 2. The outcome of both the stages is expressed in Kilolitres per Sq m (Kl/Sq m) i.e. water used per unit floor area of the constructed building. The total embodied water of the building will be given by the sum of these two amounts.

Table 1: Assessment of embodied water in the materials of the case study.

Sl.	Materials	Quantity used	Unit	EW Coefficient	Total Embodied water in material	Embodied water per unit floor area
				Kl/unit	Kl	Kl/Sq m
1	Cement	154,858	Ton	1	154,858	0.5
2	Bricks	42849.37	Cum	0.71	30423	0.1
3	Steel	38906	Ton	200-250	7781200	25
4	Aluminium	15143	Kg	0.088	1332.584	0.004
	Total embodied water of the major materials					25.604

Table 2: Water consumption for 68 months during construction.

A	B	C= A/B	D	DxC/1000
Energy consumed in 21 months	Capacity of Water Pumps	Duration of operation	Yield of bore well	Total water consumed
KWh	kW	Hours	Litres per Hour (lph)	Kl
972319.44	5.595 (7.5 H.P) &7.46 (10.0 H.P.)	24826.23	18,000 & 24,000	
	Average: 6.5275			
	For 6 pumps: 39.165 kW		21000	521350.83

4.2 Embodied water attributed by materials

As mentioned earlier, only the major stake holding materials have been considered in this study, especially as the embodied water in these materials in Indian scenario were found to be somewhat available. However, these figures either came in from more than a decade old documents or had to be calculated from mentions in studies and researches with different thrust areas. These are explained in the following text:

Cement: In an industrial documentation study of Central Pollution Control Board [13], different values were obtained on water consumption in cement industry. It was found to be varying between 0.5 Kl/ton to 1 Kl/ton. For the purpose of this study, 1 Kl/ton has been considered as embodied water foot-print of cement.

Steel: Manual on water supply and treatment [5] specifies water embodied in steel as 200-250 Kl/ton of steel manufacture. Although this is excessively high compared to that obtained in international studies, which is in the range of 40 Litres/Kg or 40 Kl/ton [8], this is the only authentic figure obtained in Indian context and, therefore, considered in this paper.

Brick: Brick manufacturing sector in India is largely informal and without much documentation on its water consumption per unit of production. A technical note of Brick Industries' Association [14] mentions water requirement as 20-30% of clay amount, which works out to be about 300 Litres/Cum of brick manufacture. Local practice, however, would certainly involve much more water, not only because of water abundance, but also owing to lesser awareness of water issues. Hence, this paper draws parallel between the consistency in concrete mix with that of brick and builds on the water-cement ratio of M20 concrete to arrive at the possible water quantity in brick manufacture [15]. This works out to be 714 Litres/Cum, more than double the previous figure. The total brick volume used in the building fabric was obtained from the number of bricks purchased by the Project.

Additionally, in the interest of research, a fourth material - Aluminium has also been included as its embodied water was available from International reports [8], though its contribution, as expected, is insignificant.

4.3 Water consumption during actual construction

The data of electrical expenses on account of on-site water pumping during construction period was available for some 68 months. This monetary value was converted into energy consumed in Kilo-Watt-Hour (kWh) by dividing it with the unit energy rate, considered here to be 3.91 INR as per available contemporary data and the energy consumed by the water pumps was calculated. Total energy equivalent for water resource use was found to be 972319.44 kWh for these 68 months. As mentioned earlier, the construction water was supplied by the six on-site bore wells. The capacities of on-site water-pumps, as obtained from site sources, were 7.5 H.P and 10.0 H.P. The quantity of water consumed during the process of construction was, thus, calculated based on the known discharge capacity or the yield of the pumps. This Water foot-print calculation for rest of the construction period i.e. 81 months was extrapolated from the results obtained. The running of water-pumps for approximately 12 hrs a day also validated the total number of pump operation hours obtained from this assessment.

Thus the water consumption for 68 months comes to around 521350.83 Kl for the building construction. Assuming uniform rate of pump operation throughout the construction period, the total Water Consumption (W_C) for the 81 months of construction works out to be 621020.84 Kl. The total built-up area of the building group being 310173.22 Sq m, embodied water (W_A) per unit area constructed is $W_A = W_C$ / Built-up area = 2 Kl/ Sq m. A previous study had estimated the construction water use of a three storied 1150 Sq m building of RC construction to be 1 Kl per Sq m [15] in the same region and therefore, the present finding appears reasonably acceptable.

4.4 Results and discussion

Inferring from the above calculations, the embodied water of the major building materials together worked out to be 25.604 Kl/Sq m while the water consumed during the actual construction was 2 Kl/Sq m. Hence, the total embodied water in a typical urban construction in Calcutta, India was found to be 27.604 Kl/ Sq m of built area, compared to 20 Kl/ Sq m of Gross Floor Area as found by the research team in Australia. The present finding for the Indian building is about 38% higher than buildings in Australia. In other words, water consumed for the construction of the entire building translates into 8562021.56 Kl of water, equivalent to a cube of about 205 metre side and the amount of water needed by nearly 34 families with five members each for one whole year at 138 litres per capita per day (lpcd). In terms of share, the contribution of steel is the highest at about 97.6% and surpasses all other materials like cement and brick. The water consumption during the actual construction process was found to be only 8% of total embodied water of the materials together. This shows that water consumption at material production stage i.e. the industries have to be more conscious about usage of fresh water and look for ways to bring this down through reuse and recycling. The emerging field of water governance may focus on this area and initiate drives that will achieve this goal. In future, water may prove to be an even more crucial challenge than energy, having far-reaching implications in areas of climate change, human habitation, health and economics.

5 Conclusion

Since water is everybody's business and any water disaster will affect population cutting across socio-economic classes, water mapping in building industry is not only significant to understand the dynamics of water resource consumption in one of the key sectors of urbanization, but also to select and decide on the most appropriate strategy to monitor and optimize this quantity for global sustainability and water management issues. This paper presented a study that intended to assess the possible range of water amount locked in the building fabric as its embodied water. It considers a live case study in the fast growing city of Calcutta and adopts a simple methodology to conduct this assessment. The result indicates the embodied water to be about 27.6 Kl per Sq m of built up area of a multi-storied apartment building of steel and RC construction. It has also assessed the percentage contribution of steel- the single-most water intensive material as well as that of the actual construction process in the total embodied water foot-print of the case-study. However, it is understood that one case-study with limited information on embodied water of building materials is not enough. Further research explorations into water foot-print estimation of individual materials and multiple case studies for different climatic zones of India will be required to reach substantial accuracy. In practice, thoughts and action towards water metering in industries as well as at construction sites may be imperative in immediate future to map and monitor this essential resource for over-all sustainable development.

Acknowledgements

The author is grateful to All India Council for Technical Education (AICTE) and Department of Science & Technology, Govt. of India for their financial support in carrying out this study.

References

[1] Vision 2050: Key messages; World Business Council for Sustainable Development, www.wbcsd.org/
[2] National Action Plan on Climate Change, Govt. of India, Prime Minister's Council on Climate Change. Online. http://pmindia.nic.in/Pg01-52.pdf
[3] National Water Policy-India, Govt. of India, Ministry of Water Resources. http://mowr.gov.in/writereaddata/linkimages/nwp20025617515534.pdf

[4] United Nations Environment Program (UNEP) and Division of Technology Industry and Economics (DTIE), Eco-house Guidelines, pp. 5-6, 2006.
[5] Manual on Water Supply and Treatment, Central Public health and Environmental Engineering Organization, Government of India, pp. 13, 1999.
[6] Crawford, R. & Treloar, G., An assessment of the energy and water embodied in commercial building construction, *Australian Life Cycle Assessment Conference (4th : 2005 : Novotel, Sydney, N.S.W.)*, pp. 1-10, 2005.
[7] McCormack, M., Treloar, G., Palmowski L. & Crawford R., Modelling direct and indirect water requirements of construction. Online. www.informaworld.com/smpp/content~db=all?content=10.1080/09613210601125383
[8] Water, Sustainable Concrete, sustainableconcrete.org.uk/concrete/constituents_of_concrete/water.aspx
[9] Rating System for Water Efficient Fixtures: A Way to Sustainable Water Management in India, Centre for Science and Environment, New Delhi, 2010.
[10] Green buildings and climate change, www.siliconrepublic.com/green-tech/item/17858-green-buildings-and-climate
[11] Construction Industry Development Council, India: India Country Report, 2005-06.
[12] Kishore, K., Save environment with green construction, Civil Engineering Portal. Online. www.engineeringcivil.com/mix-design-with-Superplasticizers.html
[13] Comprehensive Industry Document Series; Central Pollution Control Board, India. Online. http://www.ecacwb.org/node/142
[14] Technical Notes on Brick Construction; The Brick Industry Association, Online.www.gobrick.com/BIA/technotes/t9.pdf?CFID=13100589&CFTOKEN=21091890
[15] Bardhan, S., Carbon Foot-Print Studies and Sustainable Architectural Concepts for Coastal Eco-Tourism, unpublished Doctoral thesis, Jadavpur University, 2008.

The deterioration of basin and sediment overlapping in the infrastructure of the irrigation districts in Mexico

R. Lomeli[1] & N. Alvarez[2]
[1]*Instituto Mexicano de Tecnología del Agua, México*
[2]*Comisión Nacional del Agua, México*

Abstract

The surface dedicated to agriculture in Mexico is of approximately 21 million hectares, of which 6.4 million count as irrigation infrastructure, and 3.5 million (54%) correspond to 86 great areas called Irrigation Districts. The irrigated farming generates more than half of the national agricultural production. One of the main problems of the infrastructure maintenance is the accumulation of sediments in the distribution and drainage networks, caused fundamentally by insufficient actions of soils and water conservation in the basins of the irrigation dams. The present work analyzes the impact of the catchment basins degradation in the accumulation of sediments in the Irrigation Districts.
Keywords: sediment removal, irrigation districts maintenance, catchment basins, soil and water conservation works.

1 Introduction

In Mexico, the area dedicated to agricultural activity varies between 21 and 24 million hectares annually. The surface with infrastructure for irrigation is 6.4 million hectares, México ranks sixth place in the world, in terms of irrigation areas made of 86 Irrigation Districts with 3.5 million hectares and 39482 irrigation units of small irrigation with 2.9 million hectares [1–3].

In addition there are called Temporal Tecnificado districts that directly depend on the rainfall, with no irrigation infrastructure, but using different techniques and works to reduce the damage caused by the occurrence of heavy and prolonged rains or drought-, to improve the rain and moisture gathering in farming lands.

WIT Transactions on Ecology and the Environment, Vol 144, © 2011 WIT Press
www.witpress.com, ISSN 1743-3541 (on-line)
doi:10.2495/ECO110091

The agricultural production of the Irrigation Districts contributes directly in the development of the country because it represents one third of the national food production, and involves over half a million of irrigation users. These products represent about 60% of the value of the domestic production.

The infrastructure of Irrigation Districts, are made up as follows [4]:

- 174 storage dams
- 394 diversion dams
- 14490 wells
- 266 pumping plants
- 8000 miles of main canals and 23000 miles of secondary canals. 47.3 % of the canals are lined. 6000 miles are collectors drains and 13200 miles are secondary drains.
- 43100 miles of roads, of which 51% are earth road, 41% with revetment and the rest are paved.
- 3031 units of machinery and equipment for infrastructure maintenance

The following is a summary table of the classification of canals and drains, according to their main characteristics which are the bottom and the depth of water, as shown in figure 1.

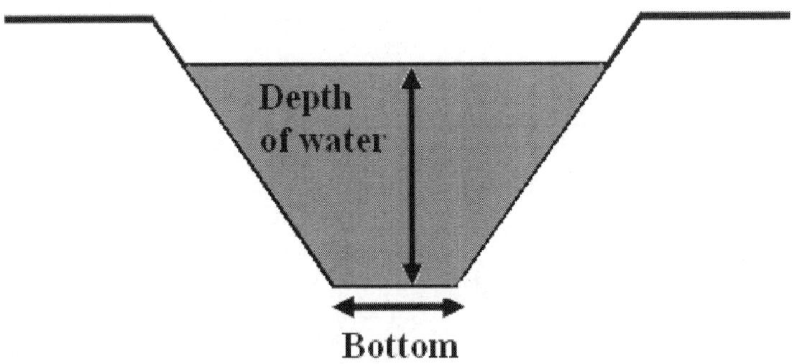

Figure 1: Cross section of canals and drains.

The infrastructure maintenance (conservation in Mexico) is an action to be performed periodically and permanently in order to preserve its state, under conditions closer to the original project.

To maintain in optimal conditions the infrastructure is essential to provide a good irrigation service, that is, to deliver water to crops timely and sufficient, which will get an optimum crop production.

The development of surface irrigation in Mexico is limited, and currently, there are few sources of water economically available, therefore, to maintain or improve the productivity of the Irrigation District is priority for optimal maintenance of infrastructure

The maintenance of the infrastructure is basic to improve the operation of the irrigation districts and to fulfill with the irrigation schedule. In Mexico to make a quick and economical field works, we use manpower, heavy weight machinery as long reach hydraulic excavator (Figure 2) and equipment as backhoe loaders.

Figure 2: Long reach excavator removing sediments in an irrigation canal.

2 Problematic

The investment in infrastructure for irrigation is high, water to one hectare of land (dam construction, infrastructure, etc.) has a cost of over 5000 €, the infrastructure rehabilitation requires approximately 1000 €, while the maintenance, has a cost of about 18 €, making it extremely profitable to perform the work of preserving the infrastructure of the irrigation districts to obtain the sustainable water use matter.

The main maintenance concepts are:

- Sediments removal in canals and drains.
- Tracking, forming and coating roads replacement.
- Repair of structures and mechanisms.
- Weed Control (aquatic and terrestrial).
- Canals slabs replacement.
- Dams, wells and pumping plant preventive maintenance

It should be noted that annually the irrigation Districts removes about 5.2 million cubic yards of sediments in distribution networks and more than 7.8 million in the drainage network that means more than 13 million cubic meters per year total.

In the following data from Organismo de Cuenca (Basins Agencies) are showed.

The main causes of sediment accumulation are:

- Uncontrolled water from natural channels into canals and drainage networks.
- Irrigation water with suspended solids.
- Inadequate protection of the canal slopes with grassland.
- Tail water.

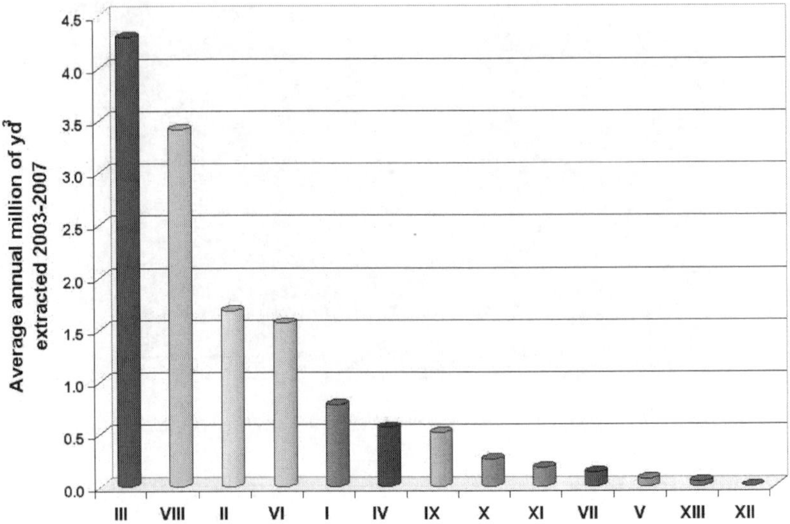

Figure 3: Average annual million of yd^3 extracted 2003-2007 distributed by Basin Agency.

Most of the Irrigation Districts source is storage dams however there is not enough soil and water conservation works in their basins, which have caused soil erosion in the highlands. Water particles of such soil move down and deposits in the storage reservoirs, reducing the dams usefulness and later producing the accumulation of sediments in the distribution network. These sediments also facilitate the development of weeds.

Figure 4: Distribution of the budget for general maintenance work.

Figure 5: Unprotected slopes cause accumulation of sediments in canals.

The 23% of the budget in the irrigation districts is used for sediments remove, the distribution of the budget of maintenance works is presented in figure 4.

To prevent sediments problems in the slopes of canals and drains of D and E types, (Table 1) in 90% of canals and 69.3% of drains, the National Water Commission in coordination with the Mexican Institute of Water Technology, developed and transferred the technology of using light weight equipments for weed control and allow the growing of grass to protect the slopes, and also facilitate the removal of aquatic weeds in canals and drains. It should be noted that currently there are over 250 light weight equipment of Irrigation Districts.

Table 1: Characteristics and percentages within the national territory of canals and drains.

Type	Bottom (m)	Depth of water (m)	Canals (%)	Drains (%)
E	Less than 2	Less than 1.2	76.2	40.4
D	Between 2 and 4	Between 1.3 and 1.7	13.8	28.9
C	Between 4 and 6	Between 1.8 and 2.4	5.4	16.9
B	Between 8 and 10	Between 2.5 and 3.0	1.4	8.0
A	Between 10 and 20	More than 3.0	3.2	5.8

3 Investment in maintenance

Under normal conditions, about the 50% of the Irrigation Districts budget should be invested in maintenance programs, in order to ensure the optimal state of the irrigation systems.

Figure 6: Light weight equipment in an irrigation canal.

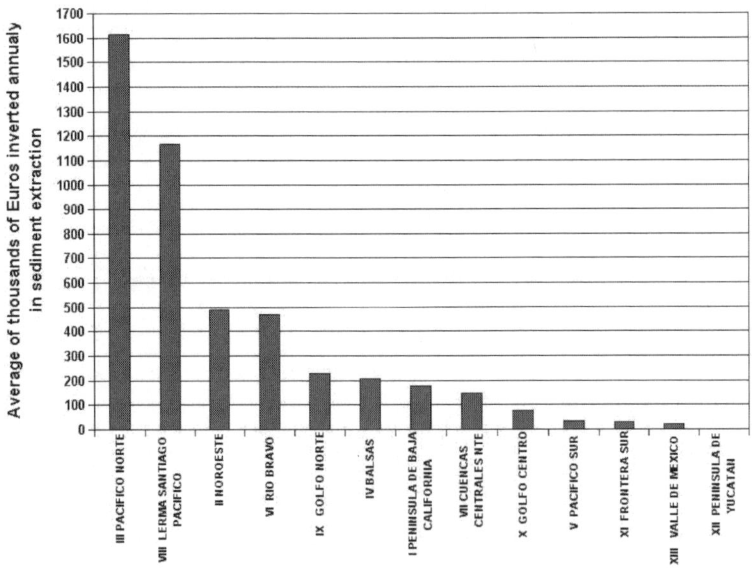

Figure 7: Average budget used in sediments extraction by Agency basin.

The Pacifico Norte Agency Basin invests more than 1´600,000 € in sediments extraction because of the storage dams basins are among the largest in the country and are deteriorated. Below is the Agency Basin Lerma Santiago Pacific Basin, which is supplied with water from the Lerma River, one of the largest in the central basin of the country which is heavily damaged.

4 Conclusions

The lifetime of most storage dams has decreased significantly due to the sediments accumulation caused by inadequate soil and water conservation works in the catchment basins.

Irrigation Districts in Mexico have upgraded their equipment and machinery to remove sediments and to protect the slopes of canals and drains at lower costs. One fifth of the Irrigation District budget is used to extract sediments. The average maintenance cost is 110 €/mile, in the distribution network and 157 €/mile in the drainage network.

5 Recommendations

Enhance the works of soil and water conservation in the impounding basins of the dams of the irrigation districts.

Promote the modernization and technical activities in the irrigation districts, to increase conduction water efficiency and avoid the risk of causing damage to the soil level plots.

References

[1] Water National Commission, Efficient use of water and electricity Coordination, *Small irrigation units organized and unorganized.* 1999

[2] Water National Commission. *Irrigated areas and volumes of water distributed in the irrigation districts*. 2009

[3] Water National Commission. *Statistics of irrigation districts*. 2009

[4] Water National Commission. *Inventories of infrastructure, machinery and equipment of the irrigation districts*. 2010

Top-bioclimate conditions associated with the natural occurrence of two Amazonian tree species for sustainable reforestation in the State of Para, Brazil

L. G. Martorano[1], D. C. A. Monteiro[2], S. Brienza Jr.[1], L. S. Lisboa[3], J. M. do Espírito Santo[4] & R. F. Almeida[5]
[1]*Embrapa Eastern Amazon, Brazil*
[2]*"Luis de Queiroz College of Agriculture"/USP, Brazil*
[3]*ITTO/Embrapa Project, Brazil*
[4]*Forest Development Institute of the State of Para, Brazil*
[5]*UFRA/EMBRAPA of Para, Brazil*

Abstract

The plants' answer to the top-bioclimate conditions is important for mapping occurrences and indications of preferential areas for the development of the forest species. This study aimed to georeference the natural occurrence of Taxi-branco (*Sclerolobium paniculatum* Vogel) and Parica (*Schizolobium parahyba* var. *amazonicum* (Huber ex Ducke) Barneby) integrated with top-bioclimate information to indicate plantations for rehabilitation of degraded landscape. The data regarding natural occurrence of species was investigated in the literature, the herbaria of Embrapa Eastern Amazon and the Paraense Emilio Goeldi Museum and demonstrative units in the Eastern Amazon. Normal climates from thirty year series with spatial resolution 3 km x 3 km were specialized in Arcgis 9.3 and exported to TerraView 3.2, to create the cellular space for integration of variables in TerraME and to generate top-climate maps and species' natural occurrence. As the Taxi-branco occurs in upland and lowland it was observed that the species prefers areas of predominantly top bioclimate with water deficit between 150 to 250 mm, and elevations below 200 m altitude, next to the river, and the Parica in about 300 m altitude, and water deficit below 180 mm. The knowledge of these zones should enhance the degree of confidence on the areas

WIT Transactions on Ecology and the Environment, Vol 144, © 2011 WIT Press
www.witpress.com, ISSN 1743-3541 (on-line)
doi:10.2495/ECO110101

indicated to the development of strategies for the genetic resources conservation and sustainable management of threatened tree species.

Keywords: preferential areas, top-bioclimate zoning, degraded landscape.

1 Introduction

In the Amazon, slash and burn practices, intense logging, expansion of agriculture and public policies of colonization and occupation can be identified as deforestation promoters in the region. The advance of deforestation may be linked to factors such as land tenure, extensive farming, illegal logging, weakness in law enforcement, lack of awareness of the population to report any irregularity, lack of commitment of the organizations that benefit from illegal products from native forest, illegal maneuvers that prevent agents that cause direct and indirect predatory deforestation from being held legally responsible for their acts and others. Depending on the magnitude of these the impacts can be intensified: loss of biodiversity, increased soil degradation, reduced carbon stocks in forest biomass, increasing emissions of greenhouse gases into the atmosphere and growth of the impoverishment of populations living in the Amazon.

The Brazilian Amazon has more than 5 million km^2 with different climate conditions, and tends greatly towards agricultural expansion. Evaluating the results released by PRODES (Program for Monitoring the Brazilian Amazon Forest by Satellite), provided by the National Institute for Space Research INPE [1], it is observed that the eastern part of Para is the most deforested, indicating that there is strong human pressure with loss of forest cover. The agricultural expansion and illegal loggers have led to the extermination of populations of several species throughout the country (Carvalho [2]).

In areas already deforested in the Amazon, the agricultural research institutions have been studying and encouraging the adoption of production systems that adopt conservation practices, and bringing social, environmental and economic benefits to the region. Forest plantations of species with potential for energy, timber, veneer and plywood are presented as alternatives to silvicultural landscape restoration and maintenance of ecosystem services by the forest canopy. In this sense, it is recommended the inclusion of native species with fast growth, high biomass production, high economic value, supply of certified wood in areas of preferred top-climate [3–5], recommended to make arrangements in: agroforestry systems (Kato and Kato [6]); livestock-crop-forest; bragantino system, forest enrichment in areas protected by law, vegetation corridors, and other actions to redesign landscapes.

According to Filho *et al.* [7], forest management is the part of forestry science that deals with the set of principles, techniques and standards, which aim at organising, sorting and controlling the productivity and efficiency of the forest, keeping the structure of the forest and its ecosystem services, the maintenance of biological diversity and socio-economic development of a region. It is noteworthy that it is important to identify the type of intended use of the species, and market conditions favourable to their ecophysiological growth stages.

Species can meet energy demands in isolated communities, both in raw material supply, the biomass as a source of renewable energy and mitigating emissions of greenhouse gases (Souza and Santos [8]).

Alvino *et al.* [9] found that of the 103 species native raised in a secondary forest with 30 years of age, in the Bragantina area in Para, 98% of species have commercial value, while 33% fall into the categories of wood for rural constructions, 30% with high commercial value, 9% for wood production, 9% with low commercial value, 7% in the fruit category, 5% as medicinal species, 4% as timber for craft and 1% with potential resiniferous. Only 2% of the species studied did not fit into these categories. The potential of native timber in the region should be further investigated, since, according to Bentes-Gama *et al.* [10] one of the greatest obstacles to their use and effectiveness of market acceptance is the lack of more information about the ecophysiological species, technological characteristics and amount in forests subject to management (Scolforo *et al.* [11]).

In this work, we tried to focus our studies on two species native to the Amazon, the Taxi-branco (*Sclerolobium paniculatum* Vogel) and Parica (*Schizolobium parahyba* var. *amazonicum* (Huber ex Ducke) Barneby), which have the potential to produce energy and timber, respectively. The Taxi-branco (Caesalpinaceae) has potential energy, rapid growth, and fixed nitrogen in the soil and has high biomass production with financial returns in the short and medium term by providing high-quality coal products. According to Lorenzi [12] it is a semi deciduous, heliophytic, selective xerophytes plant and pioneer with an average density of 0.70 g cm^{-3} and has also shown potential for dye and honey varieties in dry regions (Silva Jr [13] and Thompson [14]). The Parica has softwood and lightweight, density of 0.40 g cm^{-3} with high market competitiveness when producing laminated wood and sheets of plywood, pressed and sintered, and its pulp is of good quality and suitable for the manufacture of paper and also has excellent resistance (Pereira *et al.* [15]).

In the city of Dom Eliseu, Para and its surroundings, the average productivity of Parica ranges from 25 to 30 m^3 ha^{-1} yr^{-1} and the development of regional technologies provided greater efficiency, quality and standardization of rolled Parica regularity in the flow sale and consolidation of its supply chain in the state. Therefore, the objective was to evaluate the natural occurrence of these species associated with top-climate conditions, focusing on the productive context of sustainable management in the state of Para.

2 Material and methods

The data on the natural occurrence of the Taxi-branco and Parica was obtained from citations in the herbarium records of Embrapa Eastern Amazon, the Goeldi Museum and international databases [16], in addition to landscape data. Technical visits were made to areas of natural occurrence of these species and to plantations. Visits were made to the field, both in the city of Bragança and Dom Eliseu to evaluate natural occurrences and forest plantations.

On the field trip to the city of Bragança the technical team was composed of parabotany, foresters, undergraduate students and agrometeorologiests. Botanical material was collected to make up the collection in the herbarium of EMBRAPA, recording the geographic position with the aid of GPS (Global Position System). It also, visited Taxi-branco crops farmers in the area of demonstration units in the project "Conservation and restoration of degraded landscape in units of family farming"-INOVAGRI / Embrapa Eastern Amazon/ ITTO/ABC. This project (PD. 346/05) assesses the progress of crops in mixed systems with forest species in a farm of 13 farming families. We analyzed the production of wood in different periods of harvesting areas averaged 0.36 ha and arrangements with 150 to 200 trees to establish indicators of reclamation or degradation pathway.

To record the natural occurrence of Parica the technical trip was made to Dom Eliseu and its surroundings, where arrays were mapped to identify areas of maintenance of genetic material capable of promoting expansion programs of forest plantations in the state of Para were also used data Census of Agriculture was also used available at the IBGE [17] website in the cities, regions to identify the production of wood and its derivatives.

In generating the top-climate maps we used data relief (m), air temperature (°C), rainfall (mm) and soil water deficit (mm). Water balances were estimated, considering the soil water capacity exceeding 300 mm (Thornthwaite and Mather [18]). The top-climate letters were prepared with spatial resolution of 3 km x 3 km (3,840 points), specializing in ArcGIS 9.3 and exported to TerraView 3.2, creating the cell space for integration of variable and the model in the TerraME, fig.1.

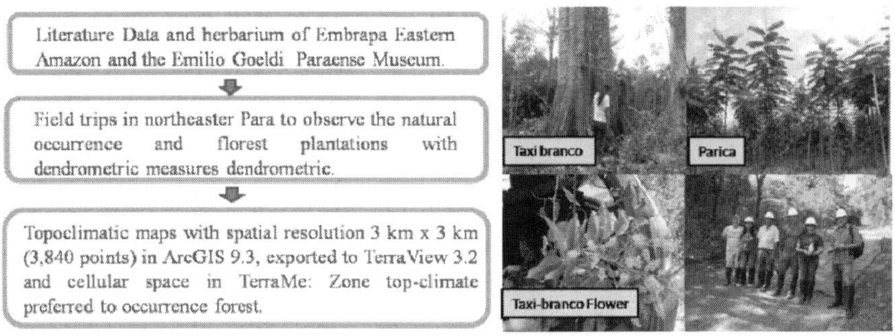

Figure 1: Methodology diagram.

3 Result and discussion

Observing the average annual data it appears that in the state of Para rainfall varies from 1,658 to 3,105, fig 2, and Bragança varies from 2,925 to 2,429 mm. During the month of March the wettest (470 mm) and October, the dry season, with monthly averages less than below 6 mm, indicating a reduction of supply of rainwater to plants. The wettest quarter runs from February to April and the dry season from September to November, and the highest water deficits in soil in

November (110 mm). The relative humidity has an annual average of around 85% to 90% in March and 76% in October.

The top-thermal annual conditions in the state of Para averages ranging from 23.3 to 27.5°C and Bragança ranging from 26.2 to 26.4°C with a maximum altitude of 100 m, fig 3. From August to December, temperatures are higher (26.7 to 27.5°C). Joining the field information available in literature and found that the Taxi-branco flowering in northeastern Para occurs in December, the rainy season in the region, influenced by the Intertropical Convergence Zone effects.

Field observations confirm that the preferred ranges indicated by Brienza Jr. *et al.* [3], reinforcing the responses of species to top-bioclimate conditions. In the field dendrometric measurements in individual Taxi-branco, found near the town of Benevides, adults with DBH of 50 cm and height of 12 m were found. When considering the data obtained by Kanashiro and Yared [19] that there are increments in *S. paniculatum* 2.5 cm DBH in forest plantations in the Amazon Basin, we infer an age of 20 years to plan mapped by the team. The trees were all found in the range indicated by preferential Brienza Jr. *et al.* [3], reinforcing the responses of species to weather conditions and climate.

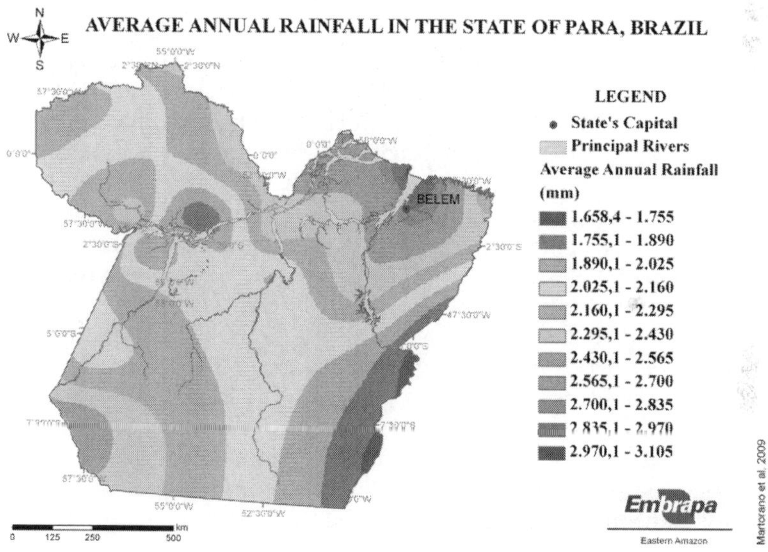

Figure 2: Annual rainfall in the state of Para, Brazil.

In Para, the areas where Taxi-branco occurred naturally were mapped, in the range of annual rainfall from 2160.0 to 2700.0 mm and Parica in areas between 2295.0 to 2430.0 mm. In Dom Eliseu city the least rainy quarter is from June to August, and the wettest runs from February to April. In terms of thermal top observed values between 25.0 to 25.8°C and altitudes in the city can reach 400 m. The annual relative humidity varies from 77 to 83% in the city as a

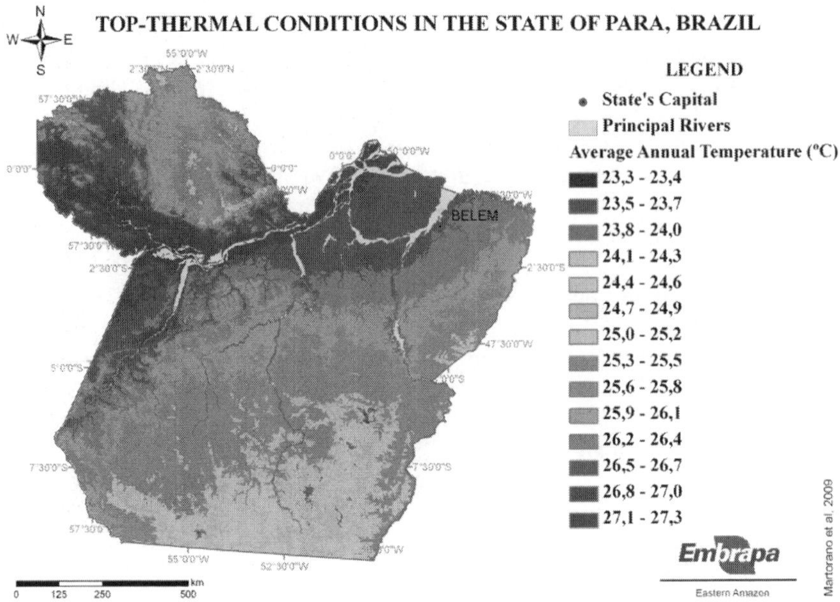

Figure 3: Top-thermal annual conditions in the state of Para, Brazil.

whole. It is observed that the lower values of relative humidity occur in October and range from 74 to78%. The moisture in the wettest month is from 88 and 90%.

As the Taxi-branco occurs in upland and low land it was observed that the species prefers areas of predominantly top bioclimate with water deficit between 150 to 250 mm and elevations below 200 m altitude next to the river and the Parica in about 300 m altitude and water deficit below of the 180 mm Table 1. Despite the low number of individuals from Taxi-branco in the region investigated, the data shows that the top bio-climatic information can help indicate the preferred areas for planting this species in the northeastern region.

The knowledge of these zones should enhance the confidence degree in the areas appointed for the development of strategies for the genetic resources conservation and sustainable management of threatened tree species, fig. 4. This evidence corroborates the studies by Lima *et al.* [20] that the species occurs in a wide range of climate conditions.

Table 1: Forest species and top-climate conditions in the State of Para.

Forest Species	Altitude (m)	Water Deficit (mm)
Sclerolobium paniculatum (Taxi-branco)	0 a 200	150 a 250
Schizolobium parahyba var. *amazonicum* (Parica)	0 a 300	<180

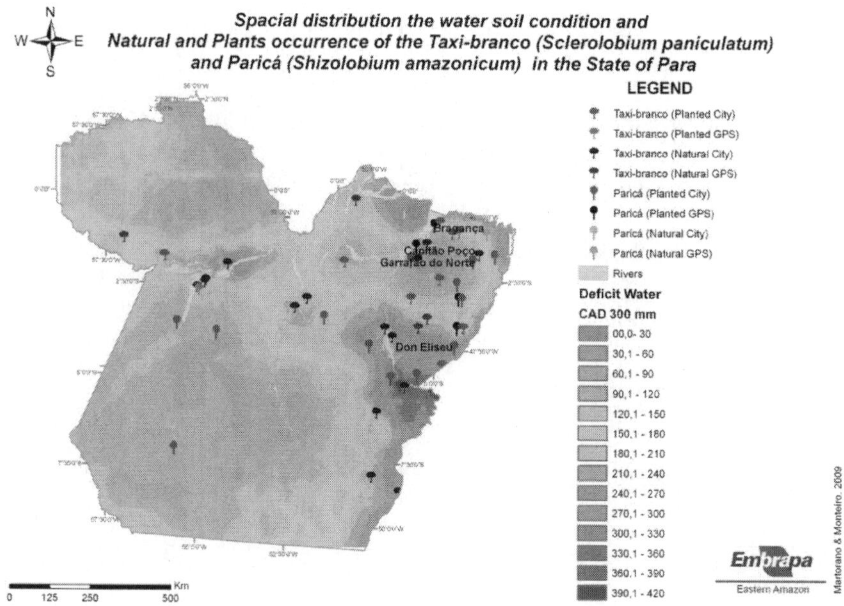

Figure 4: Tracks the soil water deficit and occurrence (natural and plantations) and Taxi-branco and Parica in Para.

Besides their wood producing and energetic potentials, these plants – which are native and largely found in the Amazon region – also produce pollen and nectar, and they are excellent to join the honey producing region in Para. In this way, the climatic change indication, such as years when the El Niño and the La Niña are in effect, will subside the honey producing strategies, because they directly associate the flowering season and the rain availability in the region.

Studies carried out by Embrapa Amazônia Oriental show that the Parica flowers are visited by numerous insects, and the more important for their polinization are medium to large bees (15 to 25mm). Among the Meliponíneos, the most common were the Trigona, the Scaptotrigona and the Melipona bees. Regarding the Taxi-branco, honey is found in small amounts, but very concentrated (containing medium amount of sugar: 82%). Among the Meliponíneos, the most frequent were the Scaptotrigona and the Melípona (Embrapa [21]).

The preferred top-bioclimate zones of these species enhance the insertion strategies in altered areas which have lost much of their forest covering to the productive area, enabling the rural producers add other income sources to theirs, reducing the pressure in the areas.

Information made available by IBGE [17] found that in the northern region, only the states of Para and Amapa appear as suppliers of firewood products and roundwood products, producing in that year in Para 80 000 m^3 of fuelwood in

Para, 1.5 million m^3 of roundwood for pulp and 716,000 m^3 for other purposes, financial support totaling around R$ 143 million, from forestry products in the region.

From 2006 to early December 2010, Sisflora [22] Para was extracted and sold about 305,000 m^3 of Taxi-branco (*Sclerolobium paniculatum* Vogel) and Parica (*Schizolobium parahyba* var. *amazonicum* (Huber ex Ducke) Barneby), accounting for around 15 million, the contribution of these two tree species was about 10 million dollars, adding jobs and income to the productive sector of the wood in the state, fig. 5.

With regard to economic aspects, the information presented in, fig. 4 emphasises the importance of identifying top-climate areas preferred as indicators of potential areas for placement within these arrangements in forest in deforested areas.

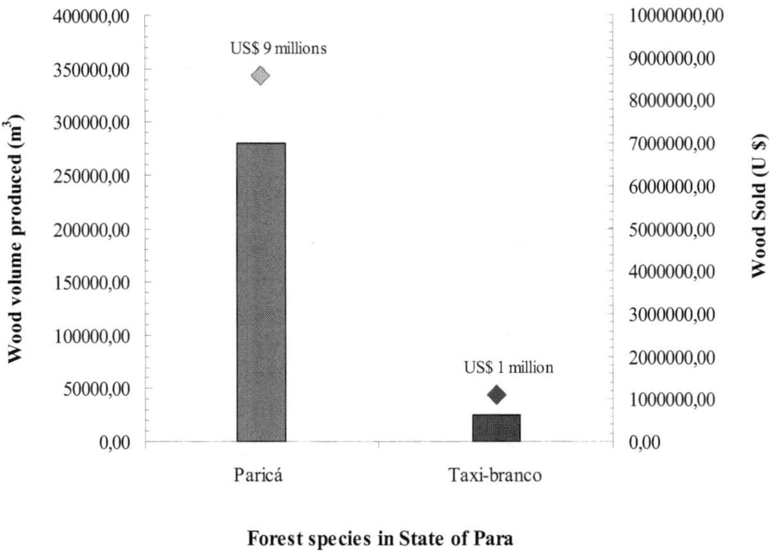

Figure 5: Average volume of production of Taxi-branco and Parica from 2006 to early December 2010 in the state of Para.

The data provided by INPE [1] to indicate that the state of Para lost nearly 20% of its forest cover. However, analyzing the percentage of deforestation in the state, it is noted that about 8% of that total is due to deforestation in northeaster Para, fig. 6. The transformation of forested lands by human actions represents one of the great forces in global environmental change and one of the great drivers of biodiversity loss with implications to life on this planet.

Fig. 7 shows that between 2005 and 2009 the deforestation in Bragança, Garrafão do Norte and Capitão Poço was around the 80%. The use of these areas with forest plantations is a reduction in pressure on the forest, helps mitigating greenhouse gases, in addition to promoting supply of raw material to the

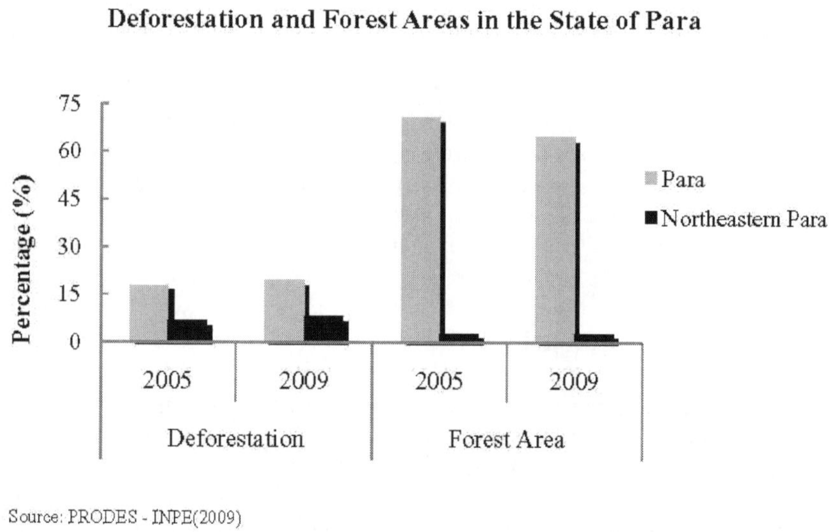

Figure 6: Percentage (%) of deforestation and Forest Areas in the State of Para (2005 and 2009 years).

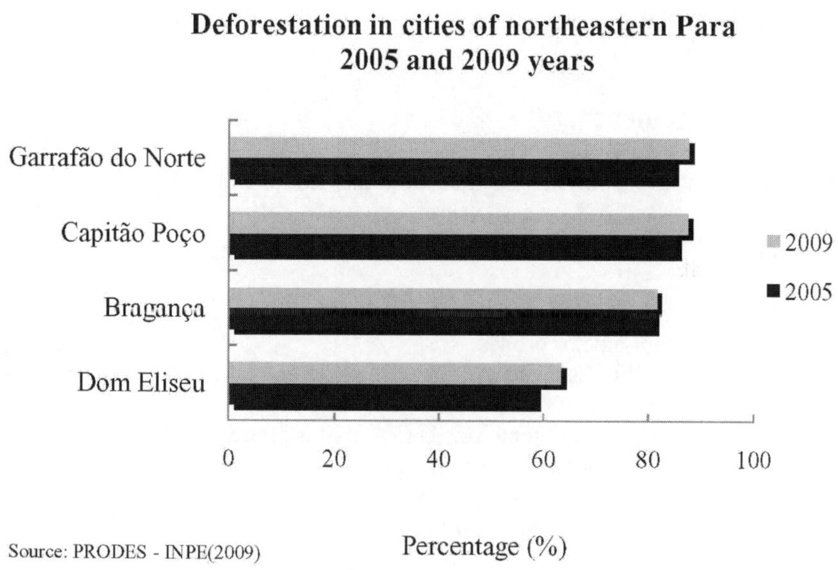

Figure 7: Deforestation percentage (%) in the northeast of Para (2005–2009).

productive sector of wood, veneer, plywood, charcoal and other forest products. Also in fig.7, it may be inferred from the apparent deforestation that it is due to the silvicultural system used in commercial forestry plantations in Dom Eliseu.

In this context, in these cities, since 2007, the potential energy of the Taxi-branco in the recovery of degraded landscape, systems of mixed forest species, with different periods of harvest, were planted by 13 families of farmers. The average size areas is 0.36 ha and the used silvicultural model, where there is the mixing 150 to 200 trees of *Sclerolobium paniculatum* Vogel (Taxi-branco); *Schizolobium parahyba* var. *amazonicum* (Huber ex Ducke) Barneby (Parica) and other species.

4 Conclusion and outlook

The top-climate conditions show that there are areas with potential for expansion of forest plantations Taxi-branco and Parica, especially in areas with severe loss of vegetation cover as the northeast Para. Based on the data obtained it is recommended the planting of Taxi-branco and Parica in order to recover the degraded landscape with characteristics similar to top-climate areas presented in this paper.

Techniques of dynamic modelling and geomatic assist in identifying areas top-bioclimate are indicated to the development of strategies for the genetic resources conservation and sustainable management of threatened tree species.

The knowledge of these zones should enhance the confidence degree for establishment of forest plantations in the restoration of sustainable landscapes and the Amazon.

Acknowledgement

S. Brienza Jr. was supported by a grant from the Emílio Goeldi Museum, Brazil

References

[1] INPE. Instituto Nacional de Pesquisas Espaciais, Prodes, 2005-2009. http//www.obt.inpe.br/prodes/

[2] Carvalho, P.E.R. Espécies Arbóreas Brasileiras. 1a edição. Colombo, PR Embrapa Florestas, 2003. Vol.1.

[3] Brienza Junior, S.; Monteiro, D.C.A.; Martorano, L.G.; Lisboa, L.S.; Maestri, M.P.; Schuler, A. Energetic potential of the *Sclerolobium paniculatum* Vogel (Taxi-branco) and its preferential top climatic condition in Eastern Amazon, Brazil. In: Conference on International Research on food security, Natural, Resource Management and Rural Development. HAMBURG: Alemanha, 2009.

[4] Martorano, L. G.; Brienza Junior, S.; Monteiro, D. C. A.; Lisboa, L. S.; Câncio, O.N.; Martorano, P.G.; Espirito Santo, J.M.do. Condições topobioclimáticas associadas à ocorrência de Taxi-branco (*Sclerolobium*

panuculatum Vogel) e Paricá (*Schizolobium parahyba* var. *amazonicum* (Huber ex Ducke) Barneby) preferenciais para implantação de plantios florestais no Estado do Pará. In: XVIII Reunião Brasileira de manejo e conservação do solo e da água, 2010, Terezina, PI. Novos Caminhos para Agricultura Conservacionista no Brasil. Terezina : Piauí, 2010.

[5] Monteiro, D. C. A.; Martorano, L. G.; Brienza Junior, S.; Lisboa, L. S.; Ferreira, M.do.S.G.; Martorano, P.G. Condições Topobioclimáticas preferenciais para plantios energéticos de Taxi-Branco (*Sclerolobium Paniculatum* Vogel) na Amazônia Oriental Brasileira. In: 62° Reunião Anual da Sociedade Brasileira para o Progresso da Ciência, 2010, Natal-RN. Ciências do Mar: Heranças para o Futuro. Natal : Rio Grande do Norte, 2010.

[6] Kato, M.S.A; Kato, O.R. Preparo de área sem queima, uma alternativa para a agricultura de derruba e queima na Amazônia Oriental: Aspectos Agroecológicos. In: Seminário sobre manejo de vegetação secundária para a sustentabilidade da agricultura familiar na Amazônia Oriental. 2000, Belém. Anais...Belém, 2000, p.35-37.

[7] Filho, A.C.; Manzi, A.; Santos, J. dos; Rocha, R.M.; Higuchi, N. 2004. A Floresta e o Clima. In: A Floresta Amazônica e suas Múltiplas Dimensões: uma proposta de educação ambiental. P. 81. INPA/CNPq. 2004. 146 p.

[8] Souza, R.C. R.; Santos, E.C.S. Incentivo ao uso de Biomassa para a geração de eletricidade na Amazônia. III Congresso Brasileiro de Regulação dos serviços públicos concedidos. ABAR-Associação Brasileira de Agências de Regulação. Gramados, RS, 2003.

[9] Alvino, F. de O.; Silva, M. F. da; Rayol, B. P. Potencial de uso das espécies arbóreas de uma floresta secundária, na Zona Bragantina, Pará, Brasil. Acta Amazônica. v. 35, n. 4, p. 413-420, 2005.

[10] Bentes-Gama, M. de M.; Scolforo, J.R.S.; Gama, J.R.V.; Oliveira, A.D. Estrutura e valorização de uma floresta de várzea alta na Amazônia. CERNE, 8 (1): 88-102.2002.

[11] Scolforo, J. R. S. ; Peres, J. F. M. ; Melo, J. M. ; Oliveira, A. D. ; Camolesi, J. F. ; Borges, L. F. R. ; Acerbi Junior, F. W. . Estimativas de volume, peso seco, peso de óleo e quantidade de moirões para a candeia (*Eremanthus erythropappus* (DC.) MacLeish).. CERNE (UFLA), v. 10, p. 87-102, 2004.

[12] Lorenzi, H. Árvores Brasileiras: Manual de identificação e cultivo de plantas arbóreas nativas do Brasil. v.2. São Paulo: Editora Plantarum. 1998.

[13] Silva Júnior, M. C. da. 100 árvores do Cerrado: guia de campo. Editora Rede de Sementes do Cerrado. Brasília, 2005. 278 p.

[14] Thompson, R. M. Estimativa de Volume, biomassa e carbono para *Esclerolobium paniculatum* Vogel. 64p.(Dissertação de Mestrado) UnB-Brasilia, 2009.

[15] Pereira, A. P., Melo, C. F. M. de, Alves, S. M. O Paricá (*Schizolobium amazonicum*) características gerais da espécie e suas possibilidades de aproveitamento na indústria de celulose e papel. Revista do Instituto Florestal, São Paulo, v. 16 A, n 2. p. 1340-1344, 1982.

[16] GBIF – Global Biodiversity Information Facility. http://data.gbif.org /species

[17] IBGE. Instituto Brasileiro de Geografia e Estatística. Senso Agropecuário 2006. http//www.ibge.gov.br

[18] Thornthwaite, C.W.; Mather, J.R. The water balance. Publications in Climatology. New Jersey: Drexel Institute of Technology, 104p. 1955.

[19] Kanashiro, M.; Yared, J. A. G. Experiências com plantios florestais na Bacia Amazônica. In: Universidade Federal do Paraná. O desafio das florestas neotropicais. Curitiba: UFPR, 1991. p. 117-137.

[20] Lima, R. M B. de; Azevedo, C. P de; Rossi, L. M. B; Garcia, L. C. Crescimento de *Sclerolobium paniculatum* Vogel na Amazônia, em função de fatores de clima e solo. Tese...UFPR, Curitiba, 2004.

[21] EMBRAPA - http://mel.cpatu.embrapa.br/plantas.

[22] Sisflora/SEMA. http//monitoramento.sema.pa.gov.br/sisflora/

Section 3
Environmental indicators

Biomes, ecosystem services and emergy: is there a relationship?

L. Coscieme, N. Marchettini, S. Bastianoni & F. M. Pulselli
Dept. of Chemistry, University of Siena, Italy

Abstract

Ecosystems and human systems are interconnected through a complex series of dynamic relations. In particular, natural functions of ecosystems generate flows of services continuously. To represent adequately the importance of ecosystems for human well-being, the concept of Ecosystem Services (ES) was defined. A complete review of methods and values for the assessment of ES from an economic viewpoint was provided by Costanza et al. in 1997. However, in ES valuation (ESV) several values are neglected or not calculated and estimation methods rely on strictly economic and subjective practices.

In order to have a physical reference upon which an objective (and complementary) evaluation can be based, we used the concept of emergy. Emergy is a thermodynamic entity based on energy quality and hierarchy.

Life on Earth is organized in an ordered series of energy transformations from one type to the next; these types do not have the same ability to do work. Emergy can be defined as the quantity of (solar) energy that is used up directly or indirectly to make a product or a service. An accounting system based on emergy can be implemented to evaluate production processes, ecosystems and nations.

In this paper the emergy of different biomes was calculated and translated into money by means of a specific "environmental" emergy-to-money conversion factor given by the ratio of the renewable emergy supporting the entire biosphere (emergy baseline) to the value of the world ES previously proposed. The results show that the new economic value is higher than classical ESV values.

Keywords: ecosystem services, thermodynamic indicators, Emergy evaluation.

1 Introduction

The natural functioning of an ecological system generates an interconnected series of processes that directly or indirectly produce goods (such as food) and services (such as climate regulation or soil formation) for the human society. At the same time these processes sustain the equilibrium of the system itself and the equilibrium of the entire Biosphere [1]. Ecosystem functions regenerate natural capital and produce renewable flows which in part converge in human systems (often after various transformations). In this way, natural originated goods and services are not considered as an integrated part of economic systems, but they are simply utilized [2, 3]. Because ES are not fully and adequately quantified in terms comparable with economic services and manufactured capital, they are often given too little weight in political decisions [4]. By recognizing ES, and the value provided to humans, ecological economists must develop indicators of value which can be used in decision-making [3, 5]. An adequate evaluation of ES can guide management policies and sustainable behaviors, for a new harmonization of the anthropic cycles with the natural cycles.

Several works made an estimation of the economic value of ES at various scales [1, 4, 6]. The "classic" approach to ESV is principally based on indirect economic methods, used to internalize the flow of services into Market [7]. The lack of a physical common basis to the ESV techniques cause a huge variability in the results, also for the same service provided by similar ecosystems (i.e. wetlands value in Costanza et al.[4] vs. wetlands value in Schuyt and Brander [6]) [1, 14, 15]. There is a high heterogeneity in the ESV techniques applicable on a territorial system, and the results are principally influenced by the valuation method. The economic evaluation causes also a total disjunction between function and service, which causes a non-effective protection of the ecosystems [8] (for a more complete review of the critics moved to the classic ESV see also Sagoff [16] and Pearce [17] note 1).

To study and evaluate properly the importance of the environment for humans it is necessary to adopt instruments common at both the anthropic and biotic sphere. Relations between man and environment can be represented by flows of matter and energy (withdrawal of resources and discharge of wastes). Accordingly, a thermodynamic-based approach seems to be the more appropriate to understand how to drive the management of natural and anthropic systems toward sustainable dynamics. The powerful use of the economic language can be maintained to dialogue properly with decision makers. However it has to be used only after an evaluation of the ecosystems, based on thermodynamic indicators [18, 19].

2 Methods

In this paper the emergy evaluation of 16 biomes, as listed in Costanza et al [4], is presented.

Emergy was introduced by H.T. Odum during the 1980s as a tool of environmental (but not only environmental) accounting [9–12]. On the basis of a thermodynamic hierarchy of energy, starting from solar energy, Odum's research has provided a measure of the environmental work necessary to generate an item or a flow.

Emergy is defined as the quantity of solar energy directly or indirectly necessary to support a given system and its level of organization. The emergy of all inputs to a system is calculated in terms of solar emjoules (sej) by means of suitable conversion factors called transformities (expressed in sej/J), or specific emergy (expressed in sej/g or other units). Emergy represents a measure able to evaluate the convergence of matter and energy (several inputs) to a system on a common basis. In the emergy analysis of all the 16 biomes considered in this paper, only renewable inputs are included, with the exception of some semi-natural or totally anthropic systems, like croplands and urban systems, which exist and provide services to humans using also non renewable inputs like fertilizers and fossil fuels.

Emergy theory offers also important tools that enable a clear communication of the physical evidences of the interactions between environmental and human systems. In particular the Emergy/Money Ratio is able to convert emergy units into "economic terms familiar to most people" (and vice versa).

The Emergy/Money Ratio is usually calculated as the ratio of Total Emergy to the GDP of a nation in a given year and it is used for all the evaluations of territorial subsystems or productive systems within the same nation. In this paper, the concept of "environmental" Emergy/Money Ratio was applied in the evaluation of ecosystem services.

3 Results and discussion

A very comprehensive bibliographic research has been necessary to associate an average emergy value to all the16 biomes (as presented in Costanza et al. [4]). In spite of the high variability found in emergy literature, we think this exercise is important because it represents a rather complete collection of emergy values for different biomes. For some biomes there was an insufficient number of published studies: for them, namely desert and tundra, we have estimated an emergy value using the National Environmental Accounting Database developed by Sweeney et al. [13]. A summary of biomes emergy flows is shown in Table 1.

The highest emergy flows are related to "ecotonal" zones and more heterogeneous biomes. A very high emergy is related to coastal zones due to the high energy of waves and tides. In Emergy literature the role of these inputs in the development of coastal and intertidal ecosystems is currently controversial.

On these energetic inputs, ecosystems base all their internal transformation cycles. The dynamics of these cycles depends on the biota composition and status, and on abiotic configurations and relations with living organisms. On this basis, the system works and produces outputs that could be used directly or indirectly by humans.

Table 1: Emergy flows of the 16 biomes.

Emergy of different biomes	
Biome	*Emergy flow* **(sej ha^{-1} yr^{-1})**
Open Ocean	***3.88E+14***
Estuaries	8.84E+16
Seagrass_Algae beds	1.93E+19
Coral Reefs	2.94E+19
Shelf	1.93E+19
Tropical Forest	4.52E+16
Temperate_Boreal Forest	1.77E+15
Grass_Rangelands	7.45E+16
Tidal Marsh_Mangroves	1.09E+16
Swamps_Floodplains	3.21E+15
Lakes_Rivers	4.97E+16
Desert	1.55E+14
Tundra	2.01E+15
Ice_Rock	3.68E+15
Cropland	2.71E+16
Urban	1.47E+18

3.1 An environmental emergy-to-money ratio for evaluating ecosystems and services

It is possible to define an indicator of the ecosystem conversion of energy inputs into services for humans. To do that, we define a specific "environmental" emergy-to-money ratio given by the sum of independent inputs supporting the entire biosphere cycles (emergy baseline, equal to 15.83 x 10^{24} sej/yr [12]) divided by the value of the world ES previously proposed [4] (this value is between a minimum of 18 x 10^{12} €/yr and a maximum of 62 x 10^{12} €/yr). The result is a global range of environmental emergy-to-money conversion factor between 2.57 x 10^{11} sej/€ and 8.68 x 10^{11} sej/€. This ratio represents how much emergy (as a proxy of environmental work) is needed for the global ecosystem to exist and provide ecosystem services (measured in money terms). Using these ratios we can calculate brand new values for the 16 biomes (Table 2), dividing the emergy flow of each biome for the minimum and maximum values of the environmental conversion factor. These values correspond to the amount of emergy necessary to support one unit money of ES that we can expect from a given biome. A range of economic values attached to each biome can be

Table 2: Classic ESV results and emergy-based values for 16 biomes.

Values of ecosystem services for 16 biomes			
Emergy Baseline (sej yr^{-1})	**World ecosystem services value (Costanza et al [4][a]) (€ yr^{-1})**		**Environmental EMR (sej €$^{-1}$)**
1.58 x 10^{25}	1.82 x 10^{13}		8.68 x 10^{11}
	6.15 x 10^{13}		2.57 x 10^{11}
Biome	**Costanza et al [4][a] (€ yr^{-1})**	**Emergy-based value (min) (€ yr^{-1})**	**Emergy-based value (max) (€ yr^{-1})**
Open Ocean	287	4.47 x 10^{2}	1.51 x 10^{3}
Estuaries	26019	1.02 x 10^{5}	3.44 x 10^{5}
Seagrass_Algae beds	21657	2.23 x 10^{7}	7.51 x 10^{7}
Coral Reefs	6924	3.38 x 10^{7}	1.14 x 10^{8}
Shelf	1835	2.23 x 10^{7}	7.51 x 10^{7}
Tropical Forest	2288	5.21 x 10^{4}	1.76 x 10^{5}
Temperate_Boreal Forest	344	2.04 x 10^{3}	6.90 x 10^{3}
Grass_Rangelands	278	8.58 x 10^{4}	2.90 x 10^{5}
Tidal Marsh_Mangroves	11385	1.25 x 10^{4}	4.22 x 10^{4}
Swamps_Floodplains	22315	3.69 x 10^{3}	1.25 x 10^{4}
Lakes_Rivers	9684	5.72 x 10^{4}	1.93 x 10^{5}
Desert	0	1.79 x 10^{2}	6.03 x 10^{2}
Tundra	0	2.31 x 10^{3}	7.81 x 10^{3}
Ice_Rock	0	4.23 x 10^{3}	1.43 x 10^{4}
Cropland	105	3.12 x 10^{4}	1.05 x 10^{5}
Urban	0	1.69 x 10^{6}	5.71 x 10^{6}

[a] Costanza et al.'s [4] values are adapted to € 2010

determined. These values are derived from a sort of re distribution of classic ES values, on the basis of renewable environmental inputs they consume to exist.

4 Conclusion

This paper has presented an application of thermodynamic tools (basically using emergy) in ESV. In fact, emergy evaluation can support ESV with a physical basis and contributes to reduce the high heterogeneity of the classic economic evaluation method. A comprehensive bibliographic research, and some original calculations, enabled a first estimation of natural emergy flows to 16 types of ecological system.

The transformation of natural and free inputs (basically sun, rain, geothermal heat, tides and waves) into economically valuable outputs, can be measured by a new indicator of the ecosystem work done to provide services to humans. An "environmental" emergy-to-money ratio enabled to estimate a set of new ES values for the 16 biomes. These values are higher than classical ESV values and can be used as a complement of them.

As nations (countries and communities) and economic systems become more and more (apparently) dependent on industrial and technological products, the mayor role of the environment for human well-being becomes less clear to the society [20]. Indicators of the work of the environment for our well-being can help to reform collective ecological knowledge. Maintaining the use of economic language to express the value of ES will permit to define the environmental conservation as a conservation of value, and not as a limit to economic development and to the increase of well-being. Using emergy, and other very informative holistic thermodynamic indicators [19, 21], to reach that economically expressed result, will make clear the direct relationship between the value of ES and the work of the biosphere.

References

[1] de Groot R.S., Wilson M.A., Boumans R.M.J. A typology for the classification, description and valuation of ecosystem functions, goods and services. *Ecological Economics*, **41**, pp. 393-408, 2002.

[2] Daily G.C. (Ed.) Nature Services: Societal Dependence on Natural Ecosystems. Island Press, Washington, D.C., 1997.

[3] Limburg K.E., O'Neill R.V., Costanza R., Farber S. Complex systems and valuation. *Ecological Economics*, **41**, pp. 409-420, 2002.

[4] Costanza R., d'Arge R., de Groot R.S., Farber S., Grasso M., Hannon B., Limburg K.E., Naeem S., O'Neill R.V., Paruelo J., Raskin R.G., Sutton P., van den Belt M. The value of the world's ecosystem services and natural capital. *Nature*, **387**, pp. 253-260, 1997.

[5] Millennium Ecosystem Assessment. *Millennium Ecosystem assessment.* Island Press, Washington, D.C., 2005.

[6] Schuyt K., Brander L. *The economic value of the World's wetlands*. WWF Living Waters: Conserving the Source of Life. Gland, Switzerland, 2004.

[7] Spangenberg J.H., Settele J. Precisely incorrect? Monetising the value of ecosystem services. *Ecological Complexity*, **7**, pp. 327-337, 2010.

[8] Peterson M.J., Hall D.M., Feldpaush-Parker A.M., Peterson T.R. Obscuring Ecosystem Function with Application of the Ecosystem Services Concept. *Conservation Biology*, **24**, pp. 113-119, 2009.

[9] Odum, H.T. Self organisation, transformity and information. *Science*, **242**, pp. 1132-1139, 1988.

[10] Odum, H.T. Environmental Accounting. Emergy and Environmental Decision Making. John Wiley and Sons. New York, 1996.

[11] Odum H.T., Brown M.T., Brandt-Williams S. Introduction and Global Budget, Folio #1. *Handbook of Emergy Evaluation.* Center for Environmental Policy, University of Florida, Gainesville, USA, 2000.

[12] Odum H.T. Emergy of Global Processes, Folio #2. *Handbook of Emergy Evaluation.* Center for Environmental Policy, University of Florida, Gainesville, USA, 2000.
[13] Sweeney S., Cohen M.J., King D., Brown M.T. Creation of a global emergy database for standardized national emergy synthesis. In *Emergy Synthesis 4: Theory and application of emergy methodology* (Brown M. ed.), Gainesville, FL, pp. 23.1-23.15, 2007.
[14] Wilson M.A., Carpenter S.R. Economic valuation of freshwater ecosystem services in the United States 1971-1997. *Ecological Applications*, **9**, pp. 772-783, 1999.
[15] Turner M.G., Odum E.P., Costanza R., Springer T.M. Market and Nonmarket Values of the Georgia Landscape. *Environmental Management*, **12**, pp. 209-217, 1988.
[16] Sagoff M. Can We Put a Price on Nature's Services? *Report from the Institute for Philosophy and Public Policy*, Vol. 17, 1997.
[17] Pearce D. Auditing the Earth: The Value of the World's Ecosystem Services and Natural Capital. *Environment*, **40**, pp. 23-28, 1998.
[18] Odum H.T., Odum E.P. The Energetic Basis for Valuation of Ecosystem Services. *Ecosystems*, **3**, pp. 21-23, 2000.
[19] Jorgensen S.E. Ecosystem services, sustainability and thermodynamic indicators. *Ecological Complexity*, **7**, pp. 311-313, 2010.
[20] Pilgrim S.E., Cullen L.C., Smith D.J., Pretty J. Ecological Knowledge is Lost in Wealthier Communities and Countries. *Environmental Science & Technology*, Vol. 42, pp. 1004-1009, 2008.
[21] Tilley D.R. *Emergy Basis for Forest Systems*. University of Florida, 1999.

Spatial and temporal trends of heavy metal accumulation in mosses in Estonia

H. Kösta & S. Liiv
Department of Environmental Education,
Tallinn Botanic Garden, Estonia

Abstract

The main sources of air pollution with heavy metals in Estonia are the country's oil shale fired power stations, the chemical and the building materials industry and road traffic. The principal power stations and most of heavy industry are concentrated in the north-eastern part of the country. The relative deposition of Cd, Cr, Cu, Fe, Ni, Pb, V and Zn in Estonia has been monitored since 1989 by the method of bioindication, using the mosses *Pleurozium schreberi* or *Hylocomium splendens* as indicator species. The selection of sampling sites and the collection of moss samples is carried out in accordance with the requirements of the International Cooperative Programme on Effects of Air Pollution on Natural Vegetation and Crops. In 1989, heavy metals concentrations in mosses were the highest in samples from North-East Estonia. By 1995, the surroundings of Tallinn, the capital city of Estonia, had become the second most polluted area in Estonia, probably due to increased traffic density and the related increase in air pollution. In 2005–2008, the changes of the concentrations of most heavy metals in mosses remained within or close to local variability in all studied regions. During 1989–2006, the decrease of heavy metals concentrations in Estonian moss samples has been in correlation with the drop in power production as well as with the decrease of emission of pollutants due to the implementation of environmentally friendlier technologies. Due to the introduction of unleaded petrol in Estonia, the concentration of lead in moss samples has decreased by up to 60% during 1989–2008.

Keywords: atmospheric pollution, emission sources, heavy metals, mosses as biomonitors.

WIT Transactions on Ecology and the Environment, Vol 144, © 2011 WIT Press
www.witpress.com, ISSN 1743-3541 (on-line)
doi:10.2495/ECO110121

1 Introduction

The widespread carpet-forming mosses *Pleurozium schreberi* and *Hylocomium splendens* possess excellent properties for monitoring the atmospheric deposition of heavy metals. There are a variety of natural and anthropogenic sources of heavy metals in the environment.

The main natural sources of heavy metals are volcanic activity, forest fires, biogenic sources and the erosion of crusty materials (Sucharova and Suchara [1]; Callender [2]). To date the most important sectors of anthropogenic emission of various common metals are the metals industry (emissions of Al, As, Cr, Cu, Fe, Zn), other manufacturing and construction industries (emissions of As, Cd, Cr, Hg, Ni, Pb), electricity and heat production (emissions of As, Cd, Hg, Ni) and road transport (emissions of Cu, Sb, Pb, V, Zn) (Fukuzaki *et al.* [3]; Ikeda *et al.* [4]; Sternbeck *et al.* [5]; Adachi and Tainosho [6]; Harmens *et al.* [7]).

In Estonia, the main sources of heavy metal emissions are thermal power stations (Eesti Power Plant and Balti Power Plant, whose combined maximum generation capacity exceeds 3 GW), the cement production plant operated by the public limited company Kunda Nordic Tsement AS and the chemical industry (the Nitrofert fertilizer plant, the oil shale processing plant of the Viru Keemia Group) – all located in North-East Estonia (NE Estonia). In addition to these, the emissions are produced also by various local industries, heating plants and traffic around major towns and on major roads (Liiv and Kaasik [8]).

A specific feature in relation to Estonia is its possession of considerable oil shale deposits. Oil shale has been mined and used in Estonia for more than 80 years – by March 2006, the country had extracted one billion tonnes of oil shale. Since the 1960s, Estonia has been the largest oil shale producer and consumer in the world. In the 1980s about two thirds of the world's oil shale output came from Estonia (Punning and Raukas [9]).

Thermal combustion of oil shale in power plants leads to increasing amounts of heavy metal emissions (some of which are radioactive) which are partly deposited in the ash fields and partly emitted into the atmosphere (Punning and Raukas [9]). In 2002, about 97% of air pollution, 86% of total waste and 23% of water pollution in Estonia came from the power industry, which uses oil shale as the main input for production process (Raukas [10]).

Even in the absence of visible pollution, trace amounts of heavy metals may accumulate in the environment and, over long periods of time, pose a major environmental and human health hazard (Walkenhorst *et al.* [11]).

In living organisms, heavy metals mostly accumulate in bone cells, kidneys and liver, where they may cause a variety of chronic diseases, toxemia and even cancer. Food chain contamination is the most important pathway for the entry of these chemical elements into the human body (Ferner [12]; Ma *et al.* [13]).

The use of widespread native terrestrial mosses as biomonitors is a well-recognized technique in studies of atmospheric contamination (Rühling and Tyler [14]; Sucharova and Suchara [1]; Holy *et al.* [15]). Mosses are easy to handle, they lack the cuticle, vascular and root systems and are capable of absorbing and fixing metallic pollutants from the atmosphere in their tissues over long periods of time (Brown [16]; Brown and Brumelis [17]; Tyler [18]; Sawidis *et al.* [19]; Thöni *et al.* [20]; Markert *et al.* [21]; Fernandez *et al.* [22]).

Monitoring the concentrations of metals in mosses provides a surrogate, time-integrated measure of the deposition of these elements from the atmosphere to terrestrial biosystems. The fact that moss biomonitoring studies are comparatively cheap permits a higher sampling density to be achieved than would be possible with conventional deposition measurements (Holy *et al.* [15]; Harmens *et al.* [7]).

The relative deposition of Cd, Cr, Cu, Fe, Ni, Pb, V and Zn has been monitored in Estonia by recording their concentrations in the epigeic mosses *Pleurozium schreberi* and/or *Hylocomium splendens* since 1989, when Estonia joined the North European air pollution monitoring programme *Atmospheric Heavy Metal Deposition in Northern Europe* (Rühling *et al.* [23]; Mäkinen [24]).

2 Methods

Every fifth year (1989/1990; 1995; 2000/2001; 2005/2006) a national moss study has been carried out in Estonia. The study scheduled for 2010/2011 is currently in progress – moss samples have already been collected from 100 sampling sites organised as a grid whose density is approximately 30 km.

In areas where pollution loads are high or critically high, a denser sampling grid has been used (Liiv and Kaasik [25]). Thus, in 1992, 1997, 2002, and 2007 moss samples were collected around two large oil shale fired thermal power stations and a cement factory in the north-eastern part of the country (NE Estonia). In 2003 and 2008, moss samples were collected along the transects originating from Tallinn (the capital city of Estonia with 403,000 inhabitants) within a 35 km radius, fig. 1.

Sampling sites were selected with a view to meeting the requirements of the standardized methodology elaborated by the European bio-monitoring project (Rühling *et al.* [23]).

The sampling of mosses was adapted to the methods of European moss study (Rühling *et al.* [26]; Atmospheric Heavy Metal Deposition in Northern Europe [27]). There was no significant interspecies difference in the uptake of Cd, Cr, Cu, Fe, Ni, Pb, V and Zn by *P. schreberi* and *H. splendens* in Estonia (Liiv and Kaasik [25]). The concentrations of these elements were determined at the Finnish Forest Research Institute (Metla) using inductively coupled plasma emission spectrometry (ICP-ES).

In order to determine which differences and trends in concentrations are significant, a local variability criterion was used. In 1996, 25 samples were collected from each of the three sampling sites (one in NE Estonia – the seat of the country's oil shale deposits and the region which the development of these deposits affects the most, one in South-East Estonia and one on the Hiiumaa island off the west coast of Estonia). The samples were analysed separately at the laboratory of the Institute of Ecology of the University of Lund (Sweden) by means of the ICP-ES method. The local variability for each of the sites was calculated by computing the standard deviation of the respective dataset,

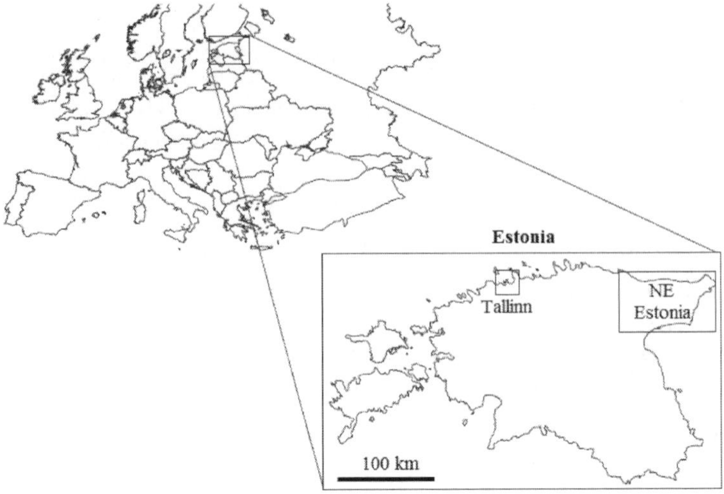

Figure 1: Locations of sampling areas with a higher density grid: NE Estonia and the surroundings of Tallinn, in 1989–2008 (Liiv and Kaasik [8]).

multiplying it by two and adding the standard error of the analytical method used. The local variability values used in this article are the largest of the corresponding figures for the three sampling sites. These values are: Cd – 0.054 μg/g, Cr – 0.84 μg/g, Cu – 0.89 μg/g, Fe – 235 μg/g, Ni – 0.59 μg/g, Pb – 3.0 μg/g, V – 1.11 μg/g, Zn – 6.95 μg/g (Liiv and Kaasik [8]).

In order to characterize the results of the moss analyses performed, a set of boxplot charts – one for each of the heavy metals whose deposition was measured in the study – was created, showing minimum, maximum and median values along with the first and the third quartile of the concentrations.

3 Results

Cd - Since 1989, the concentrations of Cd in Estonia's mosses have decreased considerably throughout the country. In 1989–2004, the median values of such concentrations in NE Estonia dropped by up to 48%. In 2005–2008, the median values of Cd concentrations in moss samples were almost identical in all three sampling regions: the entire Estonia (Estonia), NE Estonia and the outskirts of Tallinn (Tallinn), fig. 2.

Cr - Since 1989, the concentrations of Cr found in mosses in Estonia have decreased by up to 92%. In NE Estonia the concentrations have fallen by 59%. In 2005–2008 the concentrations of Cr in mosses in the Tallinn area were very close to those for the whole of Estonia. The values are not much higher in NE Estonia, fig. 3.

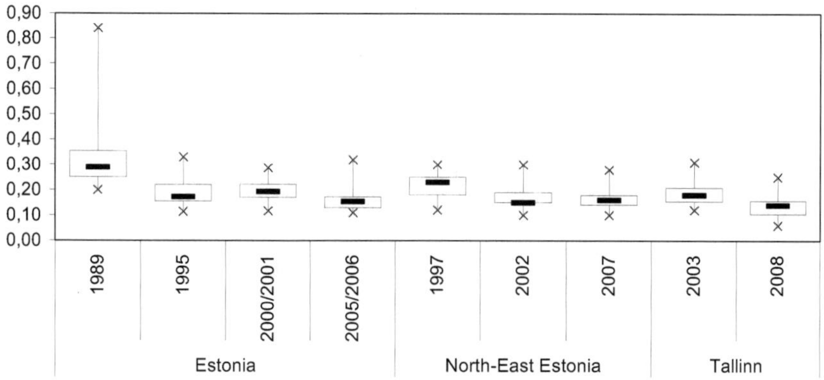

Figure 2: Minimum, maximum and median values, first and third quartile of the concentrations of Cd in mosses for Estonia (1989, 1995, 2000/2001, 2005/2006), NE Estonia (1997, 2002, 2007) and Tallinn (2003, 2008) in μg/g.

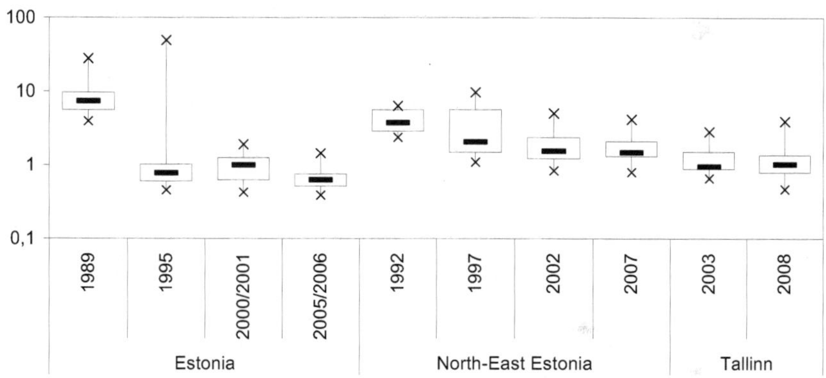

Figure 3: Minimum, maximum and median values, first and third quartile of concentrations of Cr in mosses for Estonia (1989, 1995, 2000/2001, 2005/2006), NE Estonia (1997, 2002, 2007) and Tallinn (2003, 2008) in μg/g.

Cu - In the regions with relatively high heavy metal deposition rates such as NE Estonia and the surroundings of Tallinn, the concentrations of Cu in mosses are almost identical and only show a slightly decreasing trend when the whole country is considered, fig. 4.

Fe - The concentrations of Fe in Estonian mosses have decreased by up to 70% during the last 15 years. The concentrations of Fe in mosses in NE Estonia and Tallinn are more than 2 times higher than the Estonian average, fig. 5.

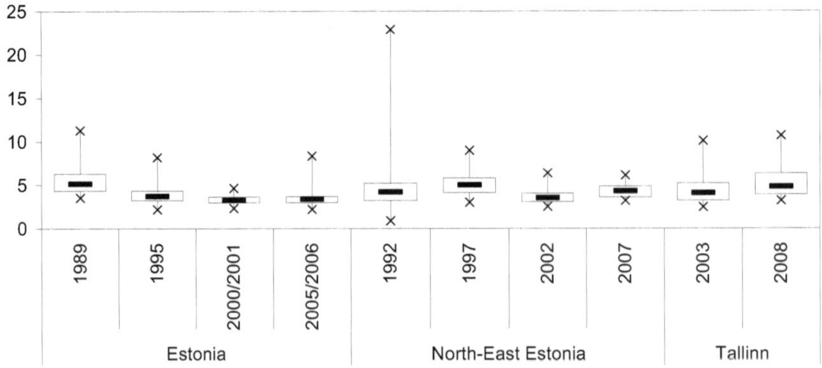

Figure 4: Minimum, maximum and median values, first and third quartile of the concentrations of Cu in mosses for Estonia (1989, 1995, 2000/2001, 2005/2006), NE Estonia (1997, 2002, 2007) and Tallinn (2003, 2008) in μg/g.

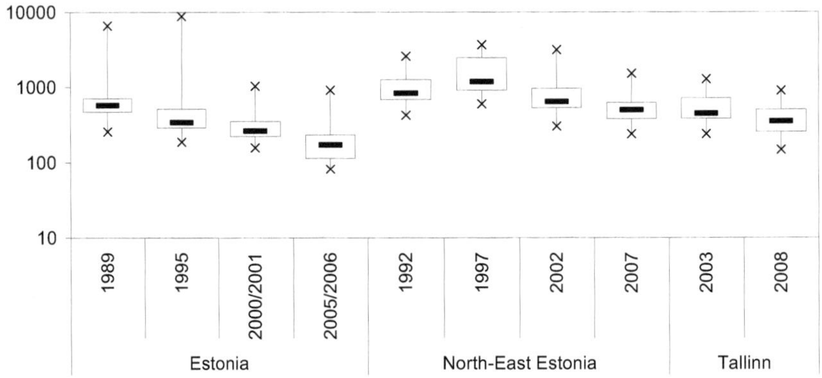

Figure 5: Minimum, maximum and median values, first and third quartile of the concentrations of Fe in mosses for Estonia (1989, 1995, 2000/2001, 2005/2006), NE Estonia (1997, 2002, 2007) and Tallinn (2003, 2008) in μg/g.

Ni - The concentrations of Ni in Estonia's mosses have decreased by up to 82%; in NE Estonia the reduction was up to 73%. The concentrations of Ni in the moss samples collected from NE Estonia in 2007 and from the Tallinn area in 2008 are very similar, and remain only slightly higher than the concentrations for Estonia as a whole measured in 2005/2006, fig. 6.

Pb - The concentrations of Pb in Estonia's mosses have decreased considerably – by up to 80% for Estonia as a whole and by 62% in NE Estonia. The concentrations of Pb in the moss samples collected from NE Estonia in 2007 are close to the concentrations found in mosses from the Tallinn area in 2008 and higher than those for Estonia in 2005/2006, fig. 7.

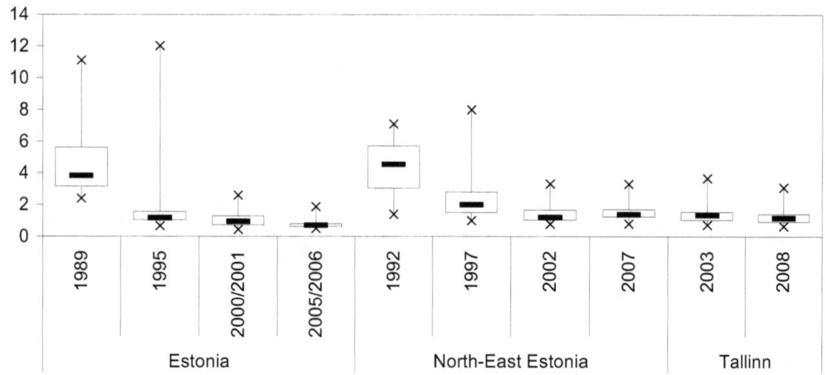

Figure 6: The minimum, maximum and median values, first and third quartile of the concentrations of Ni in Estonia (1989, 1995, 2000/2001, 2005/2006), NE Estonia (1997, 2002, 2007) and Tallinn (2003, 2008) in µg/g.

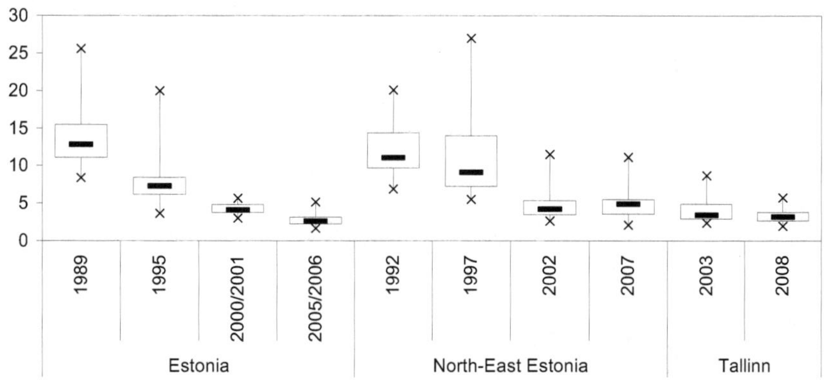

Figure 7: Minimum, maximum and median values, first and third quartile of the concentrations of Pb in mosses for Estonia (1989, 1995, 2000/2001, 2005/2006), NE Estonia (1997, 2002, 2007) and Tallinn (2003, 2008) in µg/g.

V - The concentrations of V in the mosses of Estonia have decreased by up to 76%. The concentrations found in NE Estonia in 2007 are very similar to those recorded near Tallinn, but remain higher than in the concentrations for the whole country in 2005/2006, fig. 8.

Zn - The concentrations of Zn in Estonia's mosses do not exhibit a marked shift. The changes remain within the range of local variability. The concentrations of Zn in the mosses of NE Estonia and Tallinn are nearly identical, fig. 9.

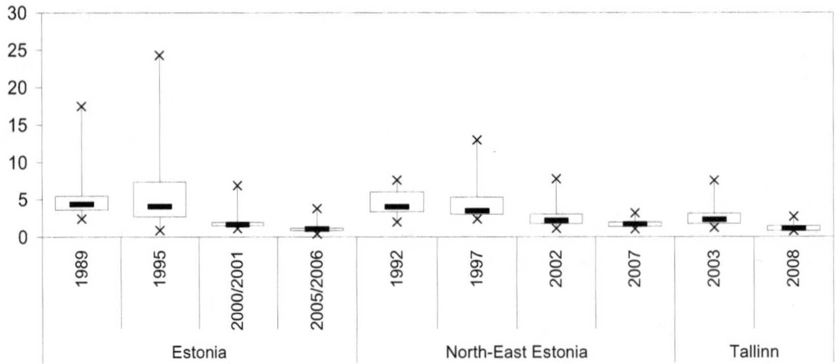

Figure 8: Minimum, maximum and median values, first and third quartiles of the concentrations of V in mosses for Estonia (1989, 1995, 2000/2001, 2005/2006), NE Estonia (1997, 2002, 2007) and Tallinn (2003, 2008) in µg/g.

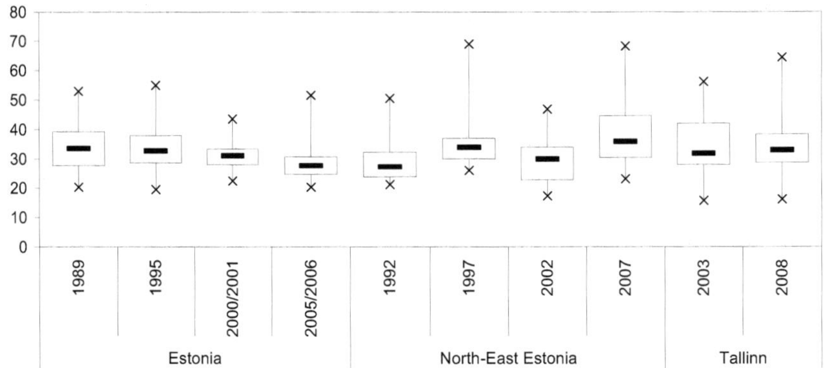

Figure 9: Minimum, maximum and median values, first and third quartile of the concentration of Zn in mosses for Estonia (1989, 1995, 2000/2001, 2005/2006), NE Estonia (1997, 2002, 2007) and Tallinn (2003, 2008) in µg/g.

4 Discussion

In 1989, the concentrations of the heavy metals investigated in the study reported here were the highest in moss samples collected from NE Estonia.

By 1995, the surroundings of Tallinn had become the second most polluted area in Estonia, probably due to the impact of increased traffic volumes and the concomitant increase in air pollution.

Traffic pollution consists of brake dust (the product of brake wear), tyre dust and fuel residuals. Fukuzaki *et al.* [3] showed that tyre dust contains the heavy metals Fe, Ni, Cu, Zn, Cd, and Pb, and contributes to airborne emissions of these elements. Zn from tyre dust is a significant pollutant in soil, air, street dust, and urban runoff (Smolders and Degryse [28]; Sadiq *et al.* [29]; Rogge *et al.* [30]; Fergusson and Kim [31]; Davis *et al.* [32]). Break dust has been recognized as a significant pollutant for Cu in the aerosols composition (Sternbeck *et al.* [5]).

In relation to the drop in power production and the implementation of environmentally friendlier technologies, the emissions of heavy metals from thermal power plants in NE Estonia have fallen remarkably since 1989.

From 1989 to 2008, the decrease in the concentrations of Cr, Fe, Ni, Pb and V in Estonian mosses has been in correlation with the decrease of emissions of solid particles from the Balti Power Plant, the Eesti Power Plant and the cement factory of Kunda Nordic Tsement AS (all in NE Estonia).

The concentrations of Zn (which is a slowly biodegradable element in the environment) have remained almost unchanged during the whole sampling period.

The decrease of concentrations of Pb in moss samples has been remarkable since 1989. The emission of Pb has decreased by 80% during 1990–2008, probably because of the introduction of unleaded fuel in Estonia.

In comparison to the rest of Europe, Estonia belongs to the group of countries with the lowest moss concentration values for Cr, Pb, Fe, Ni, V and Zn, and to the group with average concentrations of Cd and Zn.

The concentrations of heavy metals in Estonia's mosses are similar to those found in moss samples collected from the Nordic countries and from the other Baltic countries.

5 Summary and conclusions

The concentrations of Cd, Cr, Cu, Fe, Ni, Pb, V and Zn in moss samples have decreased and their spatial distribution has changed since 1989 in Estonia.

The concentrations of Cd, Cr and Ni decreased remarkably from 1989 to 1995, the concentrations of Pb in 1989–2000/2001 and the concentrations of V in 1995–2000/2001. The concentrations of Zn during the sampling period have remained within local variability or close to it.

In 1989, the highest concentrations of all heavy metals mentioned above were found in moss samples from NE Estonia. In 1995, moss samples from the region continued to have the country's highest concentrations of Cd, Cr, Fe, Ni and V, while the concentrations of Cu, Pb and Zn were the highest in the area around Tallinn. In 2000/2001, NE Estonia only accounted for the highest concentrations of Cu and Fe, the concentrations of Cr, Ni, Pb, V and Zn were the highest in the outskirts of Tallinn.

In 2005–2008, the changes of the concentrations of Cd, Cr, and Cu in all three regions (Estonia, NE Estonia, Tallinn) remained within local variability or close to it. The concentrations of Pb, Ni and V were slightly higher in moss samples collected from NE Estonia and Tallinn than in samples collected from elsewhere in Estonia. The concentrations of Fe were remarkably higher in NE Estonia than in other regions of the country.

The emission of solid particles from industrial facilities in NE Estonia has decreased considerably from 1989 to 2008. For instance, the data for the Kunda Nordic Tsement cement plant show a drop that is almost thousandfold. The concentrations of most heavy metals in moss samples across Estonia during the same period demonstrate a remarkable reduction – the concentrations of Cr, Fe, Ni, Pb and V decreased by 50–80%.

Due to the introduction of unleaded petrol, the concentrations of Pb in mosses have decreased by up to 60% during 1989–2008.

References

[1] Sucharova, J. & Suchara, I., *Biomonitoring of the atmospheric deposition of metals and sulphur compounds using moss analysis in the Czech Republic. Results of International Biomonitoring Programme 1995.* Research Institute of Ornamental Gardening Průhonice, Průhonice. 1995.

[2] Callender, E., Heavy metals in the environment – historical trends. *Environmental geochemistry*, **9**, pp. 67-107, 2005.

[3] Fukuzaki, N., Yanaka, T. & Urushiyama Y., Effects of studded tires on roadside airborne dust pollution in Niigata, Japan. *Atmospheric Environment*, **20**, pp. 377–86, 1986.

[4] Ikeda, M., Zhang, Z.W., Shimbo, S., Watanabe, T., Nakatsuka, H., Moon, C.S., Matsuda-Inoguchi, N. & Higashikawa, K., Urban population exposure to lead and cadmium in east and south-east Asia. *The Science of the Total Environment*, **249**, pp. 373-384, 2000.

[5] Sternbeck, J., Sjödin, Å. & Andre´asson, K., Metal emissions from road traffic and the influence of resuspension—results from two tunnel studies. *Atmospheric Environment*, **36**, pp. 4735–44, 2002.

[6] Adachi, K. & Tainosho, Y., Characterization of heavy metal particles embedded in tire dust. *Environment International*, **30**, pp. 1009– 1017, 2004.

[7] Harmens, H., Norris, D.A., Steinnes, E., Kubin, E., Piispanen, J., Alber, R., Aleksiayenak,Y., Blum, O., Coşkun, M., Damh, M., De Temmerman, L., Fernández, J.A., Frolova, M., Frontasyeva, M., González- Miqueom, L., Grodzińska, N., Jeran, Z., Korzekwa, S., Krmar, M., Kvietkus, K., Leblond, S., Liiv, S., Magnússon, S.H., Maňkovská, B., Pesch, R., Rühling, Å., Santamariam, J.M., Schröder, W., Spiric, Z., Suchara, I., Thöni, L., Urumov, V., Yurukova, L. & Zechmeister, H.G., Mosses as biomonitors of atmospheric heavy metal deposition: Spatial patterns and temporal trends in Europe. *Environmental Pollution*, **158**, pp. 3144-3156, 2010.

[8] Liiv, S. & Kaasik, M., Spatial and temporal variability of trace metals in mosses in Estonia. *Forestry Studies/Metsanduslikud Uurimused,* **46**, pp. 35–44, 2007.

[9] Punning, J.M. & Raukas, A., Environmental problems in the Estonian oil shale industry. *Energy and Environmental Sci*ence, **2**, pp. 723–728, 2009.

[10] Raukas, A., Opening a new decade. *Oil Shale, A Scientific-Technical Journal,* **21(1)**, pp. 1–2, 2004.
[11] Walkenhorst, A., Hagemeyer, J. & Breckle, W.S., Passive Monitoring of Airborne Pollutants, Particularly Trace Metals, with Tree Bark. *Plants as Biomonitors. Indicators for Heavy Metals in the Terrestrial Environment*, ed B. Markert, VCH, Weinheim, pp. 523–540, 1993.
[12] Ferner, D.J., Toxicity, heavy metals. *eMed Journal*, May 25, **2(5)**, pp. 1, 2001.
[13] Ma, H.W., Hung, M.L. & Chen, P. C., A systemic health risk assessment for the chromium cycle in Taiwan. *Environment International*, **10**, pp. 1016, 2006.
[14] Rühling, Å. & Tyler, G., An ecological approach to the lead problem. *Botaniska Notiser*, **122**, pp. 248-342, 1968.
[15] Holy, M., Pesch, R., Schröder, W., Harmens, H., Ilyin, I., Alber, R., Aleksiayenak, Y., Blum, O., Coşkun, M., Dam, M., De Temmerman, L., Fedorets, N., Figueira, R., Frolova, M., Frontasyeva, M., Goltsova, N., Gonzalez Miqueo, L., Grodzińska, K., Jeran, Z., Korzekwa, S., Krmar, M., Kubin, E., Kvietkus, K., Larsen, M., Leblond, S., Liiv, S., Magnússon, S., Maňkovská, B., Mocanu, R., Piispanen, J., Rühling, Å., Santamaria, J., Steinnes, E., Suchara, I., Thöni, L., Turcsányi, G., Urumov, V., Wolterbeek, B., Yurukova, L., & Zechmeister, H.G., First thorough identification of factors associated with Cd, Hg and Pb concentrations in mosses sampled in the European Surveys 1990, 1995, 2000 and 2005. *Journal of Atmospheric Chemistry*, **63**, pp. 109–124, 2009.
[16] Brown, D.H., Mineral nutrition. *Bryophyte Ecology*, ed. A.J.E. Smith, Chapman and Hall: London, pp. 383-444, 1982.
[17] Brown, D. H., & Brumelis, G., A biomonitoring method using the cellular distribution of metals in moss. *The Science of the Total Environment*, **187**, pp. 153-161, 1996.
[18] Tyler, G., Bryophytes and heavy metals. A literature review. *Botanical Journal of the Linnean Society*, **104**, pp. 231-253, 1990.
[19] Sawidis, T., Zachariadis, G., Stratis, J. & Ladoukakis, E., Mosses as Biological Indicators for Monitoring of Heavy Metal Pollution. *Fresenius Environmental Bulletin*, **2**, pp. 193-199, 1993.
[20] Thöni, L., Schnyder, N. & Krieg, F., Comparison of metal concentrations of three species of mosses and metal freights in bulk precipitation. *Fresenius Journal of Analytical Chemistry*, **354**, pp. 703-708, 1996.
[21] Markert, B., Wappelhorst, O., Weckert, V., Herpin, U., Siewers, U., Friese, K. & Breulmann, G., The use of bioindicators for monitoring the heavy-metal status of the environment. *Journal of Radioanalytical and Nuclear Chemistry*, **240(2)**, pp. 425-429, 1999.
[22] Fernandez, J. A., Ederra, A., Nunez, E., Martinez-Abaigar, J., Infante, M., Heras, P., Elias, M. J., Mazimpaka, V. & Carballeira, A., Biomonitoring of metal deposition in northern Spain by moss analysis. *The Science of the Total Environment*, **300**, pp. 115-127, 2002.

[23] Rühling, A., Brumelis, G., Goltsova, N., Kvietkus, K., Kubin, E., Liiv, S., Magnusson, S., Mäkinen, A., Pilegaard, K., Rasmussen, L., Sander, E. & Steinnes, E., Atmospheric Heavy Metal Deposition in Northern Europe in 1990. *NORD*, **12**, 41 pp, 1992.

[24] Mäkinen, A., *Biomonitoring of atmospheric heavy metal deposition in Estonia: A chemical survey of mosses in 1989*, Report no. 1/93, Ministry of Environment of Finland, 1994

[25] Liiv, S. & Kaasik, M., Trace metals in mosses in the Estonian oil shale processing region. *Journal of Atmospheric Chemistry*, **49**, pp. 563–578. 2004.

[26] Rühling, Å., Rasmussen, L., Pilegaard, K., Mäkinen, A. & Steinnes, E., Survey of Atmospheric Heavy Metal Deposition in the Nordic countries in 1985 – monitored by moss analyses. *Nord*, **21**. 44 pp, 1987.

[27] Atmospheric Heavy Metal Deposition in Europe 1995–1996. *Nord*, **15**, 66 pp, 1998.

[28] Smolders, E. & Degryse, F., Fate and effect of zinc from tire debris in soil. *Environmental Science and Technology*, **36**, pp. 3706–10, 2002.

[29] Sadiq, M., Alam, I., El-Mubarek, A.m & Al-Mohdhar, H.M., Preliminary evaluation of metal pollution from wear of auto tires. *Bulletin of Environmental Contamination and Toxicology*, **42**, pp. 743–8, 1989.

[30] Rogge, W.F., Hildemann, L.M., Mazurek, M.A., Cass, G.R. & Simoneit B.R., Sources of fine organic aerosol: 3. Road dust, tire debris, and organometallic brake lining dust: roads as sources and sinks. *Environmental Science and Technology*, 27, pp. 1892– 904, 1993.

[31] Fergusson, J.E. & Kim, N.D., Trace elements in street and house dusts: sources and speciation. *The Science of the Total Environment*, **100**, pp. 125– 50, 1991.

[32] Davis, A.P., Shokouhian, M. & Ni, S., Loading estimates of lead, copper, cadmium, and zinc in urban runoff from specific sources. *Chemosphere*, **44**, pp. 997–1009, 2001.

Sculpture to measure particulate air pollution

R. G. Wakeland
New Mexico, USA

Abstract

Make sculpture from laundry detergent bottles, found objects and salvaged construction materials. Particle air pollution collects on white fabric and plastic. This black soot-like discoloration can be compared to particle pollution quantities from USA EPA and other, international websites. K-12 curriculum derives from the sculpture itself and data collection of the particulate air pollution. Contemporary studies in education show that sculpture and visual arts used in teaching encourage reading, writing, and cognitive development.

Keywords: sculpture, particle air pollution, particulate air pollution, dust, cognitive development, teaching, reading, writing, literacy, math.

1 Effect upon nature of 21st century design

Humans by calculated design and everyday operation of factories, internal combustion engines, and non-renewable energy electric generating stations alter nature. We throw particulates into nature's pristine air. We create designer air. We measure and quantify these particulate concentrations daily on the United States of America (USA) Environmental Protection Agency (EPA) and other web sites. Created from salvaged objects, sculpture measures this particulate air pollution in a graphic, visual way, collecting particulates as black streaks and smudges on textiles and plastic surfaces. The statistics thus take on an inescapable image form. In a public forum, the sculpture confronts us with our degradation of and relationship with nature.

Art, public and private, can be analyzed for its carbon footprint. For example, sculpture composed of salvaged materials preserves air quality compared to sculpture composed of yards and meters, hectares and acres of virgin plastic sheeting and/or solid cast polymer. The lifecycle of the artwork should be factored into its impact upon the environment. Public entities are called upon to

WIT Transactions on Ecology and the Environment, Vol 144, © 2011 WIT Press
www.witpress.com, ISSN 1743-3541 (on-line)
doi:10.2495/ECO110131

support sculpture and installations constructed of recycled materials, as well as those described here to measure air pollution in public buildings and landscapes.

2 Theoretical bases and methods for using art to teach reading, writing, math and cognitive abilities

Twentieth century studies in the United States of America (USA) have shown visual arts to be effective in teaching language skills and proficiency. Lower achieving students (typically C, D and F students) scored as high as their A-B classmates on recall of visual shapes and content presented in art-appreciation classes taught by volunteers. Primarily large, colorful paintings were used. Art historical, literacy type information was included, such as name, date, and geographical location of the artist. (Epstein and Dauber [2] and Epstein and Salinas [3]).

A qualitative study was conducted at a suburban southern California, USA elementary school using visual arts strategies to teach literacy. It implemented the "Picturing-Writing" curriculum developed at the University of New Hampshire in the 1990s. The authors, two art teachers and one literacy professor, included a literature review summarizing their own and other previous studies. Among the effects of the visual imagery surveyed and studied were motivating students to read and write, a pre-writing composing device, connecting the image with poetry, dialectic response of the artist to the art object of creation, cognitive development, and ideas changing and forming during the creative process. Working on the theory that art is a "'non-linguistic' composing device that assists children in expressing their thoughts", researchers concluded "art plays a critical role as a mediating event for the compositional process" (Andrzejczak et al. [1]).

Researchers in a quantitative study compared reading comprehension in groups of children with various intervention combinations: manipulating the objects described while reading, imagining manipulation objects while reading, reading once and rereading (without any manipulation). Original and previous theory hinged on the language learner's ability to "index" words and phrases to objects or perceptual symbols (toys being manipulated). Indexing is the attachment by the reader of referent, perceptual symbols to words. Therefore, the contribution made by the association of perceptual symbols (the objects being manipulated) to the words being learned/read to accompanying/resulting reading comprehension. The common English as a second language student enigma of flawless verbal reading while comprehending nothing is recognized within first language acquisition: "To the extent that the words are not being indexed, reading becomes a meaningless exercise in word calling" (Glenberg et al. [4]).

Whether via a dump truck in a sand box or a pencil in a cup, the association of the physical world with the text, it is concluded, helps children make inferences to the real world, developmental models and perceptual symbols. Children thus understand not merely the text itself but what the text means. This also contributes to cognitive development (Glenberg et al. [4]). While the toy manipulations are not exactly sculpture apprenticeships, the physical, tactile implications for substituting sculpture for the dump truck and sand box are obvious. The vocabulary and concepts could be changed to fit the sculpture. Descriptions of previous studies in the literature encompass any kind of manipulative activity, with drawing being one. Also, visual cues are incorporated into the reported study (Glenberg et al. [4]).

Other emphasis on cognitive development to augment reading is contained in commercially available, packaged curriculum. One internationally taught course incorporates animal picture flash cards, asking for animals' names in verbal responses. It also includes practicing and developing speed in recognizing and naming objects, shapes, and colors. Activities include sorting pictures of common objects into one of three similar, corresponding abstract shapes and reproducing a design with colored shapes after a 10 second view. Incorporating also verbal command/response exercises and other techniques, this method has generally been accepted as successful. The programs tested were Cognition Enhancement Training (COGENT) and PREP. The authors call for continued studies among different demographics (Hayward et al. [5]).

Because this curriculum relies on recycled, used and everyday objects, it can be packaged by educators to send home for a weekend or summer project. Students can bring their finished sculptures, charts, graphs, and drawings back to the school. Or, suffice it to say that the home activity interaction the sculpture fosters is justification alone without the need for return of the sculpture as homework. Even science and math can be taught at home via projects such as this (Lapadat [6]).

3 The first module

The first module in this curriculum addresses the arts standard. Teachers and youth group leaders show the images of the sculpture in the appendices, facilitate discussion among students, and then students create their own sculpture.

Name and topic: sculpture to measure particulate air pollution.

Grade/content area and benchmarks, objectives: K-12 arts. Participate in the process of making and looking at works of art

Dates: minimum 2 full days or two weeks, 1 class period per day, or can be expanded. First week/day, students gather recycled materials and found objects and find one example of outdoor sculpture in the neighborhood. Second week/day making a drawing of the proposed sculpture and component parts and assemble it.

Materials and equipment: white fabric; plastic bottles and bottle caps; nylon or jute cord; found objects; clear, translucent or white plastic sheeting or Plexiglas, stapler; hole punch, scissors, washers. Parents, teachers and youth club leaders, as well as high school shop students, can do preliminary deconstruction of found objects, cutting up plastic bottles, and drilling holes.

Grouping and environment: the class listens together as leader/teacher shows images of examples such as in the appendices, and describes the project. Students contribute ideas of how to find materials, found objects, etc. to the group. Teacher facilitates this discussion. Students contribute to the group by

Figure 1: Sculpture to measure air pollution by Robin Gay Wakeland, 2010. Jute (sisal, industrial hemp) and plastic laundry detergent bottle handle and lids.

Figure 2: Sculpture to measure particulate air pollution by Robin Gay Wakeland, 2010. Found base metal wire rings, mop from construction materials salvage store, synthetic fibre.

Figure 3: Sculpture to measure particulate air pollution by Robin Gay Wakeland, 2010. Aluminium pipes and key rings borrowed from artist's booth fixtures, in situ fence and thrift store cotton fabric.

describing where they saw an outdoor sculpture nearby. Students work individually on their sculpture. Students present their finished sculpture the group and describe it and where they will install it. Students can take theirsculpture home or install it at the school. One variation is to have the class work together on one larger sculpture. This would be installed at the school, or another public building, indoor or outdoor.

Lesson sequence: teacher/youth group leaders show students the images in appendices as examples. Teacher facilitates discussion about how to collect found objects for the sculpture. Teacher leads discussion about the difference between two- and three-dimensional art. Students are given homework to find an outdoor sculpture in their neighborhood and tell the class about it. Students do drawings of how their sculpture will look and make a list of materials needed and/or components prior to assembly. Students gather materials. Parents, teachers, youth leaders and high school shop students do any drilling and working with sharp/power tools. Students assemble sculpture and describe to class how they made it and where it will be installed.

With a group collaboration, students submit drawings for a large group project. Then the students decide on how to create a group project using ideas from the individual drawings. Teacher facilitates this interaction.

Assessment: Did students grasp the idea of 3-dimensions, or did they create flat objects? Were students able to use a drawing in the creative process to conceptualize, define, and articulate their ideas? Were students able to translate their drawing into a 3-D finished project? Were students able to successfully collaborate on a large, single sculpture?

4 The second module

The second module focuses on language arts. Students compile visual images of particulate air pollution from various sources including architecture and flags in their neighbourhood. They then create verbal and written reports.

Here is the lesson plan for the language arts standard.

Name and topic: sculpture to measure particulate air pollution.

Grade/content area and benchmarks, objectives: language arts K-12. Locate and use a variety of resources to acquire information across the curriculum.

Dates: can be structured within a variety of time periods.

Materials and equipment: library, computer, internet connection, projector, digital camera (optional), paper and pencil to take notes and make drawings; word processing and slide show software.

Grouping and environment: class listens together as leader/teacher shows sculpture images, describes project and shows USA EPA flash video about particle air pollution. Students go as a group to the library. Students find books individually at library (lower grades: with help from teacher). Students take digital photos or make observations and take notes individually, except younger grades, which go as a group on a field trip to make observations outside. Students work individually on writing up their findings. Students give verbal report to group.

Lesson sequence: teachers/youth group leaders show students the sculpture images and EPA flash video. Students go to the library and find books on air pollution. Teachers help younger students. Higher grades: make annotated bibliography from library sources. Students find visual image of air pollution on internet. Students are given homework to take digital images or make observations (notes and drawings) in situ, on how particle air pollution shows up on outdoor objects (architecture, flags, etc.). This can also be accomplished as a group in a field trip. Teachers can supervise younger students with the use of a digital camera. Students compile these images and include written notes identifying the geographic location of the image, date and time, and other pertinent details. Students also include a one page written summary of their images. Students give a verbal report to the class, showing their images. Older students can produce a slide show or other digital presentation.

Here is the link for the USA EPA website and flash video:

EPA main login page
http://airnow.gov/index.cfm?action=airnow.local
flash movie about particle air pollution on EPA home page
http://www.epa.gov/airnow/pm/pm.html

Assessment: What library and web sources were used for information searches? Quantity and quality of books and images found? How well did older students understand the concept of and complete the annotated bibliography?

Figure 4: Sculpture to measure particulate air pollution by Robin Gay Wakeland, 2010. Chrome plated steel floor lamp shaft, salvaged copper pipe, Plexiglas and plastic.

Students' notes, drawings, and digital photos, as well as their written and verbal reports provide summative assessment for language proficiency.

5 The third module

The third module teaches the math standards identifying geometrical shapes and charting and graphing information. Students use their own sculpture created in the arts curriculum.

Here is the curriculum plan for the math lessons. Some corporations give support for science and math teaching. See Toshiba America Foundation Math Grants Application Form and Guidelines at http://www.toshiba.com/tafpub/jsp/recentgrants/712.jsp

Name and topic: sculpture to measure particulate air pollution.

Grade/content area and benchmarks, objectives: K-12 math. Analyze characteristics and properties of two-and three-dimensional geometric shapes. Formulate questions that can be addressed with data and collect, organize, and display relevant data to answer them.

Dates: minimum 2 weeks of regular class time, once a day; longer time span to observe particles collect on the sculpture and track EPA data. First week, students analyze geometrical shapes in the sculpture images, and in their own drawings and sculpture created in the arts lesson. Over time: make observations and/or take digital images of white fabric and plastic/Plexiglas surfaces to record particle air pollution and record EPA particle pollution for the geographical area on spreadsheets.

Materials and equipment: digital camera (optional); computer and access to internet and EPA website; computer spreadsheets software with chart/graphs; examples of 2 and 3 dimensional geometrical objects and shapes; students' finished sculptures and drawings.

Grouping and environment: class listens together as teacher/group leader shows sculpture images and compares geometrical models to the sculpture. Students individually analyze 2-D and 3-D geometric components of their own completed drawings and sculpture, and record particulate air pollution on their sculpture and enter EPA data into spreadsheets.

Lesson sequence: teachers and youth group leaders show images of the appendices and other sculpture created for the purpose, and compare geometric models to the sculpture. If possible, the teacher should obtain digital images of the students' sculpture produced in the arts curriculum and analyze it the same way for the whole class. Students then individually compare their own drawings for their sculpture, as well as the finished sculpture, with geometrical shapes. Over time: students make observations and/or take digital images of white fabric and plastic in their sculptures to record particle air pollution and record EPA particle pollution for the geographical area. Higher grades: plot data from EPA website into spreadsheet and create graphs and charts from it. Compare the particle pollution collecting on sculpture with this data.

Assessment: Have younger students mastered concepts of collecting and compiling data? Have older students mastered computer skills of entering data into spreadsheets and creating charts and graphs? Can students recognize two- and three-dimensional geometric shapes in sculpture and their own drawings and sculpture?

6 Conclusion

These strands and methods can be adapted to formal school classrooms, summer camps, and after school youth groups. The curriculum can also be packaged as a send-home project for students to complete on the weekends or on vacation. The creativity of the teachers is given free reign. The author and artist hopes teachers and youth group leaders will send their feedback, artwork, suggestions, impressions, and knowledge using this curriculum to her via email: rgwakeland4036@msn.com for inclusion in a study.

References

[1] Andrzejczak, N., Trainin, G. and Poldberg, M. (2005). From image to text: using images in the writing process. *International Journal of Education and the Arts* 6(12); http://ijea.asu.edu

[2] Epstein, J.L., and Dauber, S.L (1995). Effects on students of an interdisciplinary program linking social studies, art, and family volunteers in the middle grades. *Journal of Early Adolescence. 15,* p. 114-144.

[3] Epstein, J.L. and Salinas, K.C. (1991). *TIPS social studies and art manual and prototype presentations.*Baltimore: Center on Families, Communities, Schools and Children's Learning, Johns Hopkins University.

[4] Glenberg, A.M., Gutierrez, T., Levin, J.R., Japuntich, S., and Kaschak, M.P. (2004). Activity and imagined activity can enhance young children's reading comprehension. *Journal of Educational Psychology. 96*(3), p. 424-36.

[5] Hayward, D., Das, H., and Janzen, T. (2007). Innovative programs for improvement in reading through cognitive enhancement: A remediation study of Canadian First Nations children. *Journal of Learning Disabilities. 40*(5), p. 443-57.

[6] Lapadat, M. (1995). How parents can help us teach science. *The Hoosier Science Teacher*. 21(2):60-63.

Section 4
Sustainable development studies

Monitoring urban dynamics: the case of the Metropolitan area of Central Tuscany

R. M. Pulselli, P. Romano, V. Niccolucci & N. Marchettini
Ecodynamics Group – Dept. of Chemistry, University of Siena, Italy

Abstract

How can human activity be monitored in space and time? To answer this question it is necessary to develop new tools for investigating dynamic phenomena in territorial systems.

In this study, we elaborated data, routinely registered by a mobile telecommunication company, in order to develop a monitoring technique of population density, assuming that the intensity of mobile-phone activity in a cell (a covered area around an antenna) is directly proportional to the presence of people. In this paper, we present an application of this technique to the Metropolitan Area of Central Tuscany (Italy).

Results were represented dynamically in sequences of maps that provide a qualitative description of population behaviour in the form of spatial patterns evolving in time. A quantitative description was also provided based on the measurement of cell-phone calls intensity within given areas. Data elaborating and geo-processing allowed for combining spatial and time-series analysis and understanding the general functioning of a comprehensive geo-demographic system.

The methodology discussed here intends to give a contribution towards the implementation of bottom-up procedures in order to provide information on use and development of urban metropolitan systems in time and to improve our capacity to handle dynamic processes and thus inform planning practices and future developmental strategies.

Keywords: cell-phones, mobile telecommunication technology, geo-processing, urban dynamics.

www.witpress.com, ISSN 1743-3541 (on-line)
doi:10.2495/ECO110141

1 Introduction

Kinetics is the science that studies movement or, in general, the circulation of mass. Kinetic theory was born to explain the essential characteristics of matter, based on the hypothesis that it is made up of countless small particles in movement. In the common application to gases, this theory allows us to calculate the macroscopic behaviour of gas based on the microscopic properties (velocity, kinetic energy, etc.) of the molecules it is composed of. Chemical kinetics studies the variations of the concentration of reagents and products in time and can be established experimentally by observing, with appropriate tools, processes and chemical reactions [1].

In this paper, a monitoring technique was developed in order to investigate dynamics in urban systems, focussing on population behaviour and its variation in time. These dynamic processes, obtained through the elaboration of statistical data gathered through the technology of mobile telecommunications, were represented as a comprehensive whole in which individuals operate as the constitutive elements of a complex system and determine its spatial and temporal configuration. A bottom-up data collection was thus developed and data elaboration through geo-processing allowed us to develop a study approach for understanding the dynamic evolutionary properties of urban systems and to reveal their general functioning [2]. This is at least an investigation of kinetic processes in urban systems.

2 Materials and methods: investigating kinetics in Tuscan metropolitan area

A research aimed at developing tools for urban dynamics monitoring was first developed at the Massachusetts Institute of Technology in 2004 [3] and then elaborated by the researchers of *ecodynamics group* in Siena [4–7]. The monitoring technique was studied to observe the variations of population density in real time, and analyze its movements and dynamics in relation to the physical space of the city.

This paper presents results from a research project, namely MoTo (Mobile Toscana) funded by the General Directorate of Territorial and Environmental Policies of the Tuscan Region. Through the systematic collection of statistical data and its aggregated processing and geo-processing, MoTo developed a procedure for obtaining a real measure of population density variable in space and time. In particular, MoTo elaborated data registered by a mobile telephone company in a vast area, which is relative to the position-direction and to the activity of the antennas that transmit the signal to cell-phones. In this study, we assumed that, with a close approximation (in Italy in 2006, the percentage of stable users was registered at between 80% and 85% of the population – EITO 2007), the intensity of cell-phone activity is proportional to the number of users present in a given area, and that the variations in activity correspond proportionally to the variations in population density.

Data was registered by each individual antenna that transmits its signal over a defined area, called a cell. Coordinates and directions of the antennas allowed for the detection of the structure of the mobile phone network throughout the territory, and of the dimension-position of the single cells (Figure 1).

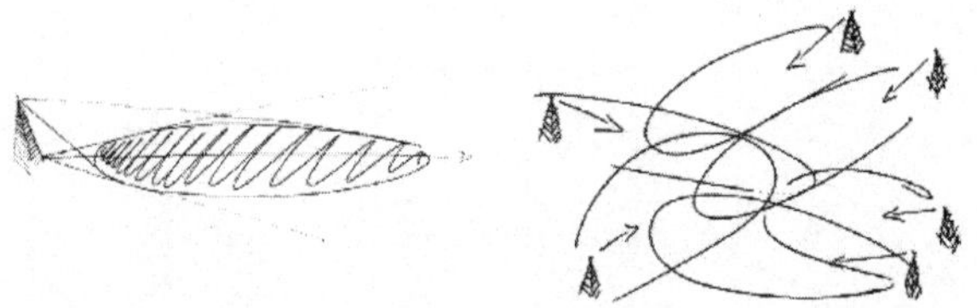

Figure 1: Scheme of antennas and cells.

Activity data from the antennas, referred to the relative cells, was processed and connected to single spatial units identified by a reference grid (500×500 m). The representation of data in the grid allowed us to visualize the intensity of activity registered in the form of an overall spatial configuration (Figure 2). The grid is extended over an area of 45×35 km. The 90×70 pixels identify a set of 6300 pixels/area defined as the *minimum spatial unit* of reference for the processing of the information on a cartographic basis.

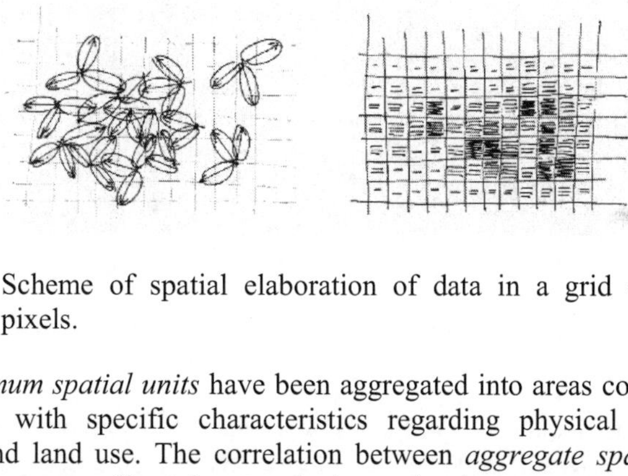

Figure 2: Scheme of spatial elaboration of data in a grid of 500×500 m pixels.

The *minimum spatial units* have been aggregated into areas corresponding to urban zones with specific characteristics regarding physical conformation, structures, and land use. The correlation between *aggregate spatial units* and ISTAT census zones allowed us to obtain results that are comparable with the Territorial Information System (SIT) of the Tuscany Region (basic cartography and ISTAT processing).

The activity of the antennas (number of calls) was registered at hourly intervals per cell sectors in subsequent segments of 550 meters along the direction of propagation of the signal. The study area is a portion of the territory of the Region of Tuscany, which includes the *metropolitan area of central Tuscany.* This includes the Florence-Prato-Pistoia plain and the zone of Empoli and extends over 1575 km² (45×35 km), with over 1,200,000 inhabitants.

Sequences of patterns with varying intensities dynamically showed the population density and its variation in space and time.

3 Results and discussion

Data geo-processing allowed us for representing dynamically the activity of population living and working in the area under study, that is for representing the geo-demographic system of the metropolitan area in Tuscany.

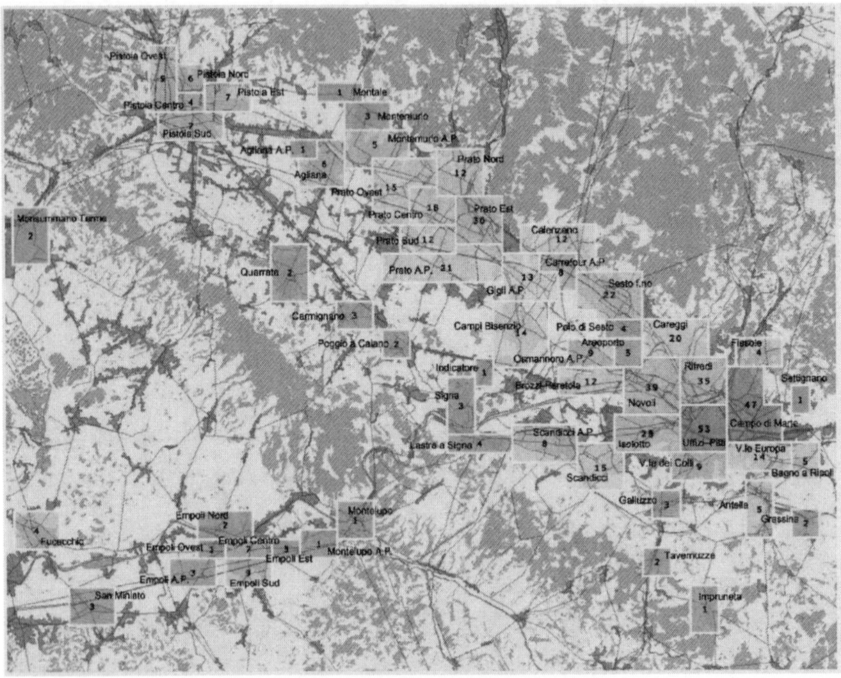

Figure 3: Map of aggregate spatial units in the metropolitan area of Central Tuscany. Values are given in thousandths.

In particular, sequences of maps were elaborated in order to show the evolution of daily activity (total 24 hours) in twelve months or the hourly activity in the 24 hours of a working day comparing different seasons. Maps show cell-phones activity in a colour-scale through *minimum spatial units*. This is also represented through values that correspond to the portion of activity in each *aggregated spatial unit* given in thousandth relative to the total activity in the area under study (45×35 km). Spatial patterns and values allowed for representing both qualitative and quantitative aspects of the human activity in the metropolitan area. In Figure 3 a synthetic map shows results in a typical working day.

The highest intensity is registered in the centre of Florence (aggregate spatial units: Uffizi-Pitti, Campo di Marte, Rifredi, Novoli), and in the areas of Sesto Fioretino and Prato (including the productive area Prato A.P.). A significant concentration is also shown in the areas of Pistoia and Empoli.

In the following diagram (Figure 4), the intensity of cell-phone activity is shown in twelve months. Values refer to the total activity in 24 hours of a working day. It is clear that the activity is almost constant during the year except for June in which it increases, probably due to the incoming of tourists at the beginning of summer season, and in July and August in which it decreases, due to the summer break, until September.

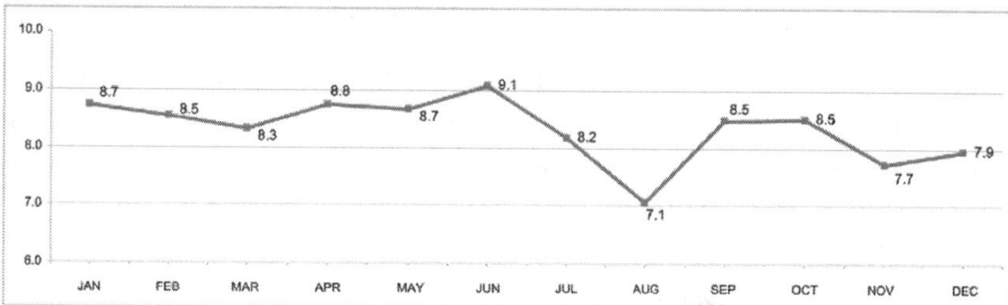

Figure 4: Diagram of total 24hours cell-phone activity during a year (2009) in the metropolitan area of Central Tuscany.

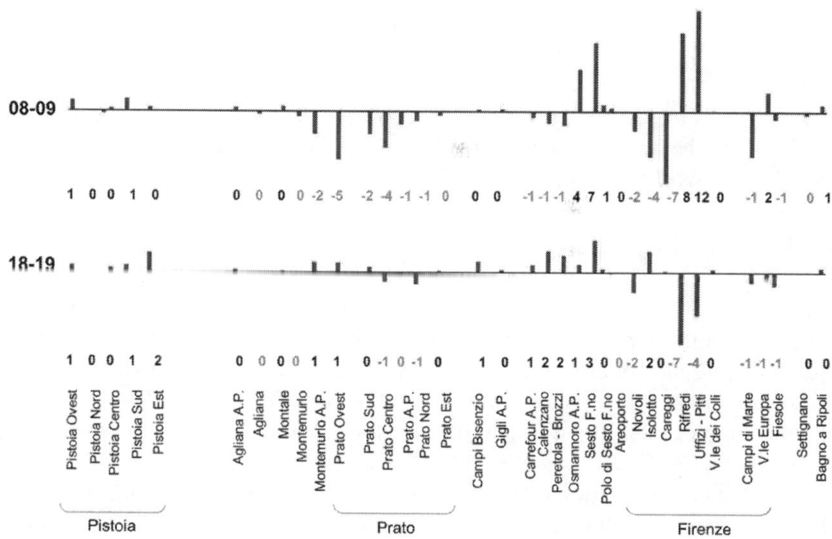

Figure 5: Diagram of cell-phone activity variation in time intervals (8-9 am and 6-7 pm) in the metropolitan area of Central Tuscany.

Other results show the mode of use in time of specific *aggregate spatial units*. In particular, among others, two aspects deal with daily life in given areas.

In Figure 5, the two diagrams show the variation of activity in the time intervals from 8 to 9 am and from 6 to 7 pm, respectively. The first diagram, that represents variation in early morning, shows an increase of activity in the city centres of Florence, Sesto Fiorentino and Pistoia, as well as in some of the productive areas (Osmannoro, Polo di Sesto, Gigli). The second diagram, that represents variation in the late afternoon, shows a decrease of activity in the areas above and an increase in residential districts in the metropolitan areas, especially around Florence (i.e. Calenziano, Campi Bisenzio, Prato Ovest, Bagno a Ripoli).

In Figure 6, curves show the intensity of activity in 24hours of a working day in five productive areas corresponding to the *aggregated spatial units* represented in the map in the right side. Results highlighted here the increase of activity during the early morning during which there are workers incoming, while activity progressively decreases during the afternoon. This shows how working hours effect the intensity of cell-phone activity, for example in the lunch break (from 1 to 2 pm) or after works.

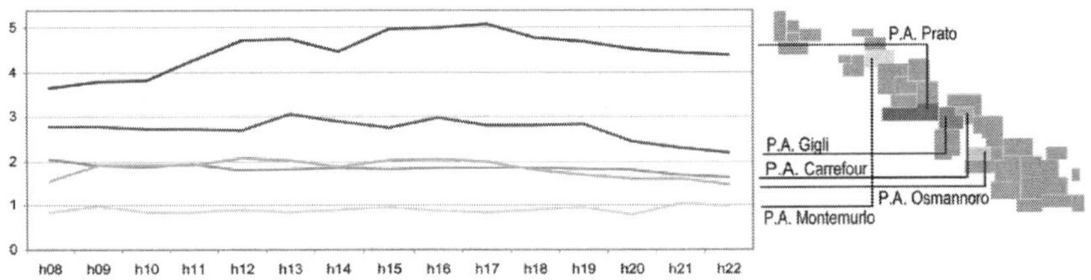

Figure 6: Time series analysis of human activity in five productive areas in the metropolitan area of Central Tuscany during 24 hours of a working day.

4 Conclusion

Cell-phone traffic data was elaborated for monitoring population density in time. Patterns of cell-phone activity, with variable intensity, were observed in order to visualize time modes of urban space use and to understand when and how long people use urban space. Results were obtained concerning a) the experimentation of a methodology for monitoring urban dynamics and b) the application to a case study, the Metropolitan Area of Central Tuscany.

a) On the one hand, this study developed a monitoring technique for analysing the geo-demographic system in a vast region based on cell-phone calls. Information provided concerns the methodology used and the procedures of data collection and its statistical elaboration, geo-processing and spatial approximation, spatial analysis and time series analysis. This work would contribute for elaborating bottom-up procedures for monitoring human activity in urban areas and providing information on their dynamics.

b) On the other hand, this technique was applied to a region in order to detect the general functioning of a wide metropolitan area and show the comprehensive behaviour of its population in time, for example in a sequence of working days. Data was elaborated relative to *minimal spatial units* in order to give a qualitative description of spatial phenomena in time through patch dynamics. Quantitative information was also provided relative to *aggregated spatial units*, in order to give a measure of the intensity of use of urban space by population. Both qualitative and quantitative descriptions were used to show the variation of activity in time and to classify different parts of the system based not only on their structure or function but also on their time of use.

Acknowledgements

The study of the Metropolitan Area in Central Tuscany, namely MoTo project, was selected from the competition for research and innovation in the territorial and environmental field, and funded by the General Directorate of Territorial and Environmental Policies of the Tuscan Region. We thank the team of the mobile telecommunication company Telecom Italia for collaborating with us and making this project possible.

References

[1] Pulselli R M, Morandi F, Tiezzi E. Integrating thermodynamics and kinetics of urban systems for regional studies. (Eds.) Brebbia C A., Hernandez S, Tiezzi E. *The Sustainable City VI, Urban Regeneration and Sustainability*. WIT Press. Southampton, UK 2010 103-111.

[2] Pulselli R M, Tiezzi E. *City Out of Chaos. Urban Self-organization and Sustainability*. WIT Press, Southampton UK 2009. (Foreword by Paolo Portoghesi).

[3] C. Ratti, R.M. Pulselli, S. Williams, D. Frenchman. Mobile Landscapes: using location data from cell phones for urban analysis. *Environment and Planning B: Planning and Design* 33(5) 727-748 2006.

[4] Pulselli R M, Ratti C, Tiezzi E. City out of chaos: social patterns and organization in urban systems. *International Journal of Ecodynamics* 1(2) 2006 125-134.

[5] Bastianoni S, Pulselli R M, Romano P, Pulselli F M. Dynamics and evolution of urban patterns: the evidence of the Mobile Landscapes project. (Ed.) Brebbia C A. *Design & Nature IV- Comparing Design in Nature with Science and Engineering*. WIT Press. Southampton, UK 2008 253-260.

[6] Pulselli R M, Romano P, Ratti C, Tiezzi E. Computing urban mobile landscapes through monitoring population density based on cell-phone chatting. *International Journal of Design & Nature and Ecodynamics* 3(2) 2008 121-134.

[7] Marchettini N, Pulselli R M, Tiezzi E.B.P. An innovative survey of urban systems dynamics: the evidence of the MoTo project. (Eds.) Brebbia C A, Hernandez S, Tiezzi E. *The Sustainable City VI, Urban Regeneration and Sustainability*. WITpress. Southampton, UK 2010 97-102.

Process sustainability prediction: a computer aided design tool for sustainable industrial development

M. Fermeglia, L. Toma & G. Longo
Department of Industrial Engineering & Information Technology, University of Trieste, Italy

Abstract

Chemical process sustainability can be estimated using different sustainability indicators. The quantitative estimation of those indicators is necessary (i) for evaluating the environmental impact of a chemical process and (ii) for choosing the best design among different available alternatives. To accomplish these goals, the computerized calculation of sustainability indicators requires the use of at least three computer tools: (i) process simulation, (ii) molecular modelling and a (iii) sustainability indicators software code. In this paper, a complete software platform, Process Sustainability Prediction Framework, integrated with process simulation programs, which support the CAPE-OPEN interfaces, is presented, discussed and applied to an example of biofuel production. The process considered is implemented in a real plant in Cuba.

Keywords: sustainability, process design, material impact, energy impact, Cape Open, process simulation, environmental impact, PSP, biofuels, bioethanol.

1 Introduction

Sustainable development has been defined as the creation of goods and services using processes and systems that are non-polluting; conserving energy and natural resources; economically viable; safe and healthy for employees, communities and consumers; socially and creatively rewarding for all working people (Veleva and Ellenbecker [1]).

In the sustainable development context, an important role is played by chemical engineering, basically for two main reasons: 1) the production of

WIT Transactions on Ecology and the Environment, Vol 144, © 2011 WIT Press
www.witpress.com, ISSN 1743-3541 (on-line)
doi:10.2495/ECO110151

chemical process plants contributes greatly to national income and is absolutely essential to modern society; thus, the society development depends on the development of the chemical process industry and vice versa; 2) many critical environmental problems can be associated directly with chemical processes or to the use of chemical products and by-products that result in waste streams to soil, water and air (Korevaar [2]). Chemical industry produces materials for multiple consumer markets, which need to be produced, used and recycled by manufacturing processes that are clean, safe and economical (Jenck et al. [3]).

In order to eliminate or reduce negative environmental problems, the environmental performance of chemical processes should be identified and quantified at an early stage of process design. Taking into account the environmental aspects at design time, an answer to "which design is more environmentally friendly, and implicitly more sustainable, for a specific chemical process?" can be provided.

The present paper is concerned with the description and application of a specific software tool, Process Sustainability Prediction (PSP) Framework.

There are several advantages deriving from the use of PSP. The most important advantage is the possibility of including the potential environmental impact criteria in the optimization of the process, on top of the traditional technical and economic criteria. Modern tools such as process simulation may result in several benefits for the design of a chemical process. These benefits include but are not limited to (i) shortened product-process development cycles, (ii) optimization of existing processes to improve energy efficiency and minimize production of waste, (iii) efficient design of new products and processes and (iv) improvements in health, safety, and environment (Fermeglia et al. [4]).

2 The evaluation of the sustainability of a process

Economy, society and environment are three basic elements of sustainability. Each of these elements is connected to, and dependent upon, the others (NJSSI [5]). It is generally acknowledged that sustainability results from a balance among the three aspects. The selection of an adequate set of indicators for sustainability evaluation is very important, for comparative analysis between different versions of a process (Sikdar [6]). With the objective of providing an easy-to-implement method for applying indicators for the purpose of analysing industrial systems for sustainability, Sikdar proposed a typology of indicators, considering the three dimensions of sustainability in three distinct hierarchical groups: (1) One dimensional (1D) indicators, which provide information about only one dimension of sustainability: economical, ecological, or social; (2) Two dimensional (2D) indicators, which provide information simultaneously about two dimensions of sustainability: socio-ecological, socio-economical, or economic-ecological; (3) Three dimensional (3D) indicators, which provide information about all three dimensions of sustainability (Martins et al. [7]).

Fermeglia et al. [8] proposed to analyse four 3D and four 1D indicators as representative for the 3D and 1D analysis respectively. The reason why the attention is focused to 3D and 1D analysis is that the former is sufficiently general and not very complicate to carry out at design time of a process and considers all possible aspects of the sustainability, while the latter is very precise, technically oriented and directly applicable to engineering practice. In the following paragraphs a brief introduction to the method is given (Fermeglia et al. [8]).

2.1 Three dimensional indicators

The Material Intensity (MI) Indicator measures the amount of non-renewable resources required to obtain a unit mass of products. It includes raw materials, solvents, and other ingredients (Martins et al. [7]). Details regarding the calculation of this indicator are given by Schwarz et al. [9]. The Energy Intensity (EI) Indicator measures the energy demands of the process. It is calculated per unit mass of products and focuses primarily on the use of non-renewable energy (Martins et al. [7]). Details about the calculation of this indicator are given by Tanzil et al. [10]. The Potential Chemical Risk Evaluation (PCRE) Indicator regards the risk associated with the manipulation, storage and use of hazardous chemical substances in the process. It is schematically represented in Figure 1.

The chemicals used in the process under study are classified depending on their relative quantity, frequency of use, danger characteristics and potential exposure. Thus, for each chemical, the designated "frequency class", "quantity class", "hazard class" and "potential exposure class" are determined according to Vincent (Vincent et al. [11]).

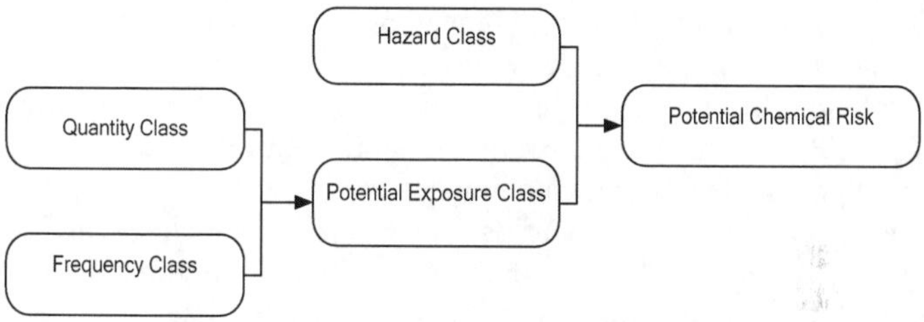

Figure 1: Potential chemical risk evaluation.

The Potential Environmental Impact Indicator is due to the emissions and discharge of hazardous chemicals to the environment. It is similar to the one for the potential chemical risk, as represented in Figure 2.

The chemicals are classified according to their relative quantity and hazard characteristics. Furthermore, the potential environmental impact depends on the transfer to a receiver medium, such as air, water or soil, which is governed by physicochemical characteristics and transfer coefficients as shown in Figure 2 (Martins et al. [7]).

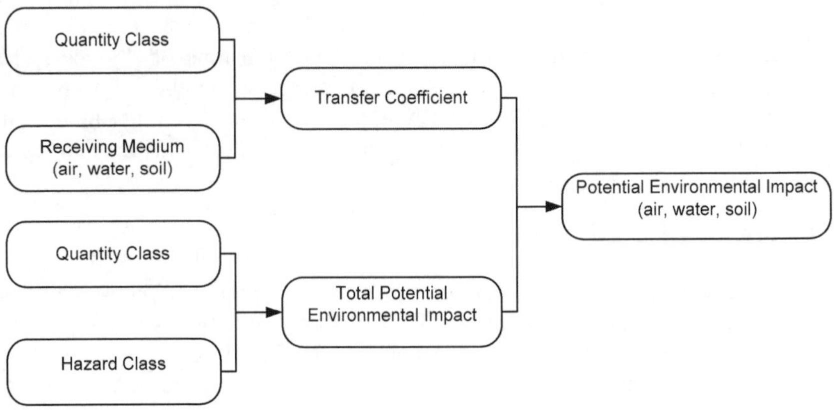

Figure 2: Potential environmental impact evaluation.

2.2 One dimensional indicator - the Waste Reduction (WAR) Algorithm

The WAR Algorithm, based on the concept of environmental impact balance, analogous to mass or energy balances, uses data from the process (stream flow-rates and mass fractions) and toxicological data to calculate the environmental impact of a chemical plant (Cabezas et al. [12]; Young and Cabezas [13]; Young et al. [14]).

Toxicological data can be grouped in different environmental impact categories: Global Warming Potential (GWP), Acidification Potential (AP), Ozone Depletion Potential (ODP), Photochemical Oxidation or Smog Formation Potential (PCOP), Human Toxicity Potential by Ingestion (HTPI), Human Toxicity Potential by Inhalation or Dermal Exposure (HTPE), Aquatic Toxicity Potential (ATP), Terrestrial Toxicity Potential (TTP). They are used in evaluation of potential environmental impact using the WAR [13].

GWP is determined by comparing the extent to which a unit mass of a chemical absorbs infrared radiation over its atmospheric lifetime to the extent the CO_2 absorbs infrared radiation over its respective lifetimes. AP or acid rain potential is determined by comparing the rate of release of H+ in the atmosphere as promoted by a chemical to the rate of release of H+ in the atmosphere as promoted by SO_2. ODP is determined by comparing the rate at which a unit mass of chemical reacts with ozone to form molecular oxygen to the rate at which a unit mass of CFC-11 (trichlorofluoromethane) reacts with ozone to form molecular oxygen. PCOP or smog formation potential is determined by comparing the rate at which a unit mass of chemical reacts with a hydroxyl radical OH to the rate at which a unit mass of ethylene reacts with OH. HTPI is calculated for a chemical if it existed as a liquid or solid at a temperature of 0°C and atmospheric pressure. As a first approximation, the lethal-dose that produced death in 50% of rats by oral ingestion (LD50) has been used as an estimate for the HTPI. HTPE was determined for that chemical if it existed as a gas a temperature of 0°C and atmospheric pressure. ATP was estimated by using toxicological data for a single, representative species of fish, Pimephales promelas (Fathead Minnows). The data for this assay comes in the form of a LC50, a lethal concentration which causes death in 50% of the test specimens. Different molecular modeling methods to evaluate the thermo-physical properties which lead to environmental impact categories are used in the PSP framework [8].

The four 1D environmental impact indexes used in WAR Algorithm and in the PSP Framework are: the total rate of impact output, the total impact output per mass of products, the total rate of impact generation and the total impact generated per mass of product. Those indexes are referred in the PSP Framework as Iout, Iout_mp, Igen and Igen_mp. The analysis of such indexes permits to select the most environmentally friendly design among different process alternatives.

3 The Process Sustainability Prediction (PSP) framework

The PSP Framework implementation has been made according to Martins and Vincent for 3D Indicators (Martins et al. [7]; Vincent et al. [11]) and to Young and Cabezas for 1D Indicators (Cabezas et al. [12]; Young and Cabezas [13]; Young et al. [14]). A general schema illustrating the main steps of the development is given in Figure 3.

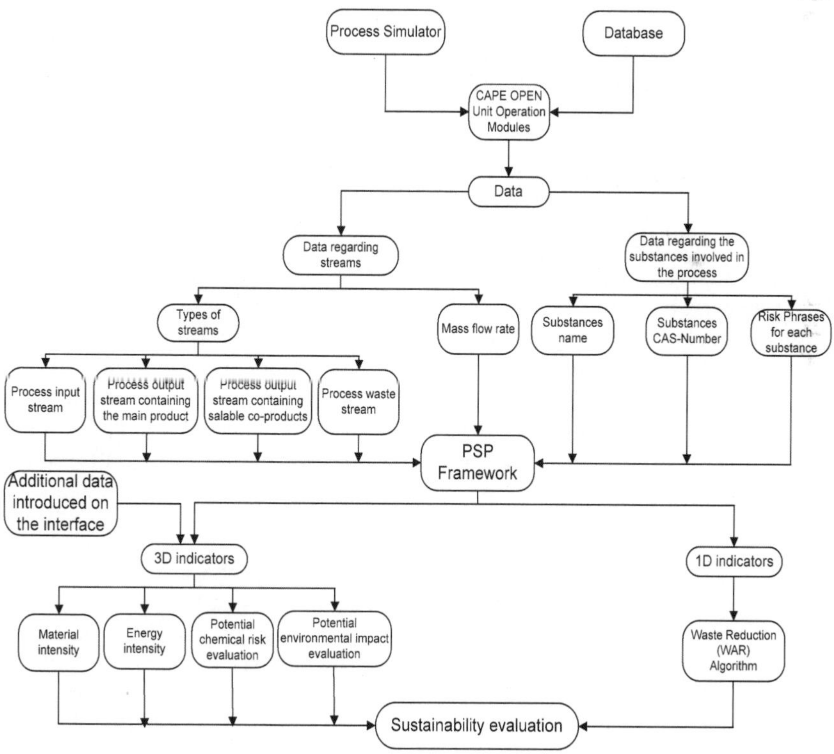

Figure 3: General schema of PSP framework development.

The two starting points for calculating the sustainability indicators are the process simulator and the database containing toxicity and risk phrases data. In a previous paper [8] we have developed methods for the estimation of toxicological properties using molecular modeling techniques. Using these methods, as well as resorting to experimental toxicological data, allowed us to generate a data base that can be used in PSP. The other issue, namely the interface to the process simulators, is addressed by developing CAPE OPEN compliant modules for accessing the data in the simulators.

The CAPE OPEN technology has been chosen since it is the only available inter platform standard in the process simulation sector. Four CAPE OPEN (CO) Modules have been developed (two CO Modules for 3D Indicators and two CO Modules for 1D Indicators).

These modules are inserted in the process flow-sheet, as any other unit operation, with the aim of extracting all the necessary data requested for further calculations (flow-rate quantities, flow-rate composition, energy demand of the process, toxicological data, risk phrases data). The CO modules in PSP are directly connected to the toxicological database (DB), containing all the necessary data for the evaluation of the indexes for about 4900 compounds.

The data obtained through the CO Modules are processed in the PSP software. Other information such as H_2O, O_2, N_2moles formed in the reaction; energy demand of the process; frequency of use and physical state of each substance involved are supplied by the user in the PSP interfaces. Details regarding the implementation of the CO modules, the development of the toxicological database, the PSP software and its applications are given by Toma [15].

4 PSP framework applications

Application of the PSP framework has been done in different contexts. Traditional chemical processes as well as innovative processes in the area of energy generation, environment protection, food processing and biotechnology have been analyzed (Toma [15], Fermeglia et al. [8]). The PSP framework, being Cape Open compliant, can be easily and directly used in any process simulator: it has been tested with PROII (Avensys-SimSci), Aspen+ (Aspentech) and CocoCofe (AmsterChem). In this paper we present an example of application of the PSP framework to a bioethanol production plant from sugar cane molasses.

4.1 Bioethanol production process: process simulation

Molasses differs from other feed stocks for alcohol production such as corn, millo and potatoes in that these plant products contain carbohydrate stored as starch. Indeed, these feedstocks must be pre-treated by enzymatic action to hydrolyze starch into fermentable sugars. In contrast, the carbohydrates in molasses are already in the form of sugars and need no pretreatment. The simplest form of sugar is glucose, which is fermented by normal distillery and brewery yeasts. Sucrose is the principal sugar contained in molasses and is readily fermentable either directly, or as its glucose and fructose components. Degradation with added water is referred to as hydrolysis. This addition of water should be taken into consideration when making any calculations of the amount of simple sugars (monosaccharides) produced from disaccharides.

In the production of cane sugar, the cane is crushed in a mill to squeeze out the juice. The juice is heated, clarified by filtration and the addition of lime (to remove cane fibers and sludge) and then evaporated to concentrate the sugar and crystallized. The syrup containing the crystals is then centrifuged to separate the crystals and the syrup residue (which still has a high content of sugar). The residue is referred to as "A molasses". It is evaporated and centrifuged again to recover more crystalline sugar; and the syrup residue is now referred to as "B molasses". The process may be repeated to yield more sugar and a "C molasses" as residue. Sugar mills normally evaporate and centrifuge a maximum of three times, but the number of treatments will depend on marketplace economics [16]. In addition to ethanol, yeast also produces other substances such as glycerol and succinic acid. Many distillers, therefore, use 95% of the 51.11% w/w figure as a realistic basis for calculating expected ethanol yields. For costing purposes, one can expect to obtain about 219.6 liters of ethanol from a ton of molasses at 46% sugars (TSAI).

Molasses fermentations may be conducted at 30–36°C, but if one is aiming for high final alcohol levels in the beer it is advisable to lower the temperature.

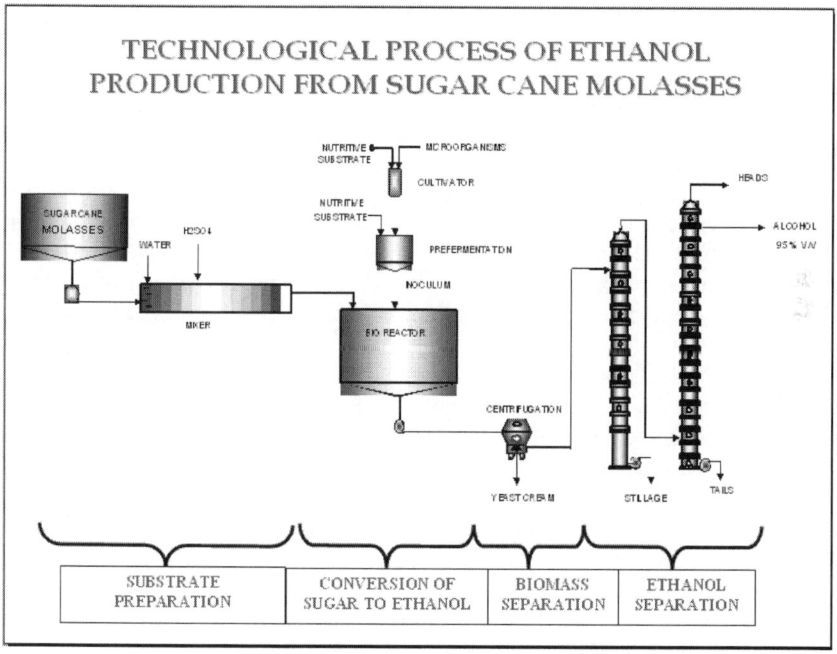

Figure 4: Bioethanol production from sugar cane molasses.

Molasses fermentations are generally much more rapid than grain fermentations and can be completed in half the time required for grain. This means that much more heat is produced per hour, which necessitates better cooling facilities than for grain fermentations. Some alcohol plants share one external heat exchanger between two fermenters.

The main problem in distillation of molasses is scale formation in the stripping column. A second problem, foaming in the distillation columns, is encountered with molasses. Foaming frequently limits the rate of distillation; and if uncontrolled it may pass from the stripping column into the rectifier and cause problems of proof, color, etc. Foaming is best controlled by continuous injection of a suitable antifoam into the mash feed line as this ensures that all the mash reaching the column is adequately dosed.

The Distillery "Paraiso", located in the central region of Cuba was selected for the simulation. The process is schematically illustrated in figure 4. The process is divided into the following steps: (i) Substrate Preparation, (ii) Conversion Sugar to ethanol (Yeast propagation and Alcoholic fermentation stage), (iii) Biomass Separation, (iv) Ethanol separation (Striping and Rectification section) [16].

The distillery has a capacity of 70 000 l of hydrated grade alcohol per day [16].

4.2 Bioethanol production process: sustainability analysis

The flow sheet of the simulation is shown in Figure 5. The diagram includes the substrate preparation, conversion sugar to ethanol (Yeast propagation and Alcoholic fermentation stage), biomass separation and ethanol separation (Striping and Rectification section).

Sensitivity studies for the process of interest have shown that the quantity of ethanol produced is sensitive to the temperature of the fermentation reactor, as anticipated in the description of the process [16]. For this reason the first case simulated corresponds to a temperature of the fermentation reactor of 46°C. In the second case the temperature of the fermentation reactor is lowered to 34°C. These two cases are examples of modification of the process conditions with no changes in the equipment schema. In the third case, the washing (final) column is not present, thus representing an example of process modification. The 3D indicators obtained using the PSP Framework are reported in Table 1.

CASE 1 shows the lowest MI and EI indicators. All three processes are continuous processes, so the frequency class for the substances present in the process is the same (permanent use; >6h/day) as well as the risk phrases, since there is no difference for the components involved in the process.

PEI has the same values for all three cases because the physical states of the substances involved in the process are the same and the receiving medium (air, water, soil) is also the same.

Therefore, due to the more favourable values of MI and EI, CASE 1 represents the best process condition in term of environmental impact among the cases considered.

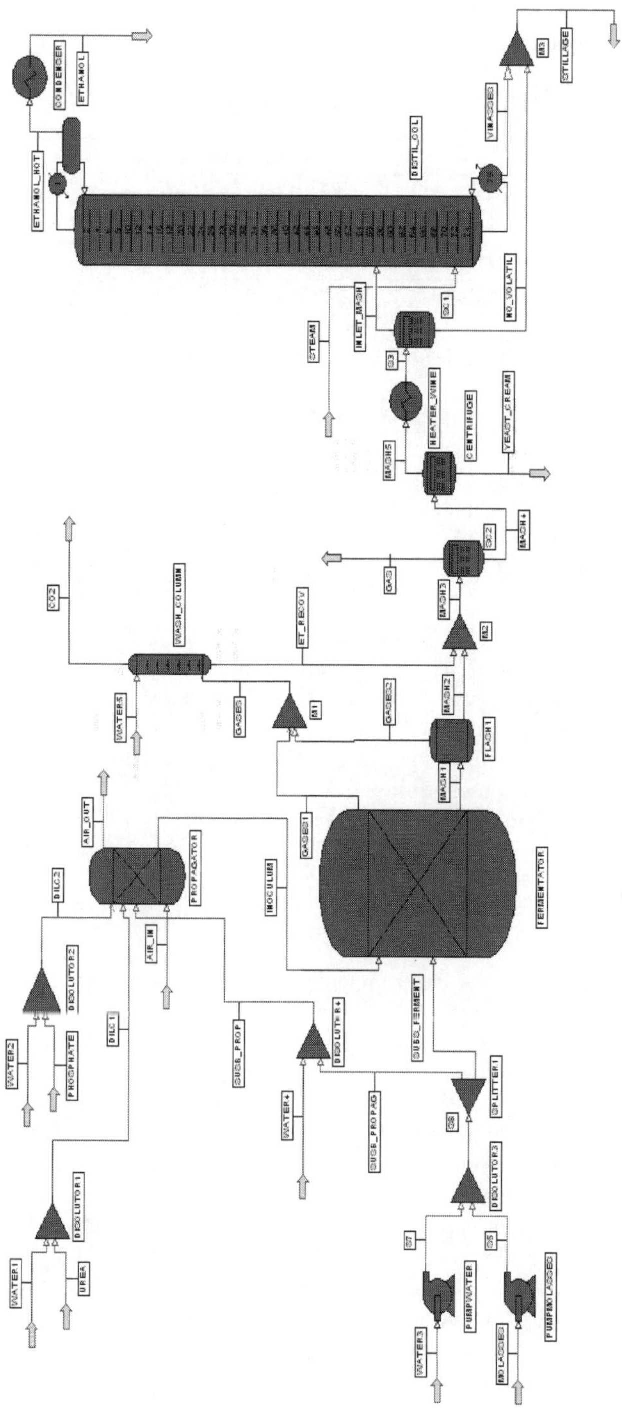

Figure 5: Flow sheet of the simulated process (PRO II).

For the same cases, WAR Algorithm, namely the 1D analysis, has been applied. The results obtained using 1D indicator, are reported in Table 2.

Table 1: 3D indicators for the different cases investigated.

CASE STUDY	3D INDICATORS			
	MI	EI	PCR	PEI
CASE 1	53125.1357	2.0566	355	8642.38
CASE 2	53129.9006	2.7069	355	8642.38
CASE 3	53364.8938	2.2137	355	8642.38

Table 2: 1D indicator (WAR algorithm) for the different cases investigated.

CASE STUDY	ENIVORONMENTAL INDEXES NAME			
	Iout (PEI/hr)	Iout_mp(PEI/kg)	Igen (PEI/hr)	Igen_mp (PEI/kg)
CASE 1	3100.47438	1.7722	-54.4912	-3.11E-02
CASE 2	3107.46802	1.7594	-47.4976	-2.69E-02
CASE 3	3123.40115	1.7902	-31.5645	-1.81E-02

CASE 3 results to have a high impact for the 3D and the 1D analysis. This is due to the absence of the scrubbing tower, and the consequent higher amount of gases emitted to the atmosphere In CASE 1 and 2 gases are recovered, thus lowering the impact of the process.

CASE 1 and 2 do not differ a lot, since the equipment are the same and the only difference is the reaction condition: CASE 2 needs a higher amount of heat at the reboiler, this justifying a higher energy consumption.

In summary, CASE 1 is the most favorable in terms of environmental impact.

5 Conclusions

A methodology to quantify the environmental impact of a chemical process is presented, discussed and applied to a process for the production of biofuels. The methodology, used in the early stages of project design, allows choosing the most environmentally friendly design among different process alternatives.

The software presented, the PSP Framework, contains the following items: (i) CAPE OPEN Modules ready to be used in any process simulator which provides CO Interfaces, (ii) a toxicological database, and (iii) various support program codes developed in Visual Basic.

The connection between the process simulator and the environmental evaluation tools has been done using a standard technique: CAPE OPEN. Four CO Modules, with different characteristics and aims have been developed in order to get automatically the necessary information from the process simulator. The modules have been tested successfully using different process simulators such as: PROII, Aspen Plus, COCO/COFE.

A relational database containing 4900 substances has been created and implemented and it is available in the PSP framework.

The ethanol production process from sugar cane molasses has been simulated using PROII process simulator. Three process alternatives are proposed and analyzed in order to choose the most environmentally friendly one.

The most environmentally friendly design is the first one, as it can be noticed from the results obtained using the 3D and 1D indicators of the PSP Framework.

In conclusion, the PSP framework proved to be a very useful decision support tool in the design phase of a process. Several different process alternatives, in terms of process conditions, process feeds and operative settings may be easily compared in terms of environmental impact. Using PSP it is possible to identify the best scenario that reduces the impact to the environment by using only 'in silico' calculations. The PSP application may lead to a new era of optimization in the process design industry and could be applied in developed, developing and emerging countries for a complete support to the global sustainability.

References

[1] Veleva V, Ellenbecker M. Indicators of sustainable production: framework and methodology. Journal of Cleaner Production. 2001; 9:519-549.

[2] Korevaar G. Sustainable Chemical Processes and Products- New Design Methodology and Design Tools. 2004. PhD Thesis- Technical University of Delft.

[3] Jenck JF, Agterberg F, Droescher M. Products and processes for a sustainable chemical indutry: a review of achievements and prospects. Green Chemistry.2004; 6: 544-556.

[4] Fermeglia M, Pricl S, Longo G. Molecular Modelling and Process Simulation: Real Possibilities and Challenges. Chemical and Biochemical Engineering Quarterly. 2003; 17: 69-79.

[5] "NJSSI" (2007). Available from: New Jersey Sustainable State Institute<http://njssi.org/about.asp?Level2ItemID=55> (last accessed October 2007)

[6] Sikdar SK. Sustainable Development and Sustainable Metrics. AIChE Journal. 2003; 49: 1928-1932.

[7] Martins AA., Mata TM, Costa CAV, Sikdar SK. Framework for Sustainability Metrics, Industrial and Engineering Chemical Research. 2007; 46: 2962 -2973.

[8] Fermeglia M., Longo G., Toma L., Computer aided design for sustainable industrial processes: specific tools and applications, AIChE Journal, 55: 1065-1078 (2009).

[9] Schwarz J, Beloff B, Beaver E. Use Sustainability Metrics to Guide Decision-Making. Environmental Protection. Available from: <http://www.bridgestos.org/CEP-July.pdf> (last accessed October 2007).

[10] Tanzil D, Ma G, Beloff B. Automating the Sustainability Metrics Approach. AIChESpring Meeting, 25-29 April 2004, New Orleans. Available from: BRIDGES to Sustainability Institute

<http://www.bridgestos.org/AIChESpring2004MetricsPPT.pdf> (last accessed October 2007).

[11] Vincent R, Bonthoux F, Mallet G, Iparraguirre JF, Rio S. Méthodologie d'Evalution Simplifiée du Risque Chimique: Un Outil d'Aide à la Décision. INRS Hyg. Secur. Travail. 2005; 195 .

[12] Cabezas H, Bare JC, Mallick SK. Pollution Prevention with Chemical Process Simulators: The Generalized Waste Reduction (WAR) Algorithm. Computers & Chemical Engineering.1997; 21: 305-310.

[13] Young DM, Cabezas H. Designing Sustainable processes with simulation: the waste reduction (WAR) algorithm. Computers and Chemical Engineering.1999; 23: 1477-1491.

[14] Young D, Scharp R, Cabezas H. The waste reduction (WAR) algorithm: environmental impacts, energy consumption, and engineering economics. Waste Management. 2000; 20: 605-615.

[15] Toma L. Computer Aided Design of Sustainable Industrial Processes. 2008. PhD Thesis- University of Padua, Italy, 2008.

[16] Domenech López F., Bioethanol production from sugar cane molassa: process simualtion and optimization, Internal report, ICS-UNIDO, Trieste, November 2005.

Tourism carrying capacity assessment and environment: the case of Crete

B. S. Tselentis[1], D. G. Prokopiou[1], Em. Gyalirakis[2] & D. Bouga[3]
[1]*Department of Maritime Studies, University of Piraeus, Greece*
[2]*Land Surveyor, Rural Planner-Engineer, Greece*
[3]*KOM 37-Architects, Greece*

Abstract

Crete is the largest island in Greece. It covers an area of 8,303 sq. km and has 170 km of beaches. Tourism supply in Crete includes more than 200,000 beds which serve more than 3.5 million international tourist arrivals per year.

Cretan environment, both natural and manmade, can sustain a polymorphic tourism product for the island. The physical and cultural environment consist of an extended coastline with many sandy beaches, picturesque cities and traditional villages, unique mountains, wetlands and lakes.

The environment plays a vital role for the sustainable development of the tourism industry and offers a unique tourist product for the island.

The objective of this paper is to evaluate the impact of tourism and other economic activities on the environment of Crete and to estimate the quality of tourism products and services offered by the island's infrastructure.

In our study we have used environmental and tourism indices to estimate the impact of tourism and other businesses in the environment of Crete, in an attempt to address the challenges and options, pertinent to formulate a novel sustainable tourist development proposal.

Keywords: tourism carrying capacity assessment, carrying capacity in tourism, study for tourism in Crete, Crete tourism and environment, tourism and environment.

1 Introduction

The World Tourism Organisation (WTO) proposes the following definition of carrying capacity: "The maximum number of people that may visit a tourist

doi:10.2495/ECO110161

destination at the same time, without causing destruction to the physical, economic, socio-cultural environment and an unacceptable decrease in the quality of visitors' satisfaction" [1]. For the Greek Islands, environment, both natural and manmade, plays a vital role for the sustainable development of the industry

Today, controlling tourist growth has become a central policy issue for the tourist trade [2], and it is noteworthy that carrying capacity assessment has become an important tool for facilitating planning and developing policy issues for the industry [3]. The purpose of this paper is to try and attain values for these indicators and integrate them into the planning process of the Greek tourist business [4].

Crete is the largest island of Greece and the fifth largest in the Mediterranean after Sicily, Sardinia, Cyprus and Corsica. It has an area of 8.162 square miles. Crete is the largest of the Greek islands, a long 260 km stretch with a width varying from 12 km to 60 km in its central part. It lies on the south of the Aegean Sea and its southern coast is on the Libyan Sea. Crete, with a population of 650,000 in year 2005, is one of the 13 regions into which Greece is divided. It forms the largest island in Greece and the second largest (after Cyprus) in the eastern Mediterranean. The island has an elongated shape: it spans 260 km (160 mi) from east to west, is 60 km (37 mi) at its widest point, and narrows to as little as 12 km (7.5 mi) (close to Ierapetra). Crete covers an area of 8,336 km^2 (3,219 sq mi), with a coastline of 1,046 km (650 mi); to the north, it broaches the Sea of Crete to the south, the Libyan Sea; in the west, the Myrtoan Sea, and toward the east, the Karpathion Sea. It lies approximately 160 km (99 mi) south of the Greek mainland [5].

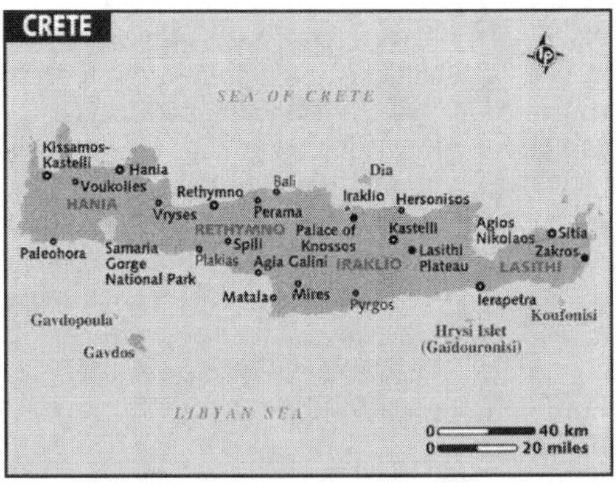

Figure 1: Map 1: Crete [6].

Crete has a unique tourist product that combines human and the natural environment, as tourism is an activity which cuts across conventional sectors in the economy. It requires inputs of an economic, social, cultural and environmental nature.

2 Population

The population of Crete is mainly concentrated in the six large towns of the island, to the capitals of the four prefectures, Chania, Iraklion, Agios Nickolaos and Rethymno and also to Ierapetra and Sitia. As far as the education is concerned, there are two universities and one polytechnic school which gather students from all over Greece. We must also point out the fact that in Crete region, there is also a factory blooming industries. The population of Crete is employed in trade, handicrafts, tourism and agricultural economy, like farming and live stock rising. Cretans who either live in the cities or the villages, all maintain the time honoured customs and manners of their tradition. Cretan traditional music and Cretan cuisine are defining their uniqueness.

Crete has the highest population growth in Greece, increasing by approximately 20%.

Table 1: Population history in Crete.

	Heraclion	Lasithi	Rethimno	Chania	Crete
1981	243,622	70,053	62,634	125,856	502,165
2001	292,489	76,319	81,936	50,387	601,131
Growth %	20.1	8.9	30.8	19.5	19.7

3 Ports and airports

Crete gathers more than 4 million tourists per year. There are many ways to reach Crete. One can get there by plane or by boat.

Ports: In Crete, there are six ports that connect Crete with the other islands and the mainland of Greece, in Heraklion, Chania, Agios Nickolaos, Kasteli Kissamou, Rethymnon and Sitia. These ports are connected with continental Greece as to some Greek islands.

Airports: Three airports are operating in Crete. Two of them are international (Heraklion and Chania) and one is national (Sitia). Thousands of charters flights

Table 2: Arrivals in Cretan Airports.

ARRIVALS IN CRETAN AIRPORTS						
	Chania International Airport		Heraclion International Airport		Sitia Airport	Total for Crete
	Arrivals - Domestic	Arrivals Abroad	Arrivals - Domestic	Arrivals Abroad	Arrivals - Greece	Arrivals Abroad
1978	153330		281469	231180		665979
1980	152482	5387	266170	386952		810991
1985	253494	31589	465721	734288	1884	1486976
1990	155157	173770	298012	926820	1992	1555751
1995	110777	334118	223649	1559204	1459	2229207
2000	254231	450258	494317	20078233	799	21277838
2005	193179	555647	448461	2003643	9032	3210003
2006	212376	660904	493760	2159452	11109	3537601
2007	248906	683669	537718	2162769	17666	3650728
2008	254659	671477	547484	2154010	20000	3647635
2009	280717	608693	553679	1952074	18916	3414079

reach Crete every year, especially during the summer season. Specifically, in Heraklion airport more than 2,000,000 tourists arrive with international flights and in Chania airport more than 1,200,000 tourists arrive every year.

Tourist seasonality in Crete concentrates most of the tourist demand from May to October. Tourism industry of the island is trying to extend the season with the support of winter time charter flights.

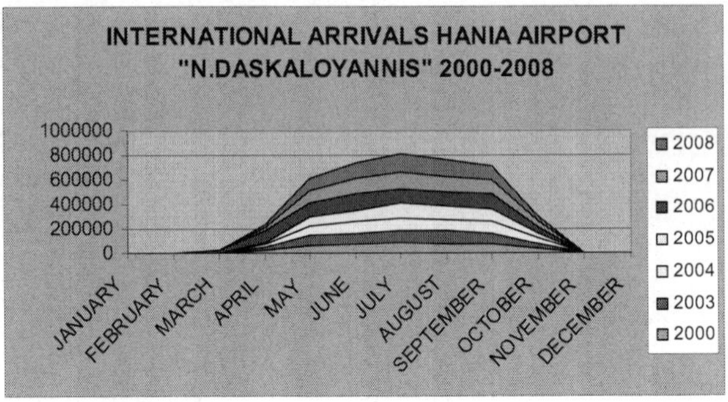

Figure 2: Arrivals in Chania Airport.

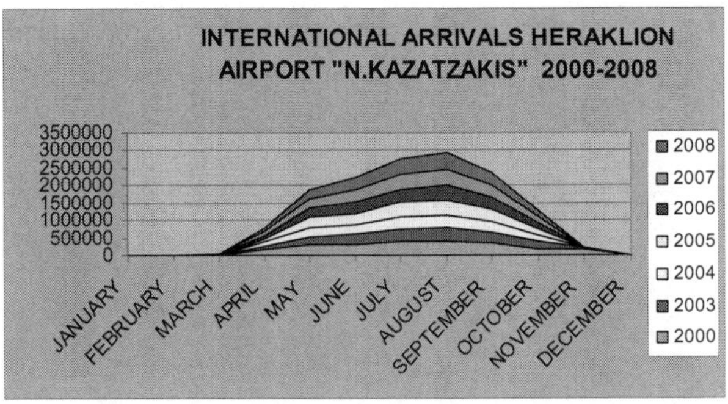

Figure 3: Arrivals in Heraclion Airport.

4 Tourist supply

Since the 1970s, mass tourism has provided Crete with increasing proportions of its GDP. Most of the physical development associated with package tourism has been concentrated on the northern coast east of Heraklion, Chania, Rethimno and Agios-Nikolaos [7]. Tourism in Crete has been as a development strategy, due to its potential for employment and income generation, infrastructural development and overall community [8]. Crete gathers more than 4 million tourists per year. Cretan tourist accommodation includes more than 200,000 beds in hotels, rooms to let, furnished apartments and traditional small hotels.

Table 3: Tourism capacity in Crete.

TOTAL TOURISM CAPACITY IN CRETE					
	2006	2010		2006	2010
PREFECTURE OF HERAKLION			**PREFECTURE OF CHANIA**		
HERAKLION (town)	2852	2646	CHANIA (town)	3353	3063
AGIA VARVARA	0	0	ACROTIRIOU	1062	997
ARCALOHORI	236	138	AN.SELINOU	19	52
ARCHANES	65	103	ARMENVN	336	403
ASTTEROUSION	151	13	VAMOU	511	606
VIANNOU	125	22	VOUKOLION	63	158
GAZI	10871	10916	GEORGIOUPOLIS	5403	5619
GOGROLAINI	0	36	EL VENIZELOU	0	0
GORTYNAS	101	79	THERISOU	63	56
GOUVES	9929	9319	INAHORIOU	0	0
EPISKOPI	0	0	KANTANOU	68	68
ZAROS	186	170	KERAMION	0	0
THRAPSANOU	0	0	KISSAMOU	892	1151
KASTELI		38	KYLYMVARIOU	1590	2128
KOFINA	0	40	KRYONERIDAS	15	46
KROUSONAS	180	0	MOUSOURON	101	101
MALIA	11478	11408	MYTHIMNIS	147	108
MIRES	97	59	NEAS KYDONIAS	10824	13295
NEA ALIKARNASOS	631	248	PELEKANOU	605	580
NIKOU KAZATZAKI	9	9	PLATANIA	8428	9610
ROUVA		0	SOYDA	99	110
TEMENOUS		0	SFAKION	339	353
TETRACHORIOU	21	21	FRE	0	0
TYLISSOU		62	ASI GVNIA	0	0
CHERSONISOS		28235	**PREFECTURE OF RETHYMNO**		
PREFECTURE OF LASITHI			RETHIMNO (town)	13287	13420
AGIOS NIKOLAOS	11954	12356	ANOGIA	50	50
IERAPETRA	4037	4238	ARKADIOU	4933	7272
ITANOU	239	275	GEROPOTAMOU	4933	5699
LAYKI		0	KOYLOYKONA	77	132
MAKRY GIALOS	582	1195	KOYRITON	0	3
NEAPOLI	30	30	LAMBIS	1472	1286
OROPEDIO LASITIOU	97	102	LAPPAION	12	12
			NIKOFOROU FOKA	304	335
			SYVRITOU		
GAVDOS		20	FOINIKA	1835	1950
			ZONIANA		

Crete is one of the most popular holiday destinations in Greece. Fifteen percent of all arrivals in Greece come through the city of Heraklion (port and airport), while charter journeys to Heraklion last year made up 20% of all charter flights in Greece. Overall, more than two million tourists visited Crete last year, and this increase in tourism is reflected on the number of hotel beds, rising by 53% in the period between 1986 and 1991, when the rest of Greece saw increases of only 25%. Today, the island's tourism infrastructure caters to all tastes, including a very wide range of accommodation; the island's facilities take in large luxury hotels with their complete facilities, swimming pools, sports and recreation, smaller family-owned apartments, camping facilities and others.

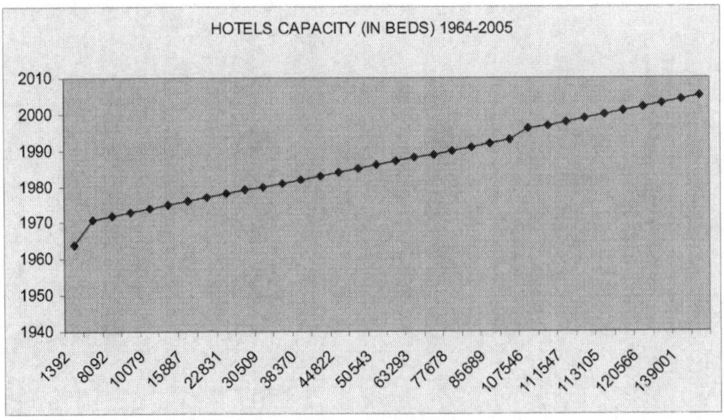

Figure 4: Hotel capacity history.

Today Crete has the infrastructure to offer its services even for athletic tourism. The combination of its numerous beaches and land will strong relief create an ideal environment [9] for the organizing of alternative tourism action, moreover, there are 69 diving centers.

Table 4: Alternative tourism enterprises, Crete [10].

ALTERNATIVE TOURISM ENTERPRISES			
HERACLION	LASITHI	CHANIA	RETHIMNO
11	3	4	8

5 Environmental indicators

The coastline is where the land meets the sea or ocean [11]. Tourist development is adding to already existing pressures in coastal areas. Population densities are increasing in the tourist coastal regions during seasonal peaks [12, 13].

There are specific touristic areas that are in more favor of tourist attention, that are settled close to Hania, Rethimnon, Iraklion, Agios Nikolaos and Sitia, the most popular cities of Crete. These cities are situated across the north coast of the island of Crete.

Furthermore, across the southern coast line, with the view of Libian Sea, there are plenty of hotels, most of them situated nearby Ierapetra – Makri Yalos, Timbaki – Messara's bay, southern Rethimno, Sfakia, and the whole west coast line of Crete (Pelekanos and Inaschorio and Gramvousa area).

The coast and its adjacent areas on and off shore is an important part of a local ecosystem as the mixture of fresh water and salt water in estuaries provides many nutrients for marine life. Salt marshes and beaches also support a diversity of plants, animals, and insects crucial to the food chain.

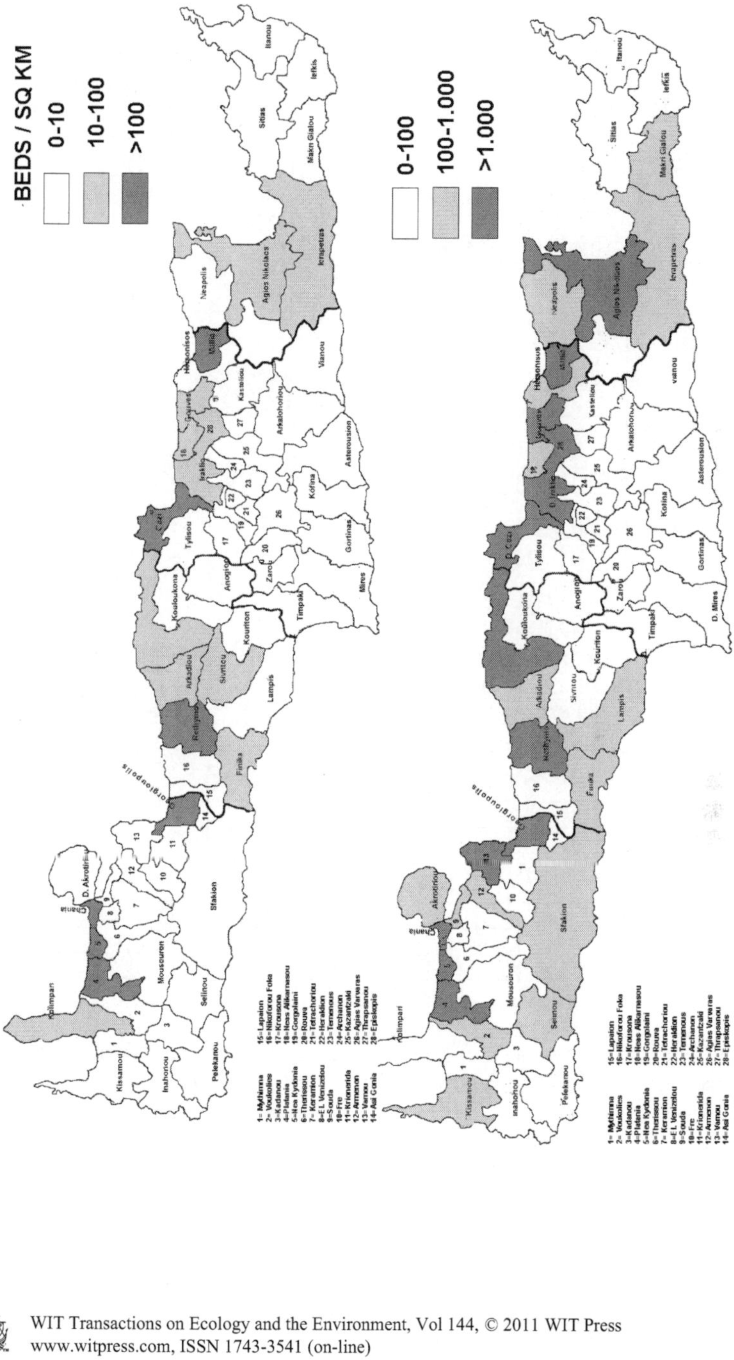

Figure 5: Map 2 and 3. Beds per sq km and beds per km of beach.

Table 5: Beach impact factor and beds per kilometers.

BEACH IMPACT FACTOR AMD BEDS PER SQUARE KILOMETRES					
	BEDS/ SQ KM	BEDS/ KM OF BEACH		BEDS/ SQ KM	BEDS/ M OF BEACH
PREFECTURE OF HERAKLION			**PREFECTURE OF CHANIA**		
HERAKLION(town)	24,26944	2205	CHANIA (town)	243,7918	3063
AGIA VARVARA			ACROTIRIOU	8,850893	433,4783
ARCALOHORI	0,580835		AN.SELINOU	0,380799	104
ARCHANES	3,268181		ARMENVN	7,288577	183,1818
ASTTEROUSION	0,063907	13	VAMOU	9,040054	1010
VIANNOU	0,099305	2,650602	VOUKOLION	2,049021	790
GAZI	113,9338	2227,755	GEORGIOUPOLIS	105,032	1123,8
GOGROLAINI	0,868349		EL VENIZELOU	0	
GORTYNAS	0,499021	34,34783	THERISOU	0,755674	
GOUVES	98,13296	2329,75	INAHORIOU	0	0
EPISKOPI	0		KANTANOU	0,930474	
ZAROS	2,367589		KERAMION	0	
THRAPSANOU	0		KISSAMOU	7,72307	230,2
KASTELI	0,308129		KYLYMVARIOU	14,18723	
KOFINA	0,276304	40	KRYONERIDAS	0,680866	
KROUSONAS	0		MOUSOURON	0,526744	
MALIA	187,8788	2852	MYTHIMNIS	1,954114	27
MIRES	0,324381	13,11111	NEAS KYDONIAS	619,6113	2215,833
NEA ALIKARNASOS	15,40564	496	PELEKANOU	3,481016	98,30508
NIKOU KAZATZAKI	0,088062		PLATANIA	243,7918	3063
ROUVA	0		SOYDA	8,850893	433,4783
TEMENOUS	0		SFAKION	0,380799	104
TETRACHORIOU	0,980255		FRE	0	0
TYLISSOU	0,473051		ASI GVNIA	0	0
CHERSONISOS	9,222133	128,2301	**PREFECTURE OF RETHYMNO**		
PREFECTURE OF LASITHI			RETHIMNO (town)	106,1239	2631,373
AGIOS NIKOLAOS	38,87564	3530,286	ANOGIA	0,442353	
IERAPETRA	10,73526	365,3448	ARKADIOU	59,10898	909
ITANOU	1,393068	45,83333	GEROPOTAMOU	29,28722	5699
LAYKI	0	0	KOYLOYKONA	0,958	
MAKRY GIALOS	7,504914	265,5556	KOYRITON	0,023859	
NEAPOLI	0,154751	150	LAMBIS	5,823326	347,5676
OROPEDIO LASITIOU	0,78476		LAPPAION	0,234687	1,875
			NIKOFOROU FOKA	3,502682	
			SYVRITOU	88,55339	
GAVDOS	1,115698		FOINIKA	14,09622	609,375
			ZONIANA		

With this indicator we analyse the pressures facing the coastal environment, as they describe the concentration of people visiting and using the facilities of the coastal area, and especially beaches. From the data presented in Table 5, Gazi 2227, Gouves 2329, 75 with tourists per km as Platanias and Nea Kydonia 2215 and Rethimno 2631 as Agios Nicolaos 3530 and Geropotamos 5699 concentrates the most serious pressures at the coasts.

Table 6: Shore and marine pollution in Heralcion [14].

	Illegal buildings	Hotels (urban waste)	Oil press (residuals)	Ships (oil residuals)	Urban waste
1997	7	1	43	5	
1998	4	2	25	1	
1999	9	7	29	0	
2000	8	15	19	7	
2001	14	6	18	6	
2002	8	12	22	4	
2003	18	6	35	1	
2004	14	9	11	5	
2005	25	7	19	5	
2006	9	5	19	2	
2007	34	4	8	4	2
2008	30	2	3	8	5

Urban waste management (solid and liquid) on Crete is characterized by lack of efficiency regarding the waste treatment in the small cities and villages. Only the large urban centres fulfil the basic requirements of modern waste management installations. It is interesting to note that other smaller settlements do not even have a complete urban waste collection network.

Table 7: Waste management in Crete.

(UC= under construction)

PREFECTURE OF HERAKLION		PREFECTURE OF CHANIA	
HERAKLION	100%	CHANIA	97%
AGIA VARVARA	UC	AKROTIRIO	NO
ARKAOLOCHORI	NO	ANATOLIKO SELINO	UC
ARCHANES	100%	ARMENON	50%
ASTTEROUSION	NO	VAMOY	NO
VIANNOU	UC	VOUKOLION	NO
GAZI	IS WORKING	GEORGIOUPOLI	NO
GOGROLAINI	UC	ELEFTHRIOU VENIZELOY	NO
GORTYNAS	NO	THERISOU	100%
GOUVES	UC	INAHORIOU	UC
EPISKOPI	50-%	KANTANOU	UC
ZAROS	NO	KERAMION	NO
THRAPSANOU	80%	KISAMOU	UC
KASTELI	25%	KOLYMVARIOU	70%
KOFINA	25%	KRYONERIDA	UC
KROUSONAS	NO	MOUSOURON	NO
MALIA	IS WORKING	MYTHIMNIS	NO
MIRES	NO	NEAS KYDONIAS	100%
NEA ALIKARNASOS	80%	PELEKANOU	NO
NIKOU KAZATZAKI	100%	PLATANIA	70%
ROUVA	NO	SOUDA	100 %
TEMENOUS	NO	SFAKION	80%

Table 7: Continued.

(UC= under construction)

PREFECTURE OF HERAKLION		PREFECTURE OF CHANIA	
TETRACHORIOU	UC	FRE	
TYLISSOU	UC	ASI GONIA	NO
TYMPAKI	50%	**GAVDOS ISLAND**	NO
CHERSONISOS	100%	**PREFECTURE OF RETHIMNO**	
PREFECTURE OF LASITHI		RETHIMNO	100%
AGIOS NIKOLAOS	90%	ANOGIA	90%
IERAPETRA	75%	ARKADIOU	100%
ITANOU	70%	GEROPOTAMOU	90%
LAYKI	45-50%	KOYLOYKONA	UC
MAKRY GIALOS	NO	KOYRITON	NO
NEAPOLI	NO	LAMBIS	UC
OROPEDIO LASITIOU	50%	LAPPAION	15%
SITIA	80%	NIKOFOROU FOKA	NO
VRACHASI	NO	SYVRITOU	NO

Table 8: Shore and marine pollution in Agios Nicolaos [15].

	Illegal buildings	Oil press (residuals)	Ships (oil residuals)	Urban waste
1998	7	1		
1999	5	3		
2000	4	1		
2001	16	4		
2002	1			
2003	9			
2004	10			
2005				
2006	15	1		2
2007	4		1	
2008	12		1	

Table 9: Shore and marine pollution in Rethimno [16].

	Illegal buildings	Oil press (residuals)	Ships (oil residuals)	Urban waste
1998	.2			
1999				
2000		1		
2001	.6	6		.1
2002	18	7		.1
2003	.4	2		12
2004	.8		1	
2005	54	7		.8
2006	.4			.2
2007	.5			.1

6 Tourism impacts

Impacts of tourism in coastal areas are: urban tourism enterprises, improper solid waste and garbage treatment, reduction of water resources, loss of free space in the towns,

Tourism can generate both positive and environmental impacts depending on how well development is planned and controlled. The principal impacts are outlined below [17]:

6.1 Negative impacts

Water pollution, air pollution, noise pollution, visual pollution, overcrowding and congestion, land use problems, environmental hazards, landslides, flooding and sedimentation of rivers, damage to historic and archaeological sites, improper waste disposal

6.2 Positive impacts

Conservation of important natural areas, conservation of important historic sites, improvement of environmental quality, enhancement of the environment, improvement of infrastructural, tourism carrying capacity assessment is the most important analytical technique to estimate the environmental pressures of uncontrollable tourist development.

7 Proposals

Tourist development indicators for Crete lead us to conclude that the tourist industry should aim at extending the tourist season to include more months, attain occupancy well over 50% for April, October, and probably more visitors with varied interests and expectations of the island.

Organization for the protection and development of environmentally sensitive areas. Due to the increased tourist demand, it is suggested that all the areas with environmental interests must be protected by a special organization [18] that, does not as yet exist now in Crete except the National Park of Samaria [19]. Natura 2000 Network areas are a step in the right direction, but must be supported with effective management schemes [20].

Extension of tourist services for the winter season months by developing and marketing novel tourist activities particular for the island (mountaineering, cultural events, conferences etc).

Completion of urban waste treatment plants and network, appropriate garbage handling (urban garbage treatment - olive oil press residuals management)

Restriction of illegal buildings at cities and coasts

Control of hotels' urban waste management systems

This concept of linking visitors with culture, nature and the environment in a harmonious way is not a new idea, but one that is now viewed on a global scale. Long-term, successful community involvement has preserved many popular rural tourist destinations such as the wine regions of Europe and the United States. Many rural communities have acted on instinct, rather than governmental directives or support, and usually with enough individual investment to achieve results [21].

Alternative tourist development.- Planning is conceptually related to sustainable development [22]. It includes approaches to deal with development and economic options, to prevent environmental damage and to involve public and stakeholders in decision-making processes. It is proposed that serious efforts have to be made in the direction of formulating viable policies and developing tools for effective implementation and control [23] as till now alternative tourism is not massively developed in Crete.

Acknowledgements

We would like to thank for their help and collaboration: Antonis Gialelakis, Neapoly, Lasithi-Crete, Dimitra Tselou and Maria Vasilaki economists, Spyros Gavanozis Hellenic Civil Aviation Authority, Greek national Tourist Organization – Heraclion Crete Branch as all the municipalities of the island.

References

[1] UNEP/MAP/PAP, 1997

[2] Coccosis H and Mexa A, ‘The challenge of Tourism Carrying Capacity Assessment’, Ashgate 2004

[3] Fernando J Garrigós Simón, Y. Narangajavana, and D. Palacios-Marqués. ‘Carrying capacity in the tourism industry’: a case study of Hengistbury Head’

[4] Zannou V, ‘Guide of Socio-Economic Studies for the Integrated Management of the Water Environment’ 1999

[5] http://en.wikipedia.org/wiki/Crete 11-12-10

[6] http://www.lonelyplanet.com/maps/europe/greece/crete/

[7] http://en.wikipedia.org/wiki/Crete 11-12-10

[8] Andriotis K., Local Authorities in Crete and the Development of Tourism-Kiousis 1984

[9] Tsokas S. , “Region of Crete”, Heraclion Crete Hotel Directory- Kalofolias publications 2005

[10] Cretan Union of Alternative Tourism

[11] The American Heritage Dictionary of the English Language: Fourth Edition. 2000. http://www.bartleby.com/61/43/C0434300.html. Retrieved 2008-12-11.

[12] Trumbic I, Tourism carrying Capacity assessment in the Mediterranean Coastal Tourist Destinations, 14 Biennial Coastal Zone Conference, New Orleans Louisiana, July 2005

[13] http://en.wikipedia.org/wiki/Coastal_zone#cite_note-0 11-12-10

[14] Heraclion Central Port Authority – Hellenic Coast Guard

[15] Agios Nicolaos Port Authority – Hellenic Coast Guard

[16] Rethimno Port Authority – Hellenic Coast Guard
[17] Lickorich L- Jenkins C, An introduction to Tourism, Butterworth Heinemann, Oxford 2000
[18] Prokopiou DG and Tselentis BS 'Proposals for sustainable development and Environmental protection for the island of Rhodes', Rhodes 2003 (in Greek)
[19] Papayanis T, 'Tourism carrying capacity in areas of ecological importance'
[20] Rose Nikki, Sustainable Tourism in Crete, excerpts from an article first printed in Stigmes Magazine, Crete
[21] Fidelis T., "Integrating environmental issues into the Portuguese planning system- 10 years of emerging challenges and persistent problems" Conference on sustainable planning and development Bologna 2005 , Wessex Institute of Technology
[22] Prokopiou DG -Tselentis BS Bousbouras D. and Toanoglou M> "Environmental impacts caused by the tourist industry in Elafonisos Island and the Neapoli district, Greece" ECOSUD 2007,Sixth International Conference on Ecosystems and Sustainable Development, Organised by: Wessex Institute of Technology, UK -The University of Coimbra, Portugal-The University of Siena, 4–6 September 2007 Coimbra, Portugal

An alternative tourism model for sustainable development in Los Cabos, Baja California Sur, Mexico

O. Arizpe & A. Gámez
Autonomous University of Southern Baja California, USA

Abstract

Baja California Sur (BCS) is the largest coastal state in Mexico and helps shape the Gulf of California. This northwestern region has a historical, cultural, and ecologic relevance that has recently been reassessed by the growing importance of tourism-related activities. Among them, eco-tourism has been acknowledged to have comparative advantage as a source of economic, environmental and social welfare, provided a process of planning, evaluation, control, and continuous adjustments is put in place. Yet as in the rest of the state territory the municipality of Los Cabos, BCS, Mexico, the most important tourist destination, has inadequate development plans. In the context of a high population and tourist pressure, the area has become extremely vulnerable. In order to counterbalance this trend and generate tools for decision makers, we characterized environmental, social and economic components in the Los Cabos municipality, emphasizing the tourism sector; public participation was included through community and government level workshops. A diagnosis was concluded, which included models and maps of economic aptitude or capacity for each micro region. As proven by our analysis, sustainable tourism could be a means for using natural and social resources in a profitable way. Decision makers and the community of Los Cabos will, nonetheless, face important challenges to strengthen the type of economic activities sustainable tourism involve versus the trend to traditional mass tourism, to secure public access to coasts and beaches, to protect landscape quality and, in general, to manage growth in the region.
Keywords: alternative tourism model, sustainable development, Los Cabos, Mexico.

doi:10.2495/ECO110171

1 Introduction

Mexico is one of the five most mega diverse countries in the world and is placed among the ten top tourist destinations in the planet. The Los Cabos region, in Baja California Sur, has a natural disposition towards tourism. The landscape is shaped by a mixture of mountains, desert and beaches; the diverse terrestrial and marine flora and fauna provide with a distinctive attraction; the Gulf of California, which bathes the state's oriental side, is deemed as the biggest aquarium of the world; and it all mixes to make Los Cabos an appealing site for visitors.

The municipality of Los Cabos, the southernmost portion of the Baja California Sur state, has become one of the fastest growing tourist destinations in Mexico. It has reached population growth rates in an average of 10% during the last decade. Tourism has mainly developed along the San José del Cabo-Cabo San Lucas corridor, and has rested on a high income tourism market that feeds on visitors from the United States and which relies on high level infrastructure to attend to this sector (Urciaga [1], SEMARNAT [2]).

An example that illustrates the importance of tourism for the economy of Los Cabos is the outstanding number of visitors it receives in comparison to the rest of the state. Beginning at the end of the 1980s and continuing through 2008, Los Cabos received 70 per cent of the 1.7 million tourists that visited Baja California Sur (Gobierno de BCS [3]). In contrast to other municipalities, its affluence is principally composed of foreign tourism. These visitors are generally 80 percent of the total visitors to the locale and reached 91 percent in 2008, which shows a long lasting trend in the region (CEI [4]). Similar to other parts of the world, growth is focused on traditional tourism, sometimes called high density tourism. The importance of growth in this activity, both at the local and international levels, has also produced its diversification. At the world level this has created an increasingly intense current in the development of tourist poles, fundamentally characterized by the search for unconventional forms in places where the human impact is minimal, and that allow for the use of conservation measures or intelligent uses of the ecosystem. This has given rise to what is called alternative tourism. Wearing and Neil [5] define this type of tourism as a series of modalities with objectives that follow from natural, social and communitarian values and that permit the enjoyment of positive and share experiences for both the hosts and the visitors. This form of tourism includes ecotourism (recreational activities associated with appreciation and encounter of nature), adventure tourism (recreational activities and sports associated with the challenges imposed by nature), and rural tourism (activities that share and interact with rural communities in all of their daily life, including social, cultural and productive endeavors).

In the Los Cabos region, particularly in the last decade, it has been repeatedly stated in different forums, that the tourism development model that is traditionally followed in the municipality and is based on sun, beaches, and high density, cannot continue over the long run and requires the study of more sustainable alternatives for development (Angeles and Gámez [6], Gámez [7]). Focus on the concentration of tourist activities and the resulting social and environmental effects was the motive for the signing of an agreement in February, 2007, between the city of Los Cabos and the Autonomous University of Baja California Sur (UABCS) calling for the formulation and adoption of a Municipal Ecology Ordinance. Built on sectorally based workshops and surveys of the public and key actors, it was noted that the socioeconomic development of the municipality should be grounded in harmony with the natural environment, and should promote scientific knowledge about the integration of social, economic and natural process in order to achieve a higher quality of life for everyone (Arizpe *et al.* [8]).

2 Development of the model

The proposed model of tourism generated in that study (Arizpe *et al.* [8]) included the following indices, indicators, and weights:

$$IT = ANAT + ACULT + SERV + INFRA$$

The Tourism index (IT) is made up of the indexes for natural attractions (ANAT), cultural attractions (ACULT), services (SERV) and infrastructure (INFRA). In order to implement weights, considerations were given to the results of the survey of the tourism sector, consultations and workshops with public participation, Delphi surveys of experts, and the existing research on the subject (Arizpe *et al.* [8]).

For traditional sun and beach tourism, the greatest weight was given to infrastructure, followed by services. Minor importance was given to natural attractions and the smallest weight was placed on cultural attractions.

$$IT = ANAT\ (0.25) + ACULT\ (0.15) + SERV\ (0.3) + INFRA\ (0.3)$$

Given the limits of traditional sun and beach tourism, a model measuring the aptitude for alternative tourism was developed using the same methodology. As will be observed, the model's components are the same but the weights are very different from the traditional tourism model. The most important value was for natural attractions (ANAT), followed by cultural attractions (ACULT), services (SERV) and infrastructure (INFRA). The following equation shows the standardized values for the alternative tourism model, and the break down of the components and values for each of the indexes:

$$ITALT = ANAT\ (4) + ACULT\ (2) + SERV\ (1) + INFRA\ (1)$$

$$\text{INDEX OF NATURAL ATTRACTIONS (ANAT)} = ANP\ (0.11) + AV\ (0.11) + AF\ (0.11) + PL\ (0.45) + LAG/BA\ (0.22)$$

ANP = Presence/absence of natural protected areas

AV = Presence/absence of natural vegetation attractions

AF = Presence/absence of wild animal attractions

PL = Number of beaches

LAG/BA = Presence/absence of Lagoons and bays

INDEX OF CULTURAL ATTRACTIONS (ACUL) = ARQ (0.33) + HIST (0.33) + PAL (0.33)

ARQ = Number of archaeological sites

HIST = Number of historical sites

PAL = Number of paleontological sites

INDEX OF TOURISM SERVICES (SERV) = SCOST (0.665) + STERR (0.335)

COASTAL TOURISM SERVICES (SCOST) = PORT + NAUT + BU + PDEP + FMAR + ZOSU

PORT = Number of port services

NAUT = Number of nautical services

BU = Number of diving services

PDEP = Number of sport fishing services

FMAR = Number of marine life viewing sites

ZOSU = Number of surfing sites

LAND-BASED TOURISM SERVICES = ACAMP + TALT

ACAMP = Number of camping services

TALT = Number of alternative tourism services

INDEX OF TOURISM INFRASTRUCTURE (INFRA) = ICOM (0.4) + IHOT (0.6)

INDEX OF COMMUNICATION ROUTES (ICOM) = PAV (0.3) + TERR (0.1) + AINT (0.3) + ANAC (0.2) + APIST (0.1)

PAV = Kilometers of paved highways

TERR = Kilometers of dirt roads

AINT = Presence/absence of international airports

ANAC = Presence/absence of national airport

APIST = Presence/absence of landing strips

INDEX OF LODGING INFRASTRUCTURE (IHOT) = CHOT

CHOT = Number of hotel rooms

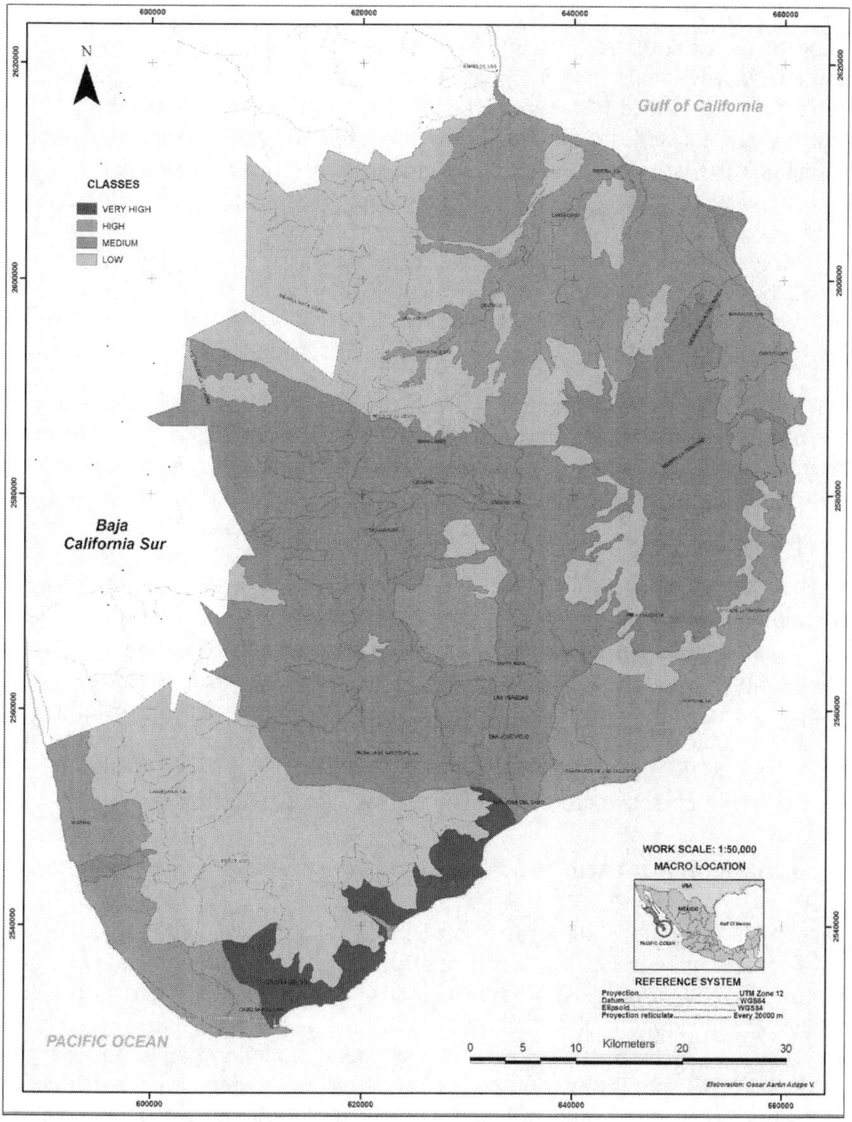

Figure 1: Extent of traditional tourism in the Municipality of Los Cabos, Mexico.

3 Results

3.1 Traditional tourism

With this model we estimated the aptitude of the high density tourism sector for Los Cabos, as shown in Figure 1. The tourism corridor Cabo San Lucas-San José del Cabo has a "very high" aptitude. The coastal zone has a "high" aptitude and some of the Environmental Units have a "medium" aptitude. The "very high" aptitude zones include only 3.1 percent, while the "high" aptitude includes 25.7 percent of the surface of the municipality. The zones of "medium" aptitude for tourism are located in the environmental units of the Sierra La Trinidad mountains in the central part of the municipality. The northern parts of the municipality, consisting of the Sierra La Laguna mountains, the environs of Sierra La Laguna, and the southwest away from the coast, have a "low" aptitude. In addition, we observed that 28.8 percent of the surface of the municipality of Los Cabos is capable of high density tourism.

3.2 Alternative tourism

Our results show that alternative tourism generates very different outcomes in the evaluation of each micro-region's aptitude. The regions with the greatest aptitude or capacity are situated in the north of the municipality. First, in the East Cape region and principally in the coastal zone of the Cabo Pulmo National Park, also in the regions of the channel of the San Dionisio-La Zorra arroyo, located in the northeast of the municipality, and finally, the Biosphere Reserve of the Sierra La Laguna and its southern reaches, which were not important areas for traditional tourism.

This strategy and methodology can be used to characterize and map all of the micro-regions in areas of interest in terms of their suitability for each type of tourism. The key is to follow the detailed procedures for evaluating and weighting precisely the components, indicators, and indexes. This methodology can be used to direct the planning and development of sustainable tourist activities with objective criteria and a formal quantitative technique.

4 Final considerations

Faced with the exponential growth of the tourism sector and related real estate development in Baja California Sur and specifically in the municipality of Los Cabos, the challenge arose to situate tourist centers in a manner that would secure a sustainable development. In this way, one understands those places with an orderly growth record, where development planning points toward the conservation of success over the long run and where the population experiences growth in its standard of living. Unfortunately, this has not occurred in any of the tourist centers of the nation, nor in those of the state, which puts at risk the continuation of growth in tourism in places that are vulnerable and lack the ability to maintain the quality of the services and natural resources that visitors expect.

Considering the obvious demonstrations of visual, economic, and social disequilibria that impose a heavy weight on governments responsible for providing services and public assistance, urban planning errors continue to be committed. These errors permit over construction, beaches are allowed to be

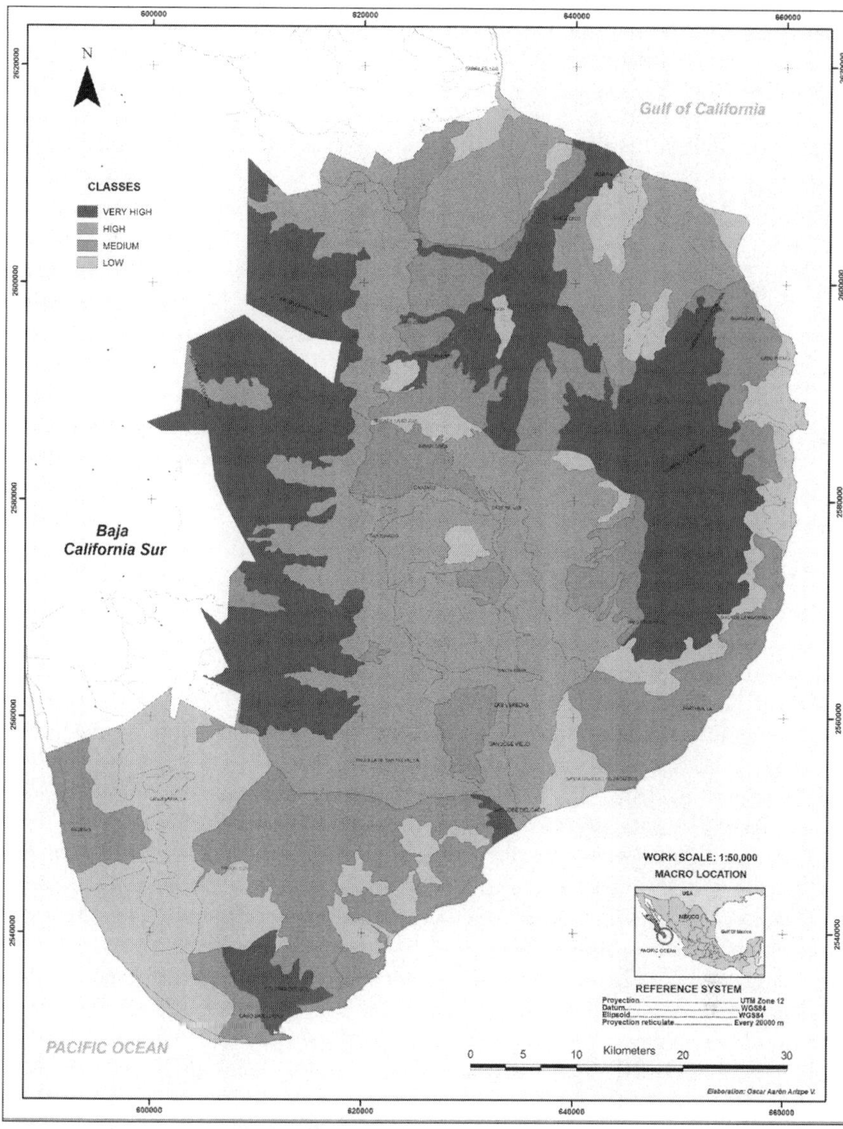

Figure 2: Suitability of alternative tourism in the municipality of Los Cabos, Baja California Sur.

become polluted while ecosystems that are emblematic are destroyed, towns become saturated with traffic, that overcrowding and the appearance of zones of misery continue to undermine the urban image, creating not only disincentives for tourists but also a deep sense of negative well-being in the resident population.

It is an unavoidable challenge for society, government, and the private sector to change this situation, so that the type and rhythm of tourism growth will not remain indefinitely. Therefore, it is a social responsibility to anticipate the future of the region with an effective form of planning. In this sense it is necessary to promote models in which not all the region is susceptible to the same type of tourism activity or with the same intensity. The development and application of the idea of the present and future suitability of tourism in the different locations of the municipality (and other regions of the state as well) will permit it to utilize the elements of anticipation, planning, and adaptation of the tourism sector to the swings in the international economy and the carrying capacity of the ecosystems.

References

[1] Urciaga García, J., Turismo alternativo. Una estrategia de desarrollo del espacio rural para Baja California Sur, *Prospectiva Económica*, **3 (5),** Julio-diciembre, pp. 177-197, 2004.

[2] SEMARNAT. Secretaría de Medio Ambiente y Recursos Naturales, *Indicadores básicos del desempeño ambiental de México: 2005,* PNUD-SEMARNAT: México, 2005.

[3] Gobierno del Estado de Baja California Sur, *IV Informe de Gobierno del Ing. Narciso Agúndez Montaño. Anexo Gráfico y Estadístico*, Gobierno del Estado de BCS: La Paz, BCS, 2009.

[4] CEI. Centro Estatal de Información, *Cuaderno de Datos Básicos 2006*, Gobierno del Estado de Baja California Sur, Secretaría de Promoción y Desarrollo Económico: La Paz, BCS, 2006.

[5] Wearing, S. Neil, J., *Ecoturismo. Impacto, tendencias y posibilidades*, Síntesis: Madrid, 2000.

[6] Angeles, M. & Gámez, A., Globalización y Desarrollo Regional. El Caso del Sector Turismo en Baja California Sur. *Globalización y regionalismo: economía y sustentabilidad,* eds. A. Ivanova & A. Guillén, Porrúa: México, DF, pp. 463-484, 2008.

[7] Gámez, A. E., El crecimiento del sector turismo en Baja California Sur. *Del saqueo a la conservación. Historia ambiental contemporánea de Baja California sur, 1940-2003*, eds. M. Cariño & M. Monteforte, UABCS-SEMARNAT-INE-CONACYT: La Paz, BCS, pp. 365-388, 2008.

[8] Arizpe, O., Fermán, J., Rivera, R., Ramírez. J. & Rodríguez, R., Evaluation and Land Use Planning Process of a High Population Growth Rate Municipality: Los Cabos, Mexico, *WIT Transactions on Ecology and Environment V*, pp. 87-95, 2008.

The importance of urban sustainable development in a flood threatened area: the case of the northern coast of El Campello (Spain)

J. Valdés-Abellán[1], A. J. Tenza-Abril[1], J. M. Ortega[1] & E. Sánchez-Camacho[2]
[1]*Departament d'Enginyeria de la Construcció, Obres Públiques e Infraestructura Urbana, Universitat d'Alacant, Spain*
[2]*Reticulares Alicante, S.L. Alacant, Spain*

Abstract

The southeast coast of the Iberian Peninsula (Spain and Portugal), and especially the Costa Blanca, is a very important tourist destination. Tourism has produced intense urban development, and in most cases, buildings and urban zones have occupied an important part of ravine basins. One of the most important climate features of this geographic area is a very intense and localized rainfall. As a consequence of these rainfalls, water levels in ravines are high.

In this work, in order to study the relationship between unsustainable urban development and flood threats, a coastal zone situated in the northern coast of El Campello (Alicante province, Spain) was chosen. Eight ravines flow into the Mediterranean Sea in this area. For each ravine, turnoff thresholds and water levels estimation were calculated for two situations: the first one is the present situation and the second one is a hypothetical initial situation, before the touristic development without urban zones in these basins.

Finally, there is a high decrease of turnoff thresholds and a rise of water levels in the studied basins. Then, it could be concluded that an unsustainable urban development produces an increase of flood threat in the particular case of the Costa Blanca, and in general in the southeast coast of Spain, due to their geographic and climate features.

Keywords: unsustainable urban development, turnoff thresholds, water levels, flood threaten, rainfall, basins, Mediterranean.

doi:10.2495/ECO110181

1 Introduction

Nowadays, floods are one of the critical hazards that human beings have to confront. These episodes constitute some of the most severe natural disasters, causing damage to property and infrastructures, often involving the loss of human lives [1]. This statement is particularly true in certain regions, such as the southeast of Spain, due to the combination of a higher frequency of occurrence of these phenomena, inherent in Mediterranean climates, together with an increased vulnerability, as a consequence of important urban development that has taken place associated with the tourist activity [2].

This study is focused on Alicante (Spain), where downpours are extremely heterogeneous [3]. Precipitation totals per storm range from 15–20 mm to 100 mm can occasionally exceed this threshold. Normally rainfall totals oscillate between 30 and 70 mm [4].

El Campello has many basins that share many characteristics with the typical Mediterranean basin with a small area and moderately steep slopes that generate short concentration times and therefore can create dangerous situations.

In the present work it has been calculated that the volume of water that can appear in the eventual rainfall which periodically takes place in this region, in two scenarios: first is the present situation and characteristics of the basin and the second one it is corresponded to the hypothetical original situation, without the urban development.

2 Methodology

2.1 Study area

The area chosen for the study is located in the coastline (see Figure 1); the municipality of El Campello has 55.27 km^2, is slightly elongated SW-NE and is situated on the Costa Blanca. It is bordered by Alicante, San Juan de Alicante, Muchamiel, Busot, Aguas de Busot, Orcheta and Villajoyosa. On its east side, El Campello faces the Mediterranean Sea along 23 km of coastline, where 10 basins flow into. This paper studies the following basins: El Amerador, Morro Blanc, Enmig, Coveta Fuma, Aguas, La Merced, Baeza and Lanuza. These basins are located between El Campello and Villajoyosa (Figure 1: 1) and all of them have common morphological characteristics.

These basins have a particular trait in common; in the last 3–4 decades they have been subject to a large urban development. This has resulted in highly vulnerable situations and has increased the risk of flooding.

El Campello is situated in the eastern part of the outer zones of the Betic mountain range, and within these in the Prebetic system. This territory is characterized by a certain lithological diversity:

(i) Jurassic and Tertiary Zone, located in the N-NW, where the mountains correspond to Jurassic outcrops. Its reliefs are less rough toward the south, were the Quaternary emerge.

(ii) Quaternary zone, for the rest of the municipality, from the foothills to the sea and south. It manifests itself as sandy deposits that appear along the Alicante's coast and other recent deposits that correspond with a colluvial and alluvial sediments close to the younger streams [5].

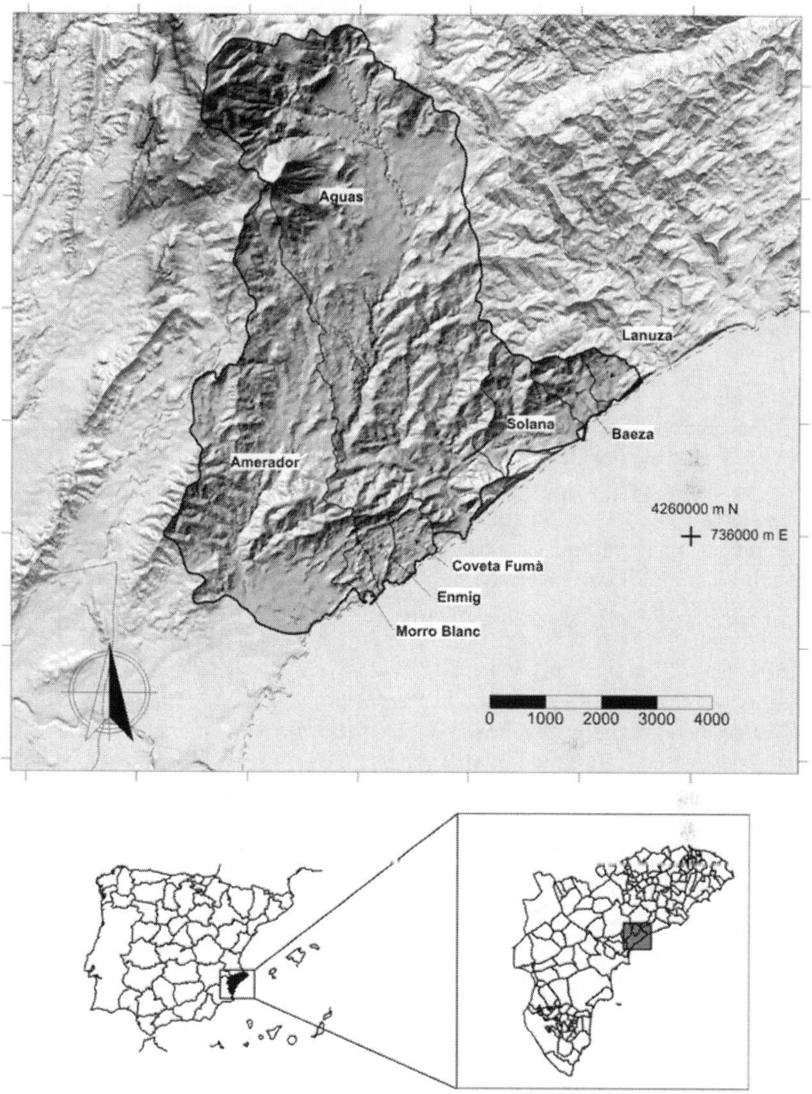

Figure 1: Location of study zone and map with basins and drainage network.

Weather can be classified as Mediterranean subtropical dry and semi-arid, covering littoral and pre-littoral lands of the south of the mountainous border.

Rainfalls are scarce, irregular and random [6]. The summer drought is extended from 3 to 5 months. The total annual days with rainfall do not exceed 60, therefore there are days with high-intensity rainstorms [7].

2.2 Basins characteristics

The analysis was developed using a 1:10,000 map edited by Cartographic Institute of Valencia, which was the starting point to define the watersheds boundary of the different basins. From this, information of size and shape of basin was calculated, information about elevation of the basins and finally hydrological information.

2.3 Rainfall assessment

Firstly the rainfall volume of considerate return period was obtained by two different methods.

The first of these methods uses information of daily rain volumes which are registered by the Spanish Agency of Meteorology (AEMET) in its broad net of meteorological stations laid around the country.

The selection of the meteorological stations to use as sources of information was based on the consideration of different aspects: the distance from the study zone, the characteristics of the relief between station and study area, the quality of the meteorological station and period of time through which records extend. After these considerations, station of Ciudad Jardin (information from 1950 to nowadays) and Relleu (information from 1962 to nowadays except one year) were considered. A statistical analysis was carried out to assess the precipitation rate associated with a return period of 100 years considering a Gumbel type I distribution of maximum values. The adjustment was validated by the Kolmogorov-Smirnov goodness-of-fit test for both stations.

The second method was based in the document "Máximas lluvias en la España peninsular" [8] edited by the Spanish Government which provides values of maximum precipitation rates by using a SQRT-ET max statistical distribution. This document was edited by the Spanish Government to provide useful information in the complete territory of Iberian Peninsula, filling the areas where does not exist meteorological stations in the vicinity.

Once both values of maximum daily rainfall by the two different methods were available, the highest of both was chosen in order to calculate the water discharge in the ravine for all eight basins.

2.4 Turnoff thresholds

Turnoff thresholds were calculated for each basin according to the Spanish roads and highways standard part 5.2-I.C. "Superficial drainage" [9], that, in turn, comes from the S.C.S. model developed by U.S. Conservation Service. Land use information was obtained through the online application GESPLAN [10], developed by regional government of Alicante. This application considers 13 types of land uses, 4 different kinds of human transformations, 4 kinds of agriculture techniques, 4 levels of natural areas and 1 kind of mining activity.

Firstly, basins turnoff thresholds were calculated for present situation. Secondly, they were calculated again for the hypothetical initial situation. For this situation, areas which nowadays are urbanized were considered as thin forest, because it is the most extended land use in the majority of studied area without urbanizing.

2.5 Water levels

Water levels in the mouth of each ravine were obtained by two different methods.

On one hand, the Temez formulation, often referred to as Rational Modified Method [9, 11], was used. This is a hydrometeorological method which gives an estimation of water discharge in the ravine from meteorological information. This method has been used in Spain for many years and its results are acceptable according to real measurements, especially in basins smaller than $50km^2$. This method only provides the peak of discharge linked to a certain return period, so users cannot obtain information about the hydrograph.

On the other hand, water levels were also calculated using HEC-HMS software [12], developed by U.S. Army. This software provides as outputs the peak of water discharge and the hydrograph linked to each return period in each of studied ravines. Different kinds of hydrographs can be obtained using HEC-HMS software, but in this study only SCS hydrograph was studied for each ravine and situation. Rainfalls in southeast of Spain are usually short and very intense. Then, in order to calculate the hydrograph for each ravine, it has been established a one-hour rainfall. This rainfall has three equal intervals with different intensities.

Assessment of these water levels in the mouth of each ravine was made for both present and hypothetical initial situations, in order to study the influence of urban development. The differences between two methods employed were also analyzed.

3 Results and discussion

In order to know the influence of urban development in flood vulnerability, turnoff thresholds and water levels were calculated for the present situation and for a hypothetic initial situation. The names of studied ravines are: Amerador, Morro Blanc, Enmig, Coveta Fuma, Aguas, La Solana, Baeza and Lanuza.

The main characteristics of ravine basins, needed to calculate turnoff thresholds and water levels, are shown in Table 1. The majority of ravine basins have a surface less than 3 km^2 and their average slope is higher than 10%. The only exceptions are the ravines of Amerador and Aguas, which present a surface of their basins between 14 km^2 and 30 km^2, and their average slopes are smaller (around 8%).

Rainfalls in 24 hours calculated for each basin are shown in Table 2.

Table 1: Main characteristics of ravine basins.

Ravine	Surface of basin (km^2)	Length of ravine (km)	Average slope of ravine (%)	Concentration time (min)
Amerador	14.72	10.30	8.5	170
Morro Blanc	0.45	0.97	9.6	27
Enmig	0.59	1.26	15.4	31
Coveta Fuma	0.82	1.06	15.5	27
Aguas	28.06	14.30	7.5	222
La Solana	2.37	3.25	11.8	66
Baeza	0.85	1.70	17.6	37
Lanuza	0.65	1.00	11.2	27

Table 2: Rainfalls in 24 hours for each ravine basin for 100 years return period.

Ravine	Rainfall in 24 hours (mm)
Amerador	184
Morro Blanc	195
Enmig	195
Coveta Fuma	195
Aguas	170
La Solana	168
Baeza	168
Lanuza	168

Turnoff thresholds have changed because parts of the basins are occupied by urban zones. As a consequence of these changes, impervious areas increase and turnoff thresholds decrease. The impervious area at present situation and turnoff thresholds for the hypothetic initial situation and present situation obtained for ravines studied are shown in Table 3.

As can be observed, in most of the ravine basins studied (except Amerador and Aguas), the percentage of impervious areas at present moment is very high, which could be a typical case of non sustainable urban development. For example, this percentage is around 60% for Coveta Fuma basin and around 50% for Morro Blanc basin. These basins have small surface, and they are close to the coast. As a consequence, they are more attractive to be transformed into urban zones by tourism.

The water levels in mouth of ravines, calculated according rational method for a 100 years return period, are shown in Table 4. Use of rational method is suggested by the Spanish Roads and Highways standards to design bridges which cross rivers and ravines.

Water levels for each ravines increase between hypothetical initial situation, without urban zones in their basin, and present situation. The highest rises have

Table 3: Turnoff thresholds and impermeable area of ravines basins.

Ravine	Impermeable area of basin (%) Present situation	Turnoff threshold Hypothetical initial situation (mm)	Turnoff threshold Present situation (mm)
Amerador	6.2	38	35
Morro Blanc	48.3	33	14
Enmig	31.1	30	18
Coveta Fuma	63.2	36	11
Aguas	1.7	65	64
La Solana	24.2	38	29
Baeza	24	42	33
Lanuza	35.6	33	19

Table 4: Water levels in the mouth of ravines obtained using a rational method for a return period of 100 years.

Ravine	Water level Hypothetical initial situation (m^3/s)	Water level Present situation (m^3/s)
Amerador	79.4	88.7
Morro Blanc	6.5	14.7
Enmig	9	11.9
Coveta Fuma	10.9	29.3
Aguas	58.4	61.7
La Solana	15	26
Baeza	7.6	12
Lanuza	7.2	15.3

been observed for ravines Morro Blanc, Coveta Fuma and Lanuza. In these cases, water levels obtained for present situation is more than twice as obtained for hypothetical initial situation. These results could mean that an unsustainable urban development produces an increase of water levels in mouth of ravines, and as a consequence, it could also produces an increase of flood threaten, mainly in urban zones placed close to the coast.

Water levels in ravine outfalls have also been calculated using HEC-HMS software, which permits to obtain the hydrograph of ravine. Rainfall temporal distribution for each ravine is shown in Table 5.

Total precipitation of this one-hour rainfall is the same than total precipitation estimated for rainfalls in 24 hours, calculated using a statistical method for 100 years return period (Table 2).

The water levels for each ravine obtained using HEC-HMS are higher than those calculated according to the rational method. This is due to the different duration of rainfalls considered for each method. Nevertheless, duration of

Table 5: Rainfalls for each ravine considered for HEC–HMS simulations.

Ravine	Time intervals (min)	Precipitation (mm)
Amerador	0-20	50
	20-40	100
	40-60	34
Morro Blanc	0-20	40
	20-40	125
	40-60	30
Enmig	0-20	40
	20-40	125
	40-60	30
Coveta Fuma	0-20	40
	20-40	125
	40-60	30
Aguas	0-20	45
	20-40	90
	40-60	35
La Solana	0-20	43
	20-40	95
	40-60	30
Baeza	0-20	43
	20-40	95
	40-60	30
Lanuza	0-20	43
	20-40	95
	40-60	30

rainfall considered for HEC-HMS simulations is very similar to real rainfall occurring in the studied zone, so it is expected that results obtained using HEC-HMS could be more realistic.

The hydrographs obtained in mouth of Ravine Amerador for present and initial situation, using HEC-HMS software, are depicted in Figure 2(A). Both hydrographs and peak water levels are very similar, because only a small part of this basin has been occupied by urban zones.

In Figure 2(B) the hydrographs of Ravine Morro Blanc are represented. In this ravine, there is an important difference between both situations studied, and the peak discharge in their mouth is considerably higher nowadays. This could be related with the unsustainable urban development in the basin during the recent years.

Hydrographs calculated in Ravines Enmig and Morro Blanc outfalls are depicted in Figures 3(A) and (B) respectively. For both ravines, the peak water level is higher at present situation than at initial situation. These results again show that a high increase of urban areas in the basins, produces greater water

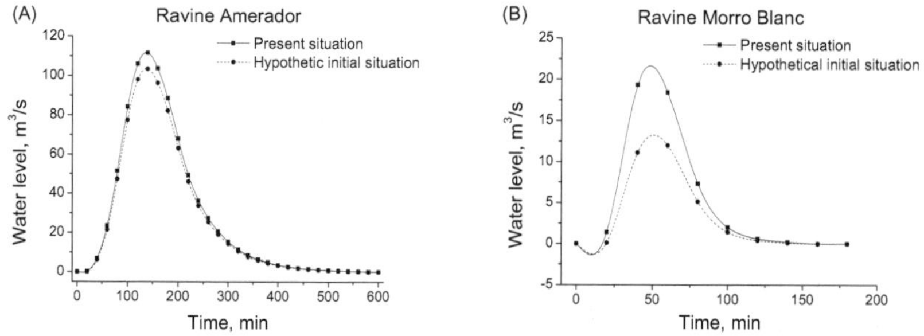

Figure 2: (A) Hydrographs obtained in the mouth of Ravine Amerador for present (continuous line with square symbols) and hypothetic initial situation (dashed line with circular symbols). (B) Hydrographs obtained in the mouth of Ravine Morro Blanc for both situations studied (the lines have the same meaning as on Ravine Amerador).

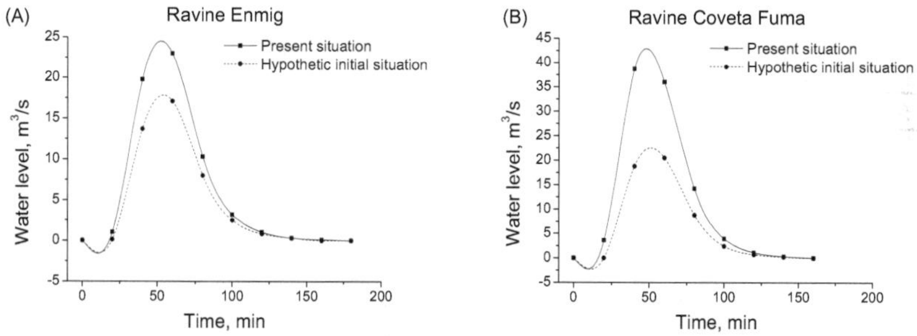

Figure 3: (A) Hydrographs obtained in the mouth of Ravine Enmig. (B) Hydrographs obtained in the mouth of Ravine Coveta Fuma. The lines have the same meaning as in Figure 2.

levels and, as a consequence, a rise of flood threaten in zones close to the mouth of ravines.

The results determined for Ravine Aguas are shown in Figure 4(A). Hydrographs obtained for this basin are very similar, because nowadays only a very small area of this basin is occupied by urban zones. La Solana basin is another example of increase of flood threaten, as a consequence of urban development, as can be observed in Figure 4(B).

Hydrographs calculated in the outfall of Ravines Baeza and Lanuza are represented in Figures 5(A) and (B). In both cases, there is an important rise of peak discharge between initial situation, without urban areas on basins, and present situation.

Finally, in Table 6 a compilation of peak water levels obtained from SCS hydrographs using HEC-HMS software for each ravine and situation studied is increase of water levels due to proliferation of urban zones, which occupy an

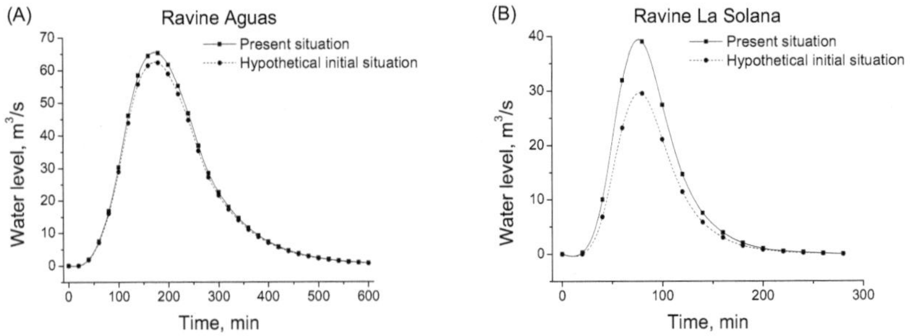

Figure 4: (A) Hydrographs obtained in the mouth of Ravine Aguas. (B) Hydrographs obtained in the mouth of Ravine La Solana. The lines have the same meaning as in Figure 2.

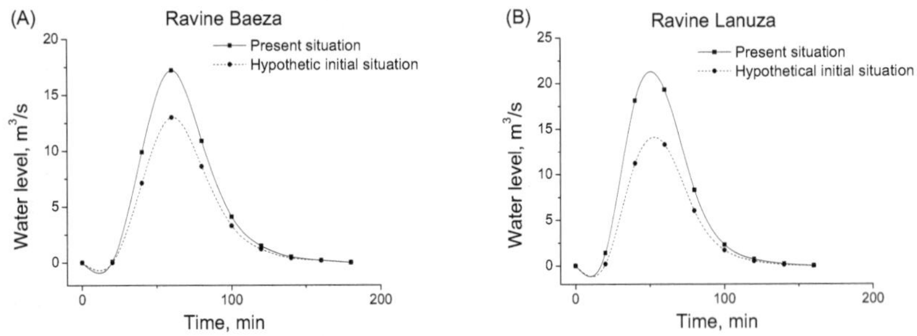

Figure 5: (A) Hydrographs obtained in the mouth of Ravine Baeza. (B) Hydrographs obtained in the mouth of Ravine Lanuza. The lines have the same meaning as in Figure 2.

Table 6: Water levels in outfalls of ravines obtained from SCS hydrographs using HEC-HMS software for a return period of 100 years.

Ravine	Water level Hypothetical initial situation (m^3/s)	Water level Present situation (m^3/s)
Amerador	103.4	112.7
Morro Blanc	12	19.4
Enmig	17.1	23.1
Coveta Fuma	20.5	39
Aguas	62.3	66.3
La Solana	29.5	39.4
Baeza	13	17.4
Lanuza	13.3	19.4

important part of the basins. This situation could provoke an important rise of flood threaten on these basins, mainly on coastline, which is where there is a higher concentration of urban areas and population.

4 Conclusions

The main conclusions that can be obtained from the results previously discussed can be summarized as follows:

- Most of basins placed in the studied zone, have a surface less than 3 km^2 and their average slope is higher than 10%.
- Touristic development has produced that part of these basins is occupied by urban zones. These changes have provoked that impermeable areas of basins increase and turnoff thresholds decrease.
- Impermeable area rate at present moment is very high in most of studied basins, which shows a typical case of unsustainable urban development.
- Peak discharge for each ravine increase between hypothetical initial situation, without urban zones in their basin, and present situation for both methods used.
- Water levels obtained using HEC-HMS software are higher than those calculated according to the rational method. These differences could be related to different duration of rainfalls considered for each method.
- Flood threaten is increased as a consequence of urban development.
- As a preliminary conclusion, it could be expected that an unsustainable urban development produces an increase of flood threaten. This statement is particularly true in case of Costa Blanca, and in general in the southeastern coast of Spain, due to their geographic and climate features.

Acknowledgement

J. M. Ortega is indebted to the Ministerio de Educación of Spain for a fellowship of the "Formación Personal Investigador (FPI)" programme (reference BES-2008-002650).

References

[1] Rozalis, S., Morin, E., Yair, Y., Price, C., Flash flood prediction using an uncalibrated hydrological model and radar rainfall data in a Mediterranean watershed under changing hydrological conditions. *Journal of Hydrology*. **394(1-2)**, pp. 245-255, 2010.

[2] Quintana, P., Ribes, A., Martin, E., Habets, F., Boe, J., Comparison of three downscaling methods in simulating the impact of climate change on the hydrology of Mediterranean basins. *Journal of Hydrology*. **383(1-2)**, pp. 111-124, 2010.

www.witpress.com, ISSN 1743-3541 (on-line)

[3] Latron, J., Gallart, F., Seasonal dynamics of runoff-contributing areas in a small Mediterranean research catchment (Vallcebre, Eastern Pyrenees). *Journal of Hydrology*. **383(1-2)**, pp. 111-124, 2010.

[4] Gil Olcina, A., Olcina Cantos, J., Rico Amorós, A.M., *Aguaceros, aguaduchos e inundaciones en áreas urbanas alicantinas*. Universidad de Alicante. 2004.

[5] Mapa Geológico de España (E1:50.000). *Hoja geológica MAGNA, nº 872-874*. IGME, Madrid. 2ª serie. 1ª edición, 1979.

[6] Ceballos-Barbancho, A., Morán-Tejeda, E., Luengo-Ugidos, M., Llorente-Pinto, J.M., Water resources and environmental change in a Mediterranean environment: The south-west sector of the Duero river basin (Spain). *Journal of Hydrology*. **351(1-2)**, pp. 126-138, 2008.

[7] Delgado, J., Llorens, P., Nord, G., Calder, I.R., Gallart, F., Modelling the hydrological response of a Mediterranean medium-sized headwater basin subject to land cover change: The Cardener River basin (NE Spain). *Journal of Hydrology*. **383(1-2)**, pp. 125-134, 2010.

[8] España Dirección General de Carreteras. *Máximas lluvias diarias en la España peninsular*. Ministerio de Fomento. Madrid, 1999.

[9] España Dirección General de Carreteras. *Instrucción 5.2-IC "Drenaje superficial"*. Ministerio de Obras Públicas y Urbanismo. Madrid, 1990.

[10] Diputación de Alicante, Ciclo Hídrico, GESPLAN, www.gesplanweb.com/alicante

[11] Temez, J.R., Extended and improved rational method. *Version of Highways Administration of Spain. Proc. XXIV Congress IAHR*, Madrid vol. A, pp 33-40. 1991.

[12] Hydrologic Engineering Center. *HEC-HMS Hydrologic Modelling System: User's Manual, Version 3.4.* US Army Corps of Engineers. Davis, 2009.

Sustainable and unsustainable agriculture in Ghana and Nigeria: 1960–2009

S. Onyeiwu, E. Pallant & M. Hanlon
Departments of Economics, Environmental Science, and Biology (respectively), Allegheny College, USA

Abstract

The agricultural sector of African economies has faced considerable challenges within the past 50 years or so. Although agricultural production on the continent rose by an annual average of 2% between 1965 and 1980 and has continued to increase by 1.8% annually since then, population growth of 2.9% per year has resulted in a per capita decline in agricultural production. From self-sufficiency in food production before the 1960s, many African countries have become net food importers, with a handful of them facing severe food shortages arising from drought, desertification, climate change and wars. In this paper we use the case of Ghana and Nigeria to explore some of the salient dynamics that have resulted in the current crisis in the agricultural sector of African economies. We argue that soil conditions, climate change, and population growth, in combination with ineffective economic policies, have contributed immensely to the sordid state of agriculture in Africa. We use historical and contemporary evidence gathered from Ghana and Nigeria during several visits to show how economic policies have interacted with biophysical and environmental factors to generate an unsustainable use of land, agricultural labor, and natural resources.

Keywords: Ghana, Nigeria, sustainable agriculture, climate, structural adjustment, environmental economics, sustainable development, Sub-Saharan Africa.

1 Introduction

Most analysts agree that African agriculture has been on an unsustainable path for the past five decades or so; however, there is no consensus on the factors responsible for this unsustainability [1–3]. Explanations of the causes of the

WIT Transactions on Ecology and the Environment, Vol 144, © 2011 WIT Press
www.witpress.com, ISSN 1743-3541 (on-line)
doi:10.2495/ECO110191

agrarian crisis in Africa have tended to be monocausal in nature, focusing on single factors like the so-called irrationality of African farmers [4], anachronistic land tenure systems/social structures [5, 6] and ineffective economic policies that generate perverse incentives for agricultural development [7]. Monocausal explanations have limited our understanding of the dynamics of African agriculture for of a number of reasons. First, they are too narrow in the sense that they ignore several factors that affect sustainable agriculture. Second, they seem to be Eurocentric by examining African agriculture on the basis of assumptions relevant to Western economies and society [8, 9]. Third, they gloss over the interconnectedness of economic and biophysical feedback loops between the various factors affecting African agriculture.

We believe a more useful approach for understanding African agriculture is an integrated methodology that explores the role of multiple factors that interact in mutually reinforcing ways. In this paper, we seek to provide a historical, biophysical, and socio-economic context for why African agriculture has become unsustainable. Evidence gathered from our visits to Ghana and Nigeria between 2001 and 2008, as well as reports by the World Bank and the FAO, are used to support our analysis.

One common thread in the conventional definition of agricultural sustainability is the notion that sustainable agriculture is a system that “makes the best use of nature’s goods and services whilst not damaging the environment” [10]. In other words, sustainable agriculture enhances the “productive values of natural, social and human capital” [11].

2 Transformation of agriculture in Ghana and Nigeria: 1960–2009

2.1 Environment

Through direct observation and evaluation of available data we have measured trends in precipitation, soil fertility and fertilizer use, forest cover, and agricultural practices. World Bank Country Environmental Analyses discovered that the cost of environmental degradation, due to both biophysical and anthropogenic factors, totalled 10% of the GDP of Ghana and 9% of Nigeria’s.

2.1.1 Precipitation

Rainfall patterns in both Ghana and Nigeria have been shifting over the past fifty years. The Ghanaian EPA estimates that rainfall for all of Ghana has been decreasing by 2.4% per decade since 1960 [12] and is predicted to decline another 10% by 2050 according to the IPCC. Similarly, the length of the rainy season in North East Nigeria has been decreasing since 1961, as has the total amount of rainfall received [13]. In the northern regions of both countries rainfall is already becoming more sporadic, intense, and less reliable [14–16]. These regions are at threat for desertification, which is already occurring in the northernmost regions of Nigeria [17].

On repeated visits to Ghana and Nigeria long-term residents told us that the beginning of the rainy season was delayed and shorter in overall duration. Additionally, farmers described rainfall as being more intense, causing greater runoff, flooding, and erosion, as well as lower rates of infiltration. Rainfall intensity is important to farmers because it can dictate success or failure of crops, whereas total rainfall gathered in collectors might not reflect this impact.

2.1.2 Forests

Multiple factors have led to the increasing rate of deforestation in Ghana and Nigeria. When Ghana received independence in 1957, forest cover measured 8.2 million hectares. In the time since, deforestation has progressed at an annual average of 65,000 hectares lost per year, leaving only 1.6 million hectares of forest cover [18] (Figure 1). Ninety-five percent of Ghana's high forest has already been logged and only 1% of what is left is within protected areas such as wildlife sanctuaries, game reserves, and sacred groves [19].

In Nigeria, areas dominated by guinea vegetation (trees, woodlands, and shrubs) decreased by 50% during the same period, with the majority of the area converting to agricultural use in the 1990s. Between 1986 and 1990, forest area declined by 3.1 million hectares and between 1990 and 1995, the amount of covered area declined by an additional 3.3 million hectares [20].

Causes of deforestation include increasing population, resulting in shorter rotations of slash and burn agriculture and increased demands for charcoal production. Several rural communities in Ghana are cutting down trees for processing into charcoal, which has become a source of income for cash-strapped rural dwellers (See section on Economy for further discussion of charcoal processing). Village women describe having to walk increasingly long distances – sometimes 12 hours – to gather sufficient fuelwood for home use.

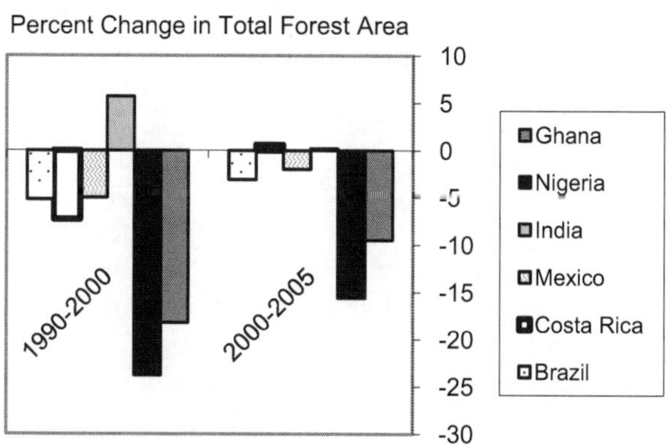

Figure 1: Deforestation in Ghana and Nigeria as compared to other, non-African nations.

2.1.3 Soil quality

Declining soil fertility is considered to be the major constraint to agricultural production even though more land is under cultivation [21, 22]. The Global

Assessment of Human Induced Soil Degradation (GLASOD) survey done in 1990 found 69% and 27% of the soils in Ghana and Nigeria, respectively, to be degraded due to, in large part, unsustainable farming practices and deforestation [23]. None of the soils found in Nigeria are conducive to productive agriculture, and many are considered the worst [17], putting farmers at an immediate disadvantage. These soils lack proper drainage and water retention, nutrient composition, and structure to allow proper root growth. As a result of population increase, pressure on land has reduced the 8-15 years natural fallow period that is required to regenerate soil fertility after 1-3 years of cropping to only 2-3 years, further reducing soil fertility [21, 22]. Soils are not adequately protected by cover crops as crop rotation is hardly practiced, resulting in fragile soils that are easily eroded, a problem exacerbated by overfarming and overgrazing. Sporadic and intense rainfalls add to these issues by inducing erosion [24, 25].

Almost all the nutrient balances in Ghana show a deficit as more nutrients are removed by harvesting or lost to erosion than are applied as fertilizers [21, 22]. This represents a loss of potential yield and progressive soil impoverishment. According to FAO estimates, cassava and yams account for almost 20% of the cropped area, but 37% of the nitrogen deficit. The highest depletion rates are in the southeast and the central west parts of Ghana, which correspond to the cassava area [26].

These nitrogen deficiencies, though, are not overcome with fertilizer application as is the case in the developed world. The application of fertilizers to soils throughout Ghana is very low, as fertilizer use has been on the decline instead of increasing as the cropped area expands. In the 1980s fertilizer imports were reduced and subsidies removed, resulting in the price being too high for most farmers (See discussion below regarding impact of structural adjustment programs of the 1980s). Fertilizer use declined from 21.9 kg of fertilizer material per hectare arable land in 1978 to 7.3 kg/ha in 1993.

2.2 Economy

Following the United Nations' declaration of the 1960s as the "First Development Decade" for developing countries, several Sub-Saharan African (SSA) countries achieved impressive growth rates that averaged 6-8% per annum. More encouraging, perhaps, is the fact that these growth rates were accompanied by a robust agricultural sector that proved to be sustainable from the economics standpoint. The agricultural sector was not only the mainstay of the Ghanaian and Nigerian economies in the 1960s, it did so without posing serious threats to these countries' ecosystems. In Nigeria agricultural exports represented about 85% of total exports in the 1960s, and contributed 60% of the GDP [27], though about three-quarters of the country's land area was under cultivation. Subsistence agriculture in both Ghana and Nigeria was based largely on the time-tested methods of shifting cultivation, crop rotation, and organic farming, without extensive use of fertilizers.

Things began to take a dramatic turn, however, in the 1970s. Agriculture's role in the Ghanaian and Nigerian economies not only began to falter, but also glided into a new trajectory that proved to be unsustainable. While agricultural value added began to decline in the 1970s, agricultural land and the area devoted to farming were on the rise - a phenomenon that signalled the onset of declining productivity in the sector. Agricultural land as a percentage of land area in Ghana rose from 51% in the late 1960s to 65% in 2005. Arable land as a percentage of land area also increased from 7% in 1969 to 18% in 2005. There were similar changes in land use in Nigeria, but they were not as dramatic as those of Ghana. For instance, agricultural land as a percentage of land area rose from 77% in 1969 to 81% in 2005, while arable land as a percentage of land area increased from 30% to 35% within the same period.

Despite increases in agricultural and arable land, agricultural value added as a percentage of GDP in Ghana dropped from 46% in 1969 to 37% in 2005. Although data on agricultural value added in Nigeria during the 1960s to 1990s are not available, data for the 2000s suggest a downward trend. Agricultural value added as a percentage of GDP fell from 40% in 2002 to 33% in 2005, while arable land per person dropped from 1 hectare to zero implying a scarcity of land for agricultural production.

2.3 Equity

Per capita income levels in Ghana and Nigeria today are not significantly different from what they were in the 1960s. Indeed, it is now believed that many Ghanaians and Nigerians enjoyed higher living standards in the 1960s than they did in the 1990s [28]. Contrary to the 1960s, Ghanaians and Nigerians are increasingly dependent on food imports to meet growing shortfalls in domestic food supply. As subsistence agriculture has not been able to generate sufficient income for rural farmers, more and more people have begun to migrate to urban areas in both Ghana and Nigeria. In 1961 76% of the Ghanaian population and

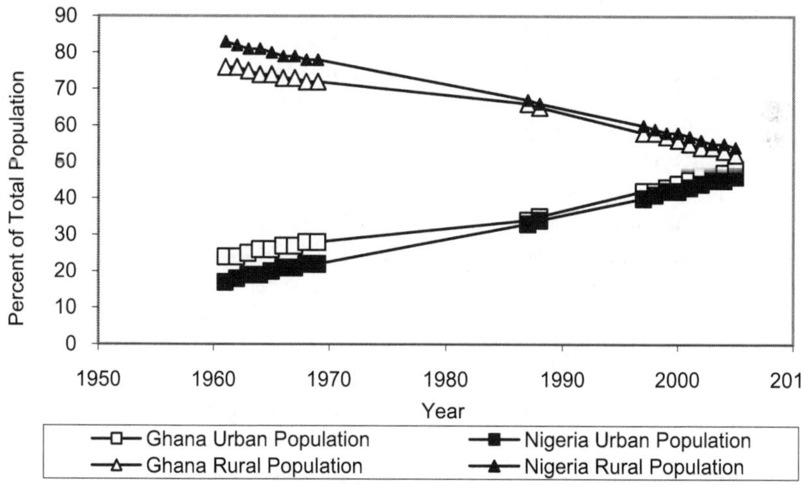

Figure 2: Rural and urban populations as a percentage of the total population in Ghana and Nigeria.

83% of the Nigerian population lived in rural areas. By 2005, these numbers had dropped drastically, to 52% for Ghana and 54% for Nigeria (Figure 2). Furthermore, of those living in urban areas in Ghana and Nigeria, 70% and 79%, respectively, were living in slums, as designated by the United Nations Human Settlements Program (UN-HABITAT).

Many analysts believe that agriculture has become unsustainable in Ghana and Nigeria because of the introduction of economic and industrial policies that are incompatible with sustainable agriculture. One of these policies is structural adjustment programs, which ostensibly were supposed to revive African economies, but have had the backwash effects of undermining the agricultural sector. We discuss the effects of structural adjustment programs in the next section.

2.4 Structural adjustment and sustainable agriculture in Ghana and Nigeria

The Lewisian assumption that Import-Substitution Industrialization (ISI) would promote industrial development and draw surplus labor away from the agricultural sector became illusory after two decades of its implementation in Ghana and Nigeria. In fact, Africa appears to have had the worst of two worlds: a world in which agriculture has become more unsustainable, and another in which the continent's industrial performance has been abysmal.

By the early 1980s, a decade aptly characterized as a "lost decade" for Africa, it became obvious that the continent was sliding over a dangerous economic cliff. To prevent deterioration in their economies, and upon the urging of the World Bank and the IMF, many African countries implemented neoliberal economic policies or Structural Adjustment Programs (SAP) in the 1980s. The introduction of SAP was expected to not only reverse this inglorious industrial development trajectory, but also to set African enterprises on a new path of efficiency, higher productivity, and international competitiveness. SAP is premised on the notion that once African countries get "their prices right" through trade liberalization, devaluation, privatization, removal of government subsidies, and reduction or elimination of budget deficits, firms will respond by reducing inefficiency, eliminating wastes, and raising total-factor productivity.

SAP is predicated on the same assumption of the Lewis model: the notion that industrial development is the key for transforming the rural sector. In addition to promoting industrial development SAP also introduced explicit policies (mainly price incentives) for strengthening the agricultural sector. A key component of structural adjustment is the removal of price controls and the abolition of marketing boards that had for many years monopolized the buying and selling of primary products. Structural adjustment and the removal of price controls were expected to boost farmers' incomes. With the attractiveness of agricultural production under SAP, farmers are expected to invest in land acquisition and improvement. In some cases, marginal land may be brought under cultivation.

Rather than promote agricultural development, SAP has exacerbated some of the problems created by ISI. One of those problems is rural-urban migration, which has now taken a new dimension. Following the liberalization of product markets under SAP, the prices of goods and services have increased significantly in both Ghana and Nigeria. For instance, the consumer price index for all items in rural areas more than doubled from 482.3 (with 1975 as a base year) in 1985, a year before SAP was introduced in Nigeria, to 1,194.6 in 1989 [29]. Rural dwellers are also now required to pay user fees for social services previously provided free-of-charge by the government, or heavily subsidized. This means that farmers have to generate additional income to cope with the increase in the cost of living. Unable to cope with the escalation in the cost of living, many young rural dwellers are migrating to urban areas to explore better job opportunities. Our visits to villages in Ghana and Nigeria show that a preponderance of rural dwellers are mainly older men and women who, by virtue of old age, are unable to migrate. They are also too frail to work long hours on the farm, thus leading to a reduction in agricultural productivity.

Following a review of rural household surveys in a number of African countries, Bryceson [3] concludes that neo-liberalism (or SAP) has exacerbated poverty in rural Africa by weakening the continent's "agrarian foundation" and by accelerating the pace of "deagrarianisation and depeasantisation" of the region. According to Bryceson [3], the removal of subsidies for agricultural inputs and social programs such as health and education, has forced rural farmers to diversify their income sources. This attempt at diversification has resulted in a shift of resources to non-agricultural activities, thus undermining agricultural production–which structural adjustment ostensibly strives to promote.

Several rural communities in Ghana appear to be coping with the escalation in the cost of living by resorting to unsustainable use of natural resources. Specifically, they are cutting down trees for processing into charcoal, which has become a source of income for cash-strapped rural dwellers. It is not uncommon to find hundreds of bags of charcoal along the Kumasi-Tamale highway, waiting to be transported onward to Accra, Cape Coast and other cities in Ghana. It is estimated that structural adjustment in Ghana caused a 4% loss in GDP just due to environmental degradation [24]. Agriculture imposed the greatest environmental costs, at 6.9% of the total or 28.8 billion cedis (U.S. $88.5 million). These costs were reflected in wind and water erosion, soil compaction, surface soil crusting and loss of soil stability and fertility, nor forgetting the indiscriminate use of fertilizers and pesticides. While these unsustainable agricultural practices had adverse effects on the environment, they also intensified the poor living conditions of a majority of Ghanaians and Nigerians [19]. In the next section, we discuss the impact of unsustainable agriculture on living conditions in a small village in Nigeria.

3 The Village of Umuluwe, Southeast Nigeria as an example of unsustainability

To really understand the impact of failing sustainability of agriculture on widening the equity gap for rural dwellers in Ghana and Nigeria we describe below the village of Umuluwe in Nigeria. In our travels in Ghana and Nigeria we found many villages like Umuluwe suffering the effects of failing biophysical factors, increased population pressure, and poor economic strategies resulting in extreme poverty and inequity.

We carried out two surveys of about 300 individuals in the small village of Umuluwe in Southeast Nigeria in 2001 and 2007. The village of Umuluwe is situated within the predominantly Christian southeast region of Nigeria, and is about 30 miles west of the regional capital of Owerri. Like most communities in Igboland, the Umuluwe is a close-knit village of about 3000 people. Except for interaction (through marriages, the church, and trade) with other neighboring villages, Umuluwe residents have limited contact with the outside world.

Umuluwe is representative of rural West African villages with difficult access from outside areas. The roads are quite a challenge for regular cars that, due to the pervasive poverty, are a luxury only very few can afford. Thcrc is no bus line, and people walk daily for hours to reach the nearby villages and the markets where their products are sold. Electricity was introduced only in the last seven years, due to the concerted effort of the community. However, because of undersupply and transmission problems, a chronic problem in Nigeria, power is on intermittently. There is no running water; the only water source is a small spring about one mile away from the center of the village. The soil has low fertility, there is no livestock and poor community members are unable to afford chemical fertilizers.

Per capita income in the village is about \$150 per annum, and the main sources of income are cash crops (palm trees and fruit), water and paving stone – the same sources of income it has relied on for at least the past five decades. Farming is mainly for subsistence, although surpluses are sometimes produced and sold at the weekly markets in neighbouring villages. Proceeds from such surpluses are used to purchase items such as meat, milk, sugar and bread. The villagers also use the proceeds to pay community dues, school fees for their children, and healthcare costs.

With support from four research assistants who are residents of the village, socio-economic data were gathered from villagers who still reside in the village and those who migrated to Obigbo–a satellite town about 120 miles from the village. Data were collected on income, savings, investment, assets, debt, remittances, occupation, education, apprenticeship training, gender, age, marital status, etc. Many of the individuals were also interviewed face-to-face in order to obtain descriptive information. While 78% of those who migrated out of the village in search of better opportunities were men, only 22% were women, which may be due to cultural constraints on the mobility of women. This may also explain why the poverty rate tends to be higher amongst rural African women. While about 56% of non-migrant villagers earned less than \$100 per year, only 6% of the migrants earned less than this amount. Conversely, while 26% of migrants earned over \$500 per year, less than 6% of non-migrants earned this amount. Income distribution is skewed in favor of men, as 21 per cent of male migrants earned more than \$5000 a year while only 5 per cent of female migrants fell within this income bracket. Likewise, 5 per cent of non-migrant males earned over \$5000, while only 1 per cent of non-migrant women did so.

We use the survey data to compute the poverty headcount index for the village. The poverty headcount index was calculated by finding the ratio of individuals who lived below the poverty line to all the individuals in the survey. The index is reported in Table 1, and it reveals a very high poverty rate in the village, with 87% of all the respondents living below $1 per day. An upward revision of the poverty line to $2 per day increases the poverty rate amongst the respondents to an alarming 94%; a rate rarely seen in most developing countries. A follow-up survey was undertaken in the same village in 2007 in order to investigate whether the poverty profile of the villagers had changed. Data on the poverty headcount index for 2007 are summarized in Table 2. As the table shows, poverty was still prevalent in 2007, but at a lower rate compared to the 2001 levels.

Table 1: Headcount index for Umuluwe* and Africa** (%): 2001 survey.

	< $1 per day	< $2 per day	< $1 per day (Female)	< $2 per day (Female)	< $1 per day (Male)	< $2 per day (Male)
Non-Migrant	93.0	97.0	98.0	99.2	85.1	94.3
Migrant	69.2	85.0	93.3	100.0	62.8	80.4
Entire Sample	87.2	94.3	97.2	99.3	76.8	89.1
Africa-rural	55.6					
Africa-urban	43.0					
Africa	52.3					

* Own calculations (based on 2000 income)
** Ali and Nwabu (2002, p. 12)

Table 2: Headcount index for Umuluwe* (%): 2007 survey.

	< $1 per day	< $2 per day	< $1 per day (Female)	< $2 per day (Female)	< $1 per day (Male)	< $2 per day (Male)
Non-Migrant	63.3	76.5	80.5	86.2	44.3	65.8
Migrant	25.5	40.0	42.9	52.4	14.7	32.4
Entire Sample	53.8	67.4	73.1	79.6	35.4	55.8

*Own calculations (based on 2006 income)

Non-income measures of poverty such as protein consumption, number of meals per day, the proportion of income spent on food, type of fuel used for cooking, savings, debt, landownership, etc. may be better markers for how really poor the villagers are. Data on some non-income measures of poverty in

Umuluwe are summarized in Table 3. Three of these measures point to the existence of chronic poverty in the village. About 75% of the respondents indicated that food accounted for the largest proportion of their expenditures, and only 33% consumes milk at least once weekly. About 24% indicated that they consume meat at least once every week. Over 60% of the respondents were in some type of debt. Over 90% used firewood as the main mode of cooking, while about half of the respondents owned no land.

Table 3: Non-income measures of poverty in Umuluwe, 2007.

Indicator	% of Sample	% Men	% Women
Consumes Milk Weekly	33	16	17
Consumes Meat Weekly	24	11	13
Food as Most Important Expenditure	75	36	39
Firewood as Main Mode of Cooking	91	42	49
Owes Money	62	27	36
Has Savings	55	34	21
Owns Land	53	37	16
Owns Goats	25	11	14
Owns Chickens	55	21	34

Source: 2006 Village Survey

4 Conclusions

Although undernourishment is not a major problem in Ghana and Nigeria (only 10% of the population in these countries are hungry), Table 3 suggests that low protein consumption is widespread. Malnourishment is of greater concern in Ghana, where protein production and consumption are very low. Ghanaians survive on a surfeit of starchy vegetables such as cassava, yam, and coco yam to provide an adequate number of calories. Compared to other non-African countries and Nigeria, the average Ghanaian receives very little protein from either animal sources or pulse consumption. Moreover, the over reliance in Ghana on starchy vegetables has led to a vitamin A deficiency, as the vitamin is found primarily in green and orange fruits and vegetables. The deficiency can lead to stunted growth and blindness, especially in children [30]. Nigerians also consume very few calories from animal products compared to other non-African countries, but their level of pulse production is equal to more advanced countries in Latin America and Asia. Ghana imports nearly no supplemental protein, further increasing the deficiency [21]. In earlier times when forests were more abundant, bushmeat was a more readily available protein source [31]. Similarly, when water was more abundant and populations were smaller, fish provided sufficient protein to a greater proportion of the population. Neither forest nor aquatic ecosystems are now able to provide adequate nourishment to the rising population. While there is not a calorie crisis in either country at the present time, the prognosis for the future is not bright, if agricultural practices do not change in either country.

Monocausal explanations have offered incomplete, and sometimes misleading, accounts of why agriculture in Africa has become unsustainable. Evidence from Ghana and Nigeria, however, suggests that agricultural sustainability has environmental, economic and equity ramifications. Specifically, the biophysical configuration of both countries has changed over the past five decades in ways that have grave implications for agricultural sustainability. Soil conditions, rainfall patterns, and forest use have all changed for the worst, making farmers in Ghana and Nigeria susceptible to declining productivity and poverty. Declining biophysical conditions in these countries have also been exacerbated by ineffective economic policies that encourage farmers to use their human and natural resources in unsustainable ways. Despite the increasing use of agricultural and arable land, Ghanaian and Nigerian rural dwellers are no better economically now than they were shortly after independence. Ghanaian and Nigerian farmers can no longer afford to continue on the path of unsustainability - a path that has not only been destructive of their natural environment, but also anathema to their economic and physiological well-being.

References

[1] Pretty, J. (1995). Regenerating Agriculture: Policies and Practices for Sustainability and Self-Reliance, London: Earthscan.

[2] Kuyek, D. (2002). "Intellectual Property Rights in African Agriculture: Implications for Small Farmers, *Genetic Research Action International,* August.

[3] Bryceson, D.F. (2004). "Agrarian Vista or Vortex: African Rural Livelihood Policies," *Review of African Political Economy,* 102: 617-629.

[4] Mortimore, M.J. and Adams, W.M. (2001). "Farmer Adaptation, Change and 'Crisis' in the Sahel," *Global Environmental Change,* 11: 49-57.

[5] Niemeijer, D. (1996). "The Dynamics of African Agricultural History: Is it Time for a New Development Paradigm?" *Development and Change,* 27: 87-110.

[6] McCusker, B. (2004). "Land Use and Cover Change as an Indicator of Transformation on Recently Distributed Farms in Limpopo Province, South Africa," *Human Ecology,* Vol. 32, No 1: 49-75.

[7] Lipton, M. (2007). "Can Small Farmers Survive, Prosper or be the Key Channel to Cut Mass Poverty?" *Tips Monitor* 38: 55-68.

[8] Ochieng, C. (2009). "Sustainable Agriculture in Africa: Towards a New Paradigm," *Project Africa 2020.*

[9] Hoben, A. (1995). "Paradigms and Politics: The Cultural Construction of Environmental Policy in Ethiopia," *World Development,* 23(6): 1007-1021.

[10] Altieri, M.A. (1995). Agroecology: The Science of Sustainable Agriculture, Boulder, CO: Westview Press.

[11] Pretty, J. (2004). "Can Sustainable Agriculture Feed Africa? New Evidence on Progress, Processes and Impact," *Environment, Development and Sustainability,* Vol.1, Nos 3-4: 253-274.

[12] Haven, B. (2004). "OneWorld Guide: Ghana." Available Online: http://uk.oneworld.net/guides/ghana/development. Accessed 15 March 2009.
[13] Hess T.M., Stephens W., Maryah U.M. (1995). "Rainfall Trends in the North East Arid Zone of Nigeria 1965-1990," Agricultural *and Forest Meteorology,* 87: 87-97.
[14] All Africa News Source (2008), "Nigeria: The Desert is fast encroaching, but Why?" Available online: http://allafrica.com/stories /200803280949.html
[15] All Africa News Source (2008), "Ghana: Climate Change threat, is country taking it serious?" Available online: http://allafrica.com/stories /200803280949.html
[16] All Africa News Source (2008), "Africa: Invest in Food crops, continent told." Available online: http://allafrica.com/stories/200808290136.html
[17] USAID (2002). Nigeria Environmental Analysis Final Report. (USAID BIOFOR IQC No. LAG-I-00-99-00013-00).
[18] Safo A. (2002). "Bleak future as Ghana's forest cover diminishes," News from Africa. Available online: http://www.newsfromafrica.org /newsfromafrica/articles/art_7890.html.
[19] Anane, M. (2007). "Towards Sustainable Development in Ghana," Voices from Africa Number 6: Sustainable Development Part 2. UN-Non-Governmental Liaison Office. http://www.un-ngls.org/documents /publications.en/voices.africa/number6/vfa6.01.html
[20] FAO (2002). "Gateway to land and water information: Nigeria national report," Available online: http://www.fao.org/ag/agl/swlwpnr/reports /y_sf/z_ng/ng.htm .
[21] FAO (2004 a.) FAOSTAT on-line statistical service http://apps.fao.org
[22] FAO (2004 b.) Gateway to land and water information: Ghana national report. Available online: http://www.fao.org/ag/agl/swlwpnr/reports /y_sf/z_gh/gh.htm .
[23] FAO. (2000). Land Resources Potential and Constraints at Regional and Country levels, FAO, Rome.
[24] World Bank (2006). Nigeria Rapid Country Environmental Analysis.
[25] World Bank (2006). Ghana Country Environmental Analysis.
[26] FAO (2005). "Fertilizer use by crop in Ghana" Available online: http://www.fao.org/docrep/008/a0013e/a0013e05.htm#TopOfPage .
[27] Cohen, R. (1981). *Labour and Politics In Nigeria,* London: Heinemann.
[28] Ayittey, G. (1999). *Africa in Chaos,* New York: St. Martins Press.
[29] Anyanwu, J.C. (1992). "President Babangida's Structural Adjustment Programme and Inflation in Nigeria," *Journal of Social Development in Africa,* 7(1): 5-24.
[30] Aguayo, V., Gamier, D., Baker, S. (2007). "Drops of Life," Published by UNICEF and Helen Keller International.
[31] Barth, E. (2008). "Catch Shares: Will a solution for fisheries work for Bushment?" Earth Trends. http://earthtrends.wri.org/updates/node/330

Biometric analysis of eucalyptus forests in the five initial years of development

A. L. T. Fernandes, T. de M. Florêncio & M. F. de Faria
University of Uberaba, Brazil

Abstract

This research is aimed at evaluating the effects of irrigation and mineral nutrition on the growth and production of eucalyptus (*Eucalyptus grandis*) cultivated in the region of Triângulo Mineiro. The experiment was conducted at the Farm School of Universidade de Uberaba, which has 850 m of altitude, in red-yellowish Latosoil. Irrigated treatments and non-irrigated treatments with different nutritional conditions (25%, 50% and 100% of the recommended doses) were compared. It was evaluated the parameters of diameter at breast height and growth in a period of five years of conducted experiment; the final productivity of wood was also determined at the end of that period. Even though results were statistically different throughout the years, at the end of the analyzed period it was observed that irrigated treatments showed the biggest biometric values as well as the biggest rates of annual medium increment. Proximity of values may have occurred due to excessive rainfall at the local, making water more available to the non-ırrıgated treatments.

Keywords: commercial wood, irrigation, fertilization.

1 Introduction

According to FAO, Brazil is the second most forested country in the world with almost 477.77 millions of hectares, and the fourth country with the biggest area dedicated to plantation forestry. However, the country lacks adequate information and functional mechanisms. The usefulness of eucalyptus is evident: virtually all parts of the tree are exploited [3].

In order to improve the production of a eucalyptus forest, one of the techniques that could be applied by growers is irrigation, which aim is the controlled supply of water to the forest species in sufficient quantity and in an

WIT Transactions on Ecology and the Environment, Vol 144, © 2011 WIT Press
www.witpress.com, ISSN 1743-3541 (on-line)
doi:10.2495/ECO110201

adequate period, complementing rainfall, as not only can irrigation benefit the growth of eucalyptus tree, but it can also reduce the cutting age and provide homogeneous development of the planted forest [1, 2].

All things considered, the present research aimed to evaluate different strategies of irrigation and mineral nutrition in the vegetative and productive development of eucalyptus cultivated in the conditions of Triângulo Mineiro, MG (Brazillian savannah).

2 Material and methods

The experiment took place in the Experimental Campus of the University of Uberaba (Farm School located at Uberaba, MG). The geographic coordinates are: Latitude 19°44′13″S, Longitude 47°57′27″W and altitude 850 m. The local climate is classified in the Köppen system as Aw hot, humid tropical and cold, dry winter. The annual rainfall is 1,474 mm and the mean annual temperature is 22,6°C. The species evaluated was *Eucalyptus grandis* cultivated with and without irrigation. Different levels of mineral nutrition (25%, 50% and 100% of the recommended dose) were applied in a fertirrigated system for those irrigated. For the non-irrigated nutrition was applied in the standard manner. Each portion had the dimensions: 60.0 m x 53.5 m. The average of the sample area was 124.6 m^2. The average flow of drippers used in the experiment after field evaluations were 2. 2L h^{-1}. Water analysis revealed it was adequate for irrigation and soil reacquired 2 tons of dolomitic lime per hectare. Plantation occurred in October, 2003. Plant spacing was 4.0 m x 1,5 m, the formulation applied at planting 08-28-16 and of coverage 20-00-20, resulting in a total of 0.25 ton. ha^{-1} and 0,2 ton. ha^{-1}, respectively. Data were collected in zigzag from the first measurable tree, corresponding to measurements from 2004 to 2008 with annual periodicity.

For field work, it was used meter tape in mm scale to measure tree's circumference 1.30 m above ground for later conversion to diameter. Haglof electric hypsometer was used to measure heights, for the lease of samples and measuring tape to measure the sample areas. Software Statistica 6.0 was used for statistical analysis of data. The volumetric calculation required the scaling of 14 trees, 7 in the irrigated sector and 7 in the non-irrigated. The choice of trees was based on diameter at breast height. The scaling was performed by collecting two orthogonal diameters at heights of sections 0.10, 0.70, 1.30, 2.0 and from this on, in every meter. Trees' heights were also collected. In order to calculate the solid volume of tree, Smalian's formula was applied:

$$V = \frac{(g_1 + g_2)}{2} \times c \qquad (1)$$

where

g_1 = base diameter

g_2 = top diameter

c = tree's height.

In order to obtain the solid volume of each tree, an average form factor was applied (0.4412).

Graphics of the collect data were made, featuring trend lines and the R^2.

3 Results and discussion

The total volume of eucalyptus in 2008 and the Mean Annual Increment (MAI) per treatment is in Table 1. Biggest values of MAI (77.6 $m^3ha^{-1}year^{-1}$) were obtained by treatment with 100% fertilized and irrigated, with total volume for

Table 1: Total volume in 2008 (m^3ha^{-1}) and Mean Annual Increment, MAI ($m^3ha^{-1}year^{-1}$) of eucalyptus.

Treatment	Statistic	Volume (m^3ha^{-1})	MAI ($m^3ha^{-1}year^{-1}$)
(E-IR-A)[1]	Mean:	376.1	77.6
	CV %	12.7	
	CI 95%	20,2	
(E-IR-B)[2]	Mean:	328.6	67.8
	CV %	22.5	
	CI 95%	35.9	
(E-IR-C)[3]	Mean:	365.8	75,4
	CV %	9.5	
	CI 95%	15.1	
(E-NI-A)[4]	Mean:	334.9	69.1
	CV %	16.4	
	CI 95%	26	
(E-NI-B)[5]	Mean:	271.8	56
	CV %	33.5	
	CI 95%	53.3	
(E-NI-C)[6]	Mean:	304.9	62.9
	CV %	20.1	
	CI 95%	32	

[1]E-IR-A: irrigated eucalyptus with 100% of nutrition.
[2]E-IR-B: irrigated eucalyptus with 50% of nutrition.
[3]E-IR-C: non-irrigated eucalyptus with 25% of nutrition.
[4]E-NIR-A: non-irrigated eucalyptus with 100% of nutrition.
[5]E-NIR-B: non-irrigated eucalyptus with 50% of nutrition.
[6]E-NIR-C: non-irrigated eucalyptus with 25% of nutrition.

that block of 376.1 m^3ha^{-1}. Lowest levels of MAI (56 $m^3ha^{-1}year^{-1}$) and total volume (271.8 $m^3ha^{-1}year^{-1}$) were observed in the treatment E-NIR-B. It is noteworthy that in Brazil a good management of production obtains a maximum of 50 $m^3ha^{-1}year^{-1}$.

Figure 1 shows eucalyptus' height with different markers referring to the annual averages and trend lines with correlation coefficients above 97%, which means data are strongly correlated. During the period studied, with lines of tendency for each treatment. In 2008 there were close height values of E-IR-A (24.16 m), E-IR-C (23.69 m) and E-NIR-A (23.25 m). Treatment E-NIR-B had the lowest levels (20.75 m)

The equations for each treatment (and each correlation coefficient) are listed below:

E-IR-A: irrigated eucalyptus with 100% of nutrition,

$$y = 13.239 \ln x + 2{,}556 \qquad R^2 = 0{,}9915 \tag{2}$$

E-IR-B: irrigated eucalyptus with 50% of nutrition

$$y = 13.805 \ln x + 1{,}5742 \qquad R^2 = 0{,}9757 \tag{3}$$

E-IR-C: non-irrigated eucalyptus with 25% of nutrition

$$y = 13.188 \ln x + 1{,}7364 \qquad R^2 = 0{,}9839 \tag{4}$$

E-NIR-A: non-irrigated eucalyptus with 100% of nutrition

$$y = 12.998 \ln x + 1{,}9664 \qquad R^2 = 0{,}9886 \tag{5}$$

E-NIR-B: non-irrigated eucalyptus with 50% of nutrition

$$y = 11.779 \ln x + 2{,}236 \qquad R^2 = 0{,}9944 \tag{6}$$

E-NIR-C: non-irrigated eucalyptus with 25% of nutrition

$$y = 11.513\, ln\, x + 2{,}0559 \qquad R^2 = 0{,}9957 \tag{7}$$

Results of Turkey's test for eucalyptus heights are in Table 2. It is observed that in 2005, 2007 and 2008 E-IR-A differed significantly from others (Table 2(A), 2(C) and 2(D)).

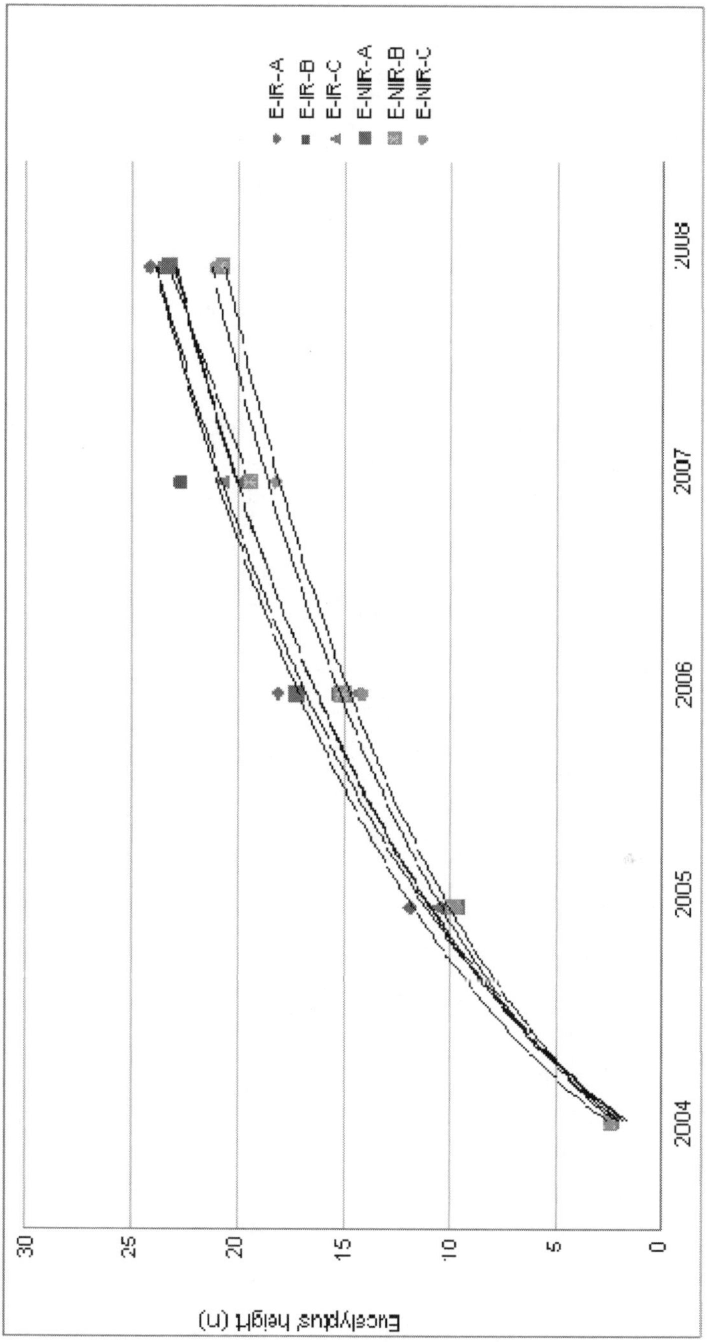

Figure 1: Eucalyptus' height (m) from 2004 to 2008.

Table 2: Results of Tukey's test for eucalyptus' height (m).

2005 (A)	
Treatment	Mean*
(E-NIR-C) [6]	9.65625 a
(E-NIR-A) [4]	9.72917 a
(E-NIR-B) [5]	9.875 a
(E-IR-B) [2]	10.17708 a
(E-IR-C) [3]	10.66667 a
(E-IR-A) [1]	11.88542 b
2006 (B)	
Treatment	Mean
(E-IR-C) [3]	14.43333 a
(E-NIR-C) [6]	14.57604 a
(E-NIR-B) [5]	15.06771 a
(E-IR-B) [2]	15.23125 a
(E-NIR-A) [4]	17.20104 b
(E-IR-A) [1]	18.0375 b
2007 (C)	
Treatment	Mean
(E-NIR-B) [5]	20.85857 a
(E-NIR-C) [6]	21.0871 ab
(E-NIR-A) [4]	22.75797 ab
(E-IR-B) [2]	23.21538 ab
(E-IR-C) [3]	23.78676 ab
(E-IR-A) [1]	24.11014 b
2008 (D)	
Treatment	Mean
(E-NIR-B) [5]	20.85857 a
(E-NIR-C) [6]	21.0871 ab
(E-NIR-A) [4]	22.75797 ab
(E-IR-B) [2]	23.21538 ab
(E-IR-C) [3]	23.78676 ab
(E-IR-A) [1]	24.11014 b

[1]E-IR-A: irrigated eucalyptus with 100% of nutrition.
[2]E-IR-B: irrigated eucalyptus with 50% of nutrition.
[3]E-IR-C: non-irrigated eucalyptus with 25% of nutrition.

[4]E-NIR-A: non-irrigated eucalyptus with 100% of nutrition.
[5]E-NIR-B: non-irrigated eucalyptus with 50% of nutrition.
[6]E-NIR-C: non-irrigated eucalyptus with 25% of nutrition.
[*]Treatment means in the same group are not statistically different from each other.

4 Conclusion

In 2006, E-IR-A was statistically equivalent to the average of E-NIR-A. In 2007, E-NIR-B had the minimum values whereas E-IR-A featured the maximum values. The biggest value of eucalyptus' height was 24.11, corresponding to E-IR-A in 2008 (Table 2D).

Irrigated treatments showed the highest biometric rates (height). Higher values of MAI were obtained by the irrigated treatment and 100% fertilized.

References

[1] Gruber, Y.B.G, Tavares, A.C.S., Detomini, E.R. e R.D. Coelho, 2006. Eucalipto irrigado tem de produzir mais: quanto é preciso aumentar a produtividade para custear cada sistema de irrigação. Agrianual: anuário da agricultura brasileira [1], 318-321.

[2] Lopes, J.L.W., Guerrini, I.A, Saad, J.C.C. e M.R. Silva, 2005. Efeitos da irrigação na sobrevivência, transpiração e no teor relativo de água na folha em mudas de Eucalyptus grandis em diferentes substratos. Scientia Forestalis [68], 97-106.

[3] Shimizu, J.Y., 2006. Pinus na silvicultura brasileira. Revista da Madeira [16], 4-14.

Negotiations over mineral extraction titles with bilateral asymmetric information: the case of developing countries

R. D Mounanga Mabiala & C. Yang
Southwestern University of Finance and Economics, China

Abstract

Why do negotiations between governments of developing countries and multinational mining companies often yield inefficient trade over acquisitions of mineral extraction titles? Adverse selection is one of the main reasons. When the terms of contracts are prepared on the basis of erroneous information, the resulting contract will be inefficient. This paper presents a theoretical model of negotiations over mineral extraction titles with bilateral asymmetric information; where a mining extraction title is negotiated between a developing country's government and a multinational mining company. When both the mine owner and the mine operator have private information at the time of trading, and when each contracting party has a continuous distribution of types, with strictly positive density function, trade inefficiency is the inevitable result. However, it is still possible to reduce the inefficiency if the agents' valuations are traded at auction, or if they resort to arbitration. Therefore, as a second-best optimal contract, two strategies remain possible; the simple double auction or the final offer arbitration. In both strategies, the average price bid amount is the optimal result of the negotiation.

Keywords: mineral extraction titles, adverse selection, contract inefficiency, second-best outcome, final offer-arbitration, double auction.

1 Introduction

A bilateral contract is an agreement in which each of the parties to the contract makes a promise to the other party. Adverse selection may arise, when both parties have private information, and then each are imperfectly informed about the characteristics of the other party.

In this paper, the complete analysis of negotiation will take into account the following requirements. Firstly, we assume that the mining firm has access to more information than the mine owner about the resource stock level, the nature of the ores extracted, the diversity of mineral composition, and the quality of the resource. Secondly, we assume that the mine owner's specific context is about his commitment. He may or may not commit himself to the concession contract proposed to the mine operator. Thirdly, we suppose that the agents' valuation for the non-renewable resource is equal to their expected mining revenue. Therefore

WIT Transactions on Ecology and the Environment, Vol 144, © 2011 WIT Press
www.witpress.com, ISSN 1743-3541 (on-line)
doi:10.2495/ECO110211

the asymmetric variable is their valuation for the mineral deposit. Finally, we assume that the arbitrator's preferred settlement is normally distributed around its mean and its standard deviation.

2 The formal framework

As Dewatripont and Bolton [1] analysed, in presence of bilateral asymmetric information, it may not always be possible to achieve trade. One assumes that the two risk neutral agents have private, independently distributed information about their valuation. However, the agent expected valuation has differentiable strict positive density. We can prove that efficient trade cannot be achieved if contracting takes place at the interim stage. We should firstly determine the set of implementable trades that satisfy interim individual rationality, and then show that efficient trading is not in this set.

2.1 The set of implementable trades

Applying the revelation principle ; the mining owner and the mining operator's problem is to choose a revelation contract. The contingent on announcement satisfies incentive-compatibility and individual rationality constraints, and maximizes the expected gain from trade; thus in any truth-telling equilibrium we have:

-from the mining owner:

$$\Pr(E(\pi_g)) = \int_{E(\underline{\pi}_f)}^{E(\bar{\pi}_f)} \Pr(E(\pi_g); E(\pi_f)) f_f(\pi_f) d(\pi_f) \quad \text{seller expected probability of trade} \tag{1}$$

$$\eta_g(E(\pi_g)) = \int_{E(\underline{\pi}_f)}^{E(\bar{\pi}_f)} H(E(\pi_g); E(\pi_f)) f_f(\pi_f) d(\pi_f) \quad \text{the expected price from the firm} \tag{2}$$

-from the mining operator:

$$\Pr(E(\pi_f)) = \int_{E(\underline{\pi}_g)}^{E(\bar{\pi}_g)} \Pr(E(\pi_g); E(\pi_f)) f_g(\pi_f) d(\pi_f) \quad \text{buyer expected probability of trade} \tag{3}$$

$$\eta_f(E(\pi_f)) = \int_{E(\underline{\pi}_g)}^{E(\bar{\pi}_g)} H(E(\pi_g); E(\pi_f)) f_g(E(\pi_g)) dE(\pi_g) \quad \text{the expected price from the state} \tag{4}$$

Thus, each agent expected payoff under the contract can be written as:

$$U_g(E(\pi_g)) = \eta_g(\pi_g) - E(\pi_g)\Pr(E(\pi_g)) \tag{5}$$

and

$$U_f(E(\pi_f)) = E(\pi_f)\Pr(E(\pi_f)) - \eta_f(E(\pi_f)) \tag{6}$$

The incentive-compatibility and interim individual-rationality constraints for both agents are then given by:

$$U_g(E(\pi_g)) \geq \eta_g(E(\pi_g)) - E(\pi_g)\Pr(E(\pi_g))\,\forall \pi_g \in \left[E(\underline{\pi}_g); E(\bar{\pi}_g)\right] \text{ State incentive constraint} \tag{7}$$

$$U_f(E(\pi_f)) \geq E(\pi_f)\Pr(E(\pi_f)) - \eta_f(E(\pi_f))\,\forall \pi_f \in \left[E(\underline{\pi}_f); E(\bar{\pi}_f)\right] \text{ firm incentive constraint} \tag{8}$$

$$U_g(E(\pi_g)) \geq 0\,\forall \pi_g \in \left[E(\underline{\pi}_g); E(\bar{\pi}_g)\right] \text{ State individual rationality} \tag{9}$$

$$U_f(E(\pi_f)) \geq 0\,\forall \pi_f \in \left[E(\underline{\pi}_f); E(\bar{\pi}_f)\right] \text{ firm individual rationality} \tag{10}$$

Assuming that agents' preferences satisfy the single-crossing condition problem, only the individual-rationality constraints for the lowest types may be binding:

$$U_g(E(\bar{\pi}_g)) \geq 0 \tag{11}$$

and

$$U_f(E(\underline{\pi}_f)) \geq 0 \tag{12}$$

The lowest expected valuation of the government is $E(\bar{\pi}_g)$ (his opportunity cost).

Each agent expected probability of trade must be monotonically increasing in his type; then by using the envelope theorem (by integrating), we obtain:

$$U_g(E(\pi_g)) = U_g\left(E(\bar{\pi}_g)\right) + \int_{E(\pi_g)}^{E(\pi_g)} \Pr(E(X_g))dE(X_g) \tag{13}$$

symmetrically,

$$U_f(E(\pi_f)) = U_f\left(E(\underline{\pi}_f)\right) + \int_{E(\underline{\pi}_f)}^{E(\pi_f)} \Pr(E(X_f))dE(X_f) \tag{14}$$

It simply means that, given monotonicity, only local incentive compatibility constraints matter, so the mining operator and the mining owner's informational values are simply the sum of the lowest buyer's expected bid and the integral of all inframarginal buyers' expected marginal value. More exactly, the mining owner's gain by pretending that he is reluctant to sell the extraction rights of the mining resource, so as to induce the mining operator to offer a higher expected price. Therefore, he would always claim to have the highest possible expected opportunity cost. To induce the government to truthfully announce a valuation; the mining operator must therefore be given some informational value, and giving him this value raises the likelihood of trade, when he announces a lower expected opportunity cost. At the same time, the mining company will always underestimate his valuation of the non-renewable resource. Thus, any firm must be given some price to be induced to tell the truth, or equivalently the likelihood of trade must be increasing in the company's announced valuation.

2.2 Agent negative virtual surplus

Efficient trade for this contract requires that trade take place with

$$\Pr(E(\pi_g); E(\pi_f)) = 1 \ \forall \ E(\pi_g) \leq E(\pi_f) \tag{15}$$

and

$$\Pr(E(\pi_g); E(\pi_f)) = 0 \ \forall \ E(\pi_f) \leq E(\pi_g) \tag{16}$$

Knowing that, the mining contract imposes the first-best trade probabilities and satisfies all incentive-compatible constraints; we show that mining contract violates the interim individual rationality constraints [1]. To prove this result, observe that (putting E1 into E5, and E5 into E13, we obtain E17):

$$\eta_g(E(\pi_g)) = E(\pi_g) \int_{E(\underline{\pi}_g)}^{E(\bar{\pi}_g)} \Pr(E(\pi_g); E(\pi_g)) f_g(E(\pi_g)) dE(\pi_g) + U_g(E(\bar{\pi}_f)) + \int_{E(\pi_g)}^{E(\bar{\pi}_g)} \Pr(E(X_g)) dE(X_f) \tag{17}$$

Symmetrically (putting E3 into E6, and E6 into E14, we obtain E18),

$$\eta_f(E(\pi_f)) = E(\pi_f) \int_{E(\underline{\pi}_f)}^{E(\bar{\pi}_f)} \Pr(E(\pi_f); E(\pi_f)) f_f(E(\pi_f)) dE(\pi_f) - U_f(E(\underline{\pi}_f)) - \int_{E(\underline{\pi}_f)}^{E(\pi_f)} \Pr(E(x_f)) dE(x_f) \tag{18}$$

Knowing that the sum of the expected payment cannot exceed the total transfer available, then [2]:

$$\int_{E(\underline{\pi}_g)}^{E(\bar{\pi}_g)} \eta_g(E(\pi_g)) f_g(E(\pi_g); E(\pi_f)) dE(\pi_g) + \int_{E(\underline{\pi}_f)}^{E(\bar{\pi}_f)} \eta_f(E(\pi_f)) f_f(E(\pi_g); E(\pi_f)) dE(\pi_f) \leq$$

$$\int_{E(\underline{\pi}_f)}^{E(\bar{\pi}_f)} \int_{E(\underline{\pi}_g)}^{E(\bar{\pi}_g)} H(E(\pi_g); E(\pi_f)) f_g(E(\pi_g)) f_f(E(\pi_f)) dE(\pi_g) dE(\pi_f) \tag{19}$$

with $H(E(\pi_g); E(\pi_f)) = \dfrac{\eta_g + \eta_f}{2}$ from the total transfer available, we can express that the mining operator expected surplus (from E18 we obtain E20):

www.witpress.com, ISSN 1743-3541 (on-line)

$$\int_{E(\underline{\pi}_f)}^{E(\bar{\pi}_f)}\int_{E(\underline{\pi}_g)}^{E(\bar{\pi}_g)} H(E(\pi_f);E(\pi_g))f_g(E(\pi_g))f_f(E(\pi_f))dE(\pi_g)dE(\pi_f) = \int_{E(\underline{\pi}_f)}^{E(\bar{\pi}_f)}\int_{E(\underline{\pi}_g)}^{E(\bar{\pi}_g)} E(\pi_f)\Pr(E(\pi_f);E(\pi_g))f_g(E(\pi_g))f_f(E(\pi_f))$$

$$dE(\pi_g)dE(\pi_f) - U_f(E(\underline{\pi}_f)) - \int_{E(\underline{\pi}_g)}^{E(\bar{\pi}_g)}\left[\int_{E(\underline{\pi}_f)}^{E(\bar{\pi}_f)}\int_{E(\underline{\pi}_f)}^{E(\pi_f)} \Pr(E(\pi_f);E(\pi_g))f_g(E(\pi_g))f_f(E(\pi_f))\right]dE(\pi_g)dE(\pi_f) \quad (20)$$

Therefore, using Fubini's theorem - proposition 1 - we obtain:

$$\int_{E(\underline{\pi}_f)}^{E(\bar{\pi}_f)}\int_{E(\underline{\pi}_g)}^{E(\bar{\pi}_g)} H(E(\pi_f);E(\pi_g))f_g(E(\pi_g))f_f(E(\pi_f))dE(\pi_g)dE(\pi_f) = \int_{E(\underline{\pi}_f)}^{E(\bar{\pi}_f)}\int_{E(\underline{\pi}_g)}^{E(\bar{\pi}_g)} \Pr(E(\pi_f);E(\pi_g))\left[E(\pi_f) - \frac{1-F_f(E(\pi_f))}{f_f(E(\pi_f))}\right]$$

$$f_g(E(\pi_g))f_f(E(\pi_f))dE(\pi_g)dE(\pi_f) - U_f(E(\underline{\pi}_f)) \quad (21)$$

The following expression

$$\int_{E(\underline{\pi}_f)}^{E(\bar{\pi}_f)}\int_{E(\underline{\pi}_g)}^{E(\bar{\pi}_g)} \Pr(E(\pi_g);E(\pi_f))\left[E(\pi_f) - \frac{1-F_f(E(\pi_f))}{f_f(E(\pi_f))}\right]f_g(E(\pi_g))f_f(E(\pi_f))dE(\pi_g)dE(\pi_f) - U_f(E(\underline{\pi}_f)) \quad (22)$$

in the adverse selection literature [2], represents the mining operators virtual expected revenue, which includes the expected revenue plus the informational expected revenue, minus the reciprocal of the value of the density function of the mining operator's preferred settlement. More precisely, the hazard rate is the cost of screening the information relative to the multinational mining firm's expected mining revenue.

Symmetrically, from the total transfer available, we can express the government expected surplus. (from E17 we obtain E23):

$$\int_{E(\underline{\pi}_f)}^{E(\bar{\pi}_f)}\int_{E(\underline{\pi}_g)}^{E(\bar{\pi}_g)} H(E(\pi_f);E(\pi_g))f_g(E(\pi_g))f_f(E(\pi_f))dE(\pi_g)dE(\pi_f) = \int_{E(\underline{\pi}_f)}^{E(\bar{\pi}_f)}\int_{E(\underline{\pi}_g)}^{E(\bar{\pi}_g)} E(\pi_g)\Pr(E(\pi_f);E(\pi_g))f_g(E(\pi_g))f_f$$

$$(E(\pi_f))dE(\pi_g)dE(\pi_f) + U_g(E(\bar{\pi}_g)) + \int_{E(\underline{\pi}_f)}^{E(\bar{\pi}_f)}\left[\int_{E(\underline{\pi}_g)}^{E(\bar{\pi}_g)}\int_{E(\pi_g)}^{E(\bar{\pi}_g)} H(E(\pi_f);E(\pi_g))f_g(E(\pi_g))\right]f_f(E(\pi_f))dE(\pi_g)dE(\pi_f) \quad (23)$$

Therefore, using Fubini's theorem - proposition 1 - we obtain:

$$\int_{E(\underline{\pi}_f)}^{E(\bar{\pi}_f)}\int_{E(\underline{\pi}_g)}^{E(\bar{\pi}_g)} H(E(\pi_f);E(\pi_g)) f_g(E(\pi_g)) f_f(E(\pi_f)) dE(\pi_g) dE(\pi_f) = \int_{E(\underline{\pi}_f)}^{E(\bar{\pi}_f)}\int_{E(\underline{\pi}_g)}^{E(\bar{\pi}_g)} \Pr(E(\pi_f);E(\pi_g)) \left[E(\pi_g)+\frac{F_g(E(\pi_g))}{f_g(E(\pi_g))}\right]$$
$$f_g(E(\pi_g)) f_f(E(\pi_f)) dE(\pi_g) dE(\pi_f) + U_g(E(\bar{\pi}_g)) \tag{24}$$

The following expression:

$$\int_{E(\underline{\pi}_f)}^{E(\bar{\pi}_f)}\int_{E(\underline{\pi}_g)}^{E(\bar{\pi}_g)} \Pr(E(\pi_g);E(\pi_f)) \left[E(\pi_g)+\frac{F_g(E(\pi_g))}{f_g(E(\pi_g))}\right] f_g(E(\pi_g)) f_f(E(\pi_f)) dE(\pi_g) dE(\pi_f) + U_g(E(\bar{\pi}_g)) \tag{25}$$

represents the mining owner virtual expected price, which includes the expected price plus the informational expected price borne by the government. More precisely, the hazard rate is the cost of screening the information relative to the government's expected price. Rearranging the last two expressions, one then obtains the participation constraint (E21+ E24 = E26):

$$U_g(E(\bar{\pi}_g)) + U_f(E(\underline{\pi}_f)) = \int_{E(\underline{\pi}_f)}^{E(\bar{\pi}_f)}\int_{E(\underline{\pi}_g)}^{E(\bar{\pi}_g)} \Pr(E(\pi_g);E(\pi_f)) \left\{\left[E(\pi_f)-\frac{1-F_f(E(\pi_f))}{f_f(E(\pi_f))}\right]-\left[E(\pi_g)+\frac{F_g(E(\pi_g))}{f_g(E(\pi_g))}\right]\right\} \tag{26}$$
$$f_g(E(\pi_g)) f_f(E(\pi_f)) dE(\pi_g) dE(\pi_f)$$

The right hand side of this equation gives the virtual surplus of the mining company and the government. This expected payoff should be nonnegative to satisfy both interim individual rationality constraints. Now one can show that the virtual surplus is negative. Substituting the probability of trade in the equation above, with the first-best trading probability one obtains a negative virtual surplus (from E26, we obtain E27)

$$U_g(E(\bar{\pi}_g)) + U_f(E(\underline{\pi}_f)) = -\left[\int_{E(\underline{\pi}_f)}^{E(\bar{\pi}_f)} \left[1-F_f(E(\pi_f))\right] dE(\pi_f) + \int_{E(\underline{\pi}_f)}^{E(\bar{\pi}_f)} \left(\left(E(\bar{\pi}_f)-E(\pi_f)\right) f_f(E(\pi_f))\right) dE(\pi_f)\right] \tag{27}$$

When the firms expected valuation rises to the governments highest expected opportunity cost, then the surplus is clearly negative (E28).

$$U_g(E(\bar{\pi}_g)) + U_f(E(\underline{\pi}_f)) = -\int_{E(\underline{\pi}_f)}^{E(\bar{\pi}_g)} \left[1-F_f(E(\pi_f))\right] dE(\pi_f) \le 0 \tag{28}$$

This result establishes that this bilateral trading problem, where both the government and the mining company have private information at the time of trading, always results in trade inefficiency when each contracting party has a continuous distribution of types, with the strictly positive density function [3].

3 The second-best contract through the double auction

The government could impose a level payment, or ask a price, which begins bargaining (which might be his highest expected amount): $H(E(\pi_g); E(\pi_f))$.

More exactly, we assume that at this asking price, he might agree to sell to the mining operator the extraction rights of the non-renewable resource. On other side, the mining operator makes a bid. In this section, one will characterize the Bayesian equilibrium of this double auction contract; then show that the simple double auction is the second best optimal contract.

3.1 Characterize the Bayesian equilibrium

One assumes that the company simultaneously names an offer price. The mining owner strategy is the price he will ask for each possible valuation. Thus, a pair strategy is Bayesian Nash equilibrium if the following two conditions hold:

$$\forall E(\pi_f) \in [0;1] \; \eta_f(E(\pi_f)) \text{ solves}$$

$$\max_f \left[E(\pi_f) - \frac{\eta_f(\pi_f) + E\left[\eta_g(E(\pi_g)) / E(\pi_f) \geq \eta_g(E(\pi_g))\right]}{2} \right] \Pr\left\{\eta_f \geq \eta_g(E(\pi_g))\right\} \quad (29)$$

$$\forall E(\pi_g) \in [0;1] \; \eta_g(E(\pi_g)) \text{ solves}$$

$$\max_g \left[\frac{\eta_g(\pi_g) + E\left[\eta_f(E(\pi_f)) / \eta_g(E(\pi_g)) \geq \eta_f\right]}{2} - E(\pi_g) \right] \Pr\left\{\eta_g(E(\pi_g)) \geq \eta_f\right\} \quad (30)$$

Assuming that their strategies are linear, and firm-government strategies are best responses to each other these strategies are indeed Bayesian-Nash equilibrium.

The linear Bayesian Nash equilibrium gives the pair strategy:

$$\left\{ \frac{2}{3} E(\pi_g) + \frac{1}{4} ; \frac{2}{3} E(\pi_f) + \frac{1}{12} \right\} \quad (31)$$

is the Bayesian Nash equilibrium; and trade will occur at:

$$\eta_f(E(\pi_f)) \geq \eta_g(E(\pi_g)) \quad (32)$$

3.2 The simple double auction as the second best optimal contract

One will characterize the second best contract and then implement the second best through the double auction.

3.3 Characterizing the second best contract

The second best ex ante optimal outcome maximizes the expected sum of utilities of the two parties, subject to incentive compatibility and interim individual-rationality constraints for each agent. As assume above the new bid or the asking price of is $H(E(\pi_g); E(\pi_f))$. Thus, Governments expected utility (from E5) could be rewritten as follows:

$$\int_0^1 U_g\left(E(\pi_g)\right)dE(\pi_g) = \int_0^1\int_0^1 \left[H\left(E(\pi_g); E(\pi_f)\right) - E(\pi_g)\Pr\left(E(\pi_g); E(\pi_f)\right)\right]dE(\pi_f)dE(\pi_g) \quad (33)$$

The mining company's expected utility equation (from E6) could be rewritten as follows:

$$\int_0^1 U_f\left(E(\pi_f)\right)dE(\pi_f) = \int_0^1\int_0^1 \left[E(\pi_f)\Pr\left(E(\pi_g); E(\pi_f)\right) - H\left(E(\pi_g); E(\pi_f)\right)\right]dE(\pi_f)dE(\pi_g) \quad (34)$$

Therefore the expected sum of utilities of the two parties becomes (E33+E34):

$$\int_0^1 U_g\left(E(\pi_g)\right)dE(\pi_g) + \int_0^1 U_f\left(E(\pi_f)\right)dE(\pi_f) = \int_0^1\int_0^1 \left(E(\pi_f) - E(\pi_g)\right)\Pr\left(E(\pi_g); E(\pi_f)\right)dE(\pi_f)dE(\pi_g) \quad (35)$$

The second-best contract is the solution to the following constraint optimization problem:

$$V = \max\left[\int_0^1 U_g(E(\pi_g))dE(\pi_g) + \int_0^1 U_f(E(\pi_f))dE(\pi_f)\right] \quad (36)$$

Subject to - proposition 2 -

$$\int_0^1\int_0^1 \left(2E(\pi_f) - 2E(\pi_g) - 1\right)\Pr(E(\pi_g); E(\pi_f))dE(\pi_g)dE(\pi_f) \quad (37)$$

When the solution holds, all incentive-compatibility and interim individual-rationality constraints are satisfied. The constraint must be binding at the optimum. Now one can implement the objective function through the double auction.

3.3.1 Implementing the second-best through the double auction

If trade occurs at an average rent level; there exists Bayesian equilibrium of this double auction contract, where the government and the firms best response functions are given by the solutions to:

$$\max_{\eta_g} \int_{\eta_g}^{1} \Big(H(E(\pi_g); E(\pi_f)) - E(\pi_g) \Big) f_{\eta_f}(\eta_f) dE(\pi_f). \tag{38}$$

and

$$\max_{\eta_f} \int_{0}^{\eta_f} \Big(E(\pi_f) - H(E(\pi_g); E(\pi_f)) \Big) f_{\eta_g}(\eta_g) dE(\pi_g) \tag{39}$$

Assuming that the mining company agrees to pay the price the government claims; it will imply that the government and the mining company have in fact the same valuation for the non-renewable resource. Thus the bid claimed by the mining owner will be paid by the mining operator. Thus the government and the mining company will try to maximize their surplus under the assumption that each party bid is distributed according to the density function. Therefore:

$$\eta_g = \frac{2}{3} E(\pi_g) + \frac{1}{4} \tag{40}$$

and

$$\eta_f = \frac{2}{3} E(\pi_f) + \frac{1}{12} \tag{41}$$

This is the agent's strategy, which is also Bayesian Nash equilibrium. The bid versus price paid is indeed the best response functions to one another, and thus form Bayesian Nash equilibrium. In this equilibrium, trade takes place at the rent level:

$$\eta_f \geq \eta_g \tag{42}$$

with

$$\eta_g = H(E(\pi_g); E(\pi_f)) \tag{43}$$

4 The second-best contract through the final offer-arbitration

This analysis follows closely Robert Gibbons' final offer-arbitration steps [4]. The government may ask for an arbitrator to regulate the litigation, as a second strategy. Indeed, the mining codes of some developing countries have legal tendencies within the framework for solving litigations. Let's assume that, the government and the mining company simultaneously make offers. The arbitrator knows but the parties do not. The parties believe that the arbitrator offer is randomly distributed according to a cumulative probability distribution, with associated probability density function.

4.1 The mining owner and the mining operator believe

The mining firms and the government beliefs are represented in the following probabilities respectively:

$$\text{Prob}\ \{\eta_f \text{ chosen}\} = \text{Prob}\ \{\eta_k < H(\eta_g;\eta_f)\} = F[H(\eta_g;\eta_f)] \tag{44}$$

and

$$\text{Prob}\ \{\eta_g \text{ chosen}\} = \text{Prob}\ \{\eta_k > H(\eta_g;\eta_f)\} = 1\text{-}F[H(\eta_g;\eta_f)] \tag{45}$$

Thus, the expected revenue amount settlement is:

$$\begin{aligned}&\eta_f \Pr ob\{\eta_k < H(\eta_g;\eta_f)\} + \eta_g \Pr ob\{\eta_k > H(\eta_g;\eta_f)\} = \\ &\eta_f F[H(\eta_g;\eta_f)] + \eta_g (1 - F[H(\eta_g;\eta_f)])\end{aligned} \tag{46}$$

Assuming that the mining company wants to minimize the expected rent settlement imposed by the arbitrator; and the government wants to maximize it; the pair of offers represents the Nash equilibrium of negotiation between the both agents; therefore

η_f^* must solve:

$$\min_{\eta_f} \eta_f F\left[H(\eta_g^*;\eta_f)\right] + \eta_g^*\left[1 - F\{H(\eta_g^*;\eta_f)\}\right] \tag{47}$$

η_g^* must solve:

$$\max_{\eta_g} \eta_f^* F\left[H(\eta_g;\eta_f^*)\right] + \eta_g\left[1 - F\{H(\eta_g;\eta_f^*)\}\right] \tag{48}$$

The price offer pair must meet the first-order conditions for these optimization problems:

$$(\eta_g^* - \eta_f^*)\frac{1}{2} f\left[H(\eta_g^*;\eta_f^*)\right] = F\left[H(\eta_g^*;\eta_f^*)\right] \tag{49}$$

and

$$(\eta_g^* - \eta_f^*)\frac{1}{2} f\left[H(\eta_g^*;\eta_f^*)\right] = \left[1 - F\{H(\eta_g^*;\eta_f^*)\}\right] \tag{50}$$

The average of the offers must be equal to the median of the arbitrator's preferred settlement [4]. Substituting the average of the offers into the first order condition for the optimization problems above, we obtain:

$$\eta_g^* - \eta_f^* = \frac{1}{f\left[H(\eta_g^*;\eta_f^*)\right]} = f^{-1}\left[H(\eta_g^*;\eta_f^*)\right] \tag{51}$$

This means, the gap between the offers must equal the reciprocal of the value of the density function of the arbitrator's preferred settlement.

4.2 The arbitrator's preferred settlement

Assuming that the arbitrator's preferred settlement is normally distributed around the mean and its variance. The normal distribution is symmetric around its mean; the median of the distribution equals the mean of the distribution. Therefore the Nash equilibrium offers that:

$$\eta_g^* = H(\eta_g^*;\eta_f^*) + \sqrt{\frac{\pi\sigma^2}{2}} \tag{52}$$

and

$$\eta_f^* = H(\eta_g^*;\eta_f^*) - \sqrt{\frac{\pi\sigma^2}{2}} \tag{53}$$

In the equilibrium, the government and the mining company's offers are centered on the expectation of the arbitrator's preferred settlement, and the gap between the offers increases with the parties' uncertainty about the arbitrator's preferred settlement.

Proposition 1

Fubini's theorem [2] states that if f *is integrable on [a;b]x[c;d], then:*

$$\int_a^b\int_c^d f(x;y)dxdy = \int_a^b\left(\int_c^d f(x;y)dy\right)dx = \int_c^d\left(\int_a^b f(x;y)dx\right)dy$$

Thus [1], $\int_{\underline{\theta}}^{\overline{\theta}} uv' = \left[uv\right]_{\underline{\theta}}^{\overline{\theta}} - \int_{\underline{\theta}}^{\overline{\theta}} u'v$ *, with* $v' = f(\theta)$ *and* $u = \int V\left[q(x)\right]dx$

so that, $$\int_{\underline{\theta}}^{\overline{\theta}}\left(\int_{\underline{\theta}}^{\theta} V\left[q(x)\right]dx\right)f(\theta)d\theta = \left[\int_{\underline{\theta}}^{\theta} V\left[q(x)\right]dxF(\theta)\right]_{\underline{\theta}}^{\overline{\theta}} - \int_{\underline{\theta}}^{\overline{\theta}} V\left[q(\theta)\right]F(\theta)d\theta$$

This is equal to, $$\int_{\underline{\theta}}^{\overline{\theta}} V\left[q(\theta)\right]\left[1-F(\theta)\right]d\theta$$

Proposition 2

The participation constraint of the second-best contract is:

$$\int_0^1\int_0^1\left\{\left(E(\pi_f)-\frac{1-F_f(E(\pi_f))}{f_f(E(\pi_f))}\right)-\left(E(\pi_g)+\frac{F_g(E(\pi_g))}{f_g(E(\pi_g))}\right)\right\}\Pr(E(\pi_g),E(\pi_f))f_f(E(\pi_f))f_g(E(\pi_g))dE(\pi_g)dE(\pi_f)$$

From the uniform distribution's properties, the cumulative distribution and the probability density function of the continuous uniform distribution are:

$$F_f(E(\pi_f))=\frac{E(\pi_f)-1}{1-0}=E(\pi_f)-1;\_and\_f_f(E(\pi_f))=\frac{1}{1-0}=1\ \forall 0\le E(\pi_f)\le 1$$

symmetrically,

$$F_g(E(\pi_g))=\frac{E(\pi_g)-1}{1-0}=E(\pi_g)-1;\_and\_f_g(E(\pi_g))=\frac{1}{1-0}=1\ \forall 0\le E(\pi_g)\le 1$$

therefore, the participation constraint becomes,

$$\int_0^1\int_0^1\left(2E(\pi_f)-2E(\pi_g)-1\right)\Pr(E(\pi_g);E(\pi_f))dE(\pi_g)dE(\pi_f)$$

References

[1] Dewatripont, M. &, Bolton, P, *Contract Theory*, MIT Press, 2004.

[2] Salanié, B. *The Economics of Contracts: A Primer,* second edition, ISBN-10: 0-262-19525-9, ISBN-13: 978-0-262-19525-6, March 2005.

[3] Myerson, R.B. & Satterhwaite, M.A. *Efficient Mechanisms for bilateral trading*, Journal of Economic theory, Volume 29, Issue 2, April , 1983.

[4] Gibbons, R.*A Primer in Game Theory,* ISBN: 0745011594, ISBN13: 9780745011592, 288 pages, paperback, Published by Pearson Higher Education, 1992.

Climate change in the Pacific: Tuvalu case-study

I. M. Madaleno
Department of Natural Sciences, Portuguese Tropical Research Institute, Lisbon, Portugal

Abstract

The Tuvalu Group is made of nine small low-lying coral atoll and reef islands, located in the South Pacific, about 1,100 km north of Fiji. With a total area of $26km^2$, it has about 11,000 residents and it is the first nation on Earth to see the rising sun every single day. The people of Tuvalu are mostly of Polynesian origin, their culture and physical type being quite homogeneous. During the month of February 2010, the Portuguese Tropical Research Institute conducted a scientific mission to the atoll of Funafuti, so as to develop an ethno-geographic study. The main objective of the project was to evaluate the Pacific people's awareness to climate change. Results have shown that about two thirds of the remote islanders do not fear the rising sea levels and trust that Divine Providence will bet on their survival. The small population of fishermen and breadfruit, taro, pulaka and coconut subsistence farmers lack economies of scale because of their remoteness yet persevere with a tranquil, slow-paced existence in a vulnerable and isolated environment where they lack resources for adequate development.

Keywords: climate change awareness.

1 Introduction

Tuvalu consists of nine islands located between 4° 36′S and 10° 45′S of the Equator and longitude 176° 8′E and 179° 52′E of Greenwich. From north to south the islands are Nanumea, Nanumanga, Niutao, Nui, Vaitupu, Nukufetau, Funafuti, Nukulaelae and Niulakita [1]. The country lies in the south-east Pacific trade wind belt and even though it is located outside the southern cyclone belt it has suffered over thirty cyclones since the 1940s, the most devastating one being the 1972 cyclone, which provoked one Funafuna construction worker,

WIT Transactions on Ecology and the Environment, Vol 144, © 2011 WIT Press
www.witpress.com, ISSN 1743-3541 (on-line)
doi:10.2495/ECO110221

interviewed in the atoll of Funafuti, in February 2010, to build an elevated wooden house for his family (see Fig. 1).

Rainfall varies from 2,700 mm per year in the northern islands to 3,500 mm in the southern subgroup, which includes Funafuti, an oceanic atoll covered with sandy and infertile coralline soils. Average temperature lies between 25° and 31°C [2]. The Tuvaluan government plea for international leadership and governmental action against climate change is well-known. Kyoto Protocol agreed on an average of 5.2% reduction of greenhouse gas emissions between 2008 and 2012 based on 1990 emission levels but so far neither this target has been met nor consensus been reached in related forum [3]. Atolls are reefs of corals 3 to 5 metres high that enclose a lagoon. At the current pace of rising sea levels, the island group is expected to disappear, fuelling concerns among Pacific peoples.

2 Methods

As extremes in temperature and rainfall become increasingly noticeable, small pacific countries are compelled to diversify their food systems in order to ensure food and nutritional security [4]. Current project was initiated in Easter Island (Rapa Nui) in 2006, in order to answer these concerns. The objectives were threefold: 1. Recovery of ancestral water and soil management practises; 2. Analysis of traditional therapies using native plant species; 3. Evaluation of pacific people's awareness to climate change. This contribution will focus mainly the last objective [5, 6].

Research used a technique designed to study individuals in their unique settings, commonly called case-study method [7]. The example of Tuvalu can be defined as a geographically delineated community as it is a homogeneous people that inhabit a small Pacific country. Fifty-eight semi-structured interviews included questions about the interrelationships among people in the community, the life of the community and concerns with the physical environment. Following ethnogeographic study tradition, observation techniques, photographic and video records were used to form a comprehensive picture of Tuvaluan culture, education, religion, farming and animal husbandry, food habits, health, water collection, arts and crafts. Literature review and archival documentation were complementary sources of information aimed at substantiating data that couldn't be obtained in the face-to-face interviewing process.

3 Results

Literature on vulnerability to natural hazards has concentrated on availability of and access to resources that people can draw upon to deal more effectively with disaster. Sustainable development discourse is now focusing the regulatory role of the local governments in mitigating and adapting concerned populations and communities to climate change. Discussion contributors such as people's perceptions of risk tend to be ignored [8, 9]. Tuvaluans awareness to climate change cannot be understood without knowledge of the Pacific island's history and traditions.

3.1 The History of the tiniest nation on Earth

Atoll societies are simple, ruled by chiefs (*aliki*) that are not just rulers but have reciprocal obligations. Communal meetings (*fatele*) are part of a philosophy of gift exchange, starting with open-handed meat consumption, followed by the exhibition of amicably rival dancing groups and the elder's advice in sounding oratory. Legends of the ancient voyagers, religious issues and socio-political problems are often voiced in these assemblies, which in Funafuti are usually organised per island, in their own meeting house (*maneapa*).

The story of the Funafuti atoll tells us that the first settler was Telematua, from Samoa, together with a woman, Futi, who stayed for good, giving her name to the island. Once atolls are quite poor, Telematua sailed in search of a land of greater fertility and where freshwater was more plentiful, landing there with his second wife Tupu. The man was forever divided between the two islands of his kingdom – Funafuti and Vaitupu – the first wife and the second, named after the travel motion (*vaai ia Tupu* or to go to see Tupu). Vaitupu is the largest island, about eight hours distant from Funafuti to the north, and it is both a table reef and an oceanic atoll with two lagoons [10]. Interestingly this bias has persisted, the capital city, Vaiaku, was established in Funafuti but any sound investment as the best schools were always settled in Vaitupu.

It is difficult to establish the time of the earliest settlement, archaeological chronology varying from 2000 years ago to the sixteenth century [11]. The earliest European settlers got there in the early 1830s and about twenty years later forced labour trade was already hitting the island group. They enslaved Tuvaluans to work in the Peruvian guano extraction sites. Historical record accounts for a decline on population by 17,000 to 3,000 residents, in a quarter of century. Missionary groups arrived in 1865. The London Missionary Society supplanted both traditional religion and other Christian faiths. Roman Catholic missions began work in 1888 but *Ekalesia Kelisiano o Tuvalu* dominates to our days.

Just before World War I broke out there were 3,300 inhabitants in the then called Ellice Group, and 25,000 in the Gilberts [12]. They were incorporated into the British Gilbert and Ellice Island Protectorate in 1892. The economic significance of the islands was related to copra production [13] and to labour provision (see table 1). Phosphate mines located on Ocean Island (Banaba, in today's Kiribati) and Nauru, absorbed Tuvaluan labourers between 1900 and 1979.

During World War II a USA navy base was established and more than 6000 Americans were based in Funafuti. They outnumbered Tuvaluans and because most local residents were transported to outer islands during the conflict, little remains from this occupation apart from the airfield. Following a 1974 referendum, Ellice Islands voted for separation from the Gilberts and the first October 1978 the group became an independent nation with the designation of Tuvalu. Few subsist of copra exporters these days, island communities are

Table 1: Tuvalu Group Demographic Evolution.

Years	Population
1850	17,000
1875	3,000
1914	3,300
1931	4,074
1947	4,487
1986	8,200
1991*	9,043
2002*	9,561
2005**	10,885

*Total accounted population during the census and not only residents as in previous countings.
** Estimated population.
Sources: Maude [1]; Grattan [12]; Munro [13]; Tuvalu [26]; WHO [19].

migration-oriented, and the sate heavily dependent on fishing licences, remittances and imports, in order to meet the most basic needs [14]. With about 11,000 residents, Tuvalu is the tiniest aid dependent country on Earth.

3.2 Nature and nurture and the Tuvaluan resilience

In an island realm there is a higher probability to endemism, in both flora and fauna, although the total number of species inhabiting it is small [15]. In the case of Tuvalu there are about 50 endemic species [16]. Pacific atolls have low agriculture potential. The soil is calcareous and sandy, so porous that any organic materials are quickly leached out, even the rainwater that generously falls. It is important to stress that reef-building corals can flourish only in clear oceanic water at 18° to 23°C, and are inhabited and destroyed by sediment or by freshwater [17].

The most abundant food native species in Tuvalu are Pandanus (*Pandanus odorantissimus and tectorius*) and Pulaka (*Cyrtosperma merkusii*) also known as giant Taro. Apart from Pulaka roots, the main staples are coconuts, locally known as Niu (*Cocus nucifera*), bananas or Futi (*Musa spp.*) and breadfruit, Mei for Tuvaluans (*Artocarpus altilis*) that also feed pigs they raise. Fruits of pandanus and breadfruit are eaten raw, either alone or with bonito, flying-fish and kingfish. The preferred drink is sweet toddy, the coconut tree flowers mixed with rainwater captured in cisterns. The inflorescences can be added to the flour into sweet bread, and distilled to produce an alcoholic drink. Fishing and coconut inflorescence collection are male tasks, whilst women prepare the family meal, and bake the bread usually in outhouse kitchens. After drinking coconut water, that for millennia has offset the scarcity of potable water, Tuvaluans eat the soft meat of the ripe nut, which constitutes an important part of the main meals. Coconut cream (*lolo*) is extracted from chopped pieces of nut meat, using the *tavaga*, and is added to the peeled and washed pulaka corms, a long lasting tradition [18]. The task takes place in the *Fale*, a structure built-up with Pandanus or Fetau wood (*Callophyllum inophyllum),* and it is a male assignment (see Fig. 2).

Figure 1: Elevated wooden house with cistern – Funafuti atoll, Tuvalu.

Figure 2: Elevated wooden structure or *Fale* where Tuvaluans extract coconut *lolo*, using the *tavaga* – Funafuti atoll, Tuvalu.

Pulaka roots are preferred to taro (*Colocasia esculenta*) because the last is a water demanding crop, which is impossible to grow on atolls, where irrigation is not recommended, as said. The giant pulaka is grown in dug pits down to the muddy horizon, because rainwater accumulates at a depth of a few metres. Vegetable waste, such as breadfruit leaves and coconut leftovers built-up the much necessary compost to cultivate this dry crop. Breadfruit, pandanus and fetau provide timber for building houses or canoes, and coconut palms covered the house roofs in ancient times. Modernity brought alien and imported building materials as well as canned fish, meat and packed rice, which became vital part of the Tuvaluan diet.

Milk is scarcely consumed, and animal husbandry is reduced to pigs and poultry. Vitamins ingested are Passion-fruit or Passinifuluti (*Passiflora edulis*), Kalampola (*Averrhoa carambola*), Guava or Kuava in Tuvaluan idiom (*Psidium guajava*), Citrus fruits, such as Laimi (*Citrus aurantiifolia*), Mango, called Mangko (*Mangifera indica*), Papaya or Olesi (*Carica papaya*), Soursop, known as Saosopu (*Annona muricata*), and Noni, pronounced Nono in Tuvalu (*Morinda citrifolia*). Nono leaf is smashed with coconut oil and applied to the skin as antiseptic for inadequate garbage disposal favours dermatological problems among Tuvaluans, because they walk barefoot or wearing slippers. In fact, according to the World Health Organisation, septic sores rank second in the communicable diseases treated in Funafuti hospital, after respiratory infections [19]. In case of fever and respiratory ailments the ancestral prescription is to wash, peel and squeeze pandanus roots and give the resulting juice as medicine, an internal application particularly recommended to children.

Lifestyle diseases such as diabetes and hypertension are common, mainly due to the excessive consumption of imported canned food, salty and sweet goods. This is principally acute in the atoll of Funafuti, where 42% of the island inhabitants live, whilst residents from outer islands develop more traditional livelihoods and are more reliant on ancestral healing practises. It was evident in the discourse of the interviewed persons, which included migrants from Nanumea (7), Vaitupu (5), Niutao (4), and from other origins with smaller representation. The sample has shown they had different approaches to food preparation and dissimilar eating habits. They also were keen to rely on natural remedies for sickness mitigation as related to born residents in Funafuti.

Common to all informants was the faith in God, the proud they were of their culture, the co-operative manner they conducted their existences within the community. Communal organisations and religious associations, respective meetings and ceremonies, were vital part of their routine, even with youngsters. Tuvaluans constitute a close-knit community where each family has its place and role to perform to the benefit of all, as I also found to be the case with Easter Island. Modernity brought some differences in lifestyle and communal duties though, for instance the ancestral duties of the family are not obligatorily passed from father to son, as in old times, for education in local schools and at the university are prioritised within the group and by the government of the islands, so as to prepare them better for future challenges.

This adaptation process is part of Tuvaluan resilience, made of brave sails men and fishermen, as much as good dancers, experienced tree climbers, skilled artisans, expected to perform better in new professions. As defined by the United Nations International Strategy for Disaster Reduction "*Resilience is the capacity of a (...) community (...) potentially exposed to hazards to adapt, by resisting or changing in order to reach and maintain a certain level of functioning* [20]". Adapt to change in a way you might be capable of enduring as a well-born and raised Tuvaluan is the Polynesian mode to show their uniqueness, their singularity.

4 Discussion

Sea-level variation in the South Pacific Ocean is difficult to assess. Average sea surface in the Western Pacific is about 40 cm higher than the corresponding surface in the Eastern Pacific because of the wind stress of the easterly Trade Winds. (...) During the El Niño-Southern Oscillation (ENSO) events, however, the differential is lost [21]. According to interviews conducted in February 2010, in the atoll of Funafuti, to a meteorologist and to the responsible director of the Tuvaluan Weather Service, data collected has shown that since 1993 the sea level has augmented 5.3 mm per year. This adds to the mentioned differential and points to non-synoptic causes. There are increased impacts on the low islands of the Pacific that cannot be disregarded, challenging science to isolate atmosphere dynamics and ocean currents from human-induced disturbances.

The people of Tuvalu I've interviewed in the atoll of Funafuti were not interested in the scientific debate. Apart from the awareness day, which takes place during the high tides, they got so used to the caprices of the ocean, to the strong Westerlies, to the heavy rainfall, to the occasional hurricanes, that the fact they observe the sea to behave strangely is just a slight detail in their lives. For generations they got used to live in the isolation of their tiny atolls. Europeans, Asians, whoever comes by is welcome. They know they won't take long to depart. Tuvaluans are fearless people.

However, fearless is not synonym to unawareness, even less of lack of interest for their islands, for their physical environments (see table 2). Tuvaluans are engaged in fighting in the international arena for their identity and for their right to exist in remote Pacific atolls, against all odds.

Table 2: Sample of the beliefs in the future of Tuvalu Islands.

Indicators	N° of Interviews	% Interviews
Belief in the disappearance of the islands	20	34,4
Disbelief on the islands' disappearance	34	58,6
No opinion	4	6,9
TOTAL	58	99,9

Source: Author Interviews, 2010.

Tuvalu case-study contrasts with Hitchings research conducted in the United Kingdom, among eleven respondents working in the financial heart of London, in 2007. They were cocooned indoors during the week, spending long hours sitting at their desks within offices that are commonly pumped full of conditioned air. They often never needed to descend to the lobby as others came and catered to their needs in food and laundry. They hurried from the car or the tube to the foyer, limiting outdoor time to a minimum so as to enjoy thermal comfort as much as possible. Because this European study group from a developed country is used to live in sanitised environments, they declared to hardly notice the changing seasons and weather patterns [22]. Results have shown there was also indifference to the challenges posed by wider climatic warming.

Of course no one in Tuvalu expects to change developed countries people's habits, even less condemn high-rise office buildings in the financial capitals of the world. Their expectation is that science and new technologies might be used to solve the global warming problem. But they hope that we lose the enduring ethnocentric perspective we have been using when we study them and let them be, too. That is why, according to Nhamo "*Developing countries have formed strong coalitions to ensure that their voices are heard as they are also the most vulnerable to climate variability and change*" [23]. As to Tuvalu, "*since the early 1990s, Pacific developing member countries have articulated their priorities for addressing climate change at a regional level through the Pacific Plan for Strengthening Regional Coordination and Integration and the Pacific Islands Framework for Action on Climate Change 2006–2015"* [24].

It is an increasingly difficult task as Manzo argues for both the messages of the iconography and the political discourse of climate change are contradictory [25]. In fact, just before, during and after the Copenhagen Conference (2009) the previously dominant danger and vulnerability signifiers have been replaced by doubt and criticism against the Intergovernmental Panel on Climate Change (IPCC). The controversy only augmented by September 2010 when Arctic Sea resources exploitation, opened by the melting ice caps, was discussed among interested players. Faith in Science was terminated? Or was it the faith in Politics that ravaged us all? Definitely distrust was established.

5 Conclusions

Discussion on the issue of climate change presented here reflects extreme positions that so far have made consensus improbable. Why change the comfort of air-conditioned houses and offices, the use of automotive means of transportation, when they were so hardly obtained through years of research, of human and financial resources investment, of technological advance, for the survival of remote islanders, who are aid-dependent?

Whilst the focus are peoples and contrasting cultures and not the ability to continue developing a wide range of livelihood options, at the same pace forever, without depleting non-renewable natural resources, there will never be an agreement as to the fact that there is (really) climate change in the Pacific, even less action and modification of lifestyles. The message from Tuvalu is: *"You believe in nothing anymore. But we believe in God. So we expect a miracle..."*

References

[1] Maude, H.E. Post-Spanish Discoveries in the Central Pacific. *Journal of the Polynesian Society,* **70**, pp. 67-111, 1961.

[2] Lane, J. *State of the Environment Report 1993.* United Nations Development Programme, Apia (Samoa), 1993.

[3] Nhamo, G. Co-Leadership in Climate Change: An Agenda to 2013 and Beyond. *Politikon,* **36(3)**, pp. 463-480, 2010.

[4] Taylor, M., Jaenicke, H., Skelton, P. and Mathur, P.N. Regional Consultation on Crops for the future: towards food, nutritional, economic and environmental security in the Pacific. APAARI, Nadi (Fiji), 2009.

[5] Madaleno, I. M. Water History in Easter Island and extreme northern Chile. *5th Water History Association Conference*. Tampere: University of Tampere (Finland), Tampere, pp. 1-9 (CD-ROM), 2007.

[6] Madaleno, I. M. Sustainable Livelihood Examples from Water Deficient Easter Island and the Lower Amazon River Floodplains. *Deutsher*

Tropentag 2007, University of Kassel-Witzenhausen and University of Göttingen, Witzenhausen, 2007. **http://www.tropentag.de/2007/abstracts/full/135.pdf**

[7] Berg, B.L. Qualitative Research Methods for the Social Sciences. Pearson, New York, 2006.

[8] López-Marrero, T. An Integrative Approach to Study and Promote Natural Hazards Adaptative Capacity: A case-study of two flood-prone communities in Puerto Rico. *The Geographical Journal*, **176 (2)**, pp. 150-163, 2010.

[9] Nolon, J.R. Climate Change and Sustainable Development: The Quest for Green Communities (Part I). *Planning and Environmental Law,* **61 (10),** pp. 3-10, 2010.

[10] Roberts, R.G. Te Atu Tuvalu, a short story of the Ellice Islands. *Journal of the Polynesian Society*, **67 (204),** pp. 394-423, 1958.

[11] Kennedy, D.G. Field Notes on the culture of Vaitupu. *Polynesian Society Memoir,* **9**, pp. 1-4, 1931.

[12] Grattan, C.H. *The Southwest Pacific since 1900.* The University of Michigan, New York, 1963.

[13] Munro, D. and Besnier, N. Plantations in the Atolls: the case of Nukulaelae. Moore, C., Leckie, J. and Munro, D. (eds.) *Labour in the South Pacific.* James Cook University of North Queensland, Townsville, pp. 178-180, 1990.

[14] Munro, D. Migration and Shift to Dependence in Tuvalu. Connell, J. (ed.) *Migration and Development in the South Pacific.* National Centre for Development Studies, Canberra, pp. 30-40, 1988.

[15] Darwin, C. *The Origin of Species by Means of Natural Selection.* Barnes and Noble Classics, New York, 2004 (first edited in 1859).

[16] Seluka, S., Panapa, T. Maluofenua, S. Samisoni, L. and Tebano, T. *A Preliminary Listing of Tuvalu Plants, Fishes, Birds and Insects.* University of the South Pacific, Tarawa (Kiribati), 1998.

[17] Spate, O.H.K. *Paradise Found and Lost.* Routledge, London, 1988.

[18] Koch, G. *The Material Culture of Tuvalu.* University of the South Pacific, Suva, 1961.

[19] WHO. *Western Pacific Country Health Information Profiles.* World Health Organisation, Manila, pp. 454-464, 2008.

[20] U.N. International Strategy for Disaster Reduction. Basic Terms of Disaster Risk Reduction. Online (http://www.unisdr.org/eng/library/lib-terminology-eng%20home.htm).

[21] Spencer, T., Geodynamics, Ocean Dynamics and Island Geomorphology in the Pacific Plate. Alexandre, J. Dapper, M. de and Symoens, J.-J. (eds) *Climatic Change and Geomorphology in Tropical Environments,* pp. 49-99, 1994.

[22] Hitchings, R. Seasonal Climate Change and the Indoor City Worker. *Transactions of the Institute of British Geographers,* **35 (2),** pp. 282-298, 2010.

[23] Nhamo, G. Co-Leadership in Climate Change: An Agenda to 2013 and Beyond. *Politikon,* **36(3)**, pp. 467, 2010.
[24] *ADB's Pacific Approach 2010-2014*. Asian Development Bank, Mandaluyong City, 2009, pp. 9.
[25] Manzo, K. Imaging vulnerability: the iconography of climate change. *Area,* **42 (1),** pp. 96-107, 2010.
[26] Tuvalu. Government of Tuvalu Demographic Profile 1991-2002. Secretariat of the Pacific Community, Noumea, 2002.

Certifications in construction: a case study comparing LEED and HQE

E. Vazquez[1], M. Miguez[2], L. Alves[3], J. Valente[3] & G. Rossi[4]
[1]*Department of Civil Construction, Urban Engineering Program, Federal University of Rio de Janeiro, UFRJ, Brazil*
[2]*Department of Transportation Engineering, Urban Engineering Program, Federal University of Rio de Janeiro, UFRJ, Brazil*
[3]*Department of Civil Construction, Federal University of Rio de Janeiro, UFRJ, Brazil*
[4]*Department of Graphics, Urban Engineering Program, Federal University of Rio de Janeiro, UFRJ, Brazil*

Abstract

Proposal: This paper presents concepts of sustainability embedded in environmental certifications applied to building construction. The article addresses the importance of certification and the benefits to the environment coming from its deployment. Building constructions consume many materials, emit many gases, and use a lot of energy and water, what point to the extreme importance of developing processes that aim to conserve natural resources, using them optimally. It is necessary to identify actions needed to integrate all the steps involved in the project life cycle, including entrepreneurs, developers, builders, manufacturers, managers of enterprises, users and society. Among these reasons, arises the certification process of the project, which helps minimizing the environmental impacts using natural resources in a more conscious way. This paper aims to present the concepts of an environmental certification focused on construction, compare two of these systems: LEED (Leadership in Energy Environmental Design) and *HQE (Haute Qualité Environnementale)*, its application methodologies application methodologies, the phases in which they apply during the life cycle of an enterprise, and their benefits. **Research Structure / Approaches**: The article is structured in three parts: a brief review on this topic historic evolution; a description of the main features of the existing certification procedures; and case studies covering a LEED certified building and

WIT Transactions on Ecology and the Environment, Vol 144, © 2011 WIT Press
www.witpress.com, ISSN 1743-3541 (on-line)
doi:10.2495/ECO110231

another with HQE certification. **Contributions**: The sustainable construction is based on actions that enable the building construction to encounter or create solutions for ecological problems, using technology, selecting the materials and their suppliers, creating buildings that meet the needs of their users and, also, the environment.

Keywords: sustainability, environmental certifications, building construction, certification LEED, Certification HQE.

1 Introduction

The Industrial Revolution began in the eighteenth century and was marked by the creation of the machine, locomotives and steam trains, exploration of coal, iron and steel. England was the pioneer in this area due to large reserves of coal and also iron ore (Gottfried [1]).

Humanity was concerned solely with production and consumption, overestimating the planet's capacity to assimilate the exploitation of natural resources, which brought consequences that eventually emerged with increasing noise pollution, environmental degradation, rural exodus and the uncontrolled growth in cities that still happens up to the present today.

In this timeline, the sustainability issue was relegated to low priority, taking shape in the mid-nineteenth century. Humanity's concern with planet development begins in the 60's, when the debates on environmental degradation started.

Civilian buildings consume many materials, emit many gases, use lots of energy and water, making a new vision extremely important to conserve the natural resources, using them optimally, taking this awareness to all parties involved in the process (from the builder up to the user) (Krygiel and Nies [2]).

Amid these reasons, arises the environmental certification process of the venture, which collaborate minimizing environmental impacts, using, in a more conscious way, natural resources resulting in many benefits.

In the Brazilian market there are two main systems of environmental certification of buildings, the LEED (Leadership in Energy Environmental Design) developed by the U.S. in 1991 and the HQE (*Haute Qualité Environnementale*) developed by France in 2002.

2 Methodology

The subject matter was developed through a comparative case study between the two certification processes, LEED and HQE, in order to identify the best practices, differences and similarities seeking a critical analysis to integrate the concepts of sustainable construction. The preparation of the article was based on literature research, articles and magazines, creating a comparative of efficient solutions for sustainability.

3 Sustainable construction

Sustainable construction is developed based on actions that enable construction to encounter and create solutions to environmental problems using technology, selecting the materials and their suppliers, creating constructions that meet the needs of their users and the environment.

To have a sustainable building is important to evaluate the site of the construction, planning every step in order to reduce the aggression to the environment before, during and after the construction. The foundation of these buildings is to study and use the constructive elements that do not harm the environment, using recyclable materials that save water, with energy efficiency, minimizing air pollution and providing better comfort for its users.

It is important not only to build sustainably, but also to prove that the building actually follows these assumptions, especially after occupation. Sustainable constructions generate fewer environmental impacts covering all stages of the building's life cycle, from the conception of the product and design, through the process of construction and use of buildings, reaching the stage of demolition. Being of great relevance the understanding of this cycle, to make decisions in advance and minimize the future effects as described in table 1 (Yudelson [3]).

Table 1: Building's life cycle phases.

PLANNING	Beginning of the life cycle of a building. Financial feasibility studies, project design and specifications and the development of constructive activities are conducted.
IMPLANTATION	Phase of construction of the building, putting into practice the projects developed.
USE	Phase contemplated by the building's use.
MAINTENANCE	Phase where the need arises for replacement of certain elements, maintenance of equipment and systems, correction of any failure of performance.
DEMOLITION	Phase when the product is no longer used.

In each phase of the lifecycle of the building there are many benefits that focusing on sustainability can bring to the environment. Designers in addition to understanding this importance need to insert these constraints in the design phase on activities that can interact with the environment, from the specification of materials to the indoor air quality during the occupation of the buildings, passing through the location of the enterprise, adopted lighting systems, air conditioning and water heating.

4 Certification in construction

Many meetings and conferences that have been held for years, tried to get the environmental quality, by new technologies, with help from organizations and creating new business models. Thus appearing tools aimed at environmental and social responsibility, such as certification in construction. The table 2 presents the history of certifying agencies worldwide.

Table 2: History of certification.

Year	Country	Certification
1990	England	BREEAM
1999	USA	LEED
2002	Australia	Green Star
2002	France	*HQE*
2003	USA	Cal-Arch
2004	Australia	Nabers
2004	USA	Energy Star

The certification in construction is an important tool that establishes a process for managing the impacts of construction on the environment, consolidating the responsibilities of all parties involved as business and environmental control agencies.

As basis for this study were used two certification systems, the LEED (Leadership in Energy Environmental Design) process, which is done by the Green Building Council Brazil, with basis on American criteria and the HQE (*Haute Qualité Environnementale*) process or AQUA (*Alta Qualidade Ambiental*) being conducted by Foundation Vanzolini with the basis on Brazilian criteria.

4.1 Certification objectives

The purpose of certification is to promote an awareness of all involved in the process, from the design phase through construction, until the user, incorporating solutions that will enable a reduction in the use of natural resources, promoting comfort and quality to its users.

A greater investment initially is presented, but has lower operating costs, valuing the property, being healthier for its users, conserving water and energy, and reducing gas emissions.

The organizations that carry out the certification provide standards and instructions for the best production of the project and evaluate if they are being followed. Incremental changes are made rather than radical ones, and methods of design and construction are used for the establishment of enterprises with high performance. The organization has a label that proves the product sustainable.

4.2 Benefits of certification

The square footage of the property, the number of rooms, location and conditions for shopping, is information presented in many ads for sales of the enterprises. Some companies are including on the ads the labels of "green buildings". A way to confirm if the property is in line with the propaganda of "green building" is whether it has any type of certification provided by some organization or its representative. At the time of purchase, to make the best choice considering the price and quality of the property is necessary to evaluate how the building was designed, not to encourage the aggression to the environment and health of its users.

The benefits brought by the certification of a building are visible in the long term. The largest impacts seek to reduce consumption of water and energy, and this factor will not always contribute at the time of purchase due to high initial cost.

Certifications are a means of valuing the enterprise in the market and there is no single standard of reference. What can occur is that a certain type of certification may not be suitable for all projects, due to its own concept.

The company is recognized in the market as environmentally friendly by linking the brand to the product, with the potential to reach new markets, reducing production costs and attracting new investment, making it easier to obtain financing.

4.3 LEED Certification

Developed by USGBC, which is a nonprofit organization with a focus on sustainability of buildings and real estate, in the United States since 1991, the system was developed for guidance and certification of sustainable constructions, internationally recognized. It confirms that the buildings were designed and constructed through strategies that aim to improve performance in terms of energy, water, reduction of CO2 emissions, better indoor quality, managing the use of natural resources and minimizing environmental impacts.

LEED is a voluntary system that can be applied to any type of construction and at any stage of the life cycle of the enterprise. This certification quantifies the level of environmental protection that an enterprise can redeem. In Brazil, it was implemented by GBC Brazil, in January 2008.

It is necessary to meet some requirements that are variable and depend on the category of certification, to obtain the score. The minimum requirements to be met at the stage of the project can earn points for certification and if not met, the project cannot be certified.

Each category of performance sums a score that defines the type of certification that will be appropriate to the enterprise. The total score determines which level of certification the project is included, which can be: certified, silver, gold or platinum.

4.4 *HQE* Certification

After the conference held in Rio, ECO 92, arose reflections in many countries. As a result of this conference, in 2002, the *HQE* (*Haute Qualité Environnementale*) was created, which is a process based on performance benchmarks established in the *Centre Scientifique et Technique du Bâtiment (CSTB)*, created in 1947, in France.

In Brazil, it is also known as process *AQUA* (*Alta Qualidade Ambiental*), adapted from the original model of the *HQE* for the Brazilian version in 2007. It was implanted by the Vanzolini Foundation, a private institution non-profit organization, formed and maintained by professors at *USP (Universidade de São Paulo)* since 1967. The process certifies that the *AQUA* development is in accordance with the requirements through independent audits.

The *HQE* is a project management process in order to obtain the environmental quality of a project construction or rehabilitation, accepting solutions that bring economies to the project.

This system has a technical reference, which is adapted to Brazil that has as characteristic to adapt to the climate of the region in which it will be deployed demanding performance results. During the process, all the Brazilian prevailing Standards are used, which are quoted in the technical reference material produced by the Vanzolini Foundation. If there is absence of some Brazilian standards, international standards may be used.

The process is structured into two instruments to assess the performance required, the enterprise management system (EMS) and environmental quality of the building (EQB).

This system is based on performance, therefore there is no score. Three performance levels are considered: good, excellent and superior. Certification requires at least three categories in excellent level, four in superior and seven in good performance level, totaling fourteen items, which will be based on audit findings, an environmental profile of an enterprise.

4.5 LEED versus *HQE*

The certification systems presented have specific characteristics and requirements each. By applying one of them in an enterprise, from the initial phase (design), it is expected that, compared to traditional buildings, it has a superior performance. The following tables are presented with the main features of both types of certification: LEED and *HQE*.

Table 2 presents a comparison for the organization of each of these systems, in which are evaluated the scope, method of application, categories evaluated and results.

Table 3: Comparison of organization of systems.

Aspects	Method of Application	Categories evaluated	Results
LEED	Attendance of mandatory and qualifier items.	Sustainable site, energy, water efficiency, materials and resources, indoor quality, innovation and design process.	Four levels: certified, silver, gold and platinum. By total score obtained.
HQE	Attendance of an environmental profile. Building certificated or not.	Eco-construction, Ecomanagement, Comfort and health.	The certification is obtained through attendance to the profile of environmental performance chosen.

Table 4: Characteristics of different methodologies.

CHARACTERISTIC	*HQE*	LEED
Model and network	Global network with local criteria, based on the French model	North American model, with global representations
Adequacy of criteria to Brazil	Yes	No
Stages of evaluation	Programming, conception e implementation	Conception
Typology of current buildings	Office buildings, schools and hotels	Core, frontage, office buildings, commercial, residential, lots, neighborhoods, hospitals, shops, schools
Scope	Environment, comfort and health	Environment, comfort and health
Expression of results	Performance profile in different themes	Overall performance level

These systems have different characteristics in methodology. It is necessary to examine which fits best for the venture that will be certified, illustrated by table 3 (Téchne [4]).

These systems have common concerns, such as saving water and energy usage, reducing emission of carbon gases, deployment in the territory, internal comfort and waste management. In the U.S. the biggest concern is the issue of energy therefore LEED corresponds to energetic leadership.

The weight percentage of each item is showed in Figure 1, by each evaluation methodologies. As noticed, LEED is not a uniform system, where a higher percentage in energy efficiency is found, focusing more on some points and less on others. Otherwise, the system *HQE* presents uniformity in the categories, considering the same importance to all items (Melhado [5]).

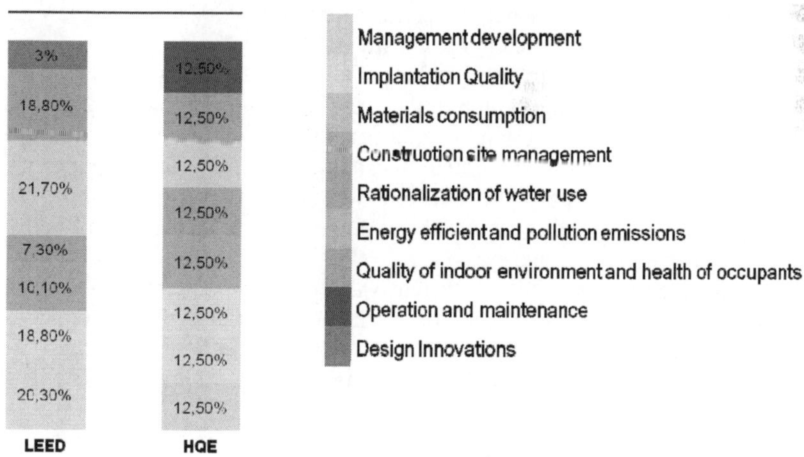

Figure 1: Comparison of the weight of the different themes.

5 Case study: certification LEED and *HQE*

The case study presents the innovations that were made in the implementation of the process of environmental certification LEED and *HQE* in different projects.

For the LEED certification system, the object of study was the Ventura Corporate Towers office building which is an enterprise composed of two towers with 36 floors, five basement floors and a garage building of five floors. It is the first certified enterprise in the gold category in Rio de Janeiro. The commercial building, Ventura Corporate Towers, was framed in the category Core & Shell which has a total of 61 points and the project received 35 points, reaching the gold level of certification.

For the *HQE* certification system, the object of study was the first certification in this system in Brazil, also a commercial building. The Leroy Merlin chose the AQUA certification because of its easiness to adapt to the Brazilian needs, developing its criteria that assess the environmental management of construction and architecture. It aimed for the quality of life for employees and customers, saving energy and water, waste disposal and maintenance, contributing to socioeconomic and environmental development of the region.

5.1 Certification LEED - Ventura Corporate Towers

Hereafter are the strategies employed, according to the concepts of sustainability required by LEED: Sustainability of space, energy efficiency, rationalization of water use, sustainable materials and indoor environmental quality in the construction site (USGBC [6]).

In the phase of construction, aiming the sustainability of the space, there was planning to ensure that precautions have been taken to avoid environmental impact, and problems with neighbors and employees. To preserve the local soil and prevent contamination with chemicals, products with low volatile organic compounds were not accepted in the construction site.

Relating to energy efficiency, the building has a glass frontage, which requires high energy consumption in the use of air conditioning. As a solution to reduce energy consumption, laminated glass 10 mm thick were used with special reflective, which have low thermal transmission of heat and provide natural lighting, thereby reducing the consumption of air conditioning. For energy management sensing light was also used, and the use of reactors and high-tech lamps that consume less energy.

In efficiency in water use, a system was created for retention and reuse of rainwater, which will be captured around the perimeter and will be conducted to a tank for reuse, where it will be pumped into the air conditioning system. Water saving devices were also used such as timers faucets, flush valves with flow control, and presence sensors in urinals.

The materials used in construction, such as steel, concrete and grout are made from 20% to 30% recycled materials and the companies that supply them were located up to 800 km of the enterprise by reducing the emission of CO_2 into the atmosphere. The wood used in construction was certified, ensuring the legality of the product. The trash generated in the phase of construction was collected, separated and received appropriate destination for each type of waste, being sent to specific recycling companies, e.g., such as drywall.

To cover the quality of the internal environment, prestressed ribbed slabs were used without intermediate pillars to ensure an interspace of 14 meters and thus allow for flexibility to internal spaces. Also, with the glass frontage and the use of tight frames that reduce air penetration, this system prevents the entry of pollutants such as carbon dioxide into the commercial rooms, ensuring good indoor air quality. To maintain air quality, smoking is prohibited inside the building, being only possible to smoke within a distance of 8 km of any entrance.

A summary of the main strategies adopted to meet the requirements of LEED certification is presented in figure 2.

Figure 2: Requirements strategy LEED.

5.2 Certification *HQE* - Leroy Merlin

Hereafter are the strategies employed, according to the concepts of sustainability required by the *HQE* certification: project design, completion of the construction, through an Enterprise Management System (EMS) to meet the performance criteria of the Environmental Quality of Building (EQB) (Vanzolini [7]).

The construction site was designed to cause little environmental impact. Through the separation of waste, woods, metals, stones and cement bags were sent for recycling. This way, the demolition stones became pavement and access of the store after recycling.

Eco-efficient materials were used, such as water-based paint that does not affect nature and discharge valve with dual flow that releases water with more or less intensity. Special urinals that use no water were also used, bringing technology and hygiene.

A project was developed to reduce the risk of flooding in the region and not waste the rain water, preserving groundwater. In addition, a reservoir of 150,000 liters was built below the parking lot, the rain water captation systems, storage and distribution. The water abstracted is used for discharge of toilets, watering gardens, outdoor and indoor cleaning. This system allows a saving of 50% of water.

In the energy scope, photovoltaic panels were installed, besides reflectors, and spots on the frontage with LEDs, which have low energy consumption. Special air conditioners were used that automatically adjust indoor temperature. These initiatives allow a saving of 17% of energy.

The floor inside the building was done in polished concrete, for easy cleaning, requiring only a damp cloth, and no use of chemicals.

Figure 3: Requirements strategy *HQE*.

The front of the store was turned to the west but this problem was solved by using materials for protection against heat, solutions provided in the phases of program and design. The frontage of the store is composed of glass and louvers that prevent heating inside the store and allow natural lighting, reducing costs on air conditioning and therefore energy.

At the entrance of the store, crawlers were adopted to provide access to disabled and facilitate, to the users, the exit of shopping carts from the store to the parking lot.

The landscaping of the store is composed by planted trees, as paubrasil and other species that attract birds, making the environment a pleasant place for employees and visitors while preserving the region's fauna. Inside the store is installed an "Ecometer" that tells all the savings made in the store in real time. In the store parking lot exist several posters that have information about the innovations made during the stage of construction, as the case of reuse of rainwater done in the shop, passing on to their customers their commitment to environmental responsibility. The store displays “eco sustainable” products for its customers, stimulating the market with the vision on environmental and economical responsibility.

A summary of the main strategies adopted to meet the requirements of *HQE* certification is presented in figure 3 (Merlin [8]).

6 Final considerations

During this work it was possible to observe that the systems presented are viable, bringing environmental and economic benefits and have similarities among themselves in relation to environmental responsibility, despite of having different methodologies.

It was noted throughout this paper that LEED is a certification system aimed at the U.S. being inflexible in some areas, especially in categories where the score for enterprise punctuates certain item or not, being fully returned to the project and not for performance the building. With this, ends up creating some obstacles in certain situations where some concepts are required, since each city has its own characteristics and problems.

The *HQE* is the opposite, it fits to the reality of the location, being more flexible and allowing choices in favor of better performance. Each project has a potential of performance and economy, and the project can justify their decisions to change due to its location, ventilation, lighting, etc.

Both systems of certification bring economic benefits to their users, besides awakening awareness on the environment. It is important that constructors and companies pay attention to these systems, because being used on a larger scale, will bring greater benefits to entrepreneurs, to users and will also be contributing to the environment.

The certification in construction is a big step to be adopted and counts with the collaboration of professionals such as designers and builders, as well as the users, having as purpose a construction that has less impact on the environment, using new resources and reuse of materials.

The current range of these systems at the national level is still small compared with the volume of construction in progress, but these organizations are taking important steps, contributing to a better quality of life, preserving the environment with ecological awareness and social responsibility.

References

[1] David Gottfried, "Green Building Design, Construction and operations", 1996
[2] Eddy Krygiel & Bradley Nies, "Green Bim", 2008
[3] Jerry Yudelson, "Green Building A_to_Z", 2007
[4] http://www.revistatechne.com.br/engenharia-civil/133/imprime77962.asp, in October2009
[5] http://www.asec.com.br/v3/docs/Doc_Encontro09_AnaMelhado.pdf, in September 2009
[6] www.usgbc.org, in July 2009
[7] www.vanzolini.org.br, in September 2009
[8] www.leroymerlin.com.br, in November 2009

Sustainability in civil construction applied in the construction site phase

E. Vazquez[1], S. Rola[2], D. Martins[1], M. Freitas[3] & L. Pinguelli Rosa[3]

[1]*Department of Civil Construction, Federal University of Rio de Janeiro, UFRJ, Brazil*

[2]*Department of Architecture and Urbanism, Federal Rural University of Rio de Janeiro, UFRRJ, Brazil*

[3]*Post-Graduation Program on the Energy Planning, Federal University of Rio de Janeiro, COPPE/UFRJ, Brazil*

Abstract

This paper presents an analysis of the theme of sustainability applied to the construction site, showing some impacts that may be imposed on the environment during the construction phase of a building and presenting some solutions to alleviate these effects. World history shows us that buildings have always been constructed to meet the immediate and basic needs of humankind. At this time of rapid growth, scant care has been taken about construction techniques or environmental conservation. Indeed, increased environmental degradation has resulted in serious risks to society and, bearing in mind that over 60% of the solid waste generated in urban centers is from construction sites, the need arises to develop techniques and solutions to mitigate these impacts. This paper seeks to present a research study and diagnosis at the construction site stage into the process of sustainability of an undertaking, with the completion of a case study. The article is structured in four parts: presentation of a brief introduction to the topic and the research methodology to be used; description of certification processes for civil construction; description of risks and mitigation of same at the construction site phase; and a case study focusing on sustainability during the construction phase of an undertaking. A sustainable undertaking is defined not only by environmental and social benefits in its use and operation, but also in its construction. This phase can cause serious damage and impacts to the neighborhood and the environment, such as depletion with excessive

WIT Transactions on Ecology and the Environment, Vol 144, © 2011 WIT Press
www.witpress.com, ISSN 1743-3541 (on-line)
doi:10.2495/ECO110241

consumption or wastage of natural resources and increased pollution caused by poor management of the construction site.
Keywords: sustainability, construction, management, construction site, environment.

1 Introduction

Modern industry coupled with technology has transformed humankind, and the consequences of the new mentality, which was heralded in with the Industrial Revolution, were felt during the twentieth and the early twenty-first centuries. They are characterized by the association of the concept of development with the increased use and consumption of the soil, as well as sources of energy and natural resources in general, and led to the formation of metropolitan agglomerations. These concentrations over time gave rise to strong environmental degradation processes Duarte [1].

World history shows us that civil construction has always existed to meet the basic and immediate needs of humankind, and at this time of rapid growth, scant attention has been given to construction techniques and environmental conservation.

There is now a pressing need to take steps to reduce or neutralize this environmental degradation. In the construction area, technical aspects that favor the construction of buildings that meet this requirement, now commonly referred to as "environmentally correct or sustainable buildings", are beginning to emerge.

Given the increase in environmental degradation, resulting in serious risks to society in general, and bearing in mind that over 60% of solid waste generated in urban centers comes from construction sites, according to Credídio [2], the need arises to develop techniques and solutions to mitigate these impacts.

Sustainable buildings have been the subject of much recent debate. This project aims to focus on sustainability during the construction phase of an undertaking. This phase has not been given the necessary attention, though it is of fundamental importance for environmental preservation. A sustainable undertaking is defined not only by the environmental and social benefits in its use and operation, but also in its construction.

2 Methodology

The theme under scrutiny was developed through an exploratory field study of the environmental impacts generated by the construction site, with the presentation of some interventions to minimize these impacts by integrating the concepts of sustainable construction. Preparation of the article was based on bibliographical research, interviews with market professionals, published articles, thereby creating a comparative study of efficient solutions for a sustainable construction site.

3 Sustainability in civil construction – certifications

Based on the current world scenario regarding the protection of the environment there has also been an increased demand to implement sustainable methods in the civil construction sector. This has led to a large increase in sustainable buildings around the world, including Brazil.

Sustainable construction is concerned with the rational use of natural resources, as the civil construction industry is one of the human activities that most impact the environment. It is responsible for 40% of the consumption of natural resources, 34% of water consumption and 55 % of wood consumption. There is also concern about the waste generated by the constructions, which is then discarded in inappropriate areas, thereby polluting the cities. No less than 67% of the total mass of solid urban waste comes from construction sites Credídio [2].

When the project for an undertaking has sustainable strategies, such that the whole cycle of the project has been planned to cause the minimum possible environmental impact, an environmental certification through some processes can be obtained.

The 'green building' certifications in developed countries concentrate on different aspects of the construction. Brazil has followed the footsteps of the United States, which foster the rationalization of resources with savings in water and energy. In this aspect, the Brazilian Committee for Sustainable Construction (CBCS) was recently established. It is a non-profit civil association that seeks to contribute to the promotion of sustainable development through the generation and dissemination of knowledge and the mobilization of the productive chain of civil construction and its customers and consumers.

4 Risks and mitigations at the construction site stage

The construction phase gives continuity to what was conceived and projected at the earlier stage. Many companies have not devoted sufficient attention to the interference caused by construction sites, which also have significant impacts, such as inconveniences to the neighborhood (noise, visual, etc.), pollution (soil, water and air), impacts of the construction site (on ecosystems, erosion, siltation, traffic, etc.) and the consumption of resources (mainly water and energy).

These interferences caused by the construction site affect society at the local level, including workers, neighborhoods and ecosystems, and at a global level, affecting society as a whole, especially in relation to pollution Cardoso and Araújo [3].

4.1 Impacts generated by the construction site

When defining the standard of the building to be constructed, a program of requirements is prepared and, based on the definition of the scope, it is possible to identify the potential risks of environmental impacts. Once the risks have been identified, it is possible to plan alternatives that address the mitigation actions for them, seeking interventions in order to attempt to increase their social and

Table 1: Potential environmental impacts during the construction phase.

Removal of buildings	Generation of a large volume of waste when the structure is of reinforced concrete with mortar finishing.
Suppression of the vegetation and erosion	Generation of a dynamic process of change in the local ecosystem, bringing risks to fauna and flora. It also causes the exposure of the lower strata that in general, are more susceptible to erosion.
Temporary constructions	Health risks, due to precarious conditions of hygiene, and worker safety, due to potential accidents.
Storage and handling of materials	Chemical contamination of the soil due to incorrect storage and handling of materials.
Circulation and maintenance of equipment	Impacts of deterioration of air quality; noise pollution, inconveniences to the community, alterations in security conditions in the neighborhood.
Consumption and wastage of resources	Accelerated depletion of mineral deposits and natural resources.
Generation of waste	High cost and increase in volumes sent to discarding areas.
Air quality	Non-renewal of air directly impacts the health of the worker, causing discomfort, damage to health, drowsiness, and loss of reaction capacity, asphyxia, or even death.

environmental performance without generating high costs in the implementation phase. Table 1 describes some environmental impacts that may occur during the construction phase. Lima and Lima [4].

5 Case study: sustainable construction sites

The case study deals with the interventions that were made in the implementation of the construction site of a hospital, in order to obtain a building site with a reduced environmental impact. This study describes the environmental impacts caused by the construction site and the strategies employed to mitigate these effects, in accordance with the concepts of sustainability required for LEED certification [5]: sustainability of the workspace; energy efficiency; rationalization of water use; sustainability of materials; and internal environmental quality at the construction site.

With regard to sustainability requirements that were established for the construction site, there are items such as: the control of erosion and sedimentation; reduction of water consumption; solid waste generated during construction; internal air quality; and the use of certified and low toxicity recycled materials. The items used at this stage also add points to the items used in the final undertaking, accumulating points and contributing to a higher level of certification.

5.1 Description of the sustainability parameters

5.1.1 Sustainable space

Relating to the first item, Table 2 shows the major impacts caused by the construction site, and the intervention that was performed.

This environmental impact is considered a pre-requisite for the Sustainable Space category. This item establishes the implementation of an Erosion and Sedimentation Control Plan and the adopted strategies of which are described hereafter.

Table 2: Impacts and interventions at the construction site.

Sustainable Space (SS)	Impacts on the surrounding area	Intervention
	Erosion and Sedimentation	Protected embankments and reduction of sediments in the air

The building has a basement, and due to the size of the plot of land it was decided to conduct the excavation for the building with embankments around the entire work area. To prevent soil erosion of the embankments and the washing away of soil during heavy rains, the use of plastic screens was adopted for their protection, as can be seen in Figure 1.

For the control of sedimentation, the construction site is regularly irrigated with water from the lowering of the water sheet. As the construction site is constantly humidified, the movement of cars inside the construction site does not generate dust.

For any sweeping in the construction site, the area is first irrigated, or else rainwater is used (for cleaning the area outside the undertaking, in this case), as can be seen in Figure 2.

Figure 1: Embankments protected with screens to avoid erosion.

Figure 2: Control of sedimentation and sweeping of the construction site using rainwater and avoiding the generation of dust.

Figure 3: View of the wheel-washers.

The trucks filled with waste leaving the site are enclosed with canvas to avoid the suspension of sediments on the journey to the external environment. Another measure to prevent the trucks from fouling the streets around the building and raising dust while circulating, was to deploy a system for washing wheels, as shown in Figure 3.

5.1.2 Water efficiency

In relation to the consumption and loss of water, Table 3 shows the impact and intervention which was performed at the construction site.

Table 3: Impacts and interventions at the construction site.

	Impacts	Intervention
Rational Water Use (WE)	Wastage of drinking water	Reduced consumption of drinking water by recycling

In this category, version 2.1 of the LEED certification does not establish any requisites regarding water consumption. Credits are based on use and occupancy of the undertaking and there are no references to the construction site. However, in a sustainable construction site, the wastage of water is also considered a crucial impact on the environment.

Therefore, in order to reduce water consumption and avoid wastage, a system to capture water from the water sheet was adopted. This water is used for routine purposes on the construction site, such as the toilet flushes (Figure 4), washing sidewalks and cleaning work, thereby avoiding the purchase from water supply trucks or the local water utility provider. To disseminate this concept, lectures are staged regularly to promote the sustainability of the construction site and a reduction in water consumption and wastage.

Figure 4: Notice regarding the use of recycled water in the bathrooms.

5.1.3 Energy and atmosphere

Regarding this item, Table 4 highlights the main impacts caused by the construction site, and the intervention that was performed.

Table 4: Impacts and interventions at the construction site.

	Impacts	Intervention
Energy and Atmosphere (EA)	Wastage of energy	Reduction in consumption of energy with more open environments

Reducing energy consumption is treated in this category as the second requisite. However, this item does not score points for reduction performed at the construction site. But, as in the former category, wasting energy on a sustainable construction site is not permitted.

This being the case, the construction site office was designed with many windows to take maximum advantage of natural lighting and ventilation, as seen in Figure 5. There is also a ventilation area between the ceiling and the roof, thereby reducing heat absorption through the roof. The electrical installation project included independent circuits, ensuring that lights are not left on in empty environments. Compact fluorescent lamps were used in all internal areas of the offices and halls of the construction site. Lectures, educational campaigns and dissemination of this concept were also offered to all employees.

Figure 5: Engineers' office on the construction site with many windows.

5.1.4 Consumption of materials and the generation of waste

Regarding the consumption of materials and solid waste generation, Table 5 highlights the main impacts and the intervention that was performed.

This category includes purchase of materials, generation and management of waste. The storage and collection of recyclable waste is the only requisite. This item is not applied to the construction site, but is addressed in certification credit 2, which refers to management of waste from construction. Credits 4.1 and 4.2, concerning the use of materials with recycled content, 5.1 and 5.2, concerning the use of regional materials, 6 and 7 relating to the use of rapidly renewable materials and certified wood, respectively, are each equivalent to 1 point in the LEED certification. The treatment given to these items is described below.

Table 5: Impacts and interventions at the construction site.

	Impacts	Intervention
Materials and Resources (MR)	Generation of solid waste Saturation of sanitary landfills Pollution from waste and transportation	Waste management Recycling of materials Using recyclable and regional materials

The waste generated on the construction site is grouped separately according to its characteristics, and is then sold for recycling. Training sessions are also staged explaining the correct use of each material in order to avoid wastage. Some of the materials that are recycled, such as scrap iron and wood, are separated and identified in certain locations around the construction site, and subsequently sold to agencies and companies that recycle (Figure 6).

Designated areas for smaller waste such as plastics and packaging of paints, separated in accordance with CONAMA resolution No. 307/2002 were also created. This material is then bundled and sold for recycling. Organic waste, like leftover food, is deposited in separate buckets, for collection by a company specializing in the treatment of organic waste.

Regarding to the consumption of materials, preference is given to companies that work with recycled materials and companies that are within a radius of approximately 500 miles from the construction site. This benefits the local economy and ensures that carbon dioxide emissions and the consumption of fuel used in freight delivery are also reduced.

The credit for the use of certified wood is not applicable to the wood used during construction, only for wood in the building, such as furniture and wooden structures. In this undertaking some wooden furniture will be used, and the wood used will have to have FSC (Forest Stewardship Council) certification, represented in Brazil by the '*Conselho Brasileiro de Manejo Florestal – FSC Brasil*,' with the guarantee that it is from reforested areas.

Although this does not count toward the certification credit score, the moulds used in the construction have reforestation certificates, and the company contracted to provide this service has an environmental support program. The moulds that arrive at the construction site are already cut and adapted to perform the services and have a thin finishing film. In addition to the project for

Figure 6: Wood separated for subsequent recycling processes.

rationalization of the construction of the moulds, this resulted in their extended use during the works and less waste.

All the steel purchased for the construction, of the AC-50 and AC-25 categories has about 70% to 85% of manufacturing components of recyclable origin.

The cement used was PC and PC-II-III, which is composed of waste such as granulated slag from blast furnaces and pozzolanic materials, resulting in lower consumption of clinker and therefore less CO_2 emissions and preservation of the deposits.

When purchasing the aluminum for the construction site, aluminum manufactured using green billet is selected, as it is composed of more than 80% of recycled raw materials.

All materials that have recycled products in their composition are accompanied by a declaration in which the supplier of the material states the percentage of recycled components in each material.

Other measures also adopted to reduce wastage include the dissemination of the specifications for the preparation of mortar to be used in the construction site on all floors and the use of plastic-based ingredients. Use of the latter reduces wastage because it has a smoother surface than concrete, and mortar made is 100% usable.

5.1.5 Internal environmental quality

Regarding this item, Table 6 presents the main impacts and the interventions made.

The requisites of this category refer to minimum air quality performance and the control of smoke generated by tobacco smoking. These two items are evaluated when the building is already in use, and are not applicable to the

Table 6: Impacts and interventions at the construction site.

	Impacts	Intervention
Internal environmental quality (EQ)	Air pollution	Reduction in carbon emissions and the production of sediment

construction site. However, they are addressed in certification credit 3, which refers to air quality during the execution of the works and stipulates the creation of an Air Quality Management Plan. Credits 4.1, 4.2 and 4.3 relate to the use of materials with low VOC. Each item is worth one point in the system.

To ensure the best air quality at the construction site, an Air Quality Management Plan was created. This plan envisages no air pollution or materials that can release this pollutant waste in the future, with some measures that are already being implemented at the construction site and other measures that will be implemented throughout the course of performance of the services, as described below.

For the cleaning procedure adopted during construction, as stated above the irrigation of surfaces to be cleaned is adopted, followed by sweeping of the indoor environments, with manual collection of larger debris.

To ensure the health of workers, smoking is not permitted on the construction site. For this purpose, a space was created for those who smoke known as the smoking area (Figure 7), a restricted and well ventilated area in the construction site.

The sequence of installation of materials is planned in order to avoid contamination of porous materials such as insulation, carpeting, ceilings and plasterboard, so that they do not disperse this pollution in the future. The storage of these materials must be protected with plastic preventing them from being contaminated.

The thermo-acoustic insulation surfaces will only be installed when all services that may release VOCs or dusts are concluded, thereby preventing damage to the materials. However, some of these materials need to be installed inside walls, as in the case of the fiberglass insulation that will be used in the drywalls. For this material, proper storage is required before execution, and at

Figure 7: Smoking area.

the moment of installation this will be done in such a way as not to expose the materials to exposure to odors.

All products that can emit VOCs or odors are sealed and properly stored in ventilated areas or sheltered from light and in their original containers to avoid leakage and consequent contamination of the environment. The furniture and carpeting will be adequately protected with plastic after its installation.

The system designed to meet the internal thermal requirements generated by the functions to be performed in the location will only come into operation when all the construction services are considered to be completed, including any final touches and rectifications due to flaws in execution. Before use, all air intakes as well as vents and diffusers will be protected by plastic sheeting that will serve as a sealant, preventing the entry of particulate matter and thereby reducing the effectiveness of the filtering elements. Furthermore, when the works are completed the ducts will be inspected and cleaned to minimize contamination

In the event that the installed fan coils need to be used during construction, temporary filters will be installed in addition to the existing filters in places where exposure to dust occurs, though all filtering elements will be replaced at the end of the works. In this case, the sites adjacent to where the works are in progress will be isolated from contamination by filters installed inside the ducts. These filters will ensure that the air conduits are kept clear.

In places where there is the need to perform heavy works that generate large amounts of dust, while the adjacent spaces remain operational, and the filtering system installed is not effective, this may be increased by using Activated Carbon or Potassium Permanganate. Building materials may not be stored in the places where the machines will be located.

To prevent the accumulation of debris and particulate materials in suspension, both the air conditioning system diffusers and the ventilation duct inlets and windows will be sealed with plastic sheeting until the completion of construction services, including any final touches and rectifications due to flaws in execution. All environments, as they are completed, will have their access doors sealed with temporary sealant in the cracks, or plastic sheeting to protect places that have no doors.

The priority in the use of construction equipment is given to those with the least impact on the premises where construction-related activities are not being performed.

6 Final considerations

With growing worldwide concern regarding the preservation of the environment, there has been a considerable amount of well-documented input from lectures and discussions by countries and organizations, which propose some guidelines to ensure sustainable development.

The civil construction industry contributes significantly to the modification of the physical space in which we live, and can radically alter the climate and produce waste and sediment that have a crucial impact on the environment. For this reason, there is a pressing need to develop strategies that mitigate or nullify these impacts.

The quest for sustainable construction has resulted in the certification of some undertakings around the world. In Brazil, the adoption of sustainable strategies is gradually being absorbed by the entrepreneurs.

As seen in this paper, the construction phase of an undertaking is responsible for the bulk of the environmental impacts. The construction site deserves special attention and all possible damage that may be generated by the construction site during construction should be carefully monitored in order to analyze what can be avoided. In urban areas, where the highest concentration of civil construction occurs, care must be taken to preserve the existing ecological environment, to avoid sudden climate change, and to curb the pollution generated, which degrades the environment and damages the neighborhood.

When all the impacts that may be caused by the construction site have been assessed, the entrepreneur must prioritize them to determine which of these impacts will be addressed and to what extent. Once that has been established, the necessary technologies and actions of a managerial nature required can be defined, establishing the resources that need to be implemented, such as equipment to be bought, professionals to be trained or hired, managerial tools to be implemented, etc, and the timeframes and costs involved.

The civil construction industry still has a long way to go on the issue of sustainability, though positive results in reducing environmental degradation and the savings generated in the consumption of resources has been a driving force behind these changes. New technologies that contribute to these reductions are more readily available on the market and with an ever decreasing differentiation in price, resulting in lower costs for a sustainable undertaking. In this case study, the increase in costs of the hospital complex will vary between 6% and 14%, the greatest impact being the investment in electrical installations and eco-efficient air conditioners.

In light of what has been presented above, it can be seen that, besides achieving the objectives originally proposed, this work can provide a contribution to the study of sustainability within the construction site.

References

[1] Duarte, Anselmo. "*Casa de Alta Tecnologia e Ecologicamente Correta*". Virtual Book, 2007. <http://www.crear.com.br/casa_ecologica.php>.

[2] Credídio, Fernando. "*Construções sustentáveis: conforto e respeito ao meio ambiente - Parte 1*". Virtual Book, 2008. http://www.ifk.org.br/construcoes_ sustentaveis:_conforto_e_respeito_ao_meio_ambiente__parte_1__344.html >.

[3] Cardosa, Francisco Ferreira, ARAUJO, Viviane Miranda. "*Levantamento do estado da Arte: Canteiro de Obras*". Tecnologia para Construção Habitacional mais Sustentável. São Paulo, 2007.

[4] Lima, Rosimeire Suzuki, LIMA, Ruy Reynaldo Rosa. "*Guia para Elaboração de Projeto de Gerenciamento de Resíduos da Construção Civil*". Série de Publicações temáticas do CREA- PR, nº 1. Paraná, 2009.

[5] LEED, "Leadership in Energy & Environmental Design – LEED", v.3. USGBC, 2009.

Sustainable carbon neutral building design: new simulation tools for teaching green buildings

S. Vassigh
Department of Architecture, Florida International University, USA

Abstract

The Building Industry is one of the major consumers of energy. In the United States buildings use 48% of the total energy consumed, significantly impacting national energy demand and contributing to global warming. The vast majority of architectural practice in the U.S. leads to construction of buildings with little concern to sustainability contributing to environmental degradation. However, as the environmental impact of buildings become increasingly recognized, the role of architects and the initial decision making process which determines the resource use and life-cycle of materials, systems, and construction processes, becomes more critical.

In order to advance sustainable development and make green building construction the standard rather than the anomaly, the education of the architects must be reconsidered. The traditional American architectural curriculum that is based on a schism between "design" and "technology" is inherently in conflict with the principal of sustainability.

Recent research in building design indicates that the most resource efficient, best performing, and environmentally sustainable buildings are designed utilizing "integrated practice," in which the various disciplines involved in building design work together at the conception of the project to improve overall building performance, reduce GHG's, and lower the costs for operation and maintenance of buildings.

Although large-scale reform of architectural curricula is a complex, ongoing, and difficult debate, producing teaching tools that can simulate integrated design can impact and promote a better understanding of sustainable practice in architecture.

WIT Transactions on Ecology and the Environment, Vol 144, © 2011 WIT Press
www.witpress.com, ISSN 1743-3541 (on-line)
doi:10.2495/ECO110251

The proposed paper will present the progress of a multi-disciplinary team of faculty who are collectively working on the completion, implementation and evaluation of a simulation software package in an interactive game format. The project teaches the concepts of "integrated design" through immersing students in a virtual world that imitates the complexity of the real world of decision-making and material choices in design. The project accomplishes this by harnessing the capabilities of simulation and dynamic modelling programs as well as powerful game engines while creating compelling and rewarding reasons for student's engagement in the learning process. The project is funded by the US Department of Education for the period of 2007-2011.
Keywords: educational tools, teaching sustainable design, augmented learning, instructional technology, green building design.

1 Introduction

The Building Industry is one of the major world-wide consumers of energy. In the United States buildings use 48% of the total energy consumed significantly impacting national energy demand and contributing to global warming [1]. Seventy-seven percent of all the electricity produced in the U.S. is used just to operate buildings. Globally, these percentages are even greater [2]. Building and construction activities consume three billion tons of raw materials each year [3]. Materials utilized in buildings tend to have a high-embodied energy, high levels of toxins and pollutants at the end of production, as well as higher levels of emissions [4].

As the environmental impact of buildings become increasingly recognized, the role of building designers and the initial decision making process which determines materials, systems, and construction processes, becomes more critical. Thus, an informed design process merging environmentally responsible practices with advanced technologies can significantly reduce the adverse impact of buildings on the environment.

A critical component of an informed design process is a clear understanding of building systems operation, interaction, and the synergetic benefits realized through their proper selection. A consideration of suitable building systems, gauging their interaction, and proposing well integrated systems can lead to producing efficient models of sustainable buildings with minimal impact on the environment [5].

The project presented in this paper reports on the progress of a multi-disciplinary team of faculty who are collectively working on the completion, implementation and evaluation of a simulation software package in an interactive format. The project is funded by the US Department of Education for the period of 2007-2011. The project teaches the concepts of "integrated design" through immersing students in a virtual world that imitates the complexity of the real world of decision-making and material choices in design. The project aims to accomplish this by harnessing the capabilities of simulation and dynamic modeling programs to create interactive and compelling reasons for student engagement and learning.

2 Project approach

The project is developed based on a few pedagogical principles: First, an interactive educational format is critical for engaging students in the process of learning. Students should be empowered to learn at the level of their individual ability while receiving support for their activities. Second, any meaningful attempt to advance technology education in the architectural curriculum cannot be pursued in isolation from the design process. Finally, understanding systems integration is a critical component of designing sustainable, carbon neural, and resource efficient buildings.

The educational content of the project is composed of two interdependent modules of Simulation Environment and [eco]Learning Interface. The following section is a detailed explanation each of these components.

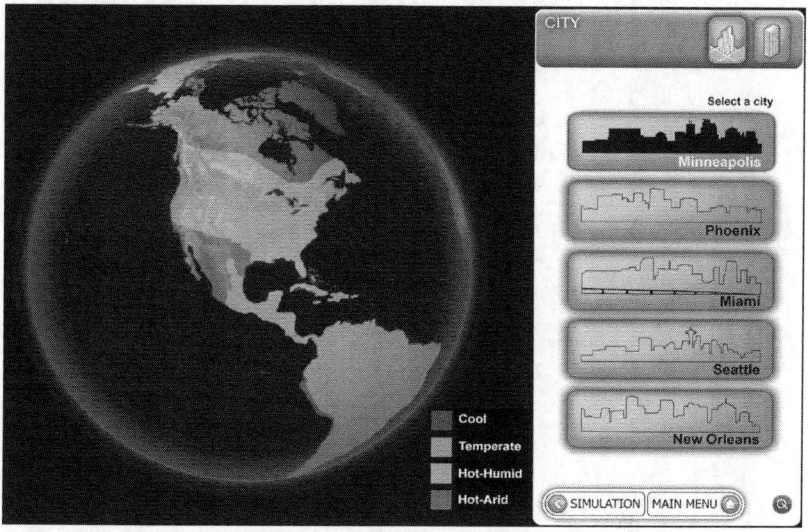

Figure 1: The Simulation Environment Interface

2.1 Simulation environment

The simulation environment is the core component of the project developed to engage students in a series of scenario-driven building design projects that make a case for sustainable design and construction. These scenarios take on a variety of ecologic and economical issues such as squandering natural resources, fuel and energy cost, and climate change. They take place in a variety of building sites and contexts allowing the students to select the venue for their learning. In each case the scenario lays out a challenge for composing a building through balancing the demanding requirements of site and climate, form and function, energy and sustainability, and building systems. Each scenario is introduced with a movie explaining the details of the required activity. Once the student has selected a mandate he/she will move to the user interface and start to assemble a building.

The simulation environment provides access to a library of building tools that contains a wide range of preconfigured building components to be used to assemble a building. Navigating through these choices enables the student to investigate and learn about each building component and their relevant properties. Each building component is tagged with a series of metrics and attributes which are imported to the interface when the component is selected. All selected choices are all stored in the library of tools ready for assembling the building and running the analysis function. During the assembly, the student can organize, move, group and edit the selected components. Specific components of the Simulation Environment are the following:

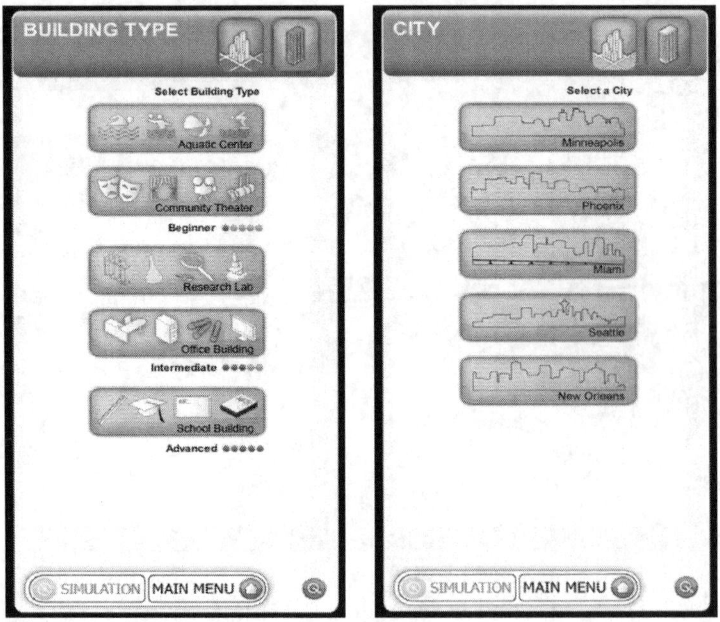

Figure 2: Menu for scenario selection.

2.1.1 Scenarios

Scenarios will provide a choice of architectural programs such as Community Center, Research Laboratory, School Building, etc. Once a choice is made the next menu will allow the student to select one of the four climatic conditions of hot and arid, hot and humid, temperate, and cool as the context for their building assembly. Once the initial selections are made, a brief movie will launch a scenario with a narrated story and three dimensional graphics introducing the entire project. The movie will include full textual and visual information of the site, the surrounding environment, topography, climatic conditions, and a detailed square footage requirement of the building program. Once the user is given the mandate, the building assembly begins.

2.1.2 Library of building blocks

Library of Building Blocks includes a number of pre-arranged computer generated models of various building components including floor templates and program elements that are used to make a building assembly. Floor templates are the initial step for creating floor plans. They indicate the shape and the exterior boundaries of the floor area, and once a selection is made the floor plan can be laid out. Program elements compose the functional components of a building such as, rooms, public spaces, dining areas, stairs, elevators and etc. The student can investigate each arrangement by placing it on the main window and examining how a particular arrangement responds to the site and the climate conditions. For example, the student can compose a horizontal arrangement of the building mass and place it in various orientations on the site to examine its energy efficiency in relation to solar orientation. In a hot and humid climate, a horizontal arrangement of open spaces such as offices which are extended in the east-west direction will provide cross ventilation of air from a southern summer breeze, reducing the need for mechanical ventilation. This arrangement will also maximize the utilization of natural day light. The feedback from this simulation environment will allow the student to investigate these and many other interactions of the building volume in response to the climate and the site. Understanding these complex interactions are a critical component of sustainable design. These interactions will be analyzed with simulation algorithms and exhibited for view in both numerical and graphical format.

2.1.3 Building envelopes

Library of Building Envelope Elements is a number of detailed computer generated models of building envelope systems such as walls, roofs, floors and windows. Each element is modelled as modules of 8ft x 12ft panels and carries properties of every material used in its composition. For example, a cavity wall is modelled with brick at the exterior surface a layer of air; rigid insulation, flashing, a layer of vapour barrier, concrete masonry, and painted gypsum board in its interior surface. Each material's physical properties or attributes such as specific weight, thermal resistance, embodied energy, and percentage of recycled materials is embedded within the choice and is imported for future performance analysis. The student could use the imported values from the preconfigured data set or enter another value to modify the design conditions. The attributes of each component will be added and editable for analysis once the building is completed.

2.1.4 Building systems

Library of Building Systems Components includes computer-generated models of building systems such as Structural Systems, Heating Cooling and Ventilation (HVAC), Alternative Energy Systems, and Lighting Systems. The student can browse the library of systems and investigate a variety of choices. For example, the HVAC Systems selection allows the student to study a series of systems such as Fan Coil Units, Variable Air Volume Units, Water Source Heat Pumps, and Packaged Terminal Air Conditioners.

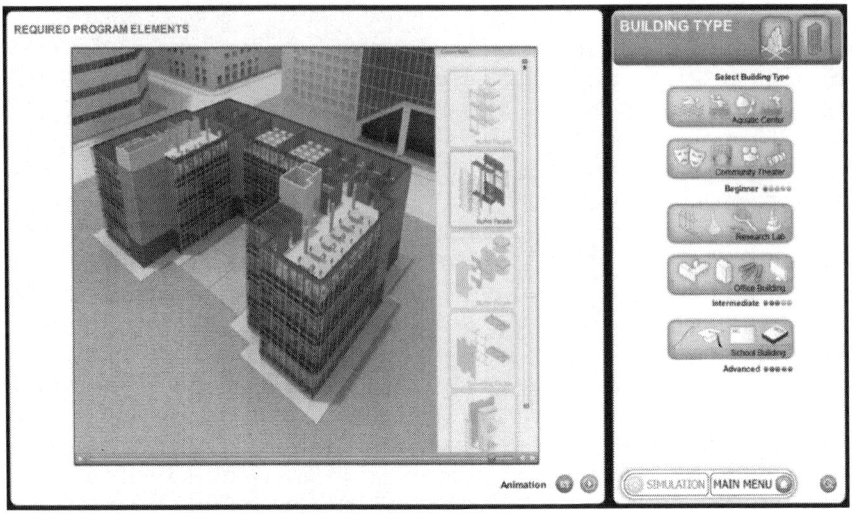

Figure 3: Screen shot showing an assembled building.

2.1.5 Analysis

Once a building is completed the student can run a few different types of analysis. The main analysis is the solar radiation simulation. This is designed to provide a better understanding of performance of the building fenestration and envelope systems. The results of the analysis provide the students with an understanding of how the location, orientation, window selection, shading, lighting and exposure will affect building heat and cooling load. The other analysis includes a summary of embodied energy, embodied water, percentage of recycled material, and an estimate of materials cost used in assembling the building. This information will allow the student to understand the impact of their choices and tradeoffs in cost versus environment.

2.2 [eco]Learning interface

The second component of the project is the [eco]Learning Interface which contains the pedagogical component of the project. It is an interactive learning tool developed to advance the education of climate responsive and ecologically sustainable buildings design and construction. [eco]Learning Interface is composed of lessons in six content areas including building form, building envelopes, climate control systems, renewable energy, natural and artificial lighting, and landscape

Each of these content areas is subdivided into learning modules that introduce the subject matter and investigate the related sustainability issues and strategies. Eco[Learning] interface harnesses the capabilities of advanced media to help users to visualize and engage with concepts that may otherwise be too ambiguous or difficult to comprehend in the traditional book format. The interface utilizes hyperlinked text, interactive diagrams, animations, analytical data and graph exhibits to provide an augmented learning environment that is accessible for users with a various learning styles and needs.

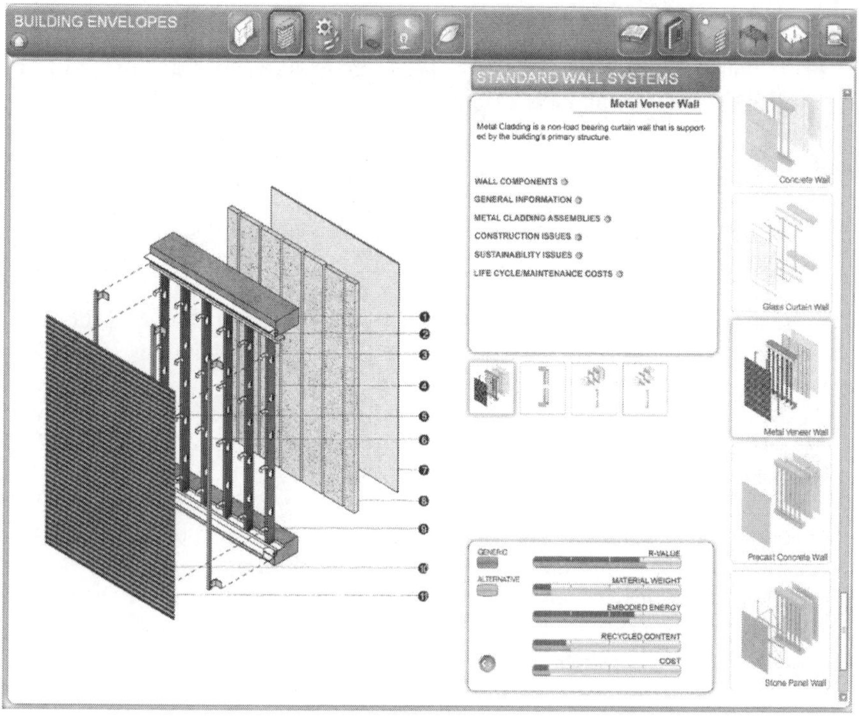

Figure 4: Screen shot showing [eco]learning interface.

This interface can be accessed from the Simulation Environment or used as an independent tool for teaching building sustainability concepts and principles. When accessed from the simulation environment, the interface provides lessons and information to help students make educated selection and compare alternatives with adequate knowledge.

The project is in its last year of funding. Once completed, it will be formally tested and evaluated, and the results will be disseminated to schools of Architecture across the nation.

3 Closing remarks

Given the size of the built environment as energy and materials consuming sector, raising students' awareness and technological competency to design efficient and sustainable and carbon neutral buildings can have a significant impact on the architectural profession as well as the national energy consumption. Though large-scale reform of architectural curricula is a complex, ongoing, and difficult debate, changing the pedagogy and delivery system of the educational content is the critical next step. The current advances in our technological capacity and the informational technology have created new opportunities to reform the traditional classroom. These technologies can help to change the role of the student from a passive observer to an active participant. The project discussed here is an example of producing teaching tools that harness digital capabilities of advanced graphic media and simulation applications for improving the sustainability education.

References

[1] Prowler, Don. "Whole Building Design Guide. National Institute of Building Sciences(NIBS)."August,07, 2008.http://www.wbdg.org/wbdg_approach.php

[2] Architecture 2030, "Architecture 2030 Will Change the Way You Look at Buildings." http://architecture2030.org/the_problem/buildings_problem_why

[3] D.M. Roodman and N. Lenssen, A Building Revolution: How Ecology and Health Concerns are Transforming Construction, Worldwatch Paper 124, Worldwatch Institute, Washington, D.C., March 1995, p. 5.

[4] Huovila, Pekka and United Nations. Environment Programme. Sustainable Consumption and Reduction Branch. Buildings and Climate Change, Status Challenges and Opportunities. Illustrated ed. UNEP/Earthprint, 2007.

[5] Whole Building Design Guide Aesthetics Subcommittee, "Engage the Integrated Design Process."October,30,2010. http://www.wbdg.org/design/engage_process.php

Financial sustainability of the waste treatment projects that follow PPP contracts in Greece: a formula for the calculation of the profit rate

A. Sotirchos, A. C. Karmperis, K. Aravossis & I. Tatsiopoulos
Sector of Industrial Management and Operational Research,
Department of Mechanical Engineering,
National Technical University of Athens, Greece

Abstract

This paper examines the initial budget estimation process for the implementation of waste treatment projects through Public-Private Partnership (PPPs) in Greece. The study, having as strong theoretical foundation the project's financial analysis that is included in the cost benefit analysis methodology, evaluates the financial sustainability of a waste management project. Initially, the budget estimation process that is currently used under the conventional procurement of public projects in Greece is followed and the results demonstrate that the specific process, where the investor's Profit Rate (PR) takes specific values, cannot be used in the case of a PPP. Furthermore, a new formula is developed, which calculates the minimum value of the private sector's PR, in order to determine the existence of positive cash flows during the operational phase of the project and to ensure the partnership's financial sustainability. The new formula can be a useful tool to the public decision-makers in Greece, because it helps them to evaluate the financial sustainability of the waste treatment projects that follow PPP contracts, during the conceptual phase.

Keywords: cost benefit analysis, financial analysis, financial sustainability, profit rate, Public Private Partnerships, waste treatment projects.

1 Introduction

Nowadays, it is common across countries that governments follow Public Private Partnership (PPP) contracts and collaborate with the private sector for the

WIT Transactions on Ecology and the Environment, Vol 144, © 2011 WIT Press
www.witpress.com, ISSN 1743-3541 (on-line)
doi:10.2495/ECO110261

purpose of delivering a project or a service that is traditionally provided by the public sector. Due to the fact that PPP projects have been rapidly increased over the last decades, there are many researchers that attempt to improve the operation of PPP projects, by identifying their key aspects [1]. Especially in the waste treatment (WT) sector, there are 398 projects with private participation that are reported in the World Bank Group's database for the period 1990 to 2009 [2], while the investments are estimated in US$ 15,878 million. Furthermore, other authors refer to the implementation of large WT projects in different countries, such as China, India, Indonesia, Thailand, Vietnam, Turkey, Mexico, Malaysia and Australia [3–5]. These projects are mainly implemented by following the Build-Operate-Transfer (BOT) type, which is the most used contract type in PPPs [6]. However, a common point between different PPP contract types is that the PPPs are long-term agreements for the provision of high quality infrastructure, products or services [7], while there are significant technical, legal and economic risks that are shared between the public and private sectors [8].

In Greece, these contract types are presented in various forms, many decades ago [9] and are mainly implemented recently, under the specialized form of concessions for the construction of 3 major projects of significant and national importance; the Rio - Antirio Bridge, the Attiki Odos Highway and the International Airport of Athens [10]. For these projects, the approval by the public authorities was a cumbersome process since the legal framework in Greece was highly segmented. Currently, the specific legal and institutional framework for PPPs in Greece is established in 2005 under the Law 3389. According to this Law, two institutions are developed, the special secretariat for PPP and the inter-ministerial committee, which execute the initial assessment process for a proposed PPP project. In 2006, the manual for the implementation of projects and services through PPP is issued by the special secretariat [11], while the "Green Paper on PPPs" (EU Committee, 2006), containing 22 questions relevant to the issues of partnership contracts is sent by the European Commission to Member States [12]. Specifically, between the 52 PPP projects that are already approved by the Greek inter-ministerial committee, there are two environmental projects for the implementation of WT units in the region of Western Macedonia and the municipality of Rafina respectively. However, due to the fact that in Greece the PPP market opened quite recently, while the PPP [13] and the WT projects [14, 15] require complicated procedures, it is concluded that the WT projects that are considered to be implemented through PPPs in Greece, include a high degree of complexity [16]. Thus, a tool that can be used by the decision-makers during the feasibility stage of a WT PPP project, in order to ensure the partnership's financial sustainability, will be very useful.

2 Public procurement framework in Greece

According to the European Community and the national law, during all phases of public projects, there have to be in act the principles of equal treatment, isonomy, transparency, proportionality, public interest protection, private rights protection, free competition and environmental protection. Public procurement rules mean that the procedure should follow one out of three routes, which are the open procedure, the restricted procedure and the negotiated one [17]. The open procedure is distinguished due to the promotion of transparency and in most cases the international competitive bidding process is followed [18], where the contracting authorities are obliged to use as a key criterion for the award, either the lowest price or the most economically advantageous tender. However, due to the fact that the lowest price criterion itself does not ensure the value for money of the PPP projects, as it does not take into account the project's technical and quality parameters, several key factors for the PPP procurement process are suggested by the literature. Indicatively, Araujo [19] and Zhang [20, 21] suggest some specific criteria for the concessionaire selection, while Zhang [22] identifies 21 specific best value contributing factors, in order to choose the right private sector partner. On the other hand, many researchers highlight crucial issues that arise in the public procurement for environmental projects [23] and especially for WT projects [24, 25]. Generally, several existing models can be used either in the tender evaluation [26, 27] or in the negotiation process of the PPP projects [28]. However, the application of these models in the procurement process for a WT PPP project requires the estimation of the project's budget by the public sector, during the feasibility stage of the project. Furthermore, the operational phase of the PPP contracts is too long, 20 to 30 years [29], so it is very difficult to model the initial budget estimation process. Main objective of this paper is to develop a formula which can be used for the calculation of the Profit Rate's (PR) minimum value on behalf of the Special Purpose Vehicle (SPV), so as to ensure the project's financial sustainability.

3 Evaluation of the waste treatment PPP projects

Generally, a WT project's evaluation process is implemented through the Cost Benefit Analysis (CBA), the life cycle assessment or the multi-criteria analysis [30]. However, the CBA is distinguished, as it can be used for the evaluation of alternative types of WT projects [31, 32]. The calculation formula for the WT projects that are implemented through PPP contracts in Greece, which is developed here, has as theoretical foundation the CBA guide for investment projects, issued by the European Commission (EC) [33]. A remarkable point is that the submission of a project proposal which includes a CBA is mandatory for all large projects that are co-financed by the European Community for the 2007-2013 period, where large are considered the environmental projects whose budget are over 25 million euros. According to the EC's guide, the financial analysis is a basic step in the CBA. The main purpose of the financial analysis for a PPP project is the use of the project cash flow forecasts, in order to calculate the net return indicators for the SPV, through the annual payments by the public sector. Further, the financial analysis includes the evaluation of the project's financial sustainability, which is crucial in all types of investments.

3.1 Financial indicators

The financial indicators which are calculated in the financial analysis, is the Net Present Value (NPV) and the Internal Rate of Return [34]. Specifically in PPP projects, the performance of the private capital is implemented through the calculation of the financial NPV of the private capital NPV (Kp). Furthermore, the investment's profitability is demonstrated by the Profitability Index (PI), which is calculated by the inflows in present value (PV), divided by the outflows in PV respectively. In PPP projects, the SPV's PI takes into account all the net inflows divided by the outflows, which are the initial capital for the infrastructure development plus the capital that will be invested annually, e.g. the payment of the subcontracting agreements, or the loan repayments, etc.

3.2 Financial sustainability

Additionally, some key points that are included in the financial analysis of the PPP projects is the bank loan's interest rate as well as the interest rate, i.e. the inflation, that is used for the adjustment of the cash flows in future value (FV), because both have an impact on the annual net cash flows, which are calculated through the annual inflows minus the outflows of the SPV. We mention that a project is classified as financial sustainable, when it does not incur the risk of running out of cash in the future. Therefore, the financial sustainability's criterion is the annual cash flows to be positive for the SPV during the project's operational phase.

4 Case study: financial analysis of a waste treatment project that follows a BOT contract in Greece

In order to achieve the desired development of a formula for the calculation of the PR, an illustrative case of a BOT project in Greece is considered and specifically is evaluated in the financial sustainability's criterion. The project is an investment in the field of WT, for the reuse of well purified waste water for multiple purposes after an intensive tertiary treatment process. It takes place in Greece and includes the construction of wastewater treatment facilities for a city of 200,000 residents.

4.1 Assumptions

The cash flows arising in different years are calculated by the discount rate following the discount cash flow analysis (DCFA) [36], while only the financial inflows and outflows are considered, i.e. the value added tax, the depreciation of fixed assets and other accounting tools are not counted for the sake of simplicity. The time horizon that is selected to run the analysis consists of a 20 year operational phase, plus 1 year as the implementation phase for the infrastructure's construction.

4.1.1 Initial investment, expenditures and revenues

The initial investment cost for the construction of the WT facilities, is estimated at €48,000,000, which does not include general costs or profits. This was originally divided into equal flows and distributed annually in the project's operational phase, i.e. €2,400,000 per year in PV, while this cost is funded by a bank agency through loan. The annual operational cost for the 1st year of the operational phase is estimated in €3,600,000 in PV, where also there are no general costs or profits. These costs met by the private contractor as investments outflows, while inflows are the annual payments from the public sector. Furthermore, the maintenance cost for the 1st year is estimated in 0.4% of the construction budget, i.e. €220,000 in PV. Other assumptions that are used in the financial analysis, is the 6.65%, as the interest rate of the loan for the SPV, while the annual repayments are also fixed and the loan's amortization period is equal to the operational phase, i.e. 20 years.

4.1.2 Discount rate

Additionally, the discount rate that is suggested by the EC for the investments implemented in the Euro-zone during the programming period 2007–2013 is 5% for a long term. Notwithstanding the above rate may vary, depending on the type of investment, e.g. in PPP projects [33]. This case study uses the 7.5% as the discount rate, which is normal to cases of PPP projects in the Greek market.

4.2 Financial indicators

All data are adjusted in FV following the inflation, which is considered 3.0%. In the financial analysis, the process flowchart for the calculation of the NPV is illustrated in the following Figure 1. The NPV and the PI evaluation indicators are calculated on behalf of the SPV, who is the contractor and operator of the investment. Specifically, in the case that the NPV takes a negative value and/or the PI<1, then the SPV should reject the project, otherwise he should select it. In these calculations, the total capital managed by the involved participants is taken into account, i.e. the SPV, the public sector and the bank loan.

4.3 Consideration of financial scenarios for different values of the PR

In order to select the best proposal for the estimation of the annual payments by the public sector, there are two scenarios tested, where the initially estimated costs are increased by 18% or 28% respectively, according to the PR designated by the legislation of public construction projects in Greece.

Generally, the project's budget estimation process specifies that these rates are used in the increasing of the costs that arise by the uniform tariffs work of the government's public works department, which contain relevant articles, prices of materials and works. Indicatively, the analytical costs of construction, electromechanical and other works are calculated and these costs are increased by 18% for projects financed by the public investment budget and by 28% in other cases (National Law, 1985; 2008), where these rates correspond to general costs and profit. A last step is to increase the rate of unplanned work of the

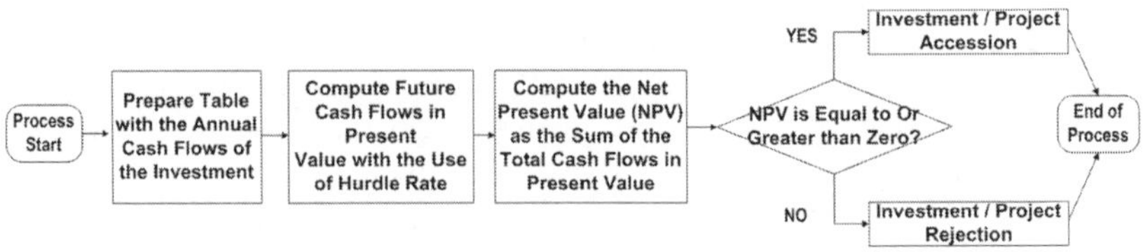

Figure 1: Process flowchart for the estimation of the Net Present Value.

project and the percentage of the value added tax. In the present case study, the resulting calculations of the NPV and the PI for both the scenarios of 18% and 28% respectively, are presented in Table 1. Additionally, the results arising through the evaluation of these scenarios in the financial sustainability criterion are also included in Table 1, while the diagrams depicting the cash flows in FV are illustrated in Figures 2 and 3 respectively. We mention that the PI and the PR take different values, because the cash flows that are adjusted with the PR are in FV and the relative cash flows for the calculation of the PI are discounted in PV.

Table 1: Project's financial sustainability.

Profit Rate:	Profitability Index	Net Present Value	Financial Sustainability		
			Lack of Liquidity	Switching Point	Availability Capital
18.00%	1.01417	1,350,000	0-7 years	8th year	8-20 years
28.00%	1.10012	9,530,000	0-2 years	3rd year	3-20 years
30.20%	1.11903	11,330,000	-	-	0-20 years

The evaluation of the diagrams in Figures 2 and 3 focuses on the existence of the switching point, where total inflows are equal to outflows and there is a change in the project's financial liquidity. In the present case, the first scenario where the estimated costs are increased with the PR of 18%, there is availability of capital from the 8th year until the end of the project life, but there are negative cash flows for the private sector by entering the operational phase, which last for 7 years, when there is a switching point. Furthermore, the second scenario, where the estimated costs are increased with the PR of 28%, there is also a lack of liquidity by entering the operational phase for 2 years, when there is a switching point. According to these results, both the scenarios are evaluated us financial unsustainable. Conclusively, the WT projects that follow PPP contracts in Greece, where the repayment of the bank loan is at a fixed rate with fixed payments, it is inefficient to follow the same process with the conventional public project procurement process, in order to estimate the project's budget, i.e. to increase the estimated costs with a PR of 18% or 28% respectively. The above results suggest the necessity to develop a formula for the calculation of the minimum price of the PR, so as to exclude the negative cash flows during the project's operational phase and to ensure the project's financial sustainability.

5 Development of a formula for the calculation of the PR

In order to develop a formula for the calculation of the minimum price of the PR in a WT PPP project, where the bank loan is repaid with equal payments throughout the project's operational phase, the following mathematical demonstration is used. The annual flows for the SPV are calculated through the next formula:

$$\text{f(annual) (t)} = \text{FV(inflows) (t)} - \text{FV(outflows) (t)} \tag{1}$$

where (t) = (1, 2, 3, 4,..., 20].

5.1 Total outflows

Let a denote the initial capital that is funded through the loan, p the interest rate and n the amortization period in years. Therefore, the FV of outflows for the loan repayment varies according to the following function, $h(t)$. Following the method of progressive amortization, the annual payments of a loan with a fixed interest, are calculated by the formula:

$$h(t) = (\text{interest} + \text{amortization}) = \frac{a}{\sum_{1}^{n} \frac{1}{(1+p)^{n}}} \tag{2}$$

Let c denote the operational costs for the 1st year of the operational phase, which is increased with a rate r per year, so the FV of this cost $g(t)$ is calculated with the formula:

$$g(t) = c \times (1+r)^{\mathrm{t}} \tag{3}$$

5.2 Total inflows

Let $R(t)$ denote the FV of the SPV's inflows, which arise through the annual payments by the public sector. These payments are increased following the interest rate r, as shown in formula:

$$R(t) = (\text{1st payment}) \times (1+r)^{\mathrm{t}} \tag{4}$$

Payment of the 1st year is estimated through the initial invested capital a, which is divided equally to the years of the operational phase v, plus the annual operating costs, multiplied by the (1+PR):

$$(\text{1st payment}) = (\frac{a}{v} + c) \times (1+PR) \overset{(4)}{\Rightarrow}$$

$$R(t) = (\frac{a}{v} + c) \times (1+PR) \times (1+r)^{\mathrm{t}}, \mathrm{t} = (0,1,2,...,20] \tag{5}$$

Formulae (1), (2), (3) and (5) give:

$$(1) \overset{(2),(3),(5)}{\Leftrightarrow} \mathrm{f}_{(\text{annual})}(\mathrm{t}) = R(t) - g(t) - h(t) \Rightarrow$$

$$\mathrm{f}(\mathrm{t}) = (\frac{a}{v} + c) \times (1 + PR) \times (1+r)^{\mathrm{t}} - c \times (1+r)^{\mathrm{t}} - \frac{a}{\sum_{1}^{n} \frac{1}{(1+p)^{n}}} \qquad (6)$$

As shown by Formula (6), in the case that the loan's amortization period is equal to the operational phase, then $n=t$. According to the financial sustainability criterion, annual cash flows for the SPV should be positive. Further, there is: $\mathrm{f}(1)<\mathrm{f}(2)<...<\mathrm{f}(n)$. Thus, Formula (6), for the f (1) > 0, results in:

$$\mathrm{f}(1) = (\frac{a}{v} + c) \times (1 + PR) \times (1+r) - c \times (1+r) - \frac{a}{\sum_{1}^{n} \frac{1}{(1+p)^{n}}} > 0 \Rightarrow$$

$$PR > \frac{c \times (1+r) + \frac{a}{\sum_{1}^{n} \frac{1}{(1+p)^{n}}}}{(\frac{a}{v} + c) \times (1+r)} - 1 \qquad (7)$$

where: a = capital funded through loan, v = operational phase (in years), c = annual operating costs, r = interest rate (inflation), p = loan's interest rate (fixed) and n =loan's amortization period (in years).

The above Formula (7) gives the desired result, as it calculates the minimum value of the PR, so as to exclude negative cash flows during the operational phase of the project. In the specific case study, the result that is calculated by the formula is PR = 30.20%, as presented in Table 1. Furthermore, the sensitivity of the minimum PR value in the above variables is estimated with the use of this formula. Indicatively, the results of the sensitivity analysis are illustrated in Figure 4, where is presented the minimum value that the PR has to take, according to a change of the variables' values in a range of (-15%, +15%), in order to achieve positive cash flows throughout the operational phase of the project for the SPV and to ensure the project's financial sustainability.

6 Conclusions

This research, having as a strong theoretical foundation the financial analysis approach, as it is described in the EC's guide to CBA for investment projects, examines the budget estimation process that is used under the conventional public procurement framework for a WT project that follows a PPP contract in

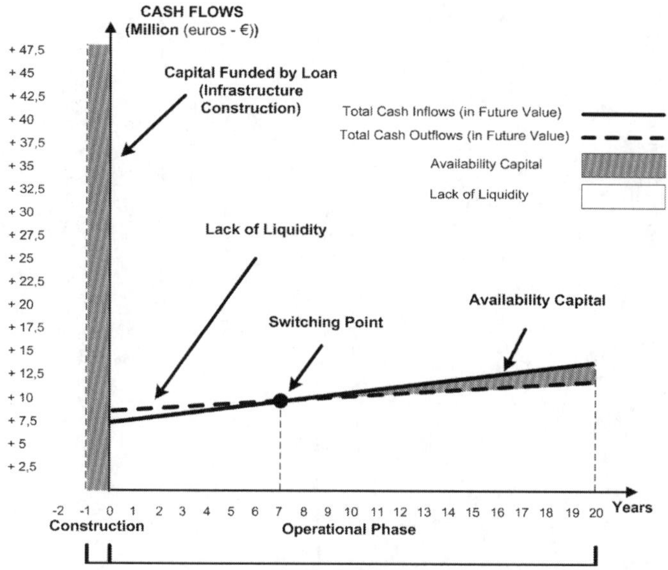

Figure 2: Cash flow diagram with payments adjustments at the PR = 18%.

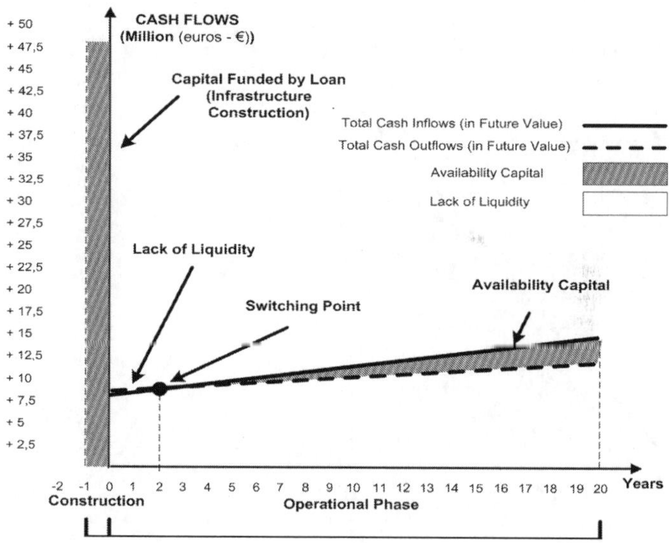

Figure 3: Cash flow diagram with payments adjustments at the PR = 28%.

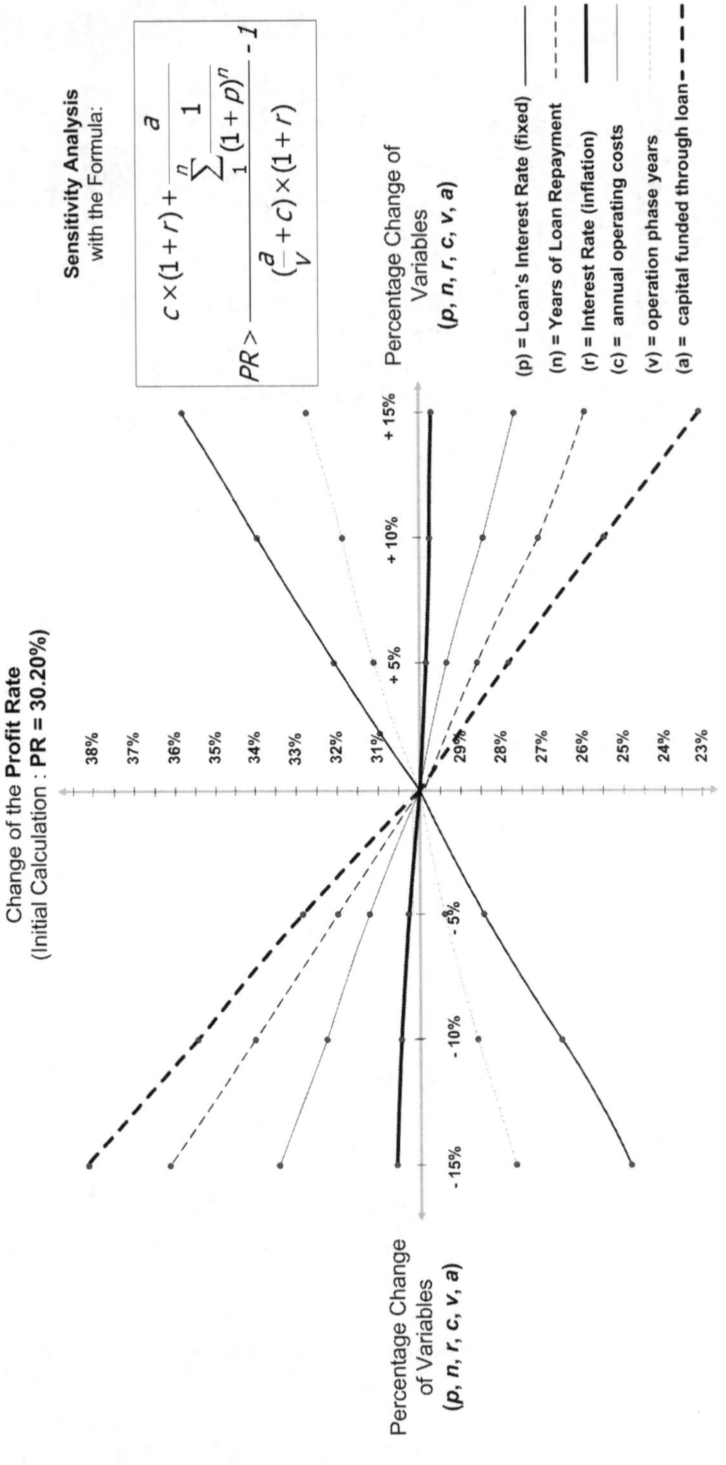

Figure 4: Sensitivity Analysis of the Profit Rate's minimum price.

Greece. Specifically, the financial analysis of a BOT project is presented and two scenarios for the annual payments are evaluated, with a PR of 18% and 28% respectively. The results demonstrate that the specific PRs cannot be used for the PPP projects in Greece. Furthermore, taking into consideration the financial sustainability criterion, a new mathematic formula is developed. With the use of the specific formula in the case study, it is demonstrated that avoiding of liquidity shortage and thus the increasing of the financial sustainability in a 20 year time horizon, is achieved with the annual payments adjusted to a minimum PR value of 30.2%. Moreover, this formula is used in the implementation of the sensitivity analysis for the PR's minimum value with the six variables which act upon it. This formula can be a useful tool, not only for the public sector during the conceptual phase of a WT PPP project, but for all stakeholders, while it can be further used in other type of investment projects, during the initial stage.

References

[1] Tang L., Shen Q., Cheng E. W.L., A review of studies on Public–Private Partnership projects in the construction industry, *International Journal of Project Management*, doi:10.1016/j.ijproman.2009.11.009.

[2] Pongsiri, N., Regulation and public private partnerships, International Journal of Public Sector Management, 15(6), pp. 487-495, 2002.

[3] Grimsey, D. & Lewis, M.K., Public Private Partnerships, the Worldwide Revolution in Infrastructure Provision and Project Finance (Chapter 1). The nature of partnerships, Edward Elgar: Cheltenham, pp.5-9, 2007.

[4] Yescombe, E.R., Public-Private Partnerships: Principles of Policy and Finance (Chapter 2). PPPs-For and Against, Published by Elsevier Ltd: Oxford, pp.15-28, 2007

[5] Haarmayer, D. & Mody, A., Financing Water and Sanitation Projects – The Unique Risks, The World Bank Group, Finance, Private Sector and Infrastructure Network, 151, pp. 1-4, 1998.

[6] Zhang, X-q. & Kumaraswamy, M.M., Hong Kong experience in managing BOT projects, Journal of Construction Engineering and Management, 127(2), pp. 154-162, 2001.

[7] Karmperis, A.C., Sotirchos, A., Aravossis, K & Tatsiopoulos, I., Funding evaluation model for the implementation of wastewater treatment projects through Public Private Partnerships. Environmental Economics and Investment Assessment III, vol. 131, eds. K. Aravossis & C.A. Brebbia, WIT Press: Southampton, pp. 147–160, 2010.

[8] Ng, A. & Loosemore, M., Risk allocation in the private provision of public infrastructure, International Journal of Project Management, 25(1), pp. 66–76, 2006.

[9] Verveniotis, P.C., Implementation of Projects with the Exploitation Concession-BOT Contracts: Construction and Arbitration, Sakkoulas: Athens and Komotini, pp. 78-79, 1993 (in Greek).

[10] Roumboutsos, A. & Anagnostopoulos, K.P., Public–private partnership projects in Greece: risk ranking and preferred risk allocation, Construction Management & Economics, 26(7), pp. 751-763, 2008.
[11] Ministry of Finance, Guide for the Implementation of PPP in Greece, Athens, 2006, http://www.sdit.mnec.gr/en/infopoint/implementation/
[12] European Union Committee, Green Bible for Public Private Partnerships, Brussels, 2004.
[13] Alonso-Conde, A.B., Brown, C. & Rojo-Suarez, J., Public Private Partnerships: Incentives, risk transfer and real options. Review of Financial Economics, 16(4), pp. 335-349, 2007.
[14] Linke, P. & Kokossis, A., Advanced process systems design technology for pollution prevention and waste treatment, Advances in Environmental Research, 8(2), pp. 229-245, 2004.
[15] Seadon, J.K., Integrated waste management-Looking beyond the solid waste horizon, Waste Management, 26(12), pp. 1327-1336, 2006.
[16] Martimort, D. & Pouyet, J., To build or not to build: Normative and positive theories of public–private partnerships, International Journal of Industrial Organization, 26(2), pp. 393-411, 2008.
[17] European Union, Co-ordination in the contracting process for Public Projects: Procurements and Services, Directive 18, 2004.
[18] Asian Development Bank, Procurement Guidelines, Ch. II, pp. 18-19, 2001.
[19] Araujo, AR., Procurement in Privately Provided Infrastructure Projects Financed by the World Bank, World Bank, pp. 16-17, 1998.
[20] Zhang, X., Concessionaire Selection: Methods and Criteria, Journal of Construction Engineering and Management, 130(2), pp. 235–244, 2004.
[21] Zhang, X., Criteria for Selecting the Private-Sector Partner in Public-Private Partnerships, Journal of Construction Engineering and Management, 131(6), pp. 631–644, 2005.
[22] Zhang, X., Public Clients' Best Value Perspectives of Public Private Partnerships in Infrastructure Development, Journal of Construction Engineering and Management, 132(2), pp. 107–114, 2006.
[23] Geng, Y. & Doberstein, B., Greening government procurement in developing countries: Building capacity in China, Journal of Environmental Management, 88(4), pp. 932-938, 2008.
[24] Lozier, J.C., Smith, G., Chapman, J.W.& Gattis, D.E., Selection, design, and procurement of a demineralization system for a surface water treatment plant, Desalination, 88(1-3), pp. 3-31, 1992.
[25] Francis Rillaerts, F., Concessions in the water sector, Desalination, 124(1-3), pp. 13-17, 1999.
[26] Hatush, Z. & Skitmore, M., Contractor selection using multi-criteria utility: an additive model, Building and Environment, 33(2-3), pp. 105-115, 1998.
[27] Pongpeng, J. & Liston, J., TenSeM: a multi-criteria and multi-decision-makers' in tender evaluation, Construction Management and Economics, 21(1), pp. 21-30, 2003.

[28] Liou, F-M. & Huang, C-P, Automated Approach to Negotiations of BOT Contracts with the Consideration of Project Risk, Journal of Construction Engineering and Management, 134(1), pp. 18-24, 2008.
[29] Grimsey, D. & Lewis, M.K., Are Public Private Partnerships value for money? Evaluating alternative approaches and comparing academic and practitioner views, Accounting Forum, 29(4), pp. 345–378, 2005.
[30] Çelik, B., Girgin, S., Yazıcı, A., Ünlü, K., A decision support system for assessing landfill performance, Waste Management, 30(1), pp. 72-81, 2010.
[31] Morrissey, A.J., Browne, J, Waste management models and their application to sustainable waste management, Waste Management, 24(3), pp. 297-308, 2004.
[32] Lavee, D., A cost-benefit analysis of a deposit–refund program for beverage containers in Israel, Waste Management, 30(2), pp. 338-345, 2010.
[33] European Commission, Guide to Cost – Benefit Analysis of Investment Projects, Directorate General Regional Policy, Brussels, 2008.
[34] Chiang, Y.H., Cheng, E.W.L. & Lam, P.T.I., Employing the NPV-Consistent IRR Methods for PFI Contracts, *Journal of Construction Engineering and Management*, **136(7)**, pp. 811-814, 2010.

Sustainable development indicators for integrated coastal management: definition area and spatial properties

J. Kaulins, R. Ernsteins & I. Kudreņickis
Department of Environmental Management, University of Latvia, Latvia

Abstract

When designing a sustainability indicator system (SIS) within the integrated coastal management (ICM) development process, there are still to be encountered a number of difficulties in the constructing of the system both in the indicator selection process and later when analysing the system and assessing its degree of correspondence to the direct objective of application, reflecting the set of planning targets and accounting for the specifics of the particular coastal territory. During research and development work in the coastal municipalities in Latvia, particularly, in Saulkrasti municipality, there was developed municipal level SIS, being locally discussed and acknowledged by experts and municipality specialists. Saukrasti SIS was developed, based on the adaptation of the known theoretical approaches and previous national case studies research done. This practice based experience also gave an inspiration for a number of theoretical generalisations about coastal indicators definition area, their spatial properties and, especially, algorithmic scheme for designing sustainability indicators.

Keywords: indicators, algorithm, coastal zone, integrated management, sustainability capitals, spatial, governance

1 Introduction to indicators definition

In the literature on measuring sustainable development, a number of sustainability indicator definitions by different authors and institutions can be found being based mainly on two approaches: conceptual and functional [1].

As the name suggests, conceptual definitions are derived from the indicator concept itself and are important for understanding it. According to FAO definition [2], an indicator is defined as a *variable, pointer, or index related to a*

WIT Transactions on Ecology and the Environment, Vol 144, © 2011 WIT Press
www.witpress.com, ISSN 1743-3541 (on-line)
doi:10.2495/ECO110271

criterion. Its fluctuations reveal the variations in those key elements of sustainability in the ecosystem, the fishery resource or the sector and social and economic well-being. The position and trend of an indicator in relation to reference points or values indicate the present state and dynamics of the system. A more general explanation is given as by Garcia and Staples [3]: *indicators are pointers that can be used to reveal or monitor conditions and trends in the fisheries sector and the marine environment.* Similar definitions can also be found by Slocombe [4], Fletcher et al, [5], OECD, [6], however one of the most comprehensive definitions is offered by Maureen Hart: "*An indicator is something that helps you understand where you are, which way you are going and how far you are from where you want to be. A good indicator alerts you to a problem before it gets too bad and helps you recognize what needs to be done to fix the problem. ... They allow you to see where the problem areas are and help show the way to fix those problems.*" [7].

All of the above definitions, however, answer to the question of „Why do we need indicators?" rather than explains what it actually is. Functional definitions explain what exactly the indicator does and how it differs from a simple parameter or measurement. This type of explanations is provided by Garcia et al [3], Smeets and Weterings [8], Hak et al., [19]). From the definitions analysed, the following one, approved by the United Nations Council on Sustainable Development in 2001, has been selected as the most comprehensive:

Indicators for sustainable development are in order to: 1. translate physical and social science knowledge into manageable units of information that can facilitate the decision-making process; 2. help to calibrate and measure progress towards sustainable development goals; 3. provide early warning to prevent damage; and 4. communicate ideas, thoughts and values (cited from Brown, Reyntjens [1]).

However, when constructing a specific indicator system, the practical use of this definition is cumbersome, as it fails to give a clear answer as to whether the selected value corresponds to indicator specifics or not. To receive such an answer, the indicator definition needs to answer to the following questions:

1) what values can serve as indicators, 2) what do these values characterise, 3) what is their role in the governance system, 4) what are the limitations of indicator functions, 5) what formal qualities distinguish an indicator from other values that can be measured, 6) what is the significance of the measurable parameter for a given management system. The definition cited earlier [1] answers to the first three of these questions and partially to number six as well.

Based on the above, we find the role of indicators in the hierarchical structure of management elements (Figure 1), which is built in line with the stages of the planning implementation cycle. Manageable units = measurable units; it means that an indicator must be expressed by numeric values. The same point also indicates to the principal role of indicators in the management system. Points 2 and 3 reflect a need for a string of successive measurements. Point 4, we consider, is more a conceptual one; to some extent, this also applies to Point 3.

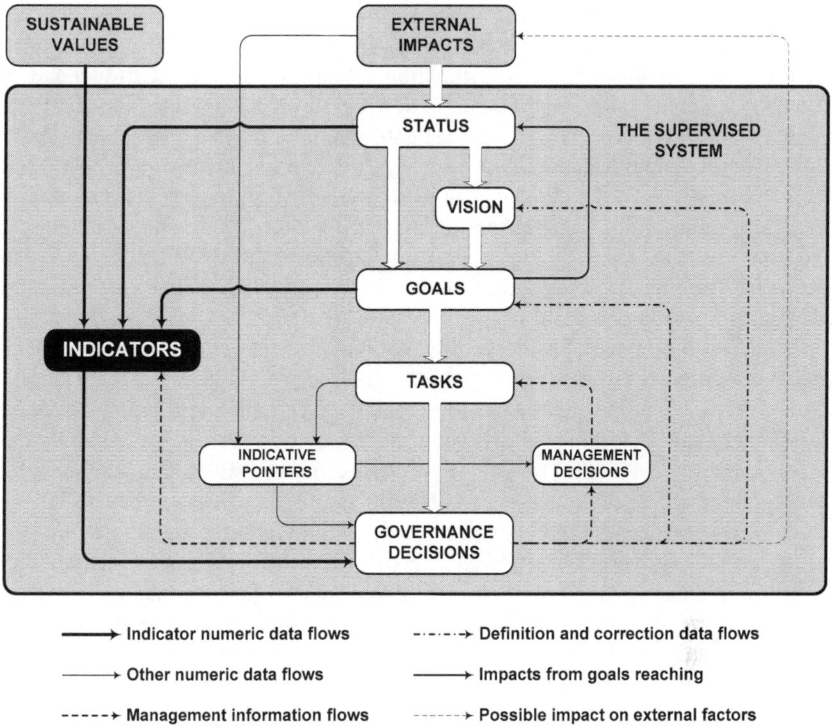

Figure 1: Indicators in planning.

Indicator applicability limits ensue from the above indirectly. A more specific description of it is provided by Sainsbury and Sumaila [10], defining that „...*an indicator that does not relate to an operational objective is not useful in this context*". More specific conditions, however, are missing. A requirement for a formal feature cannot be found in the definitions mentioned; however, we find such indications in the latest literature: *a datum or variable observed becomes an indicator only once its role in the evaluation of a phenomenon* ***has been established*** [11]. In fact it means that relevant management decision is required. This also follows from the need for financial and human resource allocation for obtaining (and often accumulating) the data, doing the calculations and preparing the reports, which can take place only based on management decision.

2 Definition area of sustainability indicators

Indicators reflect the current situation as a point of reference on the one hand, and as advancement towards a strategically set planning goal on the other. If we consider a goal as a numerically defined value, it is easy to understand that it is located on a scale and can be corrected both within the current planning cycle and within new planning cycles to come. The tasks that are the means for reaching these goals cannot be placed on this scale, as in case of tasks, the attainment of the particular value characterises the end of the process, e.g., the use of financial resources allocated for a particular project. This also indicates to the fact that an indicator needs to function at the level of long-term goals. The need for assessing the importance of the measurable parameter ensues from practical considerations as well: the indicator method, when applied correctly, is expensive and complicated enough to use it for solving relatively generalised and long-term tasks only. To create an indicator system for the coastal sustainability long term monitoring, being based not only at the task level is very challenging and this certainly shall include then both indicators as we describe and also indicative pointers, giving additional insight into the status of particular management system, as well.

Based on the above considerations, the authors first agree on the role of an indicator within the hierarchical cycle of governance (Figure 1). The diagram shows that indicators refer to the hierarchical governance level of goals only, defining initial conditions, status with relation to goal implementation and dependence on our value system, if we speak of the values that we consider worth preserving, i.e., that are sustainability factors. Indicators may influence process governance through governance decision-making – by determining or adjusting this governance (and the goals themselves) in accordance with indicator readings. The notion of sustainable governance contains two sub-notions: sustainability as the ability of the system to preserve the defined values, and governance as influence on this system. Accordingly, the definition of sustainable governance indicator should reflect both aspects. An attempt to do it in one definition could lead to complicated and miss-interpretable construction. Therefore it shall be divided into two parts: governance and sustainability.

A governance indicator illustrates a development factor or a set of factors and helps the public and decision-makers to get an impression of and control the situation with regard to the initial conditions from which the development goals defined in development planning documents ensue at the given governance level. In other words, a parameter or a group of parameters can be defined as a **governance indicator** if it/they reflect comparatively and unequivocally the **numeric values of resulting indications concerning governance goals and changes in these values** and allow for determination of the status and trends, and which can – directly or indirectly – be influenced significantly with decisions on the given governance level only.

A sustainability indicator reflects our perspectives on the values that are to be preserved in the name of our own and future generations: a parameter or a group of parameters can be defined as a **sustainability indicator** if it/they reflect comparatively and unequivocally the **numeric values of sustainability impact factors and changes in these valu**es and allow for determination of the status and trends, and which can – directly or indirectly – be influenced significantly with decisions on the given governance level only.

A parameter which at one governance level is an indicator is not necessarily an indicator at other levels; it follows from the presence of a governance level in the definition. A governance level here means the influence area within public administration: municipal, regional, national, etc. In case it is purposeful – for the sake of overall clarity – to show the values of such a parameter, this would then be an indicative pointer. It does not possess all functions of an indicator; that is, it does not reveal the efficiency of decisions taken. And vice-versa – a parameter which is only an indicative pointer at one level may become an indicator at other levels.

A phrase "...can be defined..." indicates the need for an administrative decision for a parameter to become an indicator. When applying the above definitions, an **algorithm scheme for developing an indicator system** may be constructed (see Fig. 2.). To be able to use the resultant product as a full-fledged indicator system, all indicators and indicative pointers need to have calculation methodologies developed and specific methods of result representation indicated.

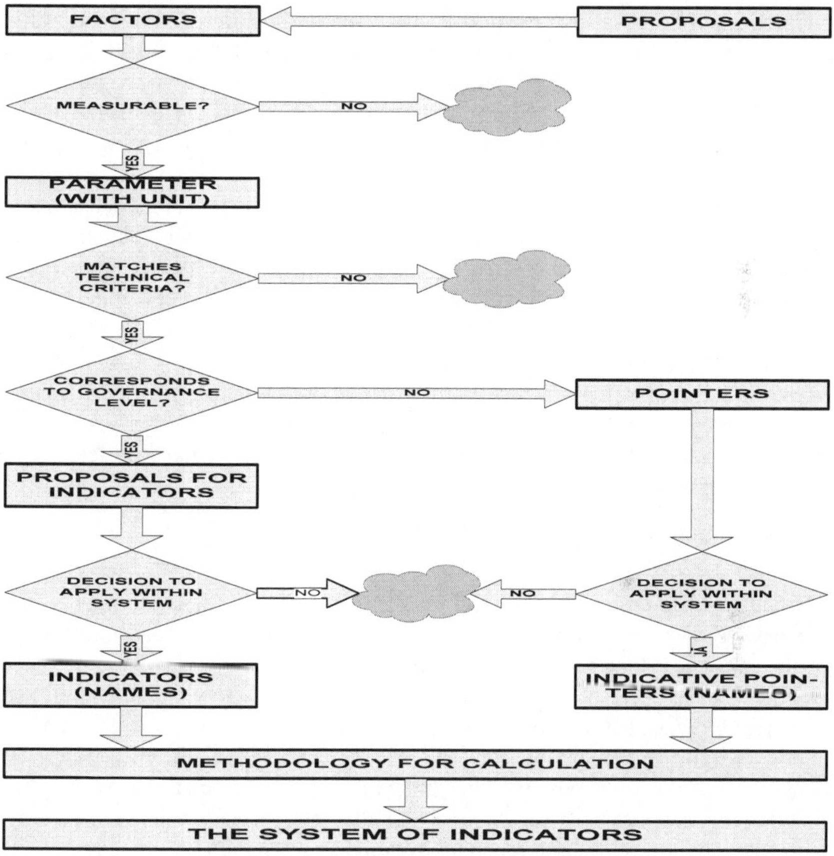

Figure 2: Algorithm for indicators.

Also, the following generic conditions are to be taken into account. Proposed factor is a parameter when it is measurable numerically. The factors proposed are in practice selected by various means: in system analysis, participatory within different target groups, following expert conclusions, through analysing existing systems as examples, etc. Also, parameter may not be used for constructing an indicator when its set of data does not meet the technical requirements regarding source data. This may not always be determined at this state; it sometimes manifests itself only when developing a methodology for calculations, or even worse – when indicator calculation is done for the first time. Parameter (or a group of parameters) becomes an indicator when it meets the technical requirements, is located within the indicator definition area and when a decision has been taken to apply it as an indicator. However, if a parameter meets the requirements but is not located within the indicator definition area, it may only be used as an indicative pointer.

From the above definitions, indicator utilisation limits, or the definition area in the respective governance system, also derive.

1. Given factor or set of factors F may serve as a territorial sustainable development indicator if it illustrates the implementation of sustainability goals defined in local governance planning documents and respective progress, and whose implementation and control falls within the competence of the given governance level. Labelling the governance level 'i' with L_i, the following law applies:

$$F \in L_i, \tag{1}$$

2. Given factor or set of factors F may not serve as a territorial sustainable development indicator if it only illustrates the values of an isolated planning task which is an intrinsic part of the set of goal-oriented measures and whose implementation applies unequivocally to a lower governance level competence:

$$F \in L_{i-n}, \tag{2}$$

where n – a degree difference in governance levels; $n \geq 1$.

3. Given factor or set of factors F may not serve as a territorial sustainable development indicator if the changes inflicted upon it by the given-level competence decisions - ΔF_{int} – are small compared to the changes inflicted by a higher-level competence decisions ΔF_{ext}:

$$\begin{cases} F \in L_{i+n} \\ \Delta F_{ext} >> \Delta F_{int}. \end{cases} \tag{3}$$

Based on the definitions and through assessment of indicator systems and their designing process, the concept of indicator integrativity was introduced, i.e., the range in which the particular indicator characterises a given governance system. The integrated management cycle planning applied in Saulkrasti municipality was based on municipal situation analysis in sustainability dimensions and on segmentation of priority integrative problem areas at dimension intersection points. The indicators were selected separately for characterising sustainability dimensions and also intergative problem areas.

By way of combining both resultant systems and assessing how the indicators refer to sustainability components, we can divide all indicators into 4 groups:

- **Sub-sectoral indicators** – describe an isolated, but governance level-specific aspect of the respective sustainability dimension,
- **sectorial indicators** – principally describe one sustainability dimension,
- **integrative indicators** – describe integrative problem areas and other processes which concerns at least two sustainability dimensions,

– **integral indicators** – describe the key, more general pointers of the governed system that characterise a given governance system in its entirety and/or compared to other similar systems.

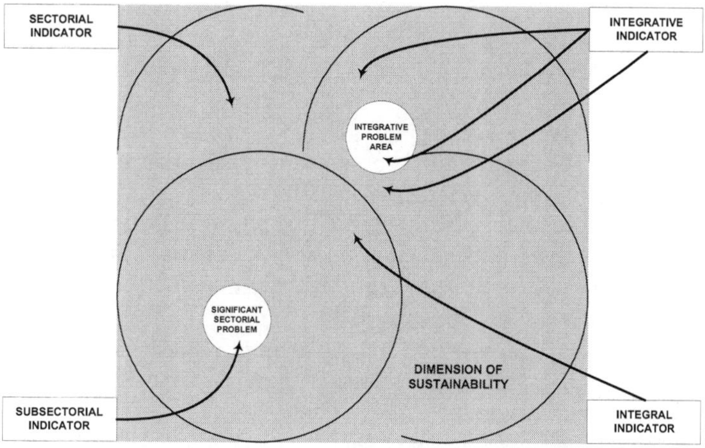

Figure 3: Indicator integrativity

The above division is to a certain degree similar to the one found in [12], but this source groups indicators according to target audience level pyramid principle and does not reflect the presence of the fourth dimension of sustainability – governance and communication environment. It can be understandable that the location of an indicator on the scale of integrativity levels generally correlates to the location in the target audience level pyramid, but this is not quite the same. Such integral indicators are fully possible which are significant or understandable only to expert audience (but required!), and an isolated sub-sectoral indicator may also characterise a very severe problem important at all levels - from the general public to experts and governance decision-makers. If both approaches are applied, it is convenient to analyse and assess the balance of the indicator system as per sustainability dimensions and to differentiate the contents of the material when drafting indicator reports and public reviews.

3 Spatial properties of coastal sustainability indicators

An indicator system for measuring coastal sustainability differs from the general case by its spatial specifics: the coastal zone is formed by a coastal line with the related set of other geospatial elements by Clark [13]. The indicator system though which coastal sustainability is assessed should therefore be able to at least differ the coastal zone from the inland and provide a comparison, to establish the origin of impact factors on the coastal status and development trends, and to create understanding of the distribution of coastal impacts within the governance territory.

Ideally, the term 'coastal zone' should apply to a territory where the specific coastal impacts can be detected, and vice versa – a territory which impacts the developments on the coast and its proximity, as these impacts:
1) may in advance be unknown precisely enough,
2) may change over time,
3) differ for different factors,
4) The specifics of spatial distribution of the data used may prevent their correct differentiation.

In practice, the term 'coastal zone' is therefore applied to a **relative** territory which (see Fig.4) – within a single system – may in addition be applied in a number of ways depending on the data character.

Based specifically on the character of data spatial distribution, the EU project DEDUCE, which aimed at developing a European coastal sustainability model based on indicator system measurements, distinguished four key types of coastal zone definition (*Marti X. et al,* [14]):

– **administratively territorial,** which mainly represents the social and demographic information, as it is accumulated mainly concerning these aspects, with the local government as the smallest unit,

– **co-ordinate character,** which reflects locations of measurements in specific geographical co-ordinates (e.g., quality of surface waters) or in an object which is small enough for the range of measurements and can relatively be taken as a co-ordinate point (port operation) or is related to the geometrical coastline (coastline dynamics, artificial coastal constructions),

– a fixed-width coastal zone formed by the coastline's **geometrical buffers**; in the specific case differentiating between the nearby coastal zone (1 km) and the distant coastal zone (10 km); in this manner, geospatial information is assessed: land use, protected territories, etc.,

– **combined**, which contains the features of administrative territorial and co-ordinate points (e.g., number and location of objects in a specific territory); in this manner, thematic sectorial information is most often reflected such as the location of eco-certified tourist residences, etc.

There should be added, that comparison zone within local governments was defined as the territories of the districts of which at least 50% lie within the coastal 50 km buffer zone, and local governments in other districts of which at least 50% lie within the coastal 50 km buffer zone.

In our opinion, an additional fifth type is worth distinguishing on the local planning level, which by its data processing methods is similar to the geometrical but is related to spatial **planning elements** rather than buffers: protected zones and other types of territories with limitations of economic activities or other special regulations concerning the use of the particular territory.

The nature of data determines not only coastal definition types but also the spatial relations to the coast by indicators themselves. Here the following cases can be distinguished:

- special coastal indicators directly characterise some values characteristic of the coast only, e.g., catch of fish, bathing water quality, artificial coastal constructions, etc.

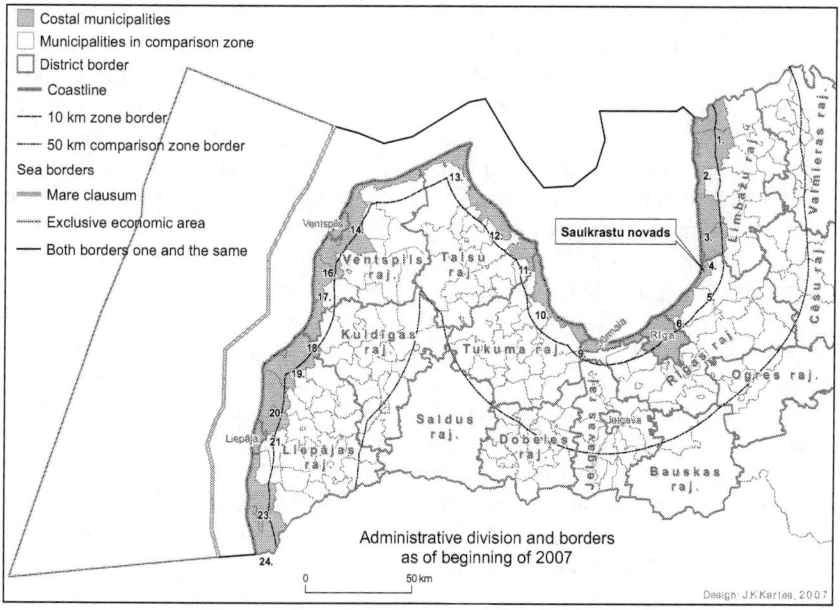

Figure 4: Coastal zone in Latvia.

- **coastal discernible** indicators which characterise elements not directly coast-specific but where a correct spatial assessment of the coast-related impacts is possible

- **coastal relatively discernible** coastal indicators, where the spatial distribution of data is "unclear", which prevents the correct determination of these impacts but our common knowledge about the territory allow for at least a qualitative assessment

- indicators **non-applicable** directly to the coast, which characterise a factor in the overall territory as a single inseparable object (e.g., number of residents, municipal budget values). However, here the coastal impacts can be assessed by mutually comparing such territories.

A strict line cannot be drawn between the discernible and relatively discernible parameters. It may in each case depend on data gathering and accumulation type. Also, when selecting the particular parameter as an indicator, we may plan changes in the data structure so as to improve coastal assessment opportunities in the periods to come.

This border also depends on the size of the local government. In national or international level indicator systems, the precision "to the municipality" is sufficient and even desirable, as in most cases easily available, safe and reliable statistical information is at hand [13,14]. In local planning, to be able to compare different parts of a territory and obtain information on coast-related impacts, more detailed elements need to be distinguished: isolated places of residence, land property (cadastre units), etc. In principle, this applies also to characterisation of spatial distribution of data origin (and availability).

The developed classification helps us understand how large and in which areas and governance levels can the coast-specific impacts and processes be, and the role of governance decisions in these, as well as allowing for a more profound assessment of the importance of the impacts established as a result of measurements. There should be added, that according to Latvia legislature 5 km zone formal border is the geometrical border of a limited economic activity zone, which may be altered based on local geographic circumstances. The coastal dune protection zone is determined according to the Protected Zone Law: in villages – 150 m, outside – 300 m in width. The coastal geometrical buffer is altered by adding specially protected biotopes which adhere to the formal protected zone.

The above approach helps select the values to be measured which – in the current data availability – reflects most fully all the coastal impacts in their different aspects, and to receive significant additional information for the interpretation of results. In addition, data sources may also be evaluated and – where possible – the degree of detailed elaboration may be improved so as to increase the indicator's spatial resolution.

The outlined approach may be fruitful not only for the coastal area but in all areas where a factor with a spatial impact is present in principle: proximity of a large city, a state border, geographical obstacles, etc.

4 Saulkrasti indicator system case study

Saulkrasti county, which is a small, relatively urbanised (for Latvian conditions) Baltic Sea coastal territory (Fig. 3), had in 2009-2010 coastal integrated development guidelines elaborated for it within the University of Latvia COBWEB project (COBWEB [15]), as the central component of municipal integrated governance. Special attention was devoted to measuring sustainability, and for this purpose, a system of indicators was developed. The system was elaborated based on the analysis of four dimensions of sustainability (natural, social and economic environment, governance and communication) and integrative problem areas as defined on their points of intersection.

Initially, over 100 indicators were proposed. However, after a selection according to the algorithm as presented in Fig. 2, the resultant system contains 55 indicators which thematically form 24 groups and reflect the status of all four dimensions of sustainability and provides overall characterisation of sectors, integrative problem areas and municipality as a whole (number in brackets show number of indicators in given thematic group and percent of total number of indicators by sustainability dimensions):

I. Nature environment (total 15, 27%): Green frame status (3), Potential loads on the environment from public utilities (5), Air quality and climate change factors (3), Surface water quality (1), Land use development (1), Nature risks (2).

II. Economic environment (total 14, 26%): Economically active people (1), Municipal budget (1),Traffic routes (2), Skulte port development (4), Tourism characteristics (6).

III. Social environment (total 13, 25%):Health care characteristics (2), Supporting for cultural environment (3), **S3** Employment and entrepreneurship

(2), Social care and social security (2), Education system characteristics (1), Social life quality (3).
IV. Governance and Communication (total 5, 9%): Activities for environment maintenance (1), Information of society about environmental events (2), Activities in nongovernmental sector (2).
V. Integral indicators (total 8, 13%)**:** Number of inhabitants (1), Area development index (1), Area attractivity index (1), Opinions of society (5).

We can see that within the system, both the traditional dimensions of sustainability are equally represented, except, governance and communication as horizontal dimension introduced as having increasingly growing role in Latvian conditions particularly, since the measurement possibilities and process itself is more time and other resources consuming as often has been based on opinion pools. It is difficult to find pointers that meet indicator requirements which characterise governance and communication; these are therefore represented to a lesser degree. This drawback, however, is compensated by the integral pointers section, which, together with the other sections, reflects the efficiency of governance perhaps most clearly, without singling out the contribution of any particular dimension.

The prevalence of integrative indicators in the system (64%) stems from the broad approach to planning, which is based exactly on such integrative perspective of seeing sustainability dimensions in their interactions. Directly integrated problem areas are decrypted by 39% of all indicators. Separate sectors (e.g., tourism) are singled out when the related issues is significant enough for the development and welfare of the entire territory. The integral indicators also include separate indicators selected to characterise particular dimensions, as these bear a considerably larger content load, but are overall designed to characterise resident attitudes and opinions. Their number is comparatively small; in case of a bigger proportion, there is a risk of obtaining too general information, which provides an insufficiently detailed picture for the purposes of practical action and decision-making. The usable data sources are mainly the information accumulated by state and municipal institutions, and opinion-polls of businessmen, NGOs and residents. In one case, the utilisation of the opportunities afforded by public monitoring is planned in one case.

Considering the role of the coast in Saulkrasti county, a conclusion may be made that the 34% of indicators bearing the load of clear coastal characteristics (special coastal and with clear resolution of coastal zone), may be insufficient for this purpose. In future, however, by means of improving data collection and accumulation in municipal institutions, this share could be enlarged towards coastal relatively discernable data.

The system has been discussed with municipal and planning experts and the wider public in seminars and focus group discussions and assessed as a practically implemental as Saulkrasti county strategic planning element, being generally balanced indicator system, which emphasizes the local specifics of Saulkrasti and includes the key general ones.

5 Conclusions

The developed indicator definitions and indicator selection algorithm allows for the careful selection of parameters that correspond to indicator meaning and purpose. This helps eliminate errors which might occur by introducing to the system parameters inappropriate or insignificant to a given governance level. The assessment of indicator spatial characteristics and classification allows for building an indicator system, in which the impacts of a coastal or other spatial factor on the respective governance system are reflected as fully as possible. When developing an indicator system for sustainability assessment in local coastal municipalities like Saulkrasti county, application of suggested above selection of indicators and the observation of indicator spatial characteristics leads to the resultant system, being balanced in terms of both reflection of sustainability dimensions as well as common territorial characteristics and description of coastal impacts. This type of an indicator system provides the opportunity to not only monitor sustainability of a territory and associated changes but also to follow the governance processes and control implementation of the strategic objectives as set in development planning documents and ensure continuous information to decision-makers and the public at large.

References

[1] Brown, J., Reyntjens, D. (2005) INDECO: Indicators – An Overview Internal Paper for Discussion. 1st April 2005

[2] Indicators for sustainable development of marine capture fisheries. FAO Technical Guidelines for Responsible Fisheries no 8. FAO, Rome. 68p. 1999

[3] Garcia, S.M., Staples, D.J. and Chesson, J. The FAO guidelines for the development and use of indicators for sustainable development of marine capture fisheries and an Australian example of their application. Ocean and Coastal Management, No 43, pp537-556, 2000

[4] Slocombe D.S. (1999) Defining goals and criteria for ecosystem-based management. Environ. Mgmt. 22: 483 – 493

[5] Fletcher, W.J., Chesson, J., Fisher M., Sainsbury, K.J., Hundloe, T., Smith, A.D.M. and Whitworth B. National ESD Reporting Framework for Australian Fisheries, Canberra, Australia, 2002

[6] OECD Environmental Indicators – Development, Measurement and Use. Reference Paper, 2003

[7] Hart M. Sustainable Measures. http://www.sustainablemeasures.com

[8] Smeets E. and Weterings R. Environmental indicators: typology and overview. European Environment Agency Technical Report no 25. EEA, Copenhagen, 1999

[9] Hak T., et al. Indicator-based evaluation of interlinkages between different sustainable development objectives. Review report on the state-of-the-art in sustainable development indicators. EU SDI project, , p.10, 2007

[10] Sainsbury K. and Sumaila U.R. Incorporating ecosystem objectives into management of sustainable marine fisheries, including 'best practice' reference points and use of marine protected areas. Reykjavik Conference on Responsible Fisheries in the Marine Ecosystem, FAO, Rome, 2001
[11] Tanguay G.A., Rajanson J., Lefebvre J.-F., Lanoie P. Measuring the sustainability of cities: An analysis of the use of local indicators. Ecological Indicators 10, 407–418, 2010
[12] Sustainable Development Indicators to monitor the implementation of the EU Sustainable Development Strategy. Brussels, 9.2.2005 SEC(2005) 161
[13] Clark R.J. Coastal zone management for the new century. Ocean & Coastal Management, Vol. 37, No. 2, pp. 191-216, 1997.
[14] Marti X. et al., Indicators Guidelines To adopt an indicators-based approach to evaluate coastal sustainable development. DEDUCE, EU Interreg III project, 98p., 2007
[15] Ernšteins R., Kauliņš J., Kudreņickis I. Sustainable Coastal Development Indicator System Studies In Latvia. Sustainable Development and Planning IV , vol.2, Southampton, WIT Press, pp.653-664, 2009
[16] Project COBWEB: http://www.hyria.fi/cobweb/about_cobweb/, last reviewed 2011.01.15

Sustainable economy of the ecological footprint: economic analysis and impacts

S. H. Shakir Hanna & I. W. Osborne-Lee
Texas Gulf Coast Environmental Data (TEXGED) Center, Chemical Engineering Department, Prairie View A&M University, USA

Abstract

The ecological footprint is an important measure in calculating the human demands and impacts on our global environment. In this respect the ecological footprint is a function of all the parameters that interact between the power of ecosystem productivity and human interactions and activities on a particular ecosystem or the demand from that ecosystem. The present paper will cover and analyze the ecosystems productivity and the human demand from the ecosystems. It will produce comprehensive analyses in measuring the possibility of capabilities of the ecosystems to provide goods and services to the human beings on our planet Earth. Further, the paper will discuss the models that can be used in measuring the sustainability of ecosystem and what we should be doing to maintain the Earth healthy ecosystems.

In this respect, the paper will assess and introduce a comprehensive model that can describe the status of our ecosystems productivities and the impacts of human population on the plant Earth. Furthermore, the paper will provide some answers to the human issues, further warning to the current trend in use and abuse of our natural ecosystems and what will be expecting from these ecosystems to provide the human needs in response to the current use of our global ecosystems. Finally, it will conclude and setup several recommendations to save our Planet Earth from the irreversible impacts of human beings on the ecosystems and our global environment.

Keywords: economic analysis, ecological footprint (EF), impacts, global biological capacity (GBC), global biological demand (GBD), global deficit capacity, global ecological capacity (GEC), sustainability index, maintenance index.

www.witpress.com, ISSN 1743-3541 (on-line)
doi:10.2495/ECO110281

1 Introduction

The term Ecological Footprint has been introduced by Wackernagel and Rees [1]; Wackernagel *et al.* [2]. However, the authors are defining Ecological Footprint (EF) as a measure of the status of the global ecosystems' ability to provide service and goods to all human beings who are living on the Earth. Further, the global ecosystems provide for the needs of all other organisms that exist on the planet and, in consequence, enable the goods and services to be provided to human beings. Therefore, the global ecosystems provide for the needs of all living beings on the Earth including the human beings.

In light of the above definition, we need to discuss in a broader sense what is actually happening on the planet Earth now. In previous ages we the human population were not faced with several aspects of shortage in resources such as water, productive soil to produce food, energy, raw materials, clean air, and so on which is now raising a question of great importance to human beings: Can the Earth continue to cover the needs of its growing human population? We must ask ourselves where we are going and how to maintain our global environment to maintain and sustain our existence. Can we maintain our existence? How can we maintain the productivity of our global ecosystems at a level to sustain our existence and standard of living in the light of our growing demands?

Warnings have been seen about the growth of human populations and their needs (Wackernagel [3], Rees [4], Lenzen *et al.* [5] and Grazi *et al.* [6]). These warnings have indicated that the growing population has affected the global ecosystems. In a recent conference of Food Security in Rome, Italy on November 16–18, 2009, the delegates of the countries attending the conference have indicted that the world is facing a shortage of food as an important one item of the global ecosystem service to human beings. Zhao *et al.* [7] have indicated that humans have had a considerable impact on the Earth associated with population increase and economic development. They indicated also that humans are faced with a series of disturbing contradictions and trends among the natural resources, the environment, and the economy, such as rising population growth, resources depletion and environmental deterioration. These trends are pointing to adverse impacts and the need for considering the importance of sustainability in maintaining globe natural resources. Otherwise, the human race will face shortage of these natural resources with serious consequences for the global economy.

Moderate United Nation Environmental Program (UNEP) [8] scenarios suggest that if current population and consumption trends continue, by the middle of the next decade, we will need the equivalent of two Earths to support us, and of course, we only have one. Turning resources into waste faster than waste can be turned back into resources puts us in global ecological overshoot, depleting the very resources on which human life and biodiversity depend.

The challenge of maintaining the Earth's ecology and sustaining the regeneration of its resources is an enormous effort that requires collaborative work between government agencies, localities and human beings themselves on a global scale. Without these efforts, global ecosystems could be facing a collapse leading to global chaos. In this respect, sustainability whether nationally, regionally, or globally requires that natural resources not be exploited beyond the renewable limits.

EF is a broad measure of resource use that highlights where consumption is exceeding environmental limits. The Best Foot Forward Company in the United Kingdom [9] uses the ecological footprint methodology to help businesses and other organizations measure, monitor and reduce their impacts.

The present paper will assess the status conditions of our global Earth, predict the Global Biological Capacity (GBC), Global Biological Demand (GBD) and introduce a comprehensive model that can describe the status of our ecosystems in term of productivity and the impacts of the human population on planet Earth. Furthermore, the paper will provide some answers to the human issues, warning on the current trend in overuse and abuse of our natural ecosystems and what should be expected in terms of the ability of the Earth's ecosystems to provide for human needs in response to the current use of our global ecosystems. The paper concludes with recommendations for avoiding irreversible impacts of human beings on the ecosystems and our global environment.

2 Material and methods

Data used in this paper were collected from different data set of series available on the web sites of the World Research Institute (WRI)-Earth-Trends [10], the World Bank [11], Food and Agricultural Organization (FAO) [12, 13], United Nation Development Program, World Wildlife Fund (WWF) [14–16] and the Global Footprint Network [17, 18]. The data were analyzed using regression, correlation, and statistical methodologies using Sigma Plot Software [19].

The Maintenance Index (MI) or (Sustainability Index) is a percentage of the total GBC (i.e. total availability or supply) in global hectares to the total GBD (i.e. consumption or demand) in global hectares from the Earth. This index explains the ability of the Earth to regenerate natural resources. In other words, it explains the status of the Earth in providing goods and services to the needs of the human population. The GBC is defined as the ability of the Earth to produce renewable natural resources in terms of global hectare/capita. GBD is defined as the resources consumption by human beings in terms of global hectare/capita. It is important to indicate the calculations of Total GBC and Total GBD are the most important factors in the calculations of indices that have been discussed in this paper. Other terms used in calculations are Cropland Footprint (CF) in billion global hectares, Grazing Land Footprint (GLF) in billion global hectares, Forest Ground Footprint (FGF) excluding fuel wood in billion global hectares, Fish Ground Footprint (FF) in billion global hectares, Total Energy Footprint (TEF) in billion global hectares, Built-up Land (BL) in billion global hectares and Global Deficit Capacity (GDC) in billion hectares. The formulae for calculating the following indices are as follows:

$$\text{MI Index} = \text{GBC} / \text{GBD} * 100 \quad (1)$$

$$\text{GBC} = \text{CF} + \text{GLF} + \text{FGF} + \text{FF} + \text{TEF BL} \quad (2)$$

$$\text{GBD} = \text{CF} + \text{GLF} + \text{FGF} + \text{FF} + \text{TEF BL consumed} \quad (3)$$

$$GDC = GBC - GBD \quad (4)$$

These estimates are calculated according to several authors of ECOTEC-U.K group [20] and Wackernagel *et al.* [21] using the following: 1) Net primary Productivity (NPP); 2) Growing crops for food, animal feed; 3) Fibber and oil; 4) Grazing animals; 5) Wood from forest; 6) Fishing from marine and freshwater; 7) Infrastructure for housing and building transportation and electrical power; 8) Burning for fuel in CO_2 emissions. These calculations depended on the governmental UN data series from the World Bank. These calculations were converted into billion hectares from the global sized hectares.

Predictions of Total GBC and Total GBD for the Earth in the next decades were calculated to show the impact of growth of the world population on the basis of annual data from a series of years 1960 to 2008 for almost 48 years of published data. The predications for the years 2009 to 2050 were analyzed using correlation and regression lines (Tables 1 and 2).

3 Importance of understanding ecological footprint

Ecological footprint as described above is the measure to analyze the human needs from the Earth in the form of resources that provide services and goods for consumption by human beings. Currently, the human population carrying capacity is exceeding the resources of the Earth [8, 20]. It is important to calculate the human population carrying capacities from each nation on this Earth against its resources and to create an Ecological Accounting System (EAS) that gives each country an ecological measure indicating how each country stands against its resources. This accounting system can be of a control of a nation and its impact on whole global environment (i.e. the maintenance indicator or parameter for the nation in global ecosystems). This is a scenario to explain how to close the gap between the Earth's resources and the human basic needs from this Earth. In general in our life we are at the level of individuals calculating our earning and expenses (i.e. income against our spending). If we have a positive balance that means we are not in debt to the others. The same will be the scenario against the Earth resources. What are the credits that each country has against its boundaries' resources exploitations? We have to think ecologically about the natural resources that we have to provide us with services, products, and the natural resources that we consume. We have to consider how are we able to save or conserve these resources for the next generations. Then we have to think about how we can trade our surplus from the resources to the others. It is imperative in this thinking process to consider the welfare of human beings over the generation of rapid wealth against humanity. At the same time, we have to regulate human behaviour in exploiting the natural resources in order to save the Earth for the next generations. The next section presents the most important factors that determine the human populations' needs and their impacts on global ecosystems of our planet Earth.

4 Impacts of the human population on food security and natural resources of the world

The human population has increased to the point that the Earth cannot support the human population with the food that is needed to survive. This conclusion can be explained in (Tables 1 and 2). In the tables, the need for each human being living on this Earth is estimated by about 2000 calories per day as an estimation of US (US Department of Health and Human Services and US Department of Agriculture, 2005) [22]. The total number of human beings who are living on this Earth is 6.69 billion people (Table 1) in the year 2008. Therefore, the daily need for the Earth's populations is about 13.85 trillion calories per day and their annual need is about 5044.30 trillion calories per year. Accordingly, what are the sources of all these calories? These calories would come from crops, meat, vegetables and others sources on the Earth's surface or below sea. The question is whether the Earth's resources can support the populations who are living on this Earth. The most useful resources from the Earth to produce the needed calories are the arable-cropped lands, forest ecosystems, water resources, grazed lands, converted as carbon footprint, fishing ground footprint, and built-up lands. It is estimated that the global land of the Earth in 2008 that can produce the food required to keep every human being in a status of normal health is requiring 2.81 global hectare/capita (i.e. the demand 18.80/6.69). However, 2008, it is estimated 1.79 hectare/capita (i.e. 12.0/6.69) is the biological capacity of the Earth to provide us with the calories as calculated from Table 1. This means that we are in shortage for producing enough calories to cover all human beings on this Earth. The current shortage in global hectare per/capita is equal to 37% in 2008 (Fig. 1). In this respect, the maintenance

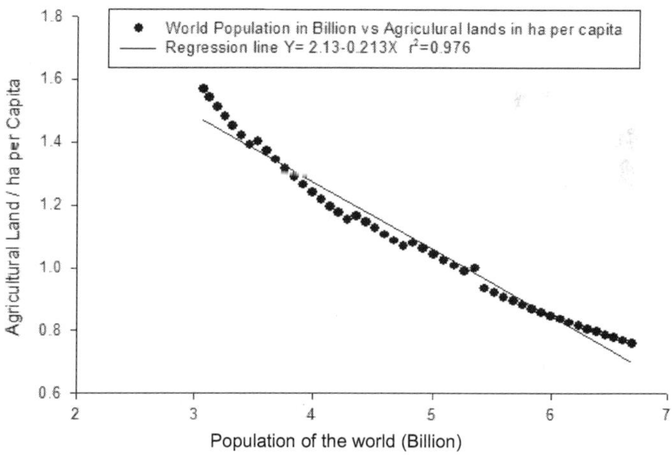

Figure 1: Relationship between the world's human population in billion and agricultural lands per capita.

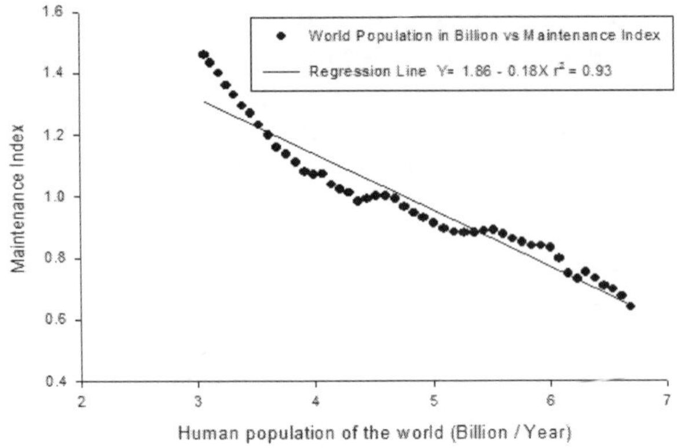

Figure 2: Relationship between the human population of the world and maintenance index (GBC in billion global ha/GBD in billion global ha).

index of the Earth as we suggested is the total GBC of the Earth/total GBD or the EF means that a decline in resources to support the existing human beings and consumption of the Earth as a percentage of the measure of how our Earth needed to be protected and to look to further conservation of the natural resources to meet the needs of human beings (Fig. 2 and Tables 1 and 2). From the regression diagram, it is expected that the when the population reaches about 9.23 billion, the maintenance index of sustainability will reach 0.19 (0.27 global hectare per capita). It is apparent from the regression model of the human population and global biological capacity resources (Fig. 3) that we can come to

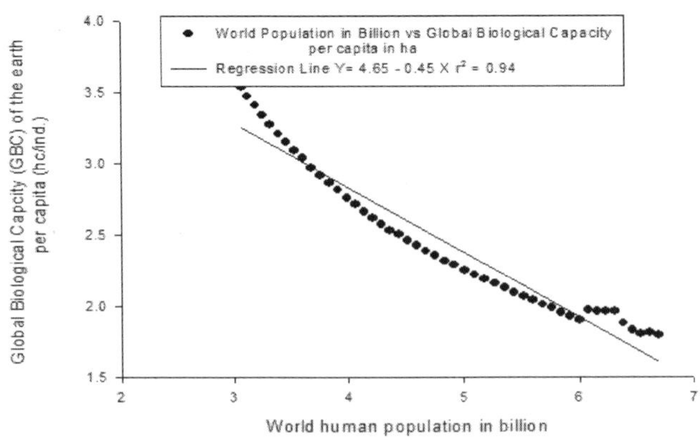

Figure 3: Relationship between the world's human population in billion and global biological capacity of the Earth per capita (ca/ind).

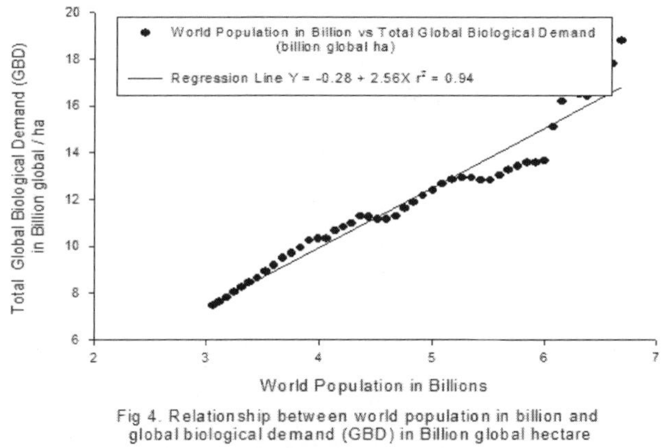

Figure 4: Relationship between the world's population in billion and global biological demand in billion global hectare.

conclude that the future of our global biological capacity on the current trend of use is about to 0.4965 hectare /capita when the population density of the globe reaches 9.23 billion in year 2050 as the projected estimates indicated in Table 2. This future generation. On the other hand, the demands from the Earth will increase to reach 23.36 global billion hectares in year 2050. This means the demands will increase to reach 2.53 global hectare/capita. Respectively, the human population will run deficit against the Earth's resources by about 2.18 hectare/capita. Also, the GDC against the surface area will reach 175% (Figs. 4, 5 and 6). We believe it is alarming that humans should control the abuse of the Earth's current natural resources.

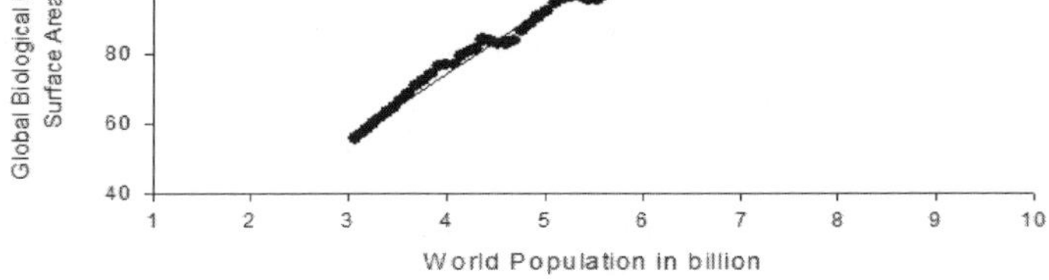

Figure 5: Relationship between the world's population in billion and global biological demand in hectare/surface areas global hectare as %.

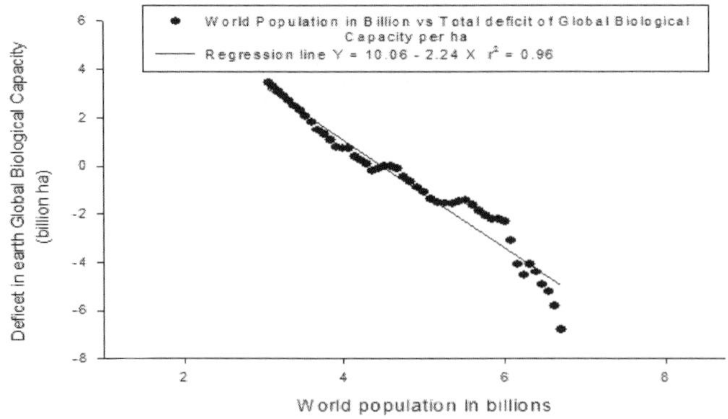

Figure 6: Relationship between the deficit in the Earth's global biological capacity in billion hectare and the world's population in billion people.

5 Economy and concept of emerging of ecological footprint

Emerging concept of ecological footprint issue has been developed over the past twenty five years since 1988 (Odum [23, 24]). This emerging concept as indicated from its name is the works of nature and the activities nature of human beings in producing products and services. On the other hand, humans are demanding from the nature the raw materials, crops, energy, services and consumption to survive. This concept can be developed in a model in economic analysis using the money factors and values of the natural resources and the service of these resources. Accordingly, to the model in the diagram (Fig. 7) below explains the flow of natural resources in the ecosystems and its relationship to the global economy. Also, this figure shows the interrelationships between human activities, their ecological footprint, maintenance of natural resources (i.e. renewable and non-renewable resources) and their impact on global economy.

The model in Fig. 7 indicates the relationships between aspects of natural resources (renewable and non-renewable), ecological footprint, human population, space and time and economy. It is important to take into consideration the factor of space and time as a value in the evaluating the economy of natural resources that we are exploiting including the rate of speed of the use of these natural resources.

According to economic analysis taking into consideration the economy model that predict the value and prospective of natural resources including the in the calculation the human impacts, it is, important to realize the relationship between the speed of increase in population worldwide and the speed of the renewability

Table 1: Global Population, Global Biological Capacity (GBC), Global Biological Demand (GBD) and Maintenance Index of Planet Earth up to the year 2008. Data are in 10-year intervals.*

Year	Global Population in billion	Surface Area in billion ha	Total Global Biological Demand (GBD) (billion global ha)	Global Biological Demand in Global Hectare/Surface Areas global hectare as %	Total Global Biological Capacity (GBC) (billion global ha)	Deficit in global billion hectare =GBC-GBD	Maintenance Index =GBC/GBD
1961	3.08	13.39	7.47	55.80	10.90	3.43	1.46
1970	3.70	13.39	9.50	70.96	11.00	1.50	1.16
1980	4.44	13.39	11.25	84.03	11.13	-0.12	0.99
1990	5.27	13.39	12.93	96.58	11.38	-1.55	0.88
2000	6.10	13.39	15.1	112.73	12.00	-3.10	0.79
2008	6.69	13.41	18.8	140.20	12.00	-6.80	0.64

* Data Sources are World Bank- FAO – WWF – Ecological Footprint Network – WRI-Earth Trends - US estimates

Table 2: Predicted Values Calculated for Global Population, Global Biological Capacity (GBC), Global Biological Demand (GBD) and Maintenance Index of Planet Earth from the year 2009 to 2050. The presented data are in 10-year intervals.

Year	Global Population in billion	Surface Area in billion ha	Total Global Biological Demand (GBD) (billion global ha)	Global Biological Demand in Global Hectare/Surface Areas global hectare as %	Total Global Biological Capacity (GBC) (billion global ha	Deficit in global billion hectare =GBC-GBD	Maintenance Index =GBC/GBD
2009	6.74	13.41	16.98	126.61	11.94	-5.04	0.65
2010	6.79	13.41	17.11	127.60	11.95	-5.16	0.64
2020	7.34	13.41	18.50	137.94	12.13	-6.37	0.54
2030	7.92	13.41	19.99	149.11	12.31	-7.68	0.43
2040	8.55	13.41	21.61	161.16	12.52	-9.09	0.32
2050	9.23	13.41	23.36	174.18	12.73	-10.63	0.20

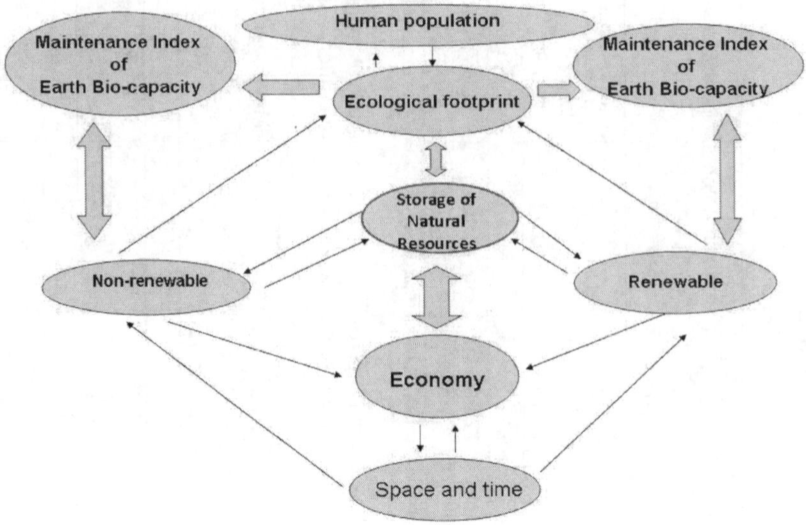

Figure 7: Diagram show the inter-relationships between natural resources, Ecological Footprint and Global Economy.

of the resources. In other words, what is the speed of growth of renewability resources against the growth of human beings? The renewability of the natural resources will become steady or experience a little growth in the next decades and in the opposite the human population will increase in alarming annual rate at 1.1%.

The EF "accounts for the flows of energy and matter to and from any defined economy and converts these into the corresponding land/water area required from nature to support these flows" Wackernagel and Rees [1].

6 The challenges facing our global ecosystem

In the light of the above analysis, the human beings are faced with a lot of challenges not only from economic point of view (i.e. unemployment, funding of projects, funding of disastrous that are happenings in the world, financial recovery …etc) but also from ecological issues. These ecological issues are:

- Lack of arable lands that are used for production of food, forestry woods, pastures and animal breading for producing milk and other meet products for human nutrition
- Conserve the wild-land habitats and to be set aside to preserve large and small preserves for the use of the next generation and to maintain the biological species diversity. In this respect, the plant species, animal species and natural ecosystems cannot utilize the Earth lands

extensively as human beings doing for exploitation the Earth natural resources in an alarming rate. Preserving the current biological species diversity of plants and animals will require leaving sizable landscape areas with light utilization from human beings or non-utilization in order to support the life cycle of the Earth.

- It is important to use the efficient technology to save our planet Earth and continue reuse the materials that we exploit from the natural resources several times to ensure the sustainability of the Earth.
- The success of our economy should be based on the ecological sustainability of our Earth. In other words, the global economy should adhere to sustainability of Global Ecological Capacity (GEC) of global Earth as a measure of equalization between the total Global Biological Demand (GBD) from the Earth in global hectare/capita and total Global Biological Capacity (GBC) of the Earth in global hectare/capita. This means that human beings will sustain the Earth ecological budget. The data in the Table 2 showed that year 1981 is year that had GBC equal to GBD.
- The most important challenge of all the above is human beings controlling themselves in sustain population numbers within the acceptable limits that the Earth can support (i.e. maintaining the ecological budget of the Earth without overloading or stressing these natural resources). This will take more efforts from the governments, non-government organizations (NGO) and individuals to sustain the Earth bio-capacity. Otherwise, the Earth returns its action (i.e. exposure to lack of food, famines, diseases, global climatic disasters on the human race...etc).
- The education is the most crucial element in the above challenges to educate the public of the issue of maintaining the Earth ecological capacity.

7 Discussion

Ecological footprint is an accountant and economical measure that can describe the balance, surplus and deficit between the global productivity (i.e. renewability of natural resources global biological capacity in global million hectare) and the human demands from the Earth resources (i.e. global biological demands or consumption in global billion hectares). Considerable attention recently to the methodologies in calculating ecological footprint and importance of this issue is discussed Zhao *et al.* [7], ECOTEC-U.K. [20] and Wackernagel *et al.* [21]. There are several scientists and researchers have considerable oppositions or weaknesses to the ecological footprint accounting ECOTEC-U.K [20]. These oppositions and concerns are 1) Measure of sustainability; 2) Only CO_2 emission calculation included in the ecological footprint rather than the other greenhouse gases; 3) Human welfare is a multi-dimensional issue and a single index over-simplifies; 4) Ecological footprint index is including quantitative rather qualitative issues; and 5) the sequestration of CO_2 in the ocean. However, in spite of these weaknesses, the authors are considering the ecological footprint as a sensitive measure that can be important to allow the decision makers to take care about the natural resources and to start thinking in matter of conserving natural resources in spite of the human pressure. From the above analysis, I found that Earth will only support in approximately 25%–30% of the total population of the globe in year 2050 and it will run a deficit of the total global biological capacity about 80% of the Earth in a conservative assumption. This conclusion will indicate an alarming to the leaders of the world, Non Governmental Organization and Government Agencies, United Nation with all its agencies to take appropriate actions to conserve world resources for the future generations. Otherwise we are facing severe competition on these natural resources (i.e. renewable and non-renewable), and none can predict the consequences of this situation. It is important that leaders of the world to concentrate on education of the population about the disastrous consequence of human population growth against the regenerative speed process of the biological capacity of the ecosystems to sustain the demand with production of these resources.

Generally, it is important to discuss the issue of energy conversion of global hectare to quantitative measure (i.e. Energy Conversion Factor per global hectare in energy units). This will be my next analysis of conversion of hectare land to energy use and then modelling for the global bio-capacity against global demands for energy from the Earth.

8 Conclusion

This paper indicates an alarming warning about the trends in Global Biological Capacity (GBC) [i.e. regenerative capacity of Earth] and Global Biological Demand (GBD) [i.e. global consumption from the Earth]. In addition, this paper shows the trends in the deficit regenerative capacity of the Earth. The current deficit regenerative capacity of the Earth is about 40% and continues to be 50%. From this study, it is found that 1981 is the year that world has a break even point between the GBC to be equivalent to GBD. In this respect, we have to monitor the world status and condition of the balance between the GDB and GBC. This will not be accomplished without the collaborative efforts of the all individuals, governmental and non governmental agencies, and global world organizations as United Nations.

References

[1] Wackernagel, M. & Rees, W., *Our Ecological Footprint: Reducing Human Impact on the Earth.* Gabriola Island, BC, and Philadelphia: New Society Publishers, 1996.

[2] Wackernagel, M., Onisto, L., Bello, P., Linares, A. C., Falfan, I. S. L., Garcia, J. M., Guerrero, A. I. S. & Guerrero, M. G. S., National natural

capital accounting with the ecological footprint concept. *Ecological Economy* vol. 29, pp. 375–390, 1999.

[3] Wackernagel, M., Ecological Footprint and Appropriated Carrying Capacity: A Tool for Planning Toward Sustainability, in *School of Community and Regional Planning.* vol. Ph.D. thesis: The University of British Columbia. Vancouver, Canada. , 1994.

[4] Rees, W. E., Ecological footprints and appropriated carrying capacity: what urban economics leaves out *Environment and Urbanisation* vol. 4, pp. 121-130, 1992

[5] Lenzen, M., Hansson, C. B. & Bond, S., On the bioproductivity and land disturbance metrics of the Ecological Footprint. *Ecological Economics* vol. 61, pp. 6-10, 2007.

[6] Grazi, F., Bergh, J. C. J. M. V. D. & Rietveld, P., Welfare economics versus ecological footprint: modeling agglomeration, externalities and trade. *Environmental and Resource Economics* vol. 38, pp. 135-153, 2007.

[7] Zhao, S., Li, Z. & Li, W., A modified method of ecological footprint calculation and its application *Ecological Modelling,* vol. 185 pp. 65–75, 2005.

[8] UNEP, A Planet in Ecological Debt. A*rendal Maps and Graphics Library*:http://maps.grida.no/go/graphic/a-planet-in-ecological-debt 2009.

[9] http://www.bestfootforward.com/

[10] WRI, EarthTrends Environmental Information, World Resource Institute: WRI - http://Earthtrends.wri.org/, 1960-2005.

[11] WorldBank, Data Series and Research, The World Bank organization. www.worldbank.org1960-2008.

[12] FAO, FAOSTAT (Food and Agriculture Organization, United Nations, Rome). 2008.

[13] FAO, FAOSTAT FAO Statistical Database FAO www.fao.org, 1960-2008.

[14] Jenkins, M., Jakubowska, J., Gaillard, V., Groombridge, B., Wackernagel, M., Monfreda, C., Deumling, D., Gurarie, E., Friedman, S., Linares, A. C., Sánchez, M. A. V., Falfán, I. S. L., Loh, J., Randers5, J. & Monfreda, C., Living Planet Report 2002, World Wildlife Fund (WWF), www.panda.org, WWF International. Avenue du Mont-Blanc, CH-1196 Gland, Switzerland 2002.

[15] Humphrey, S., Chapagain, A., Bourne, G., Mott, R., Oglethorpe, J., Gonzales, A., Atkin, M., Loh, J., Collen, B., McRae, L., Carranza, T. T., Pamplin, F. A., Amin, R., Baillie, J. E. M., Goldfinger, S., Wackernagel, M., Stechbart, M., Rizk, S., Reed, A., Kitzes, J., Peller, A., Niazi, S., Ewing, B., Galli, A., Wada, Y., Moran, D., Williams, R., Backer, W. D., Hoekstra, A. Y. & Mekonnen, M., Living Planet Report 2008, WWF–World Wide Fund, www.panda.org, WWF International Avenue du Mont-Blanc 1196 Gland, Switzerland2008.

[16] WWF—The Global Conservation Organization, H. S., http://www.panda.org/, E.F.A.A. & news_facts/publications/ living_planet _report/ footprint.

[17] Kitzes, J., Galli, A., Rizk, S., Reed, A. & Wackernagel, M., Guidebook to the National Footprint Accounts: 2008 Edition, Global Footprint Network, http://www.footprintnetwork.org/, Oakland, CA 94607-3510, USA2008.

[18] Ewing, B., Reed, A., Rizk, S. M., Galli, A., Wackernagel, M. & Kitzes, J., calculation methodology for the national Footprint accounts, 2008 Edition, Global Footprint Network. http://www.footprintnetwork.org, Oakland, CA 94607-3510, USA 2008.

[19] www.spssscience.com/sigmaplot version 8 2002

[20] ECOTEC-U.K., Ecological Footprint, European Parliament, Directorate General for Research, Directorate A, The STOA Programme March, 2001.

[21] Wackernagel, M., Schulz, N. B., Deumling, D., Linares, A. C., Jenkins, M., Kapos, V., Monfreda, C., Loh, J., Myers, N., Norgaard, R. & Randers, J., Tracking the ecological overshoot of the human economy, *Proceedings of The National Academy of Sciences of the United States of America (PNAS),* vol. 99, pp. 9266-9271, 2002.

[22] U.S. Department of Health and Human Services and U.S. Department of Agriculture, Dietary Guidelines for Americans, , 6th ed, 2005 p. 71 + IV PP + Appendix

[23] Odum, H. T., *Environment Accounting: Emergy and Environment Decision Making*. John Wiley, New York, 1996.

[24] Odum, H. T., Self-Organization, Transformity, and Information. *Science,* vol. 242, pp. 1132–1139, 1988.

Determination of the amount of carbon stored in a disturbed mangrove forest in Campeche, Mexico

R. M. Cerón, J. G. Cerón, J. J. Guerra, J. C. Zavala, L. E. Amador, E. Endañu & G. Moreno
Universidad Autónoma del Carmen, Campeche, México

Abstract

Carbon sequestration potential of mangrove forest in Campeche, Mexico, was investigated. Field work was conducted at two sites at the border of the estuary named "Terminos Lagoon" (Estero-Pargo and Bahamitas) between February and August in 2009. The mangrove forest in both sites has been subjected to anthropogenic pollution during the last years and it can be considered as perturbed. Six sampling campaigns were performed considering three climatic periods ("Norths" season, dry season and rainy season). Seasonal influence on carbon storage was evident, but also the influence of tides was significant, with the highest carbon storage during "Norths" and dry seasons (ranged from 10.63 to 37.64 Kg C m^{-2} for Estero-Pargo, and from 12.8 to 39.9 Kg C m^{-2} for Bahamitas) Carbon storage was higher in flooded soils with dominant vegetation of red mangrove in comparison with sampling zones where dominant vegetation was button mangrove. Organic matter and organic carbon content were higher at 30 cm depth during "Norths" season, when heavy rains and long periods of tidal flooding occurred with low rates of litter export that maintained anoxic conditions. Comparing these results with carbon storage data obtained in other sites, it can be suggested that neutral and sandy soils as this study area and associations of red mangrove have a good potential of carbon sequestration. Since studied mangrove individuals were from young to mature in reproductive age, it can be conclude that this potential will increase in the next years.

Keywords: carbon sequestration potential, carbon sinks, mangrove forest, wetlands, Campeche.

WIT Transactions on Ecology and the Environment, Vol 144, © 2011 WIT Press
www.witpress.com, ISSN 1743-3541 (on-line)
doi:10.2495/ECO110291

1 Introduction

The atmospheric concentration of carbon dioxide is currently increasing at the rate of 1.5 ppmv yr^{-1}. This increase of greenhouse gases has lead to an increase in the primitive temperature of the Earth and as a consequence to global climatic change. In spite of the strong inter-dependence between climate and soil quality [1], the role of soil organic carbon on global warming and its strategic importance as an alternative solution needs more recognition.

Wetlands are known to be important carbon sinks. Mangrove forest is characterized by a high productivity and a low ratio of sediment respiration to net primary production. For this reason, mangrove sediments have a high potential for a long term organic carbon sequestration. Furthermore, carbon storage in estuarine wetlands is an efficient process with a minimal release of greenhouse gases [2]. Although they cover only about 5% of the earth's surface, they contain around 40% of global soil organic carbon [3]. However, it has been estimated that about 68% of the total global area of wetlands has been lost as a result of human activities, mainly due the change of the land use, converting wetlands to farmland, forestry and urban areas.

In spite of the importance of mangrove forest as a carbon sink, the most carbon storage studies are focused to terrestrial ecosystems and little attention has been given to this ecosystems. Despite in Mexico, there are 113 Ramsar sites with a total surface of wetlands of 8,161,357 ha, and mangrove ecosystem constitutes the main wetland system, location of specific carbon sinks and their potential to sequester atmospheric carbon remains poorly defined, and there are uncertainties concerning carbon stock of wetland of local and regional scale. To complete carbon storage inventories and to reduce these uncertainties, it is necessary to obtain more information that support environmental policies to promote conservation and restoration projects as mitigation strategies. This paper reports the content of organic carbon in two sites in a disturbed mangrove forest within an important wetland in our country that has been registered as a Ramsar site [4].

2 Methods

2.1 Site description

Terminos Lagoon is located at the south western base of the Yucatan peninsula, and it is the largest estuary in Mexico. Carmen Island forms a sand barrier between Terminos Lagoon and open sea (Figure 1). Terminos Lagoon is almost completely surrounded by mangrove forest (*Rhizophora mangle, Avicennia germinans, Laguncularia racemo*sa, and *Conocarpus erectus*), covering an area nearly as large as the open-water area of the lagoon itself. Therefore, mangrove forest constitutes a significant component of the lagoon ecosystem. Maximum wind speeds range from 50 to 60 km h^{-1}, occurring during the season of winter storms named "Norths". Average annual precipitation (1680 mm) is seasonal with a dry season from February to May, and a wet season from June to October.

The period where Gulf of Mexico is subjected to the influence of cold fronts, from November to February, is named "Norths" season.

Two areas were selected for this study: a forest along Estero-Pargo, a 5.3 km long tidal channel on the lagoon side of Carmen Island and a forest at Bahamitas located at the border of the Terminos Lagoon. Both sites are within natural protected area named "Terminos Lagoon", ramsar site since 2004 (Figure 1).

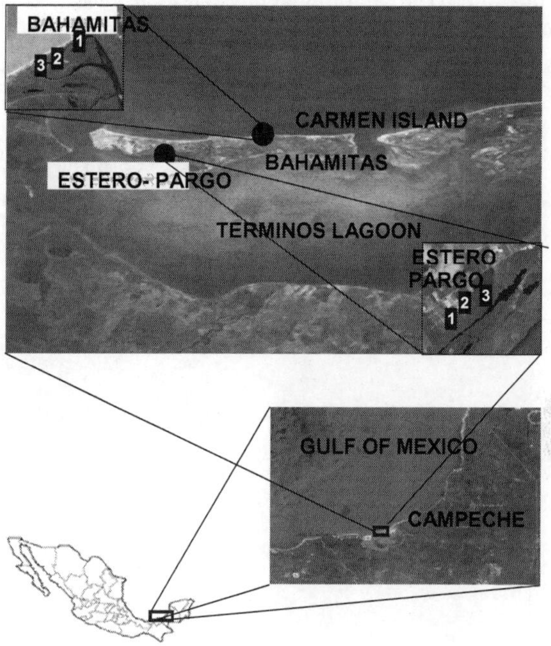

Figure 1: Location of sampling sites and sampling zones in the study area.

2.2 Sampling method and forestry inventory

Field work was conducted at two sites with three sampling zones for each one, between February and August 2009. In both sampling sites, mangrove forest is subject to anthropogenic exploitation and can be considered as perturbed. To assess seasonal changes on carbon storage, sampling campaigns were performed considering three climatic periods: "Norths" season, dry season and the beginning of the rainy season, with a total of six sampling campaigns. Not all sampling sites had comparable tidal influences during the study period, Bahamitas site was always flooded (received regular tidal inundation and was covered with fresh water during "Norths" and rainy seasons) whereas Estero-Pargo only was flooded as a result of the occurrence of rainfall during rainy and "Norths" seasons. In each site, three sampling zones of 4 m x 12 m were selected considering free access to the zone, risks, mangrove distribution by specie and disturbances (Figure 1, Table 1). Based on visual inspections of the study area, transects were established in a representative area of the mangrove forest, locating three sampling points of 1 m^2 approximately for each of three sampling zones for both sites.

Table 1 shows the location of these sampling zones and their general characteristics. In table 2, the forestry inventory for each sampling zone in both sites is shown.

Table 1: Location of the sampling zones and their general characteristics for both sites in the study area.

Sampling Site	Sampling Zone	Latitude N	Longitude W	Characteristics
ESTERO-PARGO	1	18° 39′ 02.8″	91° 45′ 46.7″	At 20 m from the border of Terminos Lagoon. Only flooded and covered by fresh water during “Norths” and rainy seasons. This site has an association of red, white and button mangrove, being white mangrove the dominant specie.
	2	18° 39′ 02.2″	91° 45′ 45.7″	
	3	18° 39′ 02.0″	91° 45′ 45.0″	
BAHAMITAS	1	18° 41′ 57.1″	91° 41′ 50.7″	Presence of wastewater and garbage. At the border of Terminos Lagoon. Always flooded. Flooded at each tide, and covered by fresh water during “Norths” and rainy seasons. This site has an association of red, white and black mangrove, being red mangrove the dominant specie.

Duplicate soil samples were collected from the ground to 30 and 60 cm depth using a 193.3 cm^3 soil sampler. The corer was carefully inserted into the soil and pushed down at 0.3 and 0.6 m. Because, of sampled soils were humid, the corer has a one-way check valve that creates a vacuum inside the corer liner as it is pushed into the soil, and when the device is pulled out of the soil it creates a suction force that retains the sample in the tube. This sampling method was used by Grossman and Reinsch [5] and by Bernal and Mitsch [6] for sampling wet soils or soils under water.

Table 2: Forestry inventory for each sampling zone for both sites in the study area.

ESTERO-PARGO													
Zone	Distribution of the species (number of individuals)				Mean diameter at breast height (cm)	Mean tree height (m)				Maximum tree height (m)			
	RM	WM	BM	BuM		RM	WM	BM	BuM	RM	WM	BM	BuM
1*	-	-	-	33	5.73	-	-	-	1.13	-	-	-	1.15
2*	-	115	-	-	4.14	-	0.93	-	-	-	1.45	-	-
3*	4	3	-	-	49.34	5.37	6.79	-	-	10.7	9.87	-	-
BAHAMITAS													
Zone	Distribution of the species (number of individuals)				Mean diameter at breast height (cm)	Mean tree height (m)				Maximum tree height (m)			
	RM	WM	BM	BuM		RM	WM	BM	BuM	RM	WM	BM	BuM
1*	5	4	15	-	9.49	2.72	1.23	2.25	-	4.06	1.49	6.39	-
2*	20	4	2	-	6.99	5.61	1.12	2.92	-	6.88	2.34	3.25	-
3*	12	5	18	-	5.31	2.49	1.49	1.77	-	3.50	4.37	3.65	-

*Stage of forest evolution for all sampling zones: Young to mature, in reproductive age. They are producing seeds every year. Stand age: 15 yr.
RM=Red Mangrove, WM=White Mangrove, BM=Black Mangrove., BuM=Button Mangrove.
- Individuals not present in the stand.

A total of 216 soil samples with replicate were taken. After extraction, each core was labelled, sealed using tight-fitting end caps, and sent for laboratory analysis [7].

2.3 Analysis method

For the laboratory analysis, the soil cores were removed from the tube. Free water was drained away and above-ground vegetation was removed. Wet mass was recorded and the cores then dried at ambient temperature to constant mass, this weight was registered to obtain the bulk density. Shells, rocks, roots and any living biomass were removed prior to grinding. Dry samples were ground and sieved to pass 2 mm particle size. Physical/Chemical parameters of each soil sample were determined according to standard procedures. pH and salinity were obtained using a 1:5 soil/water solution by electrical conductivity [8]. Soil mechanical composition (granulometric analysis) of each dried sample was analyzed by a hydrometer method [7]. Gravimetric humidity was determined according to Etchevers [9]. Total nitrogen was analyzed using the semi-micro-kjeldahl method [7]. Analysis of organic matter were made using the acid-dichromate oxidation method [7], organic carbon was analyzed according to NOM-021-RECNAT [7] and bulk density according to Gandoy [10]. Forestry inventory was performed by using a clinometer (Suunto PM-5/1520).

2.4 Estimation of carbon sequestration potential

From organic carbon content [% CO], the dry weight of soil in t ha^{-1}, the depth of sampled soil (cm), and bulk density(g cm^{-3}), the potential of sequestration of carbon (Kg C m^{-2}) was obtained as follows:

$$\text{MgC ha}^{-1} \text{ or kg C m}^{-2} = [\text{soil weight}]\ [\%\ \text{CO}]$$

where:

$$\text{Soil weight } [\text{t ha}^{-1}] = [\text{depth of sampled soil}]\ [\text{bulk density}]$$

2.5 Statistical analysis

Statistical procedure involved three stages of analysis: descriptive, comparative and relational. First, summary descriptive statistics of the carbon storage were obtained for each sampling zone, sampling point, depth and season. Mean values of the chemical and physical variables were estimated as well at these scales, these mean values were compared by using Duncan test. Secondly, hypotheses tests were conducted by two-way ANOVA in order to assess for differences between sampling zones, sampling points, between depths and between climatic seasons. All tests were conducted with SAS v. 8.0 [11]. Simple linear regression was employed to relate the carbon storage registered at each sampling zone, sampling point, depth and climatic season with the chemical and physical variables measured. Pearson regression coefficients were then tested for significance ($p<0.05$) in order to define which of the variables are significantly related with the observed values of carbon storage.

3 Results and discussion

3.1 Physical/chemical parameters

From figure 2, it can be observed that soils were more humid during Norths and dry seasons. The influence of tides was evident in both sites.

There were statistical differences between climatic seasons and sampling depths. Humidity was not statically different in both sites and a significant correlation between humidity and storage carbon was not found. Electric conductivity ranged from 0.31 to 4.07 dS m^{-1} for Estero-Pargo, and from 2.10 to 17.38 dS m^{-1} for Bahamitas. There were statistical differences between sampling depth and climatic periods. A significant correlation between electric conductivity and storage carbon was not found.

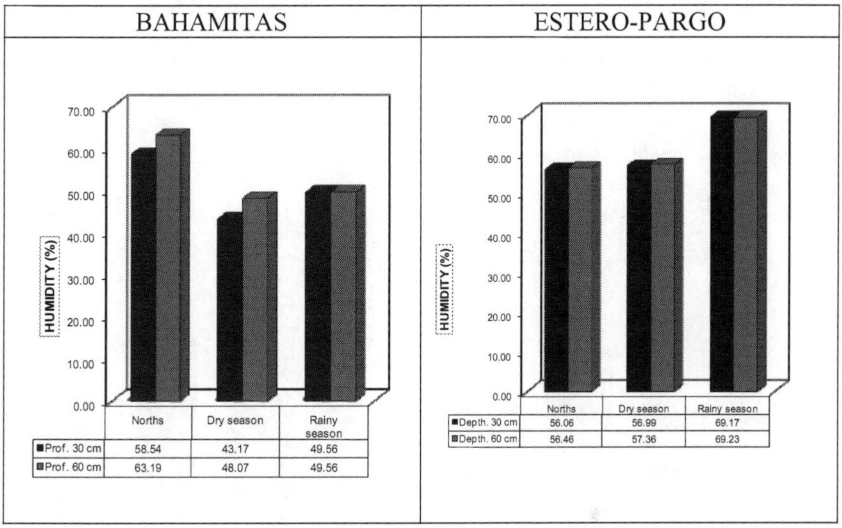

Figure 2: Gravimetric humidity for both sampling sites for each zone and for each climatic period in the study area.

Carmen Island had a sedimentary origin, for this reason, sandy soils are dominant in the study area. For both sites, soil samples showed sandy texture. In addition, pH values ranged from 6.89 to 7.12 for Estero-Pargo, and from 6.7 to 7.5 for Bahamitas, it suggests that neutral soils are dominant in the study area. There were not statistic differences between sampling zones and sampling depths. pH was not statically different in both sites and a significant correlation between pH and storage carbon was not found.

Bulk density ranged from 0.55 to 1.24 g cm^{-3} for Estero-Pargo, and from 0.57 to 1.09 g cm^{-3} for Bahamitas. In Estero-Pargo there were statistical differences between climatic periods and sampling depths. Bulk density was not statically different in both sites and a significant correlation between humidity and storage carbon was not found. On the other hand, Total nitrogen content ranged from 0.03 to 0.44% for Estero-Pargo, and from 0.02 to 0.40 for Bahamitas. There were statistical differences between climatic periods and sampling depths. Total nitrogen content was statically different in both sites considering climatic periods and a significant correlation between nitrogen content and storage carbon was not found.

3.2 Organic matter content

From figure 3 it can be observed that organic matter content was higher at 30 cm depth for all sampling zones in both sites. In study area, the long periods of tidal flooding maintained anoxic conditions (below 10 cm of depth) and high organic matter contents, it can explain the highest organic matter content at 30 cm depth. Furthermore, the accumulation process of organic matter is enhanced in sites with abundant rainfall (the heavy rains occurred during Norths season) or with a deficient drainage, it is the case of both sampling sites. During the rainy season, the accumulation of organic matter increases, however the decomposition is slow, it can explain the slightly lower values obtained for organic matter during this season, this accumulation remain until the Norths and dry seasons, resulting maximum values during these seasons.

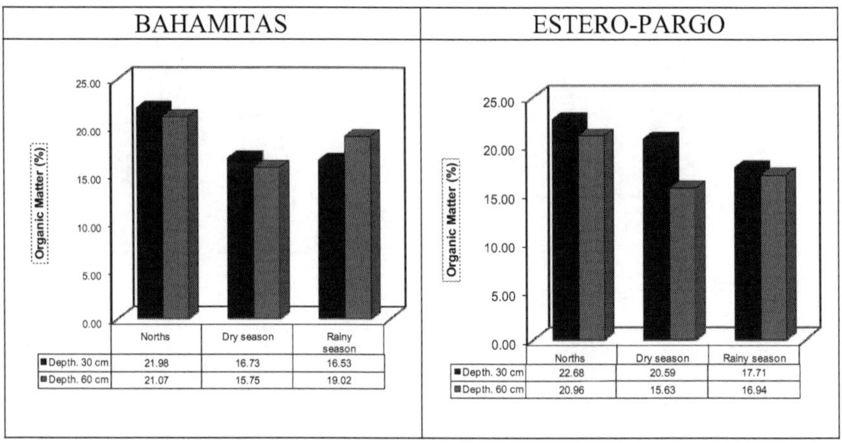

Figure 3: Organic matter content for both sampling sites for each sampling zone and for each climatic period in the study area.

Organic carbon content ranged from 2.81 to 19.7% for Estero-Pargo, and from 4.76 to 15.73% for Bahamitas. There were statistical differences between climatic period and sampling depths. Organic carbon content was statically different in both sites related to climatic season and sampling depth. A significant correlation between organic carbon content and storage carbon was found.

The highest values were found during the Norths and dry season (Figure 4). Organic carbon content decreased slightly with increasing soil depth, this pattern of organic carbon distribution is a general phenomenon in forest [12]. Seasonal variation was involved, but also the influence of tides was evident. Anoxic conditions prevailing during the Norths season, when heavy rains occurred and there was a strong influence of tides (both sites were flooded all time,) may be partly responsible for the high organic carbon concentrations. During the rainy season, water sources were mixed resulting in lower organic carbon concentrations whereas during the dry season, increased evapotranspiration concentrate salts and dissolved organic carbon which are transported vertically with percolating water.

Carbon storage ranged from 10.63 to 37.64 Kg C m^{-2} for Estero-Pargo and from 12.8 to 39.9 Kg C m^{-2} for Bahamitas. Carbon storage was greater during the Norths and dry seasons for all sampling zones studied. Climatic variation was involved, but also, the influence of tides was evident. The degree of tidal exchange influences on the amount of organic carbon accumulated.

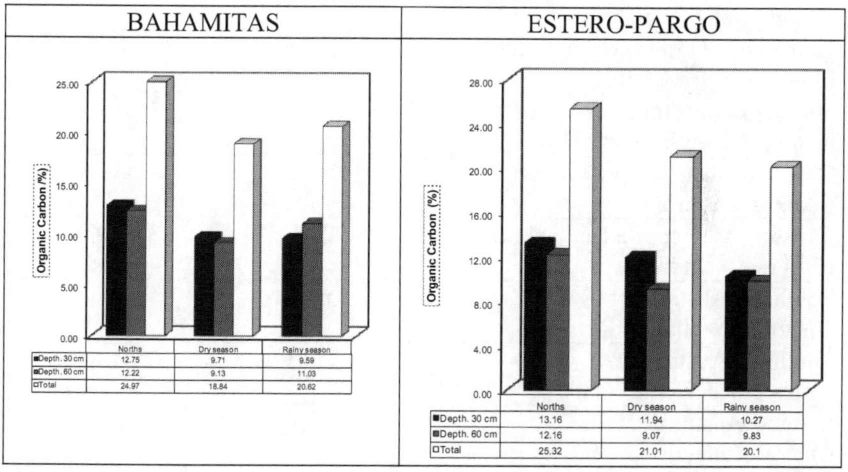

Figure 4: Organic carbon content for both sampling sites for each zone and for each climatic period in the study area.

Carbon storage is enhanced under anoxic; it can explain the highest values obtained for Norths season, when heavy rains occurred and the sites were flooded most of the time as a result of tides, because flooded wetlands provide optimal conditions for accretion of organic matter.

During the dry season, salts and dissolved organic carbon concentrates with increasing evaporation, whereas, during rainy season, pore waters were diluted with rain and run-off waters, resulting in lower dissolved organic carbon concentration, salinities and densities. In mangrove forest, the stand age is a determinant factor that influence on the amount of organic carbon in soil whatever season, Comparing our results with carbon storage data obtained in other sites (Table 3), we can suggests that sandy and neutral soils as our study area and associations of red and white mangrove have a good potential of carbon sequestration, considering that our mangrove individuals were from young to mature in reproductive age, we can expect that this potential will increase in the next years.

Table 3: Comparison of carbon storage reported by other authors.

Site/Land Use	Carbon storage (kg C m^{-2})	Reference
Gahanna woods, Ohio, USA. Isolated and forested wetland.	3.03	[6]
Old Woman Creek, Ohio, USA. Riverine flow-through wetland.	2.77	[6]
La Selva, Costa Rica. Isolated and forested wetland.	0.43	[6]
Earth University, Costa Rica. Slow-flowing slough wetland.	1.67	[6]
Palo Verde, Costa Rica. Riverine Flow-through wetland.	1.36	[6]
Southern California, USA. Coastal lagoon and wetland complex.	0.033	[13]
Okinawa Island, Japan. Pioneer Mangrove Stand.	5.73	[12]
Tabasco, Mexico. Red and white mangrove stand.	47.2-82.2	[14]
Southeast Australia Disturbed Wetland soils	6.61	[15]
Southeast Australia Undisturbed Wetland soils	11.19	[15]
Sydney, Australia Coastal wetlands	13.9	[16]
Tropical and subtropical China Marsh vegetation	40.0	[17]
Brazil, oxysol	12.0-24.0	[18]
Brazil, Agricultural soils	2.0-10.0	[19]
Campeche, Mexico. Neutral and sandy soils, Perturbed Mangrove forest. Bahamitas.	23.16	This study
Campeche, Mexico. Neutral and sandy soils, Perturbed Mangrove forest. Estero-Pargo.	23.08	This study

4 Conclusions

During this research, study area was subjected to severe cold fronts associated with intense rains, and the influence of tides in both sites was significant. Organic matter and organic carbon content decreased slightly with increasing soil depth. Carbon storage was highest during Norths and Dry seasons. In both sites, Estero-Pargo and Bahamitas were under the influence of tides and heavy rains, and were flooded the most of time. The anoxic conditions prevailed in both sites favoured the accumulation of organic matter and organic carbon in the sampled soils. In addition, accumulation of organic matter and carbon storage were also higher during the dry season, it is expected since during this season, enhanced evapotranspiration process concentrate and increase the salts and dissolved organic carbon. In the other hand, storage carbon was slightly lower during the rainy season, when pore waters were diluted with rain and run-off waters, resulting in lower dissolved organic carbon concentration, salinities and densities. Sampling zones where button mangrove is a dominant specie showed a lower carbon storage, whereas, sampling zones flooded the most of the time, with associations of red mangrove-black mangrove and red mangrove-white mangrove showed the highest carbon storage.

References

[1] Jenny, H. 1980. The soil Resource: Origin and Behavior. Springer, New York. 377 pp.

[2] Magenheimer, J.F., Moore, T.R., Chmura, G.L., Daoust, R.J., 1996. Methane and carbon dioxide flux from a macrotidal salt marsh, Bay of Fundy, New Brunswick. Estuaries 19, 139–145.

[3] Mitsch, W.J., Gosselink, J.G. 2000. Wetlands. John Wiley & Sons Inc. New York. 920 pp.

[4] Ramsar, 1999. Annotated Ramsar list of wetlands of international importance. Resolution VII.11. Conference of the contracting parties. May 1999. Bucharest, Romania.

[5] Grossman, R.B., Reinsch, T.G., 2002. The solid phase: bulk density and linear extensibility. In: Methods of soil analysis, Part 4-Physical methods. Soil Science Society of America. Madison, Wisconsin. 201-228.

[6] Bernal, B., Mitsch, W.J., 2008. A comparison of soil carbon pools and profiles in wetlands in Costa Rica and Ohio. Ecological Engineering 34, 311-323.

[7] NOM-021-RECNAT-2000. Diario Oficial, 31 de diciembre de 2002.

[8] Van Reeuwijk, L.P., 2002. Procedures for soil analysis. International Soil Reference and Information Centre and Food and Agriculture Organization of the United Nations. 6th. Technical Paper 9.

[9] Etchevers, R.J.D., Monreal, M.C., Hidalgo, M.C., Acosta, M.M., Padilla, C.J., López, R.R.M., 2005. Manual para la determinación de carbono orgánico en la parte aérea y subterránea de sistemas de producción de ladera. Colegio de Postgraduados. México.

[10] Gandoy, B.W. 1992. Manual de laboratorio para el manejo físico de suelos. Universidad Autónoma Chapingo. Departamento de Suelos. Chapingo, México.
[11] Rebolledo, H. 2002. Manual SAS por computadora. Análisis estadístico de datos experimentales. Editorial Trillas. México, D.F.
[12] Khan, M.N.I., Suwa, R., Hagihara, A., 2007. Carbon and nitrogen pools in a mangrove stand of *Kandelia obovata* (S., L.) Yong: Vertical distribution in the soil-vegetation systems. Wetlands. Ecol. Manage 15, 141-153.
[13] Brevik, E., Homburg, J., 2004. A 500 year record of carbon sequestration from a coastal lagoon and wetland complex, Southern California, USA., Catena 57, 221-232.
[14] Moreno, E., Guerrero, A., Gutiérrez, M., Ortíz, C., Palma, J., 2002. Los Manglares de Tabasco, una reserva natural de carbono. Madera y Bosques. Número especial. 115-128.
[15] Howe, A.J., Rodríguez, J.F., and Saco, P. M., 2009. Surface evolution and carbon sequestration in disturbed and undisturbed wetland soils of the Hunter estuary, southeast Australia. Estuarine, coastal and shelf science. 84, 75-86.
[16] Webb, A., 2002. Pre-clearing Soil Carbon Levels in Australia. National Carbon Accounting System Technical Report No. 12. Australian Greenhouse Office, Canberra, 204 p.
[17] Zhong, L., and Qiguo, Z., 2001. Organic carbon content and distribution in soils under different land uses in tropical and subtropical China. Plant and Soil 231,175-185.
[18] Lal, R., 2001. World cropland soils as a source or sink for atmospheric carbon. Advances in Agronomy 71, 145-191.
[19] Rabenhorst, M.C., Swanson, D., 2000. Classification of soils: Histosols. E-183:E-208 p. In: Sumner M.E., ed. Handbook of soil science. CRC Press. Nueva York, EUA.

Historical contamination and brownfield management: what does economic theory say about it?

S. Čiháková Aguilar, M. Černíková & M. Dubová
Institute of Novel Technologies and Applied Informatics, Technical University of Liberec, Czech Republic

Abstract

Residuals from past industrial, mining or military activities might represent a variety of hazards for natural ecosystems. Depending on the location of such areas, environmental assessment will be needed in order to determine if the concentration of pollutants is above the acceptable levels and can become a clear threat for human health. Furthermore, it becomes important to notice that the quality of natural resources can be directly affected. The detriment of the quality of human health and natural resources can also be translated into worse living conditions for certain inhabitants.

The redevelopment of contaminated areas where it is also difficult or impossible to identify the original polluters may incorporate an analytical background not just on legislative and technical issues. Economic theory has principles and paradigms that can explain different approaches towards the effective management of brownfields.

The scope of this work is to describe how various schools of economics focus their study on environmental problems, specifically on brownfield redevelopment. After an introductory part with definitions and a general framework, the paper presents the ideas of environmental economics, free market environmentalism, institutional economics, and ecological economics. Each part is complemented by an overview of empirical research in the field of historical contamination.

Keywords: historical contamination, brownfields, economic theory, free market environmentalism, ecological economics, environmental economics, institutional economics.

WIT Transactions on Ecology and the Environment, Vol 144, © 2011 WIT Press
www.witpress.com, ISSN 1743-3541 (on-line)
doi:10.2495/ECO110301

1 Introduction

The presence of brownfields can affect both urban and natural areas, and therefore will influence not only specific economic activities of people located in their surroundings, but also might become a hazard for natural resources and people's health. Economic theory can serve as a useful tool for understanding many relations originating as a result of brownfields: Past activities vs. present damage; remediation costs vs. benefits; brownfields vs. greenfields development; costs for nature protection vs. investment in other sectors. This part makes an overview of the schools and disciplines related to economic theory whose principles might be applied to the study, management and valuation of brownfield issues. The theoretical framework of each discipline mentioned is complemented by some results of empirical studies.

2 Environmental economics

Environmental economics can be defined as a discipline that incorporates the environment as the main subject of the economic analysis. It follows mainly the marginalistic perspective of neoclassical economics. According to this discipline, environmental problems appear due to the existence of market failures and externalities.

The analysis of externalities and distribution of resources are based on economic and quantitative terms, as a function of prices, costs and monetary benefits. These types of analyses also suggest the application of tools in order to regulate the imbalance nature vs. economy.

The traditional instruments for environmental regulation recommended by environmental economists tend to indicate "command and control" measures such as environmental regulations, and the application of taxes and tariffs on pollution, originally based in the analysis of welfare economics of negative externalities from Pigou [1].

According to the neoclassical approach environmental problems should be solved at a level that allows the maximum benefit at the least amount of abatement costs. Economically speaking, the "efficient" level of pollution is recognized at the point at which marginal damages equal marginal abatement costs. Figure 1 illustrates the neoclassical concept of pollution level efficiency. The optimal pollution level is denoted by point *OP*.

While trying to follow the way of the curve of the damages, environmental economics might want to set up a monetary expression for the lack of satisfaction that subjects of the society put for decrements of environmental quality within issues related to brownfields. This can also be seen as the calculation of the demand curve for environmental goods with information on the prices that consumers pay or are willing to pay for the goods. Environmental amenities are not bought and sold in the market, and therefore do not reflect a direct market value.

As mentioned above, neoclassical economics deals with this issue with tools approaching monetary values to non market goods. The methodologies used for

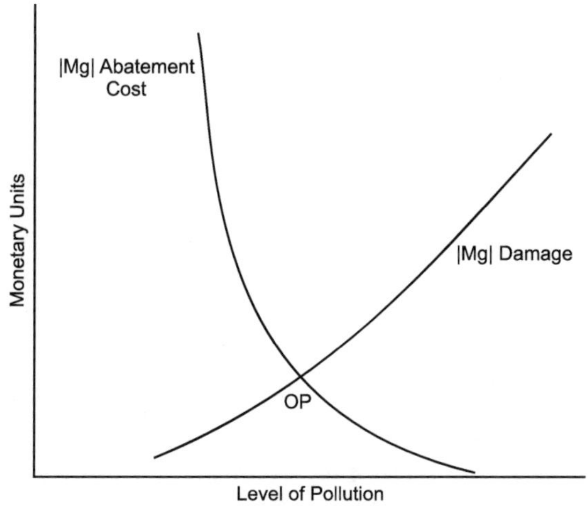

Figure 1: Optimal level of pollution. Environmental Economics Perspective (Sauer [2]).

these estimations in the branch of brownfields can vary from benefit transfer methods, to non market valuation methods such as hedonic pricing, and contingency valuation among others.

2.1 Hedonic pricing

This method suggests finding a demand curve of environmental amenities through the characteristics (qualities) that are included in an entire land property package.

As observed in different studies, valuation of damages related to brownfields tends to be done through hedonic pricing methods. Svetlik [3] presents the results of the study using the hedonic land pricing technique to identify the determinants of land prices in Monongalia County, WV, and investigates the impacts of local brownfields on residential property values. The study presents a strong negative correlation between the presence of a small dump as a brownfield and the price of housing in the town observed. The author concludes that the configuration of the landscape is a major contributor to the quality of life. Land use patterns are synonymous with location attributes, such as aesthetics, congestion, and convenience.

Hedonic pricing methods may contain limitations such as the appearance of heteroskedacidity in samples of different distributions and also of multicollinearity, thus the presence of effects which may not precisely have been caused by variables included into the model. If, for example one just includes the presence of brownfields in the proximity of a residential zone as an independent variable, and this appears to be negatively correlated with the housing prices, and forgets to include other factors, such as employment or leisure opportunities, etc., it is not valid to conclude that brownfields influence the housing market in such a zone.

Such errors should be identified, tested, clearly mentioned in the reports of results and then corrected if possible.

More applications of hedonic pricing related to brownfields try to find connections between the value of a site and the way brownfields' cleaning up is being treated. According to Greenstone and Gallagher [4], an explicit market for a clean local environment does not exist. This is the reason why the hedonic price method is commonly used to infer the economic value of non-market amenities like environmental quality to individuals. In one of his working papers he uses the hedonic pricing through housing market to develop estimates of the local welfare impacts of Superfund sponsored clean ups of hazardous waste sites (The 1980 Comprehensive Environmental Response, Compensation, and Liability Act, which became known as Superfund, gave the EPA (USA) the right to place sites that pose an imminent and substantial danger to public welfare and the environment on the National Priorities List (NPL) and to initiate remedial clean ups at those sites). The foundings of type research show that the estimated costs for the remediation of the sites are too high in correspondence to the welfare of individuals living in the sourroundings of these related sites.

3 Free market environmentalism

Some authors reject government intervention in environmental issues because of the fundamental reason that these kinds of interventions may not reflect a transparent compromise towards environmental protection. On the contrary, the forces of market and free decision-making processes could present better results for the individuals including the environment and even achieving more effective forms of protection of nature.

Free market environmentalism has adopted the basic ideas of the Austrian School of Economics, whose main representative (Hayek [5]) claims to the attention, that governmental intervention may limit freedom, which could become dangerous for the natural development of human principles in democratic societies.

Free market environmentalism does not support the idea of interventionism through taxation or any kind of environmental regulation. The solution to environmental problems should be left in the hands of a well designed system of property rights. Empirical studies on the area try to show the (in)efficiency of state regulations while trying to achieve improvements in environmental performance.

The trend of a centralized treatment of environmental issues around the management of brownfields has attracted the attention of representatives of free market environmentalism.

Empirical studies encourage the implementation of free market tools in order to achieve the solution of determinate environmental problems.

The criticism and skepticism towards the efficiency of the U.S. Superfund is one of the clearest and more recent examples of the disagreement of government interventionism within the framework of cleaning-up processes and restoration of brownfields.

In Shawn [6] the US brownfields legislation is defined to "take away states' freedom to clean up sites to their specifications" – Stroup [7] criticizes the great amount of transaction costs and thus the evident inefficiency of CERCLA in achieving its original goals· He argues that the main weakness of the program is its general and global character. The inefficiency issue has not been solved even with the attempt to lessen transactions costs with the reduction of bureaucracy through the introduction of more flexible rules for budgetary reports and controls. In other words, according to the author, federal laws have a strong potential for corruption and are also not able to solve local and specific problems related to contaminated sites. Thirteen years after, Mailman [8] presents another mechanism out of the free market framework: the U.S Bankruptcy Code. The author describes it as one of the treats for the Superfund's effectiveness for the achievement of its goals: According to ASCE and its study called the "Report Card for America's Infrastructure"' the rate of Superfund site clean-up has improved up to 2003, but it is not able to keep up with the rate at which new and potential sites are identified.

In Hamilton and Viscusi [9], health risks and population exposures are compared with the costs of the US Superfund. The conclusions of the analysis reflect excessive costs for remediation with irrelevant concrete improvements in the field of health protection, mainly on the prevention of cancer. In this study the authors point out that that governments do not necessarily come to permanent remedies when they bear the costs.

In Nugent [10], the author presents alternative ways for the development of brownfields.

His innovative idea of "cleaning soil with mushrooms" suggests that there can be always an alternative to the management of environmental goods. According to the author, it is not feasible (and also not necessary) to clean up completely one site nor all the contaminated sites, as EPA's policy intends to do. Therefore, one may look for attractive economic activities that keep people motivated to cooperate, finding the solutions that can give practical solutions and clear results for the individuals affected. Nugent says his company can gain a profit while using environmentally friendly technology on residential properties or former military sites.

4 New institutional economics

The research of new institutional economics has partially started as a reaction to welfare and mainstream economics, mainly to Pigou's conception of externalities in Coase [11]. Thanks to the concepts of Coase [12], new institutional economics (NIE) acknowledges the existence of transaction costs and their influence on economic and environmental outcomes. Instead of the idea of externalities, which Paavola and Adger [13] consider as a narrow concept, NIE is characterized by the broader concept of interdependence.

The application of NIE in brownfield management can be perceived through studies which observe and evaluate the institutional framework within specific locations. Brownfield management initiatives correspond traditionally to centralized decisions from different legislative frameworks. The cases presented below are mainly focused on the analysis of governance characteristics and information issues.

The case presented by Williams and Dair [14] reflects how generally oriented policies are not able to "encourage" companies in order to achieve "general goals". The observations made by the authors conclude that the limitations regarding the sustainability of brownfield development correspond mainly to the characteristics of the stakeholders' participation. They identified these five "main barriers" as follows:

- Stakeholders' omission of sustainability as an element of a 'successful' project.
- Inadequate participation by relevant stakeholders.
- Inappropriate timing of stakeholders' participation.
- Lack of powers to enforce the rise of sustainability options.
- Resistance to the use of sustainable technologies and materials.

It can be inferred from this study that the decision making process entered in conflict of goals definition, since engagement to achieve the general goals has not been reflected within the firms in charge of cleaning-up.

Meyer and Estrin [15] apply a combination of resource-based and transaction-cost theories in order to analyze three market entry modes for firms' expansion strategies: conventional acquisitions, greenfields, and brownfields. The authors identify brownfields as a hybrid mode of acquisition and found after observation of various studies, that firms tend to choose the acquisition of brownfield if it can provide them with already established local assets which can be combined with their competences. It becomes more attractive for companies to acquire the infrastructure of former firms so they can take advantage of certain established local characteristics and eventually "redo" the whole production concept than to risk in investments in "new land" such as greenfields, where the market is not necessarily defined. The conventional acquisition process also might represent a variety of trade-offs between both companies. The paper also suggests the following definition for brownfields "*a brownfield is a foreign acquisition undertaken as part of the establishment of a local operation. From the outset, its resources and capabilities are primarily provided by the investor, replacing most resources and capabilities of the acquired firm*". It can be inferred from the paper that the authors do not assume environmental issues as a risk factor for the acquisition of brownfields, since they suggest this entry mode as an opportunity to expand in economies where the institutional frameworks are weak (and there is therefore less environmental awareness and/or control). They also suggest undergoing a benchmark of the motivations for existing entry modes applied in different locations.

The study pursued by Bluffstone [16] gives results of empirical research in various privatized companies within five central European countries (Hungary, Lithuania, Poland, Romania and Slovakia). The goal of the study was to identify if there is a connection (and therefore a correlation) between the conclusion of a remediation process and the existence of information and plans about environmental analyses prior to and after the privatization process. The hypothesis was evaluated through econometric models. The results indicate that not only the existence of appropriate environmental information before privatization is a key issue for the development of further decontamination processes, the author found that it is also essential to combine it with environmental audit plans. One of the limitations of the analysis is that there is no data on the quality levels of the remediation, but shows also interesting additional issues about the region studied. For example, the author points out that at the time of the first (and the most) cases of privatization – in the years1990–91 – the countries included in the study had limited laws and policies for addressing environmental liabilities during the process of property transfer, which influenced that mass privatization put off environmental problems "for the future" when the owner would be forced to face the liabilities they had ignored before. The author also observed through the results that up until 1998, pollution and site contamination issues may have been handled informally.

Reports, such as the Great Lakes Commission [17], have recapitulated and observed data concerning brownfields management in specific regions. This study mentions how public participation for brownfield development has been addressed in the U.S. Great Lakes Region. It is observed that the most common public participation element in the region is notification. According to the authors, the sole fact that an announcement is placed in local newspapers is important but represents just a little piece of a comprehensive plan for public involvement.

5 Ecological economics

Ecological economics is a discipline whose origins correspond mainly to the reaction to the ideas of environmental economics. Ecological economics reject the concept of the natural environment only by means of its use and option values for human beings and thus, considers that is necessary to include not only human needs in the evaluation and management of ecosystems. It is focused on the intrinsical value of nature and therefore has an ecocentric approach. According to Costanza et al. [18] the objectives of study of this discipline are a wider understanding of the complex interrelationships between the economic, social and environmental systems in order to find solutions to conflicts existing within these systems.

Differently from free market environmentalism, ecological economics believes in the power of a centralized environmental policy which can support programs towards the achievement of the maximum protection of environmental amenities according to limits set by natural scientists. The evaluation of such actions will be developed mainly under the results of cost-effectiveness analysis. A cost effectiveness analysis is able to show the ratio at which money expenditures are related to expected results for specific areas. In other words, ecological economics might develop studies trying to find the cheapest way of achieving the desired environmental target.

The empirical work of ecological economics in the branch of brownfields includes quality assessments for projects concerning urban planning as, for example, Nijkamp and Rodenburg [19], or applications of specific systems such as GIS in Geoghegan et al. [20], suggesting wider approaches for natural valuation than the neoclassical ones (hedonic pricing).

Another close approach to the analysis of ecological economics in the brownfield area can be found in studies concerning the application of industrial ecology.

Industrial ecology is a broad, holistic framework for guiding the transformation of industrial systems from a linear model to a closed loop model that resembles the cyclical flow of ecosystems. Unlike the traditional linear model of economics, the industrial ecology model, through a systems perspective, seeks to optimize overall efficiency, ultimately eliminating all waste. Industrial ecology aims to reduce the overall environmental impact of an industrial system through the selection and optimization of the elements of the organizational system and its interaction. Industrial ecology can therefore be seen as an integrated system's approach to managing the environmental effects relating to energy and material consumption and capital generation in industrial systems. Industrial ecology encompasses the concept of cleaner production, whereby cleaner production focuses on waste reduction within a single organization. However, through an interdependent system of industrial organizations, thus modeling a natural habitat, it is hoped to achieve lower environmental impacts and a better exploitation of resources.

Eilering and Vermeulen [21] focus on the presentation of different case studies in the Netherlands about the implementation of eco-industrial parks. (An eco-industrial park is a community of companies, located in a single region, that exchange and make use of each other's by-products or energy Desrochers [22].) The study explicitly defines the design of eco-industrial parks as one of the concrete applications of industrial ecology. Additional to the traditional advantages that systems of interdependent companies and factories can create cooperation, the authors conclude after their observations that eco-industrial parks helped to achieve significant reductions of environmental burden and should be taken into account as alternative models for brownfield redevelopment.

According to Martin et al. [23], industrial ecology may focus on a two-dimensional analysis that includes the locality's stage of development and its degree of business concentration. The authors suggest promoting industrial parks in areas where development already exists and business concentration is lower. Figure 2 illustrates this idea of presenting brownfields as one of the most suitable areas for the creation of industrial parks.

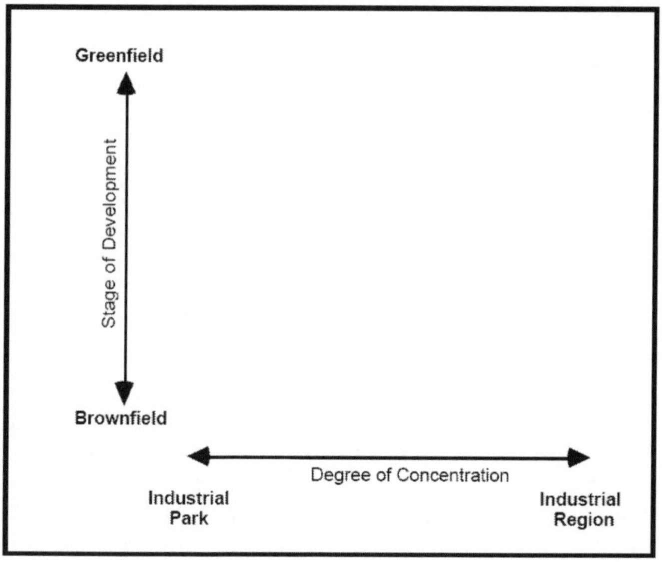

Figure 2: Industrial ecosystems (Martin et al. [23]).

6 Conclusions

One can identify notable differences in the application of economic theory towards the management of brownfields, depending on specific schools of thought. Neoclassical economics, through the study of environmental economics focuses on the utilitarian approach and tend to promote the use of command and control measures in order to achieve the "pollution optimum". Free market environmentalism does not support the government interventionism and suggests, on the other hand, market oriented solutions reflecting the freedom of decisions made by affected stakeholders. New institutional economics make assessments of the success of institutional arrangements towards the resolution of environmental conflicts among affected parties. Ecological economics underline the importance of a multidisciplinary and global approach in order to achieve not only economically efficient solutions, but also ecologically effective ones.

Even if the studies presented do not clearly expose which school they belong to, we made an effort in order to classify them according to the methods used and instruments proposed in their conclusions.

The most common supporting disciplines and techniques for the economics of brownfields encountered for the purposes of this work included cost-benefit analysis, analysis, hedonic pricing, regression analysis, cost-effectiveness multicriteria analysis, and qualitative research.

The prevailing point within the economic theory applied to brownfield redevelopment in this work was the management of conflicts. Each school tries to approach historical contamination in different ways, but for all economists environmentally risky brownfields are particular because they lack "original responsible subjects" for the pollution created in the past, which makes the problem become a "common present one".

Acknowledgement

This article was created under the state subsidy of the Czech Republic within the Research and Development project "Advanced Remediation Technologies and Processes Center" 1M0554 – Programme of Research Centers supported by the Ministry of Education.

References

[1] Pigou, A.C. *The Economics of Welfare*, London: Macmillan, 976 p. 1920.
[2] SAUER P. "Ekonomické souvislosti řešení ekologických důsledků pobytu sovetských vojsk na našem území", Acta Oeconomica Pragensia, č.2/1993.
[3] Svetlik, J. B. *Externality Effects of Local Brownfields on Residential Property Values*. Northern West Virginia Brownfields Assistance Center, West Virginia Water Research Institute Morgantown, WV. Paper presented at the 2007 Business of Brownfields Conference, in Pittsburgh, PA. 2007. Online http://www.eswp.com/PDF/Svetlik_BFpropvalues07.pdf
[4] Greenstone, M., Gallagher, J. *Does Hazardous Waste Matter? Evidence from the Housing Market and the Superfund Program*. MIT Department of Economics Working Paper No. 05-27. Online http://ssrn.com/abstract=840207. Massachusetts 2007.
[5] Hayek, F. *The Fatal Conceit: The Errors of Socialism*. Routledge. London, 194 pp. 1989. ISBN 978-0226320663.
[6] Shawn, J. S. Not Quite A Gentleman. *The Environmental Forum*. The Environmental Law Institute, Washington, D.C. 2003. ISSN 0731-5732.
[7] Stroup, R.; Townsend B. EPA's New Superfund Rule: Making the Problem Worse. Regulation: *The Cato Review of Business & Government* No. 3, Washington, D.C .1993.
[8] Mailman, R. Cleaning Up Its Act: Improving CERCLA Efficiency within the Bankruptcy Code. *27 Whittier Law Review*. Whittier CA. Vol. 27; No. 2, pp. 557-596 2005. ISSN 0195-7643.
[9] Hamilton, J.; Viscusi, W. *Calculating Risks: The Spatial and Political Dimensions of Hazardous Waste Policy*. The MIT Press .Cambridge, Mass., 340 pp. 1999. ISBN 978-0262082785.
[10] Nugent, S. *Mushrooms Meet Brownfields. A Market for Fungal Remediation*. PERC Reports. Volume 24 | Number 4. Washington DC 2006. Online http://www.perc.org/articles/article847.php
[11] Coase, R. The Problem of Social Cost. *Journal of Law and Economics*, Vol III. pp. 1-44. 1960. ISSN 00222186.
[12] Coase, R. The nature of the firm. *Economica*, Vol. IV. pp. 386-405.1937.1920. ISSN 0013-0427.

[13] Paavola J., Adger N. *New institutional economics and the environment: conceptual foundations and policy implications.* CSERGE Working Paper. University of East Anglia, Norwich 2002. Online http://www.uea.ac.uk/env/cserge/pub/wp/edm/edm_2002_06.pdf

[14] Williams K., Dair C. Five barriers to sustainable brownfield development. *Town and Country Planning*, Vol .72, No. 11. pp. 344-345. 2003. ISSN 0040-9960.

[15] Meyer, K. Estrin, S. Brownfield Entry in Emerging Markets. *Journal Of International Business Studies*. Palgrave. Hampshire. Vol. 32. No. 3, pp.575-584.2001. ISSN 0047-2506.

[16] Bluffstone, R. *Privatization and contaminated site remediation in Central and Eastern Europe: Do environmental liability policies matter?* Ecological Economics, Elsevier, vol. 63. No. 1. 2007. p. 31-34. 2007. ISSN 0924-6460.

[17] Great Lakes Commission. *Linking Brownfields Redevelopment and Greenfields Protection for Sustainable Development.* 2001. Online http://www.glc.org/bridges/9-01BridgesI.pdf

[18] Costanza, R., Cumberland J., Daly H., Goodland R., Norgaard R. *An Introduction to Ecological Economic* (Hardcover) CRC Press. 275 p. 1997.

[19] Nijkamp, P., Rodenburg, A. Success factors for sustainable urban brownfield development A comparative case study approach to polluted sites. *Ecological Economics, Vol. 40, p. 235-252.* 2002. ISSN 0921-8009.

[20] Geoghegan, J., Wainger, L., Bockstael, N. Spatial landscape indices in a hedonic framework: an ecological economics analysis using GIS. *Ecological Economics, Vol.* 23, No. 3, pp.251-264. 1997. ISSN 0921-8009.

[21] Eilering, J., Vermeulen, W. Eco-Industrial Parks: toward industrial symbiosis and utility sharing in practice. *Progress in Industrial Ecology* Vol. 1, No. 1-3. 2004. ISSN 1478-8764.

[22] Desrochers, P. Eco-Industrial Parks. The case of private planning. *In Rethinking Green. Alternatives to Environmental Bureaucracy.* Edited by Higgs and Close. The Independent Institute. Oakland. 2004. pp. 297-324. ISBN 0-945999-97-6

[23] Martin, S., Weitz A., Cushman, R., Sharma, A., Lindrooth, R., Moran, S. *Eco-Industrial Parks: A Case Study and Analysis of Economic, Environmental, Technical, and Regulatory Issues: Final Report.* Research Triangle Park, NC: Research Triangle Institute. 1996. 274 pp. Online http://www.rti.org/pubs/case-study.pdf.

Section 5
Data Mining applications
(Special session by A. Rabasa)

Identifying the most relevant issues to minimize an automated irrigation water supply in vineyards using Data Mining

A. Rabasa Dolado[1], J. M. Molina Martínez[2], H. Puerto Molina[3], J. J. Rodríguez-Sala[1], L. Santamaría Arana[1] & A. Ruiz Canales[3]
[1]*Centro de Investigación Operativa, Universidad Miguel Hernández de Elche, Spain*
[2]*Dpto. Ingeniería de Alimentos y del Equipamiento Agrícola, Universidad Politécnica de Cartagena, Spain*
[3]*Dpto. Ingeniería, Universidad Miguel Hernández de Elche, Spain*

Abstract

In order to carry out optimal daily irrigation programming for vineyard production, different variables have to be considered. Some of them are related to soil (soil moisture variation), other variables are related to crop (variety, dendrometry, leaf temperature) and finally some climatic variables are considered (reference evapotranspiration ETo and rainfall). This daily irrigation programming is traditionally based on physical–mathematical models (soil water balance, radiation, drainage among others). This is a model-driven methodology. However, in this paper a data-driven methodology is proposed. With this methodology, the daily water consumption of a crop is estimated from daily values acquired through the variables described above. This approach predicts minimal water consumption. In addition, the significant variables in daily water consumption according to crop and phenological phases are determined. The experiment was developed in a vineyard farm located in Villena, Spain. Representative daily data of five vineyard varieties (Cabernet Blanc, Merlot, Petit Verdot, Sauvignon Blanc and Syrah) were collected from 2005 to 2009. These data were subsequently analysed by Data Mining techniques that are oriented to extraction and rule systems classification. This methodology will be used in the development of applied algorithms in software for decision tools in irrigation programming for vineyard production.

Keywords: Data Mining, irrigation, vineyard.

1 Introduction

Wine production is one of the main economic activities in Villena (Alicante), in the South of the Valencian Community (Spain). Nowadays, several approved wineries in this area are experimenting with foreign varieties of vines such as Cabernet Sauvignon or Merlot, both French varieties. These varieties have been adapted to the climatology, altitude and soil conditions of this area.

In order to adapt these new varieties to these specific conditions, several Agronomical practices have to be implemented. One aspect of these practices is irrigation scheduling. Due to the specific climatology (annual rainfall, average temperature and evapotranspiration of 420 mm, 15ºC and 1,248 mm respectively), irrigation supply has to be applied in order to obtain an adequate yield.

Moreover, irrigation enables increased yields [1], though a moderate water deficit is often desirable to improve wine composition [2]. Deficit irrigation has been suggested as a strategy to improve fruit composition for premium quality wines reducing canopy vigour, increasing fruit exposure to light and reducing berry growth to avoid dilution effects [3].

Deficit irrigation consists in applying water rates to replace only part of the vine's potential evapotranspiration either during the whole season or only during some previously established phenological periods. Regulated deficit irrigation can be applied as a strategy to reduce the possible negative impact of irrigation on wine quality. In the past, Salón et al. [4] studied the response to irrigation of cv. Bobal. However, in the previously cited varieties (Cabernet Sauvignon or Merlot) the effects of different irrigation regimes on vine performance and wine quality have not been reported yet for this area.

Soil water availability is a critical factor for vine performance and wine composition. This availability is monitored by means of soil moisture gauges. The use of soil-based water status indicators have been adopted as an adequate strategy for water balance estimation.

The use of plant-based water status indicators has become very popular in recent years for studying plant-water relations and for planning more precise irrigation programmes, because it is recognized that the tree itself is the best indicator of its water status. One of these methodologies, monitoring trunk diameter variations (TDV), is widely used to detect plant water status [5, 6]. This methodology is commonly named as dendrometry.

The ultimate goal of these techniques is to provide vine growers with information about the more appropriate volumes of irrigation to apply in each phenological period, depending on the desired yield levels and wine styles.

An added structural problem of this area is the aridity of the climate and the scarcity of water resources, which is a historical and recurring characteristic in southern Europe, where during several periods in the year, water supply is not guaranteed. In order to optimize water supply and to rationalize the management of demand per vineyard an adequate irrigation scheduling could be established. This scheduling is based mainly on soil moisture gauge measurements, dendrometry and climatology. Several other variables are also related to irrigation scheduling of vineyards.

In this paper, a data-driven methodology is proposed, where the daily water consumption of a crop is estimated from the daily values acquired for the variables described above. This approach predicts minimal water consumption and determines the significant variables of daily water consumption according to crop and phenological phases. This methodology is based on experimental data from a commercial vineyard.

The experiment was developed at a vineyard situated in Villena, Spain. Representative daily data are were collected from five vineyard varieties (Cabernet Blanc, Merlot, Petit Verdot, Sauvignon Blanc and Syrah) from 2005 to 2009. These data were subsequently analysed by Data Mining techniques, which are oriented to extraction and rule systems classification.

It is possible to apply this methodology to the development of applied algorithms in software for decision tools in irrigation scheduling for vineyard production.

Daily irrigation scheduling is traditionally based on physical-mathematical models (soil water balance, radiation, drainage among others) as in model-driven methodology is proposed.

2 Problem definition and main objective

Irrigation scheduling is basically a method for calculating how much and when irrigation water has to be applied. In order to answer these questions, some data have to be collected, which are mainly soil-based, plant-based and meteorological-based. These data are processed in the water balance. Water balance is obtained by applying a mass conservation equation [7–10]:

$$ET = P + I - \Delta S - D - R \tag{1}$$

where, ET is evapotranspiration (1), P is rainfall, I is irrigation, ΔS is the variation of water stock between two dates, D is the drainage under the profile and R is the runoff [11, 12]. ET is obtained by multiplying reference evapotranspiration (ETo) by the crop coefficient (Kc). The first factor is related to meteorological data and the second depends on the crop. In the development of the experiment, daily data were obtained (ETo, P, I, ΔS). Drainage and runoff were not considered in this experiment. In addition, some data to detect plant water status were collected, as well as monitoring trunk diameter variations (TDV) or dendrometry.

In order to carry out the experiment on optimal daily irrigation scheduling for vineyard production, different variables have to be considered. Some of them are related to soil (soil moisture variation), other variables are related to crop (variety, dendrometry) and finally some climatic and agricultural variables are considered (reference evapotranspiration ETo, rainfall and irrigation dose). Finally a leaf wetness sensor is included. This is a meteorological parameter that describes the amount of dew and precipitation left on surfaces. It is only used for monitoring leaf moisture for agricultural purposes, such as fungus and disease control, for control of irrigation systems, and for detection of fog and dew conditions, and early detection of rainfall. The latter is not included for irrigation scheduling but its measurement data have been tested in order to contrast them with meteorological data.

Site description and experimental design: The experiment was carried out during the seasons from 2005 to 2009 at several plots with different varieties: Cabernet Blanc, Merlot, Petit Verdot, Sauvignon Blanc and Syrah vines (Vitis vinifera L.). These plots are situated in a valley between two hills which has sandy franc soil. These vineyards were planted in 2001 with a spacing of 2.80 x 0.90 m^2 (density is around 4,000 plant·ha^{-1}).

The irrigation system is trickle irrigation with integrated emitters spaced 0.80 cm. The average flow of the emitter is 3.85 l·$hour^{-1}$ (1.125 emitters/plant). The vineyard is near Villena (38°34′31.4″N, 0°56′35.58″W, elevation 540 m), Alicante, Spain (Figure 1).

Figure 1: Aerial view of the experimental plots (Googlemaps).

Weather conditions during the experiment were measured with an automated meteorological station at the plot and reference evapotranspiration (ETo) was calculated as in [13].

Irrigation frequency varied from 3 to 5 days per week. The previously mentioned water balance parameters were measured daily throughout the experimental period. ETo and cumulated daily rainfall (P) were measured by the automated meteorological station. Daily irrigation was measured with a pluviometer. Several FDR soil moisture gauges measured the amount applied to each monitored plant (one per variety) and at several depths (10, 30 and 60 cm deep). Using these data, the variation of water stock between two dates (ΔS) was obtained. The daily trunk diameter was measured in three plants per variety. The daily measurement was obtained from the average data in order to detect water stress in the plant. Finally, the leaf wetness sensor measured the leaf moisture every day. These last two data are not related to irrigation scheduling but are associated to plant and meteorological status.

In this case study, the attributes to be considered are the following:

- DATE: Is the date of data collection throughout the experimental period.
- SOIL_MOIST_10: Is the soil moisture measurement obtained with a FDR soil moisture probe 10 cm deep in percentage % (0-100).
- SOIL_MOIST_30: Is the soil moisture measurement obtained with a FDR soil moisture probe 30 cm deep in percentage % (0-100).
- SOIL_MOIST_60: Is the soil moisture measurement obtained with a FDR soil moisture probe 30 cm deep in percentage % (0-100).
- CF_AVG_INCR: Is the daily average trunk diameter (mm).
- LEAF_HUMIDITY: Is the leaf moisture in percentage % (0-100).
- ETo: Is the reference evapotranspiration obtained by Hargreaves method (mm).
- RAINFALL: Is the daily cumulated rainfall (mm).
- IRRIGATION: Is the daily cumulated irrigation (mm).

After the data collection, a specific data mining methodology was applied in order to detect possible wrong data and determine the weight of all data in the water balance. This methodology is used in several applications and, in this case, it was used to obtain algorithms with a low data quantity in order to apply them to a hardware design for automatic irrigation.

3 Methodology

In contrast to the classical model driven approach, where mathematical or physical models are used to describe or forecast problems, in this paper the data driven perspective defended by Solomatine [14] is suggested as the most accurate and efficient approach for classification tasks from historical data, without consideration of previous context knowledge.

Decision Trees are widely applied in this framework, where a target attribute must be classified. There are different tree implementations depending on the nature of the variables to be considered, the branch cardinality and some other constraints. In this paper the authors chose the ID3 algorithm to extract classification rules responding the structure: A1, A2, …An → C.

Ai being the categorical antecedent variables and C the categorical consequent or target variable, as shown in Table 1, in section 2.

After applying ID3 algorithm [15] with the original data set D, a rule set R is obtained. The reduction algorithm RBS [16] reduces R to R' that contains only significant rules to predict the consequent as shown in Fig.2.

Table 1: Attribute description.

Attribute	Original values	Pre-processed values
DATE	Dates typed as dd/mm/yyyyy	Jan-Apr: Winter May-Jun: Spring Jul-Aug: Summer Sep-Oct: Autumn Nov-Dec: Winter
SOIL_MOIST_10 SOIL_MOIST_30 SOIL_MOIST_60	Positive numerical values	< 8.5: Default [8.5..16.4]: OK > 16.4: Exceed
CF_AVG_INCR	Positive numerical values	< 300: Decrease (*) [-300..300]: Constant >300: Increase
LEAF_HUMIDITY	Positive numerical values in range 0..10	0..3: Low 4..6: Medium 7..10: High
ETo	Positive numerical values	[0..3]: Low (3..6]: Medium (6..10]: High >10: Very High
RAINFALL	Positive numerical values normalized	0: Null (0..5): Low [5..15): Medium [15..25]: High >25: Very High
IRRIGATION	Positive numerical values	0: Null (0..3): Low [3..10): Medium >=10: High

(*)All attributes were discretized exclusively attending expert criteria except CF_AVG_INCR obtained from the density function around Average $\pm 2 \cdot$StdDeviation.

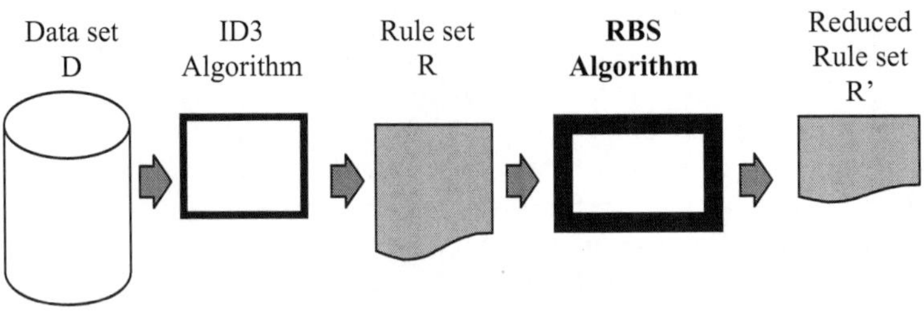

Figure 2: From Data set to Reduced Classification Rule set.

4 Computational experiment

In this section the results of the computational experiment are shown exclusively under the data analysis approach, without expert interpretation.

After the RBS algorithm had been applied to the 5 scenarios, the 5 vineyard varieties have very similar correlation between the studied antecedents and the irrigation variable, with minor differences.

Trying to minimize the water consumption, the irrigation target variable was instanced to "Null" and the most important patterns leading to no-irrigation instance are the following:

- Attributes SOIL_MOIST_10, SOIL_MOIST_30 and SOIL_MOIST_60 have a very low correlation with the target variable.
- Climatic data correlate more with Irrigation if they are considered on the previous day, instead of the current day.
- Rules in Spring and Autumn have a uniform distribution, so there are no supportable patterns to describe the vineyards behaviour in those seasons.
- The most repeated pattern in Winter is CF_AVG_INCR=Null, ETo=Low, RAINFALL=Null. With this antecedent, the IRRIGATION is NULL with confidence close to 98%.
- In Summer (with dry conditions ETo=Low, RAINFALL=Null/Low), while CF_AVG_INCR do not decrease, the Irrigation is NULL with confidence close to 92%, but when CF_AVG_INCR decreases, IRRIGATION variable turns to MEDIUM.

5 Conclusions

The data obtained for Cabernet Blanc and Merlot could be extrapolated to Petit Verdot, Sauvignon Blanc and Syrah vines. The main variation is the irrigation dose and timing.

Because climatic data correlate more with irrigation if they are considered on the previous day, instead of the current day, it is necessary to consider the automation of irrigation scheduling in order to apply the specific irrigation dose.

This methodology could be used to obtain algorithms with a low quantity of data in order to obtain cheaper irrigation hardware devices. With low data, the designed hardware spends less memory and space. It could be reached through Feature Selection techniques.

The methodology developed could be applied to improve irrigation scheduling in vineyards. From a similar field of data treatment, it is possible to adapt some irrigation practices (deficit irrigation, partial root drying, etc.) to the specific local conditions of the vineyard and use adequate water quantity and no more than this. This is one of the main objectives for water resource management in an arid zone with water scarcity.

References

[1] Williams, L.E. & Matthews, M.A., Grapevine (Chapter 34). *Irrigation of agricultural crops agronomy monograph no. 30*. eds. B.A. Stewart & D.R. Nielsen, ASA-CSSA-SSSA Madison, WI, USA pp 1019–1055, 1990.

[2] Jackson, D.I. & Lombard, P.B., *Environmental and management practices affecting grape composition and wine quality-A review*. American Journal of Enology and Viticulture 44:409–430, 1993.

[3] McCarthy, M.G., Loveys, B.R., Dry, P.R. & Stoll, M., *Regulated deficit irrigation and partial rootzone drying as irrigation management techniques for grapevines*. Deficit irrigation practices. FAO Water Reports No 22. Rome, Italy, pp 79–87, 2000.

[4] Salón, J.L., Chirivella, C. & Castel, J.R., Response of Vitis Vinifera cv. Bobal to deWcit irrigation in Requena, Spain. Water relations, yield and wine quality. American Journal of Enology and Viticulture 56:1–18, 2005.

[5] Goldhamer, D.A. & Fereres, E., Irrigation scheduling protocols using continuously recorded trunk diameter measurements. Irrigation Science 20:115–125, 2001.

[6] Naor, A., Irrigation scheduling and evaluation of tree water status in deciduous orchards. Hortic Rev 32:111–166, 2006.

[7] Faures, J.M., Goodrich, D.C., Woolhiser, D.A. & Sorooshian, S., *Impact of small-scale rainfall variability on runoff model*. Journal of hydrology. Amsterdam: Elsevier Science. B.V, 1995.

[8] Kerkides, P., Michalopoulou, H., Papaioannou, G. & Pollatou, R., *Water balance estimates over Greece*. Agricultural Water Management, 32, 85-104, 1996.

[9] Moreno, F., Cayuelas, J.A., Fernández, J.E., Boy, E., Murillo, J.M. & Cabrera, F., *Water balance and nitrate leaching in an irrigated maize crop in SW Spain*. Agricultural Water Management, 32, 71-83, 1996.

[10] Sánchez-Cohen, I., Lopes, V.L., Slack, D.C. & Fogel, M.M., *Water balance model for small scale water harvesting systems*. Journal of Irrigation and Drainage Engineering, 132 (2), 123-128, 1997.

[11] Maraux, F. & Lafolie, F., *Modeling soil water balance of a maize-sorghum sequence*. Soil Science Society of America Journal, 62 (1), 75-82, 1998.

[12] Schiller, G. & Cohen, Y., *Water balance of Pinus halepensis Mill. afforestation in an arid region*. Forest Ecology and Management, 105 (1-3), 121-128, 1998.

[13] Hargreaves, G.H. & Samani, Z.A., *Estimation of standard deviation of potential evapotranspiration. Proceedings ASCE*, Journal of Irrigation and Drainage Engineering, 114(1), 1988.

[14] Solomatine, D.P., *Applications of data-driven modeling and machine learning in control of water resources*. Idea Group Publishing. Computational Intelligence in Control, 197-217, 2002.

[15] Quinlann, J.R., *Induction of decision trees*. Machine Learning, vol. 1, 81–106, 1986.

[16] Abadía, R., Almiñana, M., Escudero, L.F., Pérez, A., Rabasa, A. & Santamaría, L., *A Rules Reduction Algorithm Based On Significance Measure*. Ed. Zanasi, Alzamora, Ebecken, Brebbia. Cádiz. WIT Press. Proc. IX Data Mining and Information Engineering. 63-72, 2008.

A study of drivers for agricultural land abandonment using GIS and Data Mining techniques

B. M. Zaragozí[1], J. T. Navarro[1], A. Ramón[1] & J. J. Rodríguez-Sala[2]
[1]*Instituto Interuniversitario de Geografía, Universidad de Alicante, Spain*
[2]*Centro de Investigación Operativa,*
Universidad Miguel Hernández de Elche, Spain

Abstract

This paper examines the problem of agricultural land abandonment through knowledge of driving factors. First, we reach out a review of the concept of abandonment in order to define the negative and positive aspects of this process in the landscape. Then we focus on a real case study in the Marina Baja region (SE Spain). Finally, we propose, from a theoretical perspective, the use of a methodology to define the most important variables for studying the abandonment processes in this region.

First of all, we performed various analyses using Geographical Information Systems (GIS) to create a geodatabase which stores the main factors generally associated with abandonment. The variables considered can be grouped into three groups: environmental (such as climate, soils or relief), socio-economic (such as demographic, accessibility or technology) and those related to cultural practices in the crop (such as irrigation or rain-fed stone walls).

Considering over a hundred variables we need to answer two questions: (1) which of the variables are really driving factors in the abandonment process of an agricultural plot? and (2) how important is each of them alone or in combination with others? The authors advocate the use of GIS for the feature creation or management and Data Mining techniques to perform the feature selection. Specifically, we propose the use of a concrete metric, aci (Attribute Correlation Index), to answer the questions raised above.

Keywords: Geographical Information Systems, Data Mining, land abandonment.

www.witpress.com, ISSN 1743-3541 (on-line)
doi:10.2495/ECO110321

1 Introduction

The abandonment of agricultural lands is an important phenomenon in many regions in the European Union. We can appreciate these processes using some indicators like the continuous reduction of the population employed in agriculture or the decline in the number of exploitations. These processes are likely to continue over the coming years as a result of the lack of viability of many of these exploitations [1].

The concept of "abandoned lands" is not completely clear, there is no single definition of land abandonment in the EU countries and this process even does not exist in a few of these countries [2]. However, we can find different interpretations in the literature describing it from a static or a dynamic point of view [3]. Following a static point of view, we can find some visible effects of the abandonment like the spontaneous re-vegetation or the progressive deterioration of fences, roads, irrigation systems and other agricultural infrastructure. When all these changes make it very difficult or impossible for the continuation of the agricultural activities, the situation is described as total abandonment or effective abandonment [4].

However, many of the farms that in theory are not viable can continue to be active for a long time after the expected period. The reasons for this are many, including transfer of ownership, search for other incomes outside the agriculture (part-time agriculture), or continuation of activities beyond retirement age [1], among others. These processes can help continue the agricultural activity but they also indicate the state of the farm prior to an expected total end of activities. The processes that lead to the total abandonment can be prolonged for an indefinite period of time. In that case, the parcels are not strictly abandoned, and that is why the situation can be described as "semi-abandonment". The semi-abandonment state is not easily perceived because the remaining activities can mask this situation. On the other hand, total abandonment is easier to locate geographically, but it includes in its definition many areas that could have been unexploited for a long time. For these reasons, in this paper we prefer an intermediate approach – we consider that an abandoned land is one that has an absence of any current land use but had a recent history of agricultural land use, normally, less in last 10 years.

A concept that helps to explain the semi-abandonment processes is "marginality". It has been traditionally associated with those lands that have low potential productivity. In case of any adjustment marginal lands are usually the first to be abandoned, so it is an important concept in order to perform abandonment estimations in a short or mid-term period. Like some biophysical factors, other structural and socioeconomic factors can be related to the marginality [5]. In Table 1 we can find some of the most common factors related to marginality and abandonment.

Abandonment is related to many processes implied in land use changes. Depending on the type of change and the region, abandonment can be considered a positive or negative change. Some positive effects linked to abandonment are

Table 1: Some identified drivers of abandonment, based on [5].

Geo-bio-physical	Socio-economic	Management
• elevation • geology • slope • fertility • soil depth • soil erosion • climate • …	• market incentives • migration • technology • industrialization • farm characteristics • farmer age • accessibility • …	• unadapted agricultural systems • land mismanagement • soil degradation • frequent flooding • overexploitation • productivity loss • …

connected to the re-vegetation of these areas with natural species adapted to the physical conditions. Due to this re-vegetation there are other positive effects caused by the abandonment process, like a better hydrological regulation and water quality, soil recovery and erosion mitigation, increased fertility, biomass increase and higher carbon sequestration. However, in Mediterranean areas this re-vegetation process can be quite slow and the pointed positive effects will arrive too late [6]. The negative effects associated to abandonment are also related to the re-vegetation process. These negative effects can be the loss of landscape heterogeneity and biodiversity, increased fire risk [7], soil erosion and desertification [8], reduction of water stocks, biodiversity loss and reduced population of adapted species or the loss of cultural and aesthetic values [9].

Once reviewed some important concepts, in this paper we consider that a better understanding of the driver factors that lead to the abandonment is a key subject, and we propose a new methodology that uses generally available geospatial data to find out the variables or drivers that lead to the abandonment of agricultural parcels (cadastral parcels).

2 A case study

2.1 Study area

In order to study the abandonment processes we have selected a Mediterranean region with a diverse geography where nowadays two socio-economic processes can be easily observed: firstly, agriculture and urbanisation intensification, and secondly, land abandonment. Now, we need a short geographical description of the selected study area in order to find out the most important factors that are structuring that region [10].

The study area is located in the southeast of the Iberian Peninsula, in the province of Alicante. Its extension covers about 578.8 squared kilometres and it is divided into a mountainous area and another coastal area. This region comprises 18 municipalities. In such a small area, the Marina Baja has many interesting contrasts. In Figure 1 you can appreciate that the altitudes range from sea level to 1558 m that is reached at the peaks of Sierra de Aitana. Between these mountains the most important river system is Guadalest-Algar (with

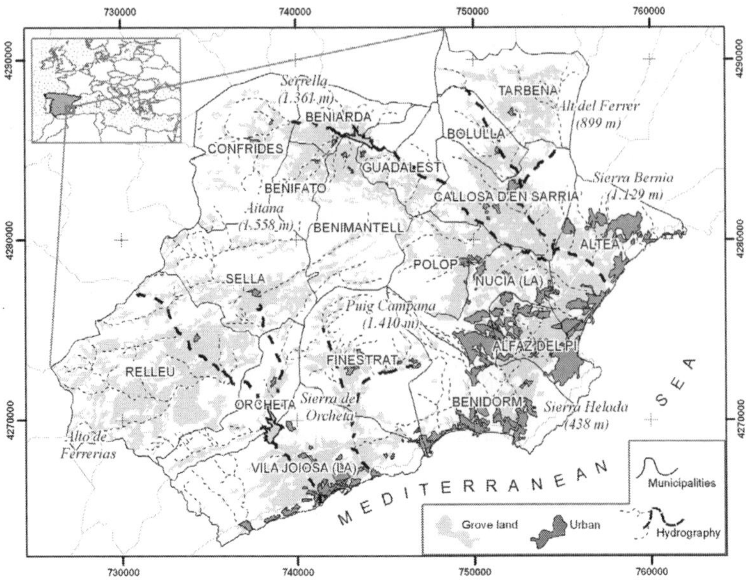

Figure 1: The Marina Baja.

modules from 0.90 to 0.78 m^3/s, respectively), but it should be noted that the resources provided by this river allow most of the irrigated crops and supply to towns. Very related to this physical distribution we can appreciate a very diverse climate with sensible differences between sea and mountain and north to south. Just to give a few climatic data we can say that average temperatures range between 17 and 18ºC and annual rainfall between 400 and 700 mm.

Before the 1960s, the Marina Baja was isolated because its abrupt topography and there were no highways or main roads as known as today but due to many socio-economic changes the Marina Baja structure has experimented a radical change. Today, over 60% of the Valencian tourist activity is concentrated here, mostly located in Benidorm, main municipality in the region [10]. These trends are still going on and producing constant land-use changes associated with two main processes: tourism/urbanisation and agricultural intensification/abandonment.

2.2 Data sources and GIS analysis

In this section we explain three steps of the GIS analysis performed for each data source: analysis of source, derivation of features and extraction to the parcel level (Object Oriented Analysis). Of course, all the data sources referred here are available for other Spanish regions and due to the INSPIRE initiative they should be available for all EU countries (Directive 2007/2/EC). This way the methodology posed here can be extrapolated to many other sites.

We acquired many different datasets that model as much as possible the abandonment drivers identified during the literature review and derived from the knowledge of the study area. These datasets include remote sensing images and altimetry datasets from the Valencian Institute of Cartography (ICV), the parcel boundaries defined by the Spanish Land Parcel Identification System (SIGPAC), and a geospatial database used for agriculture management. At last, the climatic data were obtained at the meteorological station level from the National Institute of Meteorology (INM). Of course, all these datasets needed to be pre-processed to get some useful and continuous variables. This way many of the attributes defined were derived from the previous datasets by application of different methods.

The cartographic boundaries to define the agricultural plots were obtained from the SIGPAC. In this database, each polygon represents a continuous area within a parcel, with normally a single agricultural use [11]. The total number of plots of the study area was 108,000. Directly from the parcels we calculated some shape-metrics including some very basic metrics like area or perimeter, but these metrics can be very useful because for example the shape of the area highly related to the agriculture viability [6].

The orthophotos employed were facilitated by the ICV. The images have a spatial resolution of 0.25 m/pixel and three spectral bands, including the near infrared, necessary to calculate the vegetation index. These images were acquired in 2007. To perform the feature extraction of the images we used the software FETEX 2.0 (Feature Extraction). This software was created by the CGAT (Geoenvironmental Mapping and Remote Sensing) research group, from the Department of Cartographic Engineering, Geodesy and Photogrammetry of the Universidad Politécnica de Valéncia. It is programmed in IDL 6.2 as an ENVI extension [12]. This software is designed to work with land cover databases assisting in the classification process of existing plots or even to determine land cover changes by comparing the current and the older database. Using this software it is possible to obtain over 50 characteristics describing an agricultural plot (spectral, textural and structural properties of the parcel image).

Using the same orthophotos we performed a first classification of the abandoned parcels following some criteria:

- No evidence of certain local agricultural practices
- Stonewalls are broken
- There are crop species mixed with natural vegetation
- Agricultural terraces are totally recolonized by natural species
- Presence of abandonment effects, such as erosion or wildfire
- The size of species is highly diverse
- There is bare soil but the databases indicate any agricultural use

Using the SAGA GIS software (System for Automated Geoscientific Analyses) and the ICV elevation data, we created a DEM by the Triangulation method. After that we converted the DEM to a raster format with a fine grid cell size (2x2 meters) to facilitate the feature extraction. SAGA GIS is useful for extraction of many features related to the DEM, but there are other proximity and interpolation methods that we used for preparing the other layers.

Table 2: Summary of possible features to include in this methodology.

Group / Number	Variables	Source
Shape/Plot features (6)	area, perimeter, shape index, fractal index…	SIGPAC geometry
Image features (56)	spectral features, texture features, structural features…	ICV CIR Images
DEM features (34)	Mean elevation, slope, aspect… Topographic Humidity Index…	Isohipses 1/10,000
Database features (26)	landuse03, landuse09, municipality, irrigation associations…	SIGPAC, Cadastre, Agricole cooperatives…
Climate features (60)	Precipitation, temperatures, water balance…	INM, IVIA… (Interpolation)
Soil features (12)	pH+, Organic matter…	Lab análisis (Interpolation)
Proximity features (10)	Accessibility, proximity to urban areas, to water sources…	ICV10
Aesthetic landscape (6)	Visibility, fragility…	DEM, Landuse…

Using the above-mentioned software and the methods available in there it is possible to extract hundreds of features related to the drivers of abandonment at the parcel level (see Table 2). In the methodology proposed here we must analyze this huge amount of information to get an explanation about the abandonment process in the study area, but this is high time and effort demanding, so we have selected a varied list of 10 features to test a Data Mining (DM) method that will help us to select the most important features or variables.

The first feature is the area (AREA) that can be easily calculated from the SIGPAC polygons. All the GIS programs should be able to calculate such a simple metric. We also calculated a shape index (SH_INDEX).

The easiest DEM feature we can extract is the mean elevation of each parcel (MEAN_ELEV). The crops in the study area are highly influenced by the elevation because this variable modifies the climatic characteristics decreasing the temperature range that is a limiting factor for many subtropical crops. Also based on the DEM, we extracted the Topographic Wetness Index (TWI). The TWI has been used to study different hydrological processes and biological processes based on topography [13].

Using the high resolution images available, we calculated the Normalized Difference Vegetation Index (NDVI). The NDVI is useful to distinguish artificial from natural plots, or dry crops from irrigated crops, for example. At the parcel level we could calculate many statistics relating the index to each parcel but we have selected some of the most simple: mean and standard deviation of the NDVI (MEANNDVI and DEVSTNDVI). We also have selected one of the many possible structural features we can extract from the orthoimages. The First Derivative near the Origin (FDO) is a feature based on the empirical semivariogram of the pixels that cover one parcel and it is also calculated using FETEX.

There are many climatic features we could extract or derive from the climatic data. Using the climatic data from INM (temperature and precipitation data from

1950 to 2007) we performed an interpolation as explained by [14]. Finally, we obtained monthly GIS layers for mean precipitations and temperatures (mean, max and min). However, in this study we had to limit the number of variables so we decided to calculate a climatic index used in vegetation studies. The Patterson index or Forest Productivity Index (FPI) has been widely used and adapted to different regions. Nowadays, it is still integrated in more complex methodologies to consider the climatic biomass productivity (e.g.: in [15]). This index is based in the assumption that high temperatures and strong precipitations are necessary for the biomass production and it gives an estimation of productivity in m^3 of wood/ha·year. The FPI formula is:

$$FPI = \frac{tm_{12} \cdot G \cdot P \cdot f}{12 \cdot [T_{12} - t_1]} \tag{1}$$

where P is the annual precipitation in mm, tm_{12} is the mean temperature in the warmest month, t_1 is the mean temperature in the coldest month, T_{12} is the mean of the maximum temperatures of the warmest month, G is the duration of the vegetative period (in months) and f is a sunstroke factor.

We used the SIGPAC land uses for two different time periods 2003 and 2009 (LU03 and LU09). Since they were already at the parcel level we did not need any additional process to add these two variables to the DM analysis. We also used information indicating if a parcel is irrigated or not.

Finally, we wanted to use any socio-economic variable and we selected the population density in 2009 (DENS_POP). This kind of sources is normally aggregated at the municipality level but it could be still explanatory. In these cases we apply for each parcel the value of its overlapping municipality.

The abandonment process has been studied using photo-interpretation as described before, and we have conducted some field work to validate a 10% of the total parcels in order to be sure that the abandonment is real.

3 Feature selection using DM techniques

Usually, statistical techniques provide useful tools to manage huge databases with millions of records. This is a length problem. But sometimes, as the case study proposed in this paper, the dimension problem is not in length but in width. Thus, considering 10 variables each of them acquiring only 5 possible values could imply a maximum of 100,000 rules (10^5 combinations). Although there are several decision tree algorithms with contrasted accuracy [16] that implement heuristics and pruning techniques, as ID3 for nominal attributes or C4.5 for nominal or numerical ones, sometimes a Feature Selection procedure [17] becomes necessary.

In this paper, the *aci* measure, provided by the RBS algorithm [18], is used to carry out the Feature Selection procedure. The RBS (Reduction Based on Significance) is an algorithm that calculates the critical significance and confidence boundaries to classify every rule in their specific region of significance: *A1* contains significant and discriminating rules; *A2* ontains supportable and reliable rules and *A3* contains rules with insufficient support to be classified as significant or not. All the rules outside those regions are considered as non-significant ones.

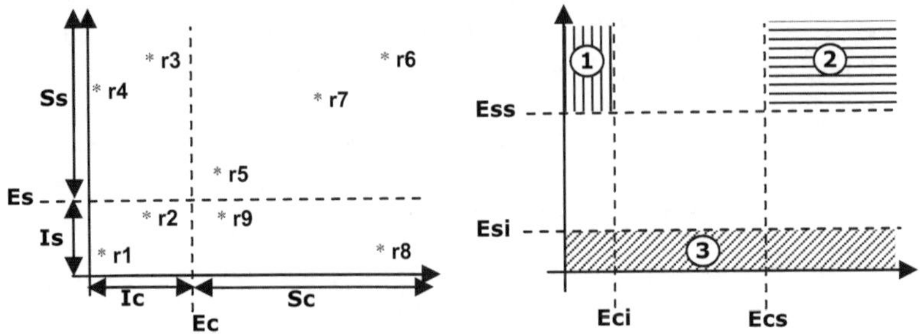

Figure 2: Example of rule distribution and significance region division obtained with RBS.

Besides, this algorithm calculates an empirical measure of variables correlation: *aci* (Attribute Correlation Index), the incidence of the attributes over the target variable.

The *aci* value associated to this Rule Set is defined as follows:

$$RS_{A\rightarrow C}.aci = \frac{RS_{A\rightarrow C}.srp}{A1 + A2 + A3} \qquad (2)$$

where $RS_{A\text{->}C}\ .aci$ is a set of rules typed as: antecedent (*A*) implying consequent (*C*). While antecedent is a nominal attribute or group of them, the consequent is a unique attribute, frequently called class attribute or target variable. For example $RS_{A\text{->}C}$ can be instanced to *A*(area, crop_type) -> *C*(abandonment), and a rule in this rule set could be (area=very low, crop_type=arable lands)->(abandonment=yes) with confidence=93.5%. $RS_{A\text{->}C}\ .srp$ is the proportion of significant rules in this set, and *Ai* the area of every significant region (*i*=1, 2, 3). High values of *aci* mean *A* is highly correlated with *C*, and vice versa.

Using this algorithm we can select the variables that are related to abandonment. However, this algorithm requires nominal variables so before applying the algorithm we performed a discretization process.

The landuse fields LU03 and LU09 were already discretized having 20 and 28 values respectively. Nevertheless, some parcels did not have any value so we redefined them as "NO_DEF".

Population Density (PD) has been classified in three groups (Low-Medium-High) to separate the crowded municipalities on the coast from the interior agricultural towns and finally the mountainous and depopulated towns.

In the most of the parcels the irrigation values (irrigated percent area of the parcel) are 0 or 100, but there are a few plots where the values remain between these two values. In those cases using the 50% limit, we have decided to use only two ranges (YES/NO) to classify the irrigation field (IRRIG).

Table 3: Discretization applied to get the nominal values.

LU03 (20 values) and LU09 (28 values)	MEANNDVI (5 values): (-inf, -0.3] => VL (-0.3 , -0.1] => L (-0.1 , 0.1) => M [0.1 , 0.3) => H [0.3 , +inf) => VH	DEVSTNDVI (3 values): (0.00 , 0.15] => L (0.15 , 0.35) => M [0.35 , 0.50) => H
FDO (4 values): (-200 , 0] => L (0 , 200) => M [200 , 400) => H [400 , +inf) => VH	SH_INDEX (4 values) (0 , 1] => L (1 , 2) => M [2 , 5) => H [5 , +inf) => VH	AREA (4 values): (0 , 2000] => VL (2000 , 10000] => L (10000 , 50000) => M [50000 , 100000) => H [100000 , +Inf) => VH
MEAN_ELEV (4 values): (0 , 200] => L (200 , 500] => M (500 , 800) => H [800 , +Inf) => VH	MEAN_FPI (5 values): (0 , 3] => VL (3 , 4] => L (4 , 5) => M [5 , 6) => H [6 , +7) => VH	MEAN_TWI (5 values): (0 , 5] => VL (5 , 10] => L (10 , 15) => M [15 , 20) => H [20 , +) => VH
IRRIG >=50 => Yes <50 => No	DP <100 => L [100 , 400] => M > 400 = H	

The other numerical values have been discretized in 3, 4 or 5 values based on the following criteria (with these criteria, the discretization is as explained in Table 3):

1. If the distribution was uniform in all its values range the variable has been divided into 5 values VL, L, M, H, VH (Very Low, Low, Medium, High, Very High) symmetrical, except the attribute "DEVSTNDVI" which has been divided into three values (L, M, H) because it has a very narrow range (between 0 and 0.5)

2. When the average of the values of a particular attribute lies far to one end, particularly to the minimum value, we have decided to discretize it in 4 ranges (L, M, H, VH), so that the labels L, M and H are for values around the mean and VH label is reserved for the more distant values above the average.

After this reclassification process we ran our own program and after 21 minutes of process time we obtained 8191 rules/group of rules ordered by the *aci* values indicating the variables or combinations that explain better the abandonment process.

Analysing all the rules with a higher *aci* value, we find that a key field in all the examples is LU03, and the second most recurring is M_FPI so we can say that both are particularly important in determining the abandonment of an agricultural parcel. The LU09 field seems that is somewhat less decisive than

LU03 as it appears in other sets of rules with lower *aci*, but in both variables we can see that the abandonment is highly associated with the "arable lands" category (those lands with a recent agricultural past but with no vegetation now). It is obvious that this definition of arable land matches in many cases with our abandonment definition, but this can not help to estimate marginality; this is why LU needs always to be combined with other variables.

The abandonment is related to a low or very low area value (less than one Hectare). The mean elevation seems also to be important. Normally, a low value in elevation is highly associated with the abandoned parcels, but this is not a direct relationship, in this study area the zones with low elevation (0-200 m) are those were we can find a higher urbanisation process. This means that these preliminary results are going to change once we include more variables in the analysis.

4 Conclusions

In this paper we have shown that nowadays we are able to generate and use a huge geodatabase for a relatively small study area. The analysis and interpretation of such a huge database could take a long time and in many cases we only would be able to get the obvious conclusions. For these reasons we have tested and confirmed that the *aci* metric/algorithm can help us to explore a very particular problem as the agricultural land abandonment.

We have conducted a test with a small number of variables and many of them were already well known and related to abandonment by the bibliography. However, we have confirmed those relationships for our particular study area. Variables obtained from the existing landuse databases (SIGPAC), the parcel area, the NDVI or the mean elevations are related to abandonment in this area. On the other hand, the climatic index (Patterson; MEAN_FPI) has not been used in previous studies, but we found that it is (in second position) one of the most important explanatory variables.

Performing the same analysis for over a hundred of new variables we will be able to explore new relationships that can help us in the study of a complex problem such as abandonment.

5 Future research

The GIS feature extraction process can be performed using different software but due to the amount of GIS layers we need to process, it is necessary to automate the whole process once the total variables are determined. In our next research we will automate the feature extraction from GIS layers.

The proposed software implementing the RBS algorithm is nowadays a prototype. Therefore in this paper a simplified pilot study is approached, considering a dataset with only 10 attributes.

For future research, a complete computational experiment (with over a hundred variables) will be developed and analyzed in detail. As the proposed algorithm needs nominal variables as input data, the discretization procedure accuracy is a critical constraint. Therefore, dynamic clustering techniques will be developed in order to perform an accurate and efficient discretization task.

References

[1] Sineiro, F., López, E., Lorenzana, R. & Valdés, B., La tipología de las explotaciones en función de su viabilidad económica y demográfica; aplicación a las explotaciones de bovino de Galicia. *Economía Agraria y Re- cursos Naturales*, **4(8)**, pp. 63-85, 2004.

[2] Moravec, J & Zemeckis, R., Cross Compliance and Land Abandonment, Deliverable D17 of the CC Network Project, SSPE-CT-2005-022727, 2007.

[3] Corbelle, E. & Crecente, R., El abandono de tierras: concepto teórico y consecuencias. *Revista Galega de Economía.* **17(2)**, 2008.

[4] Food and Agriculture Organization (FAO). The role of agriculture and rural development in revitalizing abandoned/depopulated areas. 34th Session of the European Commission on Agriculture. Riga. 2006.

[5] Rey, J., Martins, A., Nicolau, J.M. & Schulz, J.J., Abandonment of agricultural land: an overview of drivers and consequences. *CAB Reviews: Perspectives in Agriculture, Veterinary Science, Nutrition and Natural Resources,* **2 (57)**, pp. 14, 2010.

[6] Bielsa, I., Pons, X. & Bunce, B., Agricultural Abandonment in the North Eastern Iberian Peninsula: The Use of Basic Landscape Metrics to Support Planning. *Journal of Environmental Planning and Management.* **48(1)**, pp. 85-102, 2005.

[7] Lloret, F., Calvo, E., Pons, X. & Díaz-Delgado, R. Wildfires and landscape patterns in the Eastern Iberian Peninsula. *Landscape Ecology*, **17**, pp. 745 – 759, 2002.

[8] Koulouri, M. & Giourga, C., Land Abandonment And Slope Gradient As Key Factors Of Soil Erosion In Mediterranean Terraced Lands. *Catena*, **69(3)**, pp. 274-281, 2007.

[9] Otero, I., Boada, M., Badia, A., Pla, E., Vayreda, J., Sabaté, S., Gracia, C.A. & Peñuelas, J. Loss of water availability and streambiodiversity under land abandonment and climate change in a Mediterranean catchment (Olzinelles, NE Spain). *Land Use Policy* **In Press**, 2010.

[10] Peña, J., Bonet, A., Bellot, J., Sánchez, J.R., Eisenhuth, D., Hallet, S. & Aledo, A., Driving forces of land-use change in a cultural landscape of Spain. A preliminary assessment of the human-mediated infl uences (Chapter 6). Modelling Land-Use Change: Progress and Applications, eds E. Koomen, J. Stillwell, A. Bakema, H.J. Scholten. Springer: Netherlands, pp. 398, 2007.

[11] Mirón, J., Cadastre and the reform of European Union's common agricultural policy. Implementation of the SIGPAC (1). *Catastro*, 54, pp. 161-172, 2005.

[12] Ruíz, L.A., Recio, J.A., Fernández-Sarría, A. & Hermosilla, T., A Tool For Object Descriptive Feature Extraction: Application To Image Classification And Map Updating. The International Archives of the Photogrammetry,

Remote Sensing and Spatial Information Sciences, Vol. XXXVIII-4/C7, 2010.

[13] Sørensen, R., Zinko, U., & Seibert, J., On the calculation of the topographic wetness index: evaluation of different methods based on field observations. *Hydrol. Earth Syst. Sci. Discuss.*, **2**. pp. 1807-1834, 2005.

[14] Ninyerola, M., Pons, X. & Roure, J.M. A methodological approach of climatological modelling of air temperature and precipitation through GIS techniques. *International Journal of Climatology,* 20. pp. 1823-1841. 2000.

[15] Benavides, R., Roig, S. & Osoro, K., Potential productivity of forested areas based on a biophysical model. A case study of a mountainous region in northern Spain. *Ann. For. Sci.* **66 (1)** 108, 2009.

[16] Zhang, H. & Su, J. Learning Probabilistic Decision Trees For AUC. *Pattern Recognition Letters,* 27, pp. 892–899, 2006.

[17] Chen, Y-L., Hsu, W-H. & Lee, Y-H., TASC: Two-Attribute-Set Clustering Through Decision Tree Construction. *European Journal of Operational Research,* 174, pp. 930–944, 2006.

[18] Abadía, R., Almiñana, M., Escudero, L.F., Pérez, A., Rabasa, A. & Santamaría, L. A, Rules Reduction Algorithm Based On Significance Measure. Proc. IX Data Mining and Information Engineering, eds. Zanasi, Alzamora, Ebecken & Brebbia, WIT Press. Cádiz, pp. 63-72, 2008.

Section 6
Energy and the environment

Analysis of the determining parameters of energy efficiency on residential buildings in the Mediterranean climate

A. Pastor[1], F. García-Alonso[2], J. A. Reyes[2] & Y. Villacampa[2]
[1]*Technical I+D Projects of AIDICO,*
Technological Institute of Construction, Spain
[2]*Dep. de Matemática Aplicada, Escuela Politécnica Superior,*
Univ. Alicante, Spain

Abstract

In order to maintain a reasonable standard of living and comfort, society is dependent on using very high levels of energy. The challenge is to pursue sustainable development while maintaining the level of activity, transformation and progress in every productive sector. In this paper, the influence of every parameter that determinates the energy requirements of heating and cooling in residential buildings will be studied, evaluating the thermal performance of building solutions, the treatment of gaps, control and reduction of solar radiation devices, thermal bridges and ventilation, etc. In addition, there will be an analysis of the factors that optimize the performance of heating, ventilating, air conditioning and domestic hot water systems and integration of renewables in every one of them. Finally, an economic feasibility analysis of building improvements will be performed at both construction and design of facilities.
Keywords: building, energy requirements, energy consumption, CO_2 emissions, cooling, heating, hot water system.

1 Introduction

Concern about the environment and climatic change is forcing governments of the world to research capable solutions and to correct it. The consequence of this issue is the Protocol of Kyoto, in which countries of EU have engaged to obey.

WIT Transactions on Ecology and the Environment, Vol 144, © 2011 WIT Press
www.witpress.com, ISSN 1743-3541 (on-line)
doi:10.2495/ECO110331

One recommended measure to achieve the aims proposed is the encouragement of saving and energy efficiency in all sectors. Energy consumption, as much in Spain as in the EU, is mainly satisfied by fossils fuels, emitting compounds to the atmosphere that cause a big environmental impact and, although the energetic production of renewable origin is gravely worthwhile, its percentages of contribution on a global level are not representative because of its low amount.

The consequence of energy consumption, especially those of fossil origin, is one of the main agents responsible for the destruction of the environment by means of the broadcast of greenhouse effect gases. Through the gases which are contemplated in the Protocol of Kyoto, CO_2 represents three quarter parts of the total and more than 90% of this has an energy origin [5].

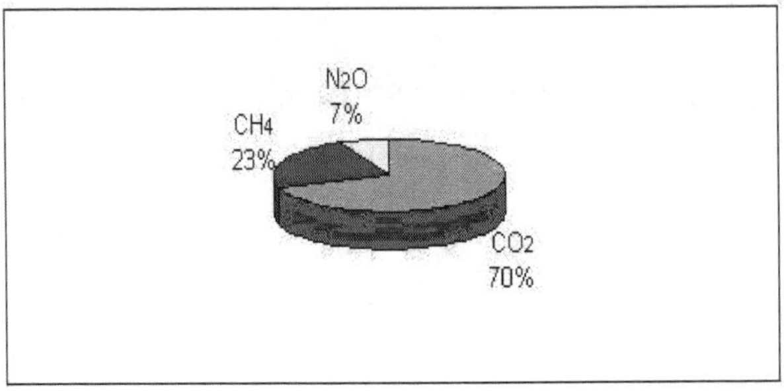

Figure 1: Percentage of polluting gas emissions.

Between the final consumers of energy, the residential sector in Spain occupies third place in importance. But, unlike what happens in the main part of Europe, this sector is found a little bit far from the consumption of industry and transport; it represents 15% of the final energy consumption [1].

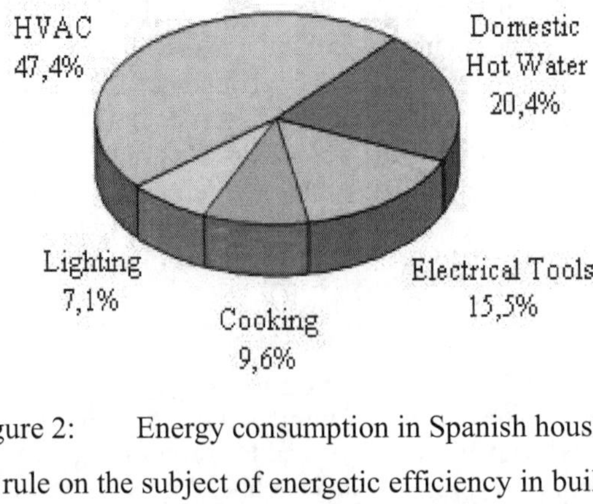

Figure 2: Energy consumption in Spanish housing.

The Spanish rule on the subject of energetic efficiency in buildings, following the European directives, requires the verification of some minimum requirements such as the calculation of energy consumption or indicators associated to this consumption [4]. Both operations require calculation tools to find the energy demand of the building, consumption of its thermal installations and the environmental impact produced by the amount of CO_2 emissions.

The relation between the energetic needs of the buildings, the associated consumption of energy that is supposed, and the CO_2 broadcast produced care related in fig. 3:

$$\frac{Energy\ Demand\ of\ Building}{\eta_{Thermal\ Systems}} = Energy\ Consumption$$

$$\Downarrow\ Energy\ Type$$

$$CO_2\ EMISSIONS$$

Figure 3: Relationship between energy parameters.

This study will determine which parameters influence the energy efficiency as the method of calculation recognized in the Spanish legislation. A detailed analysis will be carried out of a typical residential building from the Mediterranean area. Starting from these data, some modifications are proposed in order to improve its contributions and reduce CO_2 emissions in this sector.

2 Materials and methods

Figure 4 shows the procedure followed in this study, and in concentrating as much on the constructive levels as well as functional ones, the energy demand is determined. The official simulation tool is employed LIDER [2].

To evaluate the annual energy consumption of heating, refrigeration and sanitary hot water, the simulation of the thermal installations will be carried out. This procedure is done by a simulation application recognized by the Spanish legislation: Calener VyP [3]. This software allows the environment impact value to be calculated according to Fig. 3, showing the CO_2 gas emitted to the atmosphere. Once obtained, the influence of different properties of the building and its installations will be studied.

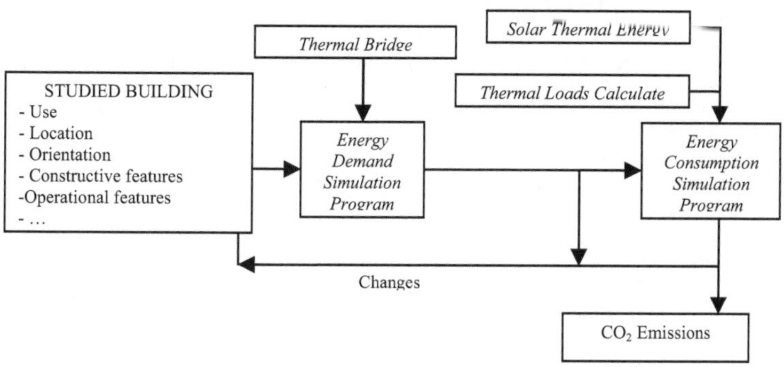

Figure 4: Project flowchart.

2.1 Study of main parameters of the energy demand

The energy demand of the building depends on its constructive and operational characteristics. Several simulations have been carried out in order to analyze the following:

- Orientations of facade.
- Thermal inertia of the exterior closings.
- U-value of the closings. Influence of thermal insulation. Conductivity and thickness.
- Treatment of gaps: Permeability of the carpentries, percentage of gaps occupied by the frame, U-value of frames and glasses, solar factor, dimensional appearances and execution…
- Treatment of the thermal bridges.
- Projection of shadows by outside elements to the building under consideration.
- Air of ventilation.

2.2 Study of the main parameters of the performance of the thermal installations

The properties analyzed at this point do not influence the energy demand. However, the energy consumption of the HVAC and production of sanitary hot water systems are directly affected.

- Type of systems.
- Power, performance and type of the equipment of thermal energy production.
- Type of energy employed by the installations.
- Solar contribution in the production of sanitary hot water.

3 Results and discussion

In view of the results obtained in successive simulations, it has been determined that some parameters have greater importance in the energetic behaviour of the building and in the performance of its installations. It will represent the values of the energetic indicators obtained in front of variations in different constructive properties and the working of the systems.

3.1 Evolution of the energetic indicators by modification of influential parameters on the energy demand

After studying each property separately and its relative importance in the demands of heating and cooling, we can proceed to see the energetic evolution of the building under consideration. It maintains the thermal systems at a constant with the aim of reducing its energy needs.

Table 1: Parameter changes in energy demand.

Change Code	Initial Building	Changed Building
1	Initial Conditions	
2	Mineral Wool in Facade (k = 0,041 W/mK)	Mineral Wool in Facade (k = 0,031 W/mK)
3	Exp. Polystyrene on Roof (k = 0,029 W/mK)	PUR on Roof (k = 0,025 W/mK)
4	5cm of Mineral Wool in Facade	10cm of Mineral Wool in Facade
5	4cm of Exp. Polystyrene on Roof	8cm of PUR on Roof
6	Double Glazing (U = 3,3 W/m^{2}K)	Low Emissive Double Glazing (U = 1,4 W/m^{2}K)
7	Glass´ Solar Factor = 0,75	Glass´ Solar Factor = 0,6
8	Metal Window Frame (U = 4 W/m^{2}K)	PVC Window Frame (U = 1,8 W/m^{2}K)
9	Permeability Carpentries 50 (m^3/hm^2)	Permeability Carpentries 3 (m^3/hm^2)
10	Solar Protection in Summer: 50%	Solar Protection in Summer:70%
11	Ventilation / Hour = 0,8	Ventilation / Hour = 0,76
12	Usual Thermal Bridges	Calculated Thermal Bridges

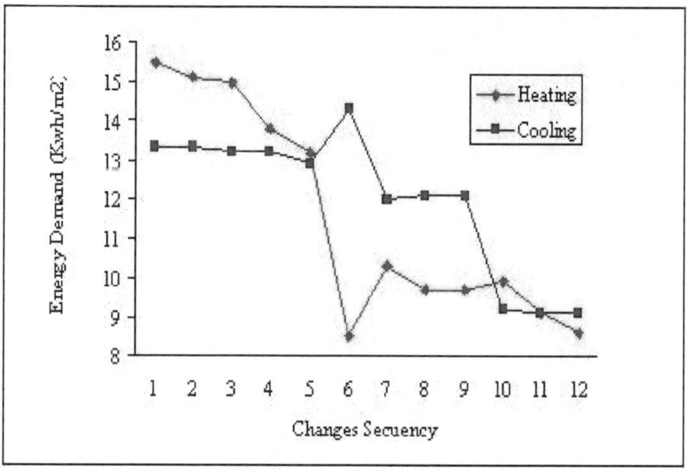

Sequence of Changes

Figure 5: Evolution of energy demands.

The demand of heating is mainly influenced by the heat transfer of the constructive elements and thermal properties. The demand of refrigeration is quite affected in front of parameters dependent on conduction and radiation heat transfer through the gaps.

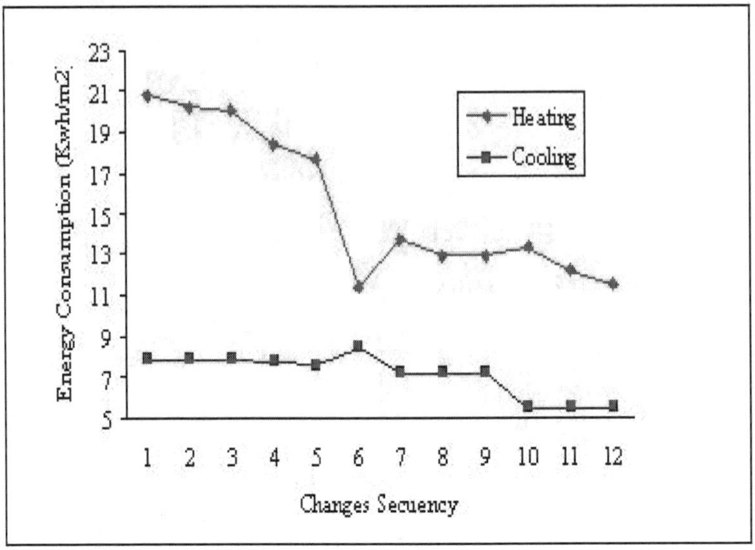

Sequence of Changes

Figure 6: Evolution of energy consumption.

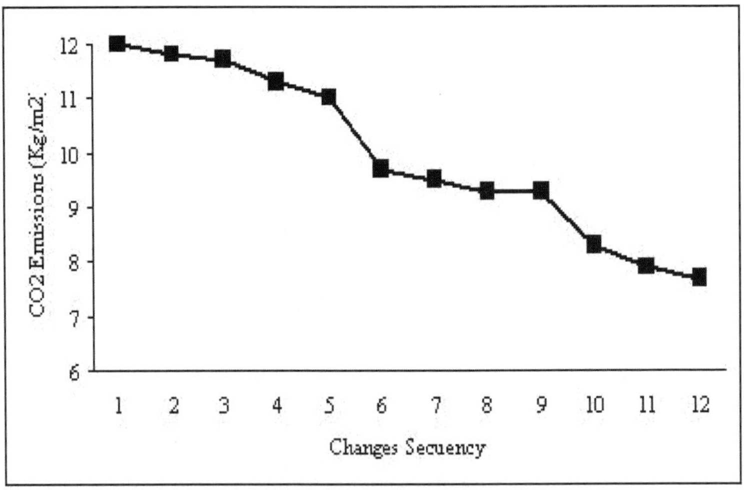

Sequence of Changes

Figure 7: Evolution of CO_2 emissions.

Improving the thermal properties of closings supposes to reduce the energy needs up to 44,5% for heating and 32% in cooling.

The energy consumption of heating reduces up to 45% of the initial value. The electrical consumption of cooling systems decreases by 30%.

3.2 Evolution of the energetic indicators by modification of influential parameters on the thermal installations

The energy consumption of different installations is evaluated by means of simulating their elements. For this, the previous variations are maintained which affect the energy demand.

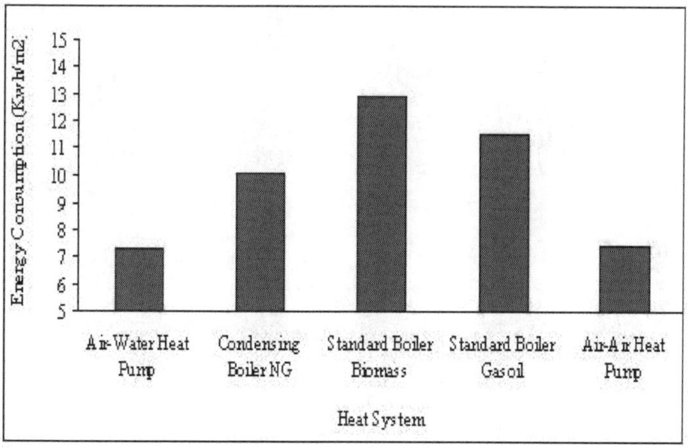

Figure 8: Energy consumption according to heating system.

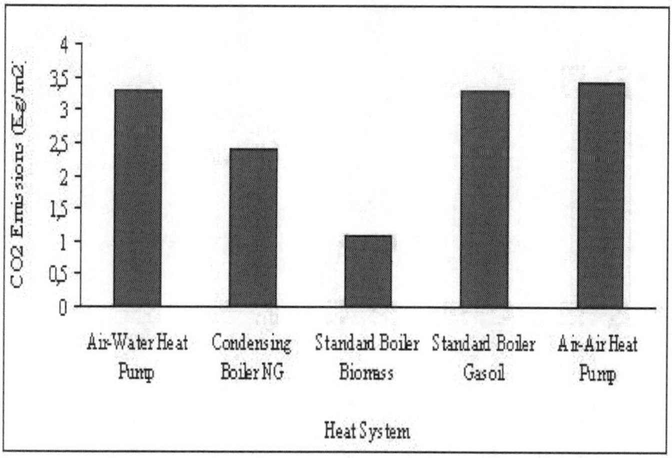

Figure 9: CO_2 emissions according to heating system.

Figures 8 and 9 show how the energy consumption depends on equipment of thermal energy production, such as the principle of working and its performance. However, the broadcasts are more influenced by the type of energy that they employ. For instance, the consumption of electricity, in spite of being minor is more polluted than other energetic sources.

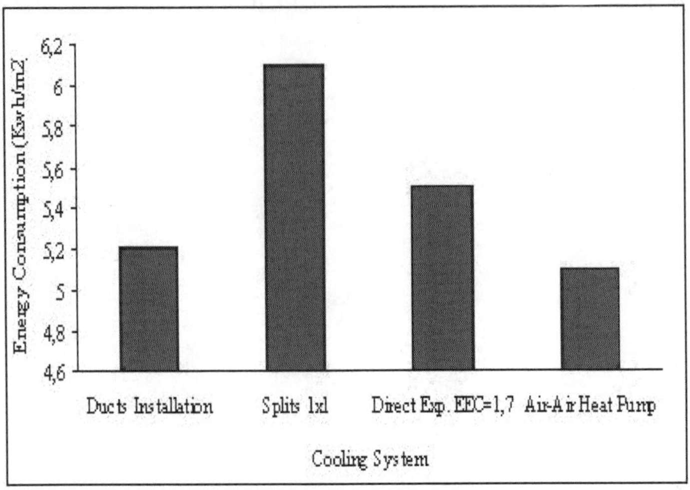

Figure 10: Energy consumption according to cooling system.

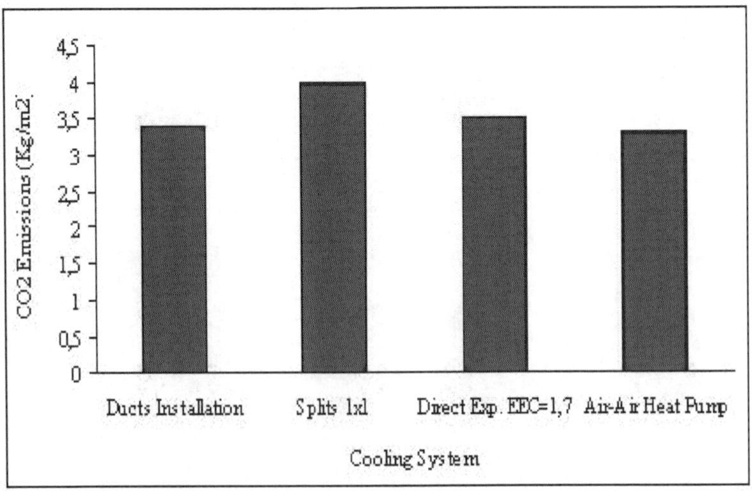

Figure 11: CO_2 emissions according to cooling system.

The cooling systems under consideration are all fed by electrical energy. Therefore, broadcasts and energy consumption will depend on the correct election of the equipment (power, performance, real needs of the buildings…). This fact involves a good engineering project.

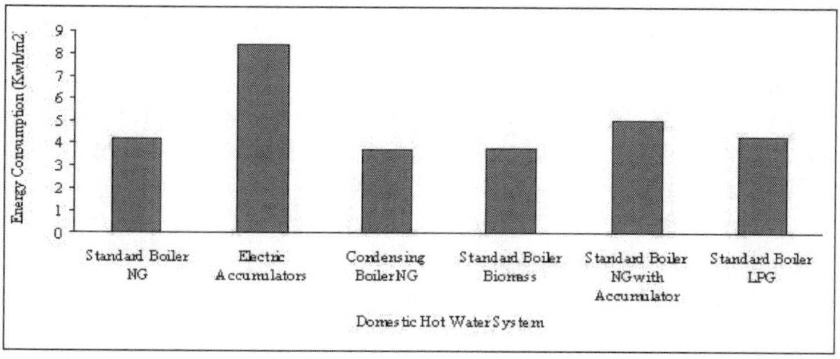

Figure 12: Energy consumption according to domestic hot water system.

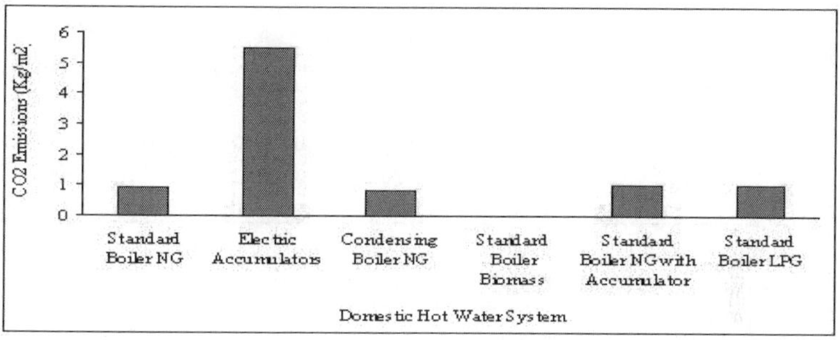

Figure 13: CO_2 emissions according to hot water system.

Figures 12 and 13 show producing hot water using electrical energy is highly inefficient and polluting. On the contrary, the employment of renewable energies, such as biomass, is the best solution from the point of view of the environment.

3.3 Studies of economic feasibility of proposals of improvement the energy efficiency on buildings

Table 2 represents the savings accumulated of a house of $90m^2$ if it installs air conditioning systems and sanitary hot water adapted to the real needs.

Table 2: Cumulative economic saves.

Type of System	Cumulative Cost (€)		
	Years		
	5	10	15
Initial Systems	1585,1	3706,4	6545,2
Recommended Systems	950,4	2222,2	3924,3
Economic Cumulative Savings (€)	634,7	1484.2	2620,9

In the calculations, 6% of inflation rate in the price of energy has been considered in order to update the annual values.

From the constructive point of view an analysis of the time needed to amortize an over cost of initial investment because of improvements in the properties of the thermal insulation installed on exterior closings of the building purposed is carried out.

The aim consists of verifying the moment in which the increase of the initial inversion is compensated with the energetic saving associated to the installation of an isolation with minor conductivity and more thickness.

$$Extra\ \ Cost_{initial}$$

$$= [(\ AnnualPayment)_{Original\ Building} - (AnnualPayment)_{\mathrm{Im}\,proved\ Building}] \cdot \sum_{year=1}^{year=n} (1+f)^{(n-1)}$$

n = 1,2,…= year of calculation, f = 0,06 = Inflation rate of the price of energy.
The approximated values of the isolation cost installed in the Spanish buildings are used to make this analysis [7].

In Fig. 14 it is observed that the time of amortization will be produced starting from 13,5 years for the beginning of the working of the building.

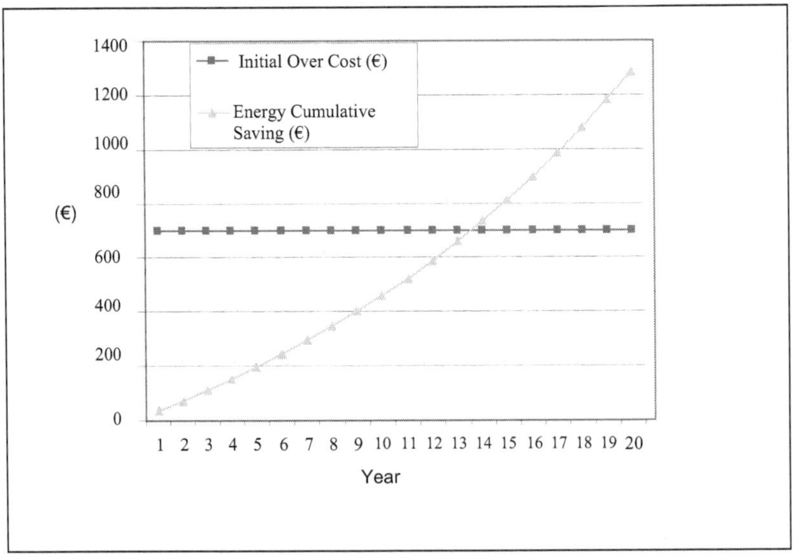

Figure 14: Payback time. Overcost of thermal isolation.

4 Conclusions

Energy efficiency on buildings is influenced from the phase of design until the execution of its closings and installations. Reducing the energy needs in the east of Spain mainly involves limiting conduction heat transfer on closings and solar radiation effects in the gaps.

Besides, the correct execution of the thermal bridges [6], the reduction of infiltrations, detailed calculation of the ventilation, thermal loads and solar protection and correct choice of equipment is necessary.

The type of energy to condition the building is essential to evaluate the environmental impact of the constructions. The improvements in their properties involve some energy saving. Therefore, increasing energy efficiency does not enter in conflict with the economic feasibility of the sustainable construction.

Acknowledgements

The author thanks F. García-Alonso, J. A. Reyes and Y. Villacampa for their technical advising and devotion in the preparation of this survey.

This work has been supported by AIDICO, Technological Institute of Construction.

References

[1] Ministry of Industry of Spain, *Libro Blanco de la Energía*, 2007.
[2] IDAE, *LIDER Tutorial*, 2006.
[3] IDAE, *Calener VyP Tutorial*, 2006.
[4] European Union, *Energy Performance of Buildings Directive*, 2002.
[5] McKenzie, T., *Temperature profoundly affects coupling of photosynthetic electron transport and* CO_2 *uptake in* Lobaria pulmonaria: *a case for measurement at field–ambient temperatures*. The Lichenologist (2001), 33: 453-455 Cambridge University Press: 10.1006/lich.2001.0339.
[6] Instituto Eduardo Torroja de la Construcción, *UNE EN ISO 10 211 (partes 1 y 2). Puentes Térmicos en Edificación. Flujos de Calor y Temperaturas Superficiales: Métodos Generales de Cálculo y Puentes Térmicos Lineales.* 1998.
[7] ANDIMAT, *Costes de Aislamiento en las Edificaciones Españolas*, 2007

Energy efficiency in South Africa: policy perspectives and the path to low carbon growth

G. Nhamo & A. Bimha
University of South Africa, South Africa

Abstract

Energy efficiency is becoming one of the key paths to transforming the South African economy into low carbon growth. This vision has been accepted by both the government and business community. To this end, the White Paper on Energy, National Energy Efficiency Strategy of 2005 (revised 2008), Energy Efficiency Accord and the National Climate Change Response Green Paper are among key policy documents. The Energy Efficiency Strategy spelt out a target of a national final energy demand reduction of 12% by 2015. Industry and Government agreed that the Energy Efficiency Strategy should be reviewed in three-year cycles. This paper reviews and assesses the energy efficiency policy landscape in South Africa and how this is assisting the nation to transform into a globally recognised low carbon economy. We conclude that South Africa is progressing well in terms of putting in place new as well as refining old pieces of legislation on energy efficiency leading to a low carbon growth path.
Keywords: energy efficiency, South Africa, low carbon growth, climate change.

1 Introduction

The International Chamber of Commerce (ICC) realizes that energy efficiency is a critical component if the world is to move sustainably to a low carbon, sustainable energy economy ICC [1]. According to the ICC, a number of benefits are associated with energy efficiency and these include: reduced production, distribution and marketing associated costs; reduced emissions and other environmental impacts; extended availability of large but non-renewable resources; improved competitiveness and overall productivity; and making energy more affordable to consumers, especially in developing countries.

WIT Transactions on Ecology and the Environment, Vol 144,
www.witpress.com, ISSN 1743-3541 (on-line)
doi:10.2495/ECO110341

Energy efficiency is very high on both the South African government and private sector policy agendas (Figure 1). The national government takes the lead in energy efficiency and is currently using its buildings as an example. Among key government policies and industry initiatives reviewed making reference to energy efficiency were: the National Climate Response Green Paper (NCCGP), Draft Carbon Tax Option (DCTO), Industrial Policy Action Plan II (IPAP II), Draft Integrated Resources Plan (DIRP), South African Renewables Initiative (SARI), Western Cape Government Green Procurement Policy (WCGGPP), National Energy Act (NEA), Long Term Mitigation Scenario (LTMS), revised National Energy Efficiency Strategy (NEES), Energy Efficiency Accord (EEA), White Paper on Energy Policy (WPEP), White Paper on Renewable energy (WPRE) and the Carbon Disclosure Project South Africa (CDP SA). From Figure 1, the revised National Energy Efficiency Strategy (NEES) of 2009 stands out as the key policy document in terms of energy efficiency.

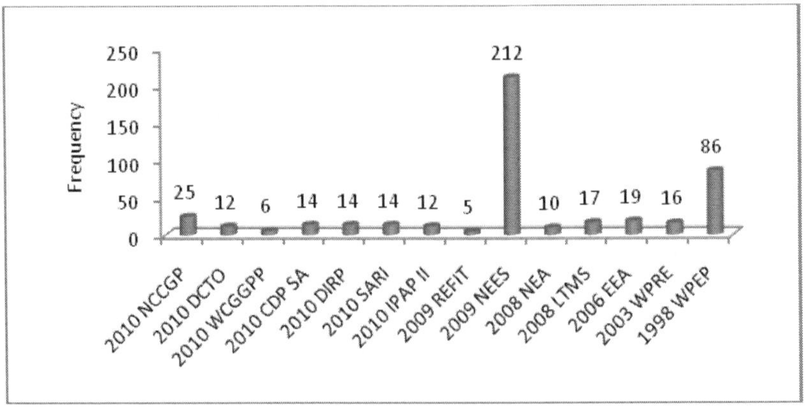

Figure 1: Energy efficiency mentioned in key South African policies.

2 Energy efficiency: an international perspective

In order to promote energy efficiency within communities, the ICC [1] identifies a number of policy directions. These include: advanced technology development, combined heat/cooling and power production, technology transfer and cooperation, fiscal instruments, energy services, building codes and standards, voluntary commitments and partnerships and consumer information. The World Economic Forum (WEF) maintains that energy efficiency should be the cornerstone to meeting the climate change goals WEF [2]. Energy efficiency will also enable humanity to meet the growing energy demand. If per capita energy demand continues unabated, then “energy demand is expected to increase by 40% by 2050, carbon emissions to increase to 34.5 GT CO_2 by 2020 and 40.2 GT CO_2 by 2030” and capital required to meet projected energy demand to 2030 will cumulatively be US$ 26 trillion (in 2008 dollars). Various policies and schemes have been put in place addressing energy efficiency. For example, the

EU aims to reduce energy use by 20% through cost-effective demand and supply measures ICC [1]. The G8 identified energy efficiency as a major element requiring action leading to reduced emissions whilst improving competitiveness, health and employment. As of 2005, 37 countries had ratified energy-efficiency labeling systems for appliances and electronic equipment (UNEP *et al.* [3]. UNEP *et al.* [3]) highlights how energy efficiency has a huge potential for creating jobs. Energy efficiency has been recognized as part of the growing green global markets. UNEP *et al.* [3] noted that the global market volume for environmentally sound technologies (products and services) was estimated at about $1,370 billion (€1,000 billion) in 2008 with a projected $2,740 billion (€2,200 billion) by 2020. Energy efficiency technologies that included appliances, industrial processes, electrical motors, insulation, etc had a share of $617 billion (€450 billion) in 2008 and this was estimated to rise to $1,233 billion (€900 billion) by 2020.

Energy efficiency has been taken seriously in China. In fact, China has been ranked the world's number one in terms of the Ernst and Young 'All Renewables Index' for November 2010 (Ernst and Young [4]). The Chinese government demands that the efficiency of pumps and fans be improved from a typical 75–80 percent in 2000 to 80–87 percent, and coal fired industrial boilers from 65 percent to 70–80 percent (both by 2010). It has also mandated a reduction, between 2000 and 2020, of energy needed per ton of steel produced from 906 kgs of coal equivalent (kgce) to 700; for aluminum from 9.9 tons of coal equivalent to 9.2 tons; and for cement from 181 kgce to 129 kgce. These measures are ambitious, but also very difficult to implement (Martinot and Junfeng [5]). In the UK there is a requirement to comply with stringent energy-efficiency agreements National Treasury [6]. The agreements are negotiated with sector associations and the results of the agreements are reviewed regularly with continued discounts relying on targets being achieved.

A Carbon Disclosure Project (CDP) 2010 report on Corporate Clean Energy Investment Trends in Brazil, China, India and South Africa (BASIC) looked at energy efficiency. In Brazil the strong regulatory requirements for the energy sector are having an effect on corporate investment and companies still wish to see good performance in energy efficiency recognised through a corporate energy efficiency index or individual energy efficiency contracts with government CDP [7]. In China high-level policy signals have been effective in stimulating investment in energy efficiency. Although there is still an appetite for more specific regulation applicable to a larger number of companies, green finance measures have been effective in promoting energy efficiency investments. Like in Brazil companies would like to see measures that provide incentives or recognition for good energy efficiency performance CDP [7].

3 South Africa's policy on energy efficiency

A number of energy efficiency policies and regulatory regimes are already in place in South Africa. These have been designed to address, especially the negative impacts of climate change. In addition, energy efficiency is seen as a tool for managing demand in a country that was plugged with blackouts in 2008. Among the noticeable policies and legislative initiatives addressing energy efficiency is the: White Paper on Energy, National Energy Act, revised National Energy Efficiency Strategy, Energy Efficiency Strategy, Industrial Policy Action Plan II, Integrated Resources Plan, National Climate Change Green Paper, Carbon Tax Option, Taxation Laws Amendment Act and the National Greenhouse Gas (GHG) emissions pledge of 2009. These and other initiatives will now be considered in the next sections.

3.1 Energy White Paper for South Africa – 1998

The White Paper DME [8] identified the use of standards and appliance labelling as the entry level interventions when implementing energy efficiency. At the time of finalising the White Paper, there were no energy efficiency standards, norms or regulations. Government had to promote energy efficiency awareness in households. It was estimated that energy efficiency could save between 10% and 20% of consumption at that time. Energy efficiency norms and standards for commercial buildings and industrial equipment as well as voluntary guidelines for the thermal performance of housing were to be established in the near future. Energy efficiency and conservation were twin pillars to an overall energy policy. Energy efficiency had to be an integral element to the Integrated Resource Planning (IRP) framework. The IRP considers both the supply and demand side options for meeting energy service requirements. The White Paper also established that South African energy consumers in industry and commerce were generally unaware of a need for and the potential of energy efficiency improvements and savings. The lack of energy efficiency expertise in industry was also identified as one of the major barriers. Government pledged to investigate possibilities of establishing appropriate institutions and build capacity for implementing energy efficiency strategies. Appropriateness of using environmental levies on energy sales would be investigated.

3.2 National Energy Act – 2008

The Act establishes the South African National Energy Development Institute whose functions in respect of energy efficiency include the need to: (i) undertake energy efficiency measures as directed by the Minister; (ii) increase energy efficiency throughout the economy; (iii) increase the gross domestic product per unit of energy consumed; and (iv) optimise the utilisation of finite energy resources RSA [9]. The Act also gives the responsible Minister power to promulgate energy efficiency standards for specific technologies, processes, appliances, devices, motor vehicles and buildings.

3.3 National Energy Efficiency Strategy – 2005 (revised 2008)

The initial Energy Efficiency Strategy was finalised in 2005 DME [10]. The Strategy spelt out a target of a national final energy demand reduction of 12% by 2015 and this has remained the same in the revised Strategy DME [11]. This is expressed as a percentage reduction against the projected national energy use in 2015. The revised Strategy of 2008 points out to a number of key sectors requiring energy efficiency. The industrial (taking up 36.2% of consumption in 2004), residential (17.9%) and mining (7%) sectors are considered the heavy users of energy DME [11]. The transport sector (taking up 25.7% of total national energy consumption in 2004) uses up to three-quarters of the country's petroleum products. The "promotion of energy efficient vehicles and those with lower emissions, building a public transport infrastructure and a travel demand management system are some of the key features of the approach adopted" DME [10]. The government makes mention of an electrical appliance labelling initiative launched to rate energy consumption of all new 'white products'. The changing of South African residents' behaviour and raise awareness about the costs and benefits of energy efficiency is deemed the most difficult area although this has been ongoing since 2005. The vision of the Strategy is to "contribute towards affordable energy for all, and to minimise the negative effects of energy usage upon human health and the environment and contributing towards secure and affordable energy for all" DME [11]. The vision will be achieved by encouraging sustainable energy development and energy use through efficient practices in the country. The Strategy will be implemented through Sectoral Implementation Plans with systems put in place to permit periodical monitoring of progress against the spelt out target to be reviewed at the end of each phase. The eight goals of the strategy are organised under the three pillars of sustainable development namely: social, environmental and economic sustainability.

The Strategy makes provision for sector programmes in the following sectors: Industry and Mining, Commercial and Public Buildings, Residential and Transport. Cross-cutting issues are also identified including: Integrated Energy Planning, Renewable Energy, Environment and Health as well as The Cleaner Fuels Programme. The strategy also dedicates a section dealing with climate change and how the DoE has established the National Designated authority to deal with CDM projects. Six barriers to energy efficiency are identified in the Strategy to include: Energy Pricing, Lack of knowledge and understanding of Energy Efficiency, Institutional barriers and resistance to change, Lack of investment confidence and practice of 'bounded rationality' DME [11]. The DME [11] notes that a few countries have set comprehensive targets for energy efficiency and these countries include Slovenia, Japan, The Netherlands and New Zealand. The South African Strategy provides for the implementation of sector programmes in a three-phase approach, timed as follows: Phase 1 - March 2005 to February 2008 (completed); Phase 2 - March 2008 to February 2011; and Phase 3 - March 2011 to February 2015. A review of the national and sectoral targets will continue to be done after the end of each phase. As the industrial and mining sectors in the country are different, it was agreed that sub-sectoral targets should be implemented. Further details regarding the targets are spelt out in Table 1.

The Energy Act will be used to implement regulations on the management, measurement, monitoring and reporting of energy efficiency DME [11]. The

DoE is to establish a formal system that will facilitate continuous updating of registered "figures related to energy efficiency; in particular, indicators for efficiency measurement" DME [11]. Such a system will involve a multi-stakeholder engagement process. A number of supporting mechanisms and instruments will be used and these include: Efficiency Standards; Appliance Labelling; Certification and Accreditation; Education, Information and Awareness; Research and Technology; Regulation; Energy Audits and Energy Management Systems. With regards to finance instruments the following will be applied: Incentives, Fee Bates, Financing the Public Sector Implementation Plan, Energy Service Companies, CDM, Voluntary and retail carbon markets, Demand Side Management and Energy Pricing. Lastly, the Strategy identifies cross-cutting and closely linked issues that include integrated energy planning, renewable energy, environment and health and Cleaner Fuels Programme.

Table 1: Energy efficiency targets.

Sector (S)/Sub-sector (SS)	*Target to 2015*
Industry and Mining Sector (S)	Final Energy Demand Reduction of 15%
Iron and Steel Industry (SS)	Improvement in energy intensity of 1% per annum
Chemical and petrochemical industries (SS)	Improvement in energy intensity of 1% per annum
Mining (SS)	Final energy demand reduction of 10% by 2015 (using an adjustable baseline)
Paper and pulp and printing industries (SS)	Improvement in energy intensity of 2% per annum
Cement industry (SS)	Improvement in energy intensity of 2% per annum
Power Generation (S)	Target of 15% reduction in "parasitic" electrical usage
Commercial and Public Building (S)	Target Final Energy Demand Reduction of 20%
Residential (S)	Target Final Energy Demand Reduction of 10%
Transport (S)	Target Final Energy Demand Reduction of9%

Source: Authors, based on DME [11].

3.4 Energy Efficiency Accord – 2005

The Energy Efficiency Accord (Accord) was adopted for implementation in 2005 following the conclusion of the Energy Efficiency Strategy of the Republic of South Africa in 2005 DME [10]. In the Accord, industry and government committed to a number of outputs and outcomes as outlined in Table 2. Industry and Government agreed that the Energy Efficiency Strategy should be reviewed regularly in three-year cycles. The review would take the form of a collaborative process between the Parties. Furthermore, the Energy Efficiency accord compels signatory industries to develop a specific energy efficiency strategy within one year of the signing of the Accord. The National Business Initiative (NBI) would act as the liaison with Government whilst Business Unity South Africa was mandated to deal with legislative issues.

Table 2: Commitments by industry and government.

Industry's commitment	Government's commitment
• Promote the development of sector specific strategies and targets; • Promote the use of Demand Side Management contracts concluded with energy suppliers, where they exist, and to negotiate additional such contracts where appropriate, towards achievement of the above target; • develop common reporting requirements for energy usage from all energy sources, taking into account, where possible, existing internationally recognised protocols for reporting such as those developed by the Global Reporting Initiative; • define industry-specific projected energy use in the future, based on Business-unusual (BAU) growth expectations; • agree on 2000 as the baseline year against which performance will be measured; • establish methodologies that will allow the baseline quantification for energy use/intensity in various sub sectors, and to take into account the need to measure specific energy intensity (providing for the differing sub sectors) rather than absolute energy use in order to promote industrial growth whilst achieving energy efficiency and recognising the energy conservation measures already in use in some sub sectors; • establish methodologies to take into account increased production so that the pursuit of improved energy efficiency does not hamper industrial growth; • develop a generic energy auditing protocol that can be adapted for use by the individual sector and company signatories; • endeavour that, where appropriate, training material prepared during the course of this accord be accredited through the relevant SETA and that the sector skills plans of the various sectors include the development of the skills necessary to sustain the commitments made in this accord; and • Where appropriate, exploit opportunities presented by energy efficiency projects to develop CDM projects.	• engage with industry signatories leading up to the three yearly interval reviews that are envisaged in the Strategy; • develop strategies, including the provision of fiscal and other incentives with signatories to incentivise the achievement of agreed sectoral targets by industry on a sector or enterprise basis; • afford signatories the opportunity to provide input to the concept and structure of future energy related strategy, regulations and standards; • build a relationship of trust and cooperation with signatories; • encourage and enable information sharing and networking activities with the private sector and to seek synergy with the greenhouse gas initiatives undertaken by the Department of Environmental Affairs and Tourism; • identify and share Best Practices (where appropriate to the South African context) among all parties and consider them as the basis for the establishment of standards; • recognise industry signatories as contributors to the ongoing process to revise the National Energy Efficiency Strategy; and • Promote CDM projects as a vehicle to achieve improved energy efficiency.

Source: Authors, based on DME [12].

3.5 Industrial Action Plan II – 2010

The Industrial Policy Action Plans (IPAPs) are developed following the adoption of the National Industrial Policy Framework (NIPF) by cabinet in January 2007 DTI [13]. The NIPF sets out the government's broad approach to industrialisation and has as one of the founding objectives the need to strengthen "building energy-efficiency standards in response to the national electricity shortage" [13]. As part of the actions in the IPAP II of 2010/11 through to 2012/13, the Department of Trade and Industry (DTI) is set to develop and strengthen national standards to support the creation or resuscitation of specific industries. The South African Bureau of Standards (SABS) will develop national standards as major milestones for the measurement and verification of energy efficiency and energy efficient electric appliances. Other national standards will be developed for solar water heaters, wind energy turbines, automotive diesel fuel and bio-diesel metal free unleaded petrol, alternative fuel vehicles, co-generation and transport of dangerous goods and globally harmonised system for classification of chemicals.

The IPAP II DTI [13] looks critically at Industrial Energy Efficiency (IEE). This has been linked to the rising electricity prices and increasing commitments in relation to carbon emission reduction. The IEE is viewed as presenting an opportunity for the establishment of an industry in relation to machinery and services which improve energy efficiency in the industrial sector. A potential area for significant increases in energy efficiency is the adjustment and/or replacement of industrial motors. To follow up on the proposed actions, an industrial energy efficiency programme will be developed. The programme will include consideration of more attractive financing models and the scaling up of the capacity of the National Cleaner Production Centre (NCPC). The IEE programme will try to counteract higher energy prices, lower emissions and create new goods and services. The IEE Programme is targeted for full implementation during Q2-Q4 of the 2010/11 cycle with the DTI taking the lead supported by the NCPC, Department of Energy, Department of Environmental Affairs and the Industrial Development Corporation.

3.6 National Climate change Response Green Paper – 2010

The Green Paper dedicates a section on energy focusing on the impact that energy efficiency, renewable energy technologies and a nuclear rollout could have on the country's carbon footprint DEA [14]. The industrial sector is identified as having great opportunities for improving energy efficiency that could be realised through improved lighting efficiency, compressed air efficiency, motor efficiency, thermal efficiency, steam system efficiency and HVAC efficiency. The Green Paper makes reference to a need to ensure that the Integrated Resource Plan for Electricity Generation (IRP) DME [15] and its future iterations take account of the peak, plateau and decline trajectory described in the LTMS DEAT [16]. The application of market-based policy measures like carbon tax is highlighted. The use of such market-based policy measures could motivate industry in implementing energy efficiency measures and investments in the development of clean technologies. It is anticipated that in future government will set "ambitious and mandatory targets for energy efficiency" DEA [14]. In addition, ongoing energy efficiency should be scaled-up. A call is made to improve energy efficiency knowledge and understanding in the various sectors through awareness campaigns, demonstration programmes, audits and education. The government also plans to work with the mining industry to increase energy efficiency in production processes. The hospitality industry is another sector mentioned. The government also wishes to support the development and implementation of energy efficiency plans by municipalities. With regards to the utilisation of waste for energy efficiency, gypsum is singled out as an example of a waste stream that could be reused in the manufacture of ceilings for low income homes leading to increased energy efficiency DEA [14].

3.7 Reducing GHG Emission: The Carbon Tax Option – 2010

The discussion document advocates for the pricing of environmental goods and services that generate huge quantities of GHGs Martinot and Junfeng [5]. The pricing must be adjusted to reflect the full costs of production and consumption and carbon taxes are part of a mechanism that can achieve this by encouraging greater energy efficiency and the use of cleaner, low carbon technologies. Tax incentives and specifically targeted government programmes for energy efficiency measures now form part of the government's policy response to climate change. Since 2008, government has implemented a range of taxes and incentives to assist it in managing emission reductions. Among such taxes and incentives are the electricity levy, income tax exemption for revenues generated from the sale of certified emissions reduction units from the CDM projects, taxation of incandescent light bulbs, the CO_2 vehicle carbon emissions tax and the proposed energy efficiency savings tax incentive. The discussion document also notes effort being undertaken by the Department of Energy to encourage energy-efficiency measures.

3.8 Integrated Resources Plan – 2010–203

The Integrated Resources Plan (IRP) includes a number of critical assumptions, among them, a need for continued investment in demand side management (DSM) initiatives to improve energy efficiency DME [15]. The IRP notes that a decline in energy intensity provides an indication of the expected improvement in energy efficiency. A scenario is presented that assumes the economy growing as per the expected economic growth without any change in energy intensity, then the expected demand will "be significantly higher (at 770 TWh by 2034)" DME [15]. Energy efficiency could bring expected demand down to 496TWh, thus, a saving of 35%. As a separate concept to DSM, energy efficiency focuses on the use of electricity by consumers and the effect of the price increases would impact on energy efficiency.

3.9 Energy efficiency in other policies

The Long Term Mitigation Scenario DEAT [16] makes reference to energy efficiency as portrayed in the National energy Efficiency Strategy of 2005 that aims to achieve a final energy demand reduction of 12% by 2015 [10]. Although not mentioning energy efficiency directly, the South African government pledge towards emissions reduction targets in 2009 has a strong bearing with regards to energy efficiency. President Jacob Zuma announced a pledge to reduce emissions by undertaking mitigation actions which will result in a deviation below the current emissions baseline of 34% by 2020 and 42% by 2025. The pledge was conditional on a fair, ambitious and effective global climate change agreement and provision of financial, technology and capacity building support from developed countries The Presidency [17].

4 Lessons learnt and associated challenges

The space accorded this paper will not permit a full audit in terms of assessing the implementation of energy efficiency initiatives in South Africa. However, a summary assessment of certain critical elements enshrined in the Energy Efficiency Accord will be done as the Accord remains a key landmark. The growth of signatories of major energy consuming companies from 30 in 2005 to the current 51 (Figure 2) can be considered a success story to the Accord NBI [18]. The Accord's monitoring and verification instrument assisted stakeholders involved in the finalisation of SABS SATS 50010: 2010 dealing with the Measurement and Verification of Energy Savings. This has also resulted in greater recognition of the importance of monitoring and verification of energy savings in the country.

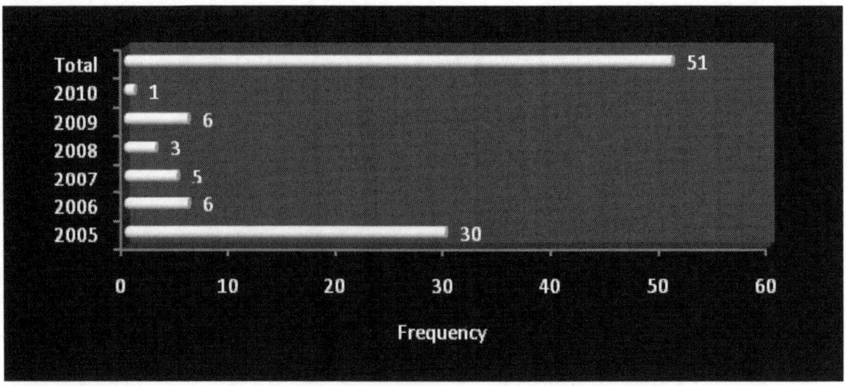

(Source: Authors.)

Figure 2: Trends in Signatories to the EEA (2005–8).

Through the Accord, both business and government agreed to utilise the CDM projects as a vehicle to achieve improved energy efficiency. Judging by the number of CDM project, energy efficiency takes up a significant portion. As of 20 December 2010, out of the 140 CDM projects submitted for consideration to the Designated National Authority, 22 were for energy efficiency, second to 26 for renewable energy DoE [19]. Although energy efficiency contributes significantly to the total number of CDM projects in the country, the overall figure of 140 projects is still low compared to those in competing countries of China (ranked first with 1100 CDM projects recorded during the same time, India with 582 and Brazil with 180 UNFCCC [20]. A 2008 report assessing progress with regards to the implementation of the Accord established that the Energy Efficiency Technical Committee (EETC) had been formed with the NBI administering the EETC and acting as the secretariat DME and NBI [21]. The EETC regularly meets to share best practices in terms of energy efficiency as outlined in the Accord. Nineteen out of the 44 signatories to the Accord in 2008 collectively consumed 24% of the national electricity consumption, a figure in excess of 56,560 GWh DME and NBI [21]. A more recent evaluation from respondents to a questionnaire indicated that the key success of the Accord has been in supporting the enhanced recognition of the importance of energy efficiency in the business sector leading to significant investment in this regard. In terms of key challenges, maintaining the momentum within member companies has been identified. Although a number of companies have been approaching the NBI wishing to sign up to the accord since Q2 of 2010, this has not taken place as the NBI is revising its approach to Energy Efficiency Accord.

5 Future road map

The starting point into the future concerns the harmonization of the definition of energy efficiency. The 1998 White Paper on energy defines energy efficiency as "a measure of the savings of energy", which is used to provide goods and services, while maintaining the desired benefits DME [8]. The 2008 National Energy Act defines it as the: "economical and efficient production and utilisation of an energy carrier or resource" RAS [9]. The revised National Energy Efficiency Strategy defines energy efficiency as "achieving the same or improved output with a reduced input of energy" DME [11]. In the draft Integrated Resources Plan DME [11] energy efficiency means "the effective use of energy to produce a given output (in a production environment) or service (from a consumer point of view), i.e. a more energy-efficient technology is one that produces the same service or output with less energy input'. What emerges from the various definitions in government policies is a clear need to harmonise the definitions.

There are also challenges associated with the institutional arrangements and capacity. The various pieces of legislation are administered in no less than 10 different government departments. These include among them the National Energy Regulator, Department of Energy, Department of Trade and Industry, Department of Public Enterprises, South African National Energy Development Institute, Department of Environmental Affairs and National Treasury. This brings bureaucratic delays, competition and inefficiencies. For example, a delay in the implementation of the Electricity Levy or any other Carbon Tax related initiative by the National Treasury holds back the implementation of all other pieces of legislation aimed at energy efficiency that are drafted and passed from other departments. With special reference to the Energy Efficiency Accord there also comes into play industry associations like the NBI and the current 51 member signatories to the Energy Efficiency Accord. The revised National Energy efficiency Strategy DME [11] has come up with a matrix that identifies primary and secondary stakeholders as per the given tasks in energy efficiency.

The CDM still remains a viable option for implementing energy efficiency in the country. There is need to promote more activity in this space, especially given that there will be a National Climate Change White Paper soon. Member companies to the Accord have clearly indicated that there is value when they work as a collective and share information in the future. Companies are looking forward to a more certain regulatory environment and more analysis of what the regulatory environment means for them. Maintaining momentum moving forward in a context of regulatory uncertainty and institutional changes within government is also a challenge facing the NBI as the key implementing agency. Looking ahead, business has indicated its wish to continue working in the energy efficiency realm with a proposal to develop a Business Network for Leadership in Energy Efficiency. Although questions have been raised in terms of the overlaps between the Energy Efficiency Accord and the CDP, the NBI as implementing agency for both initiatives realises the two initiatives are mutually supportive but different.

6 Conclusion

The paper has traversed through a plethora of energy efficiency legal documents in South Africa with most of them being in tandem with international energy efficiency trends. This is an indication that South Africa is progressing well in attaining energy efficiency which is beneficial in transforming to a low carbon economy. There is evidence of government and business' leading and participative roles in formulating and implementing energy efficiency policies. The Energy Efficiency Accord and the CDM have been key initiatives in this regard. Going forward, an integrated approach to energy efficiency institutional arrangements, capacity and legal frameworks is required. The proposal to develop a Business Network for Leadership in Energy Efficiency is a welcome development. More investment will also be required in the CDM space.

Acknowledgement

The authors wish to thank *Exxaro Resources Limited* for sponsoring the Chair in Business and Climate Change under Unisa's Institute for Corporate Citizenship.

References

[1] ICC. Energy Efficiency A world business perspective. Geneva: International Chamber of Commerce; 2007.

[2] WEF. Energy efficiency: Accelerating the agenda. Geneva: Workd Economic Forum; 2010.

[3] UNEP, ILO, IOE, ITUC. Green Jobs: Towards Decent Work in a Sustainable, Low-Carbon World. Nairobi: UNEP/ILO/IOE/ITUC; 2008.

[4] Ernst & Young. Renewable energy country attractiveness indices. London: Ernest & Young; 2010.

[5] Martinot E, Li Junfeng L. Powering China's development: The role of renewable energy. Washington DC: Worldwatch; 2007.
[6] National Treasury. Reducing Greenhouse Gas Emissions: The Carbon Tax Option. Pretoria: National Treasury; 2010.
[7] CDP. Corporate clean energy investment trends in Brazil, China, India and South Africa. London: Carbon Disclosure Project; 2010.
[8] DME. White Paper on the Energy Policy of the Republic of South Africa. Pretoria: Department of Minerals and Energy; 1998.
[9] RSA. National Energy Act (No. 34 of 2008). Cape Town: Government Printers, 2008.
[10] DME. Energy Efficiency Strategy of the Republic of South Africa. Pretoria: Department of Minerals and Energy; 2005.
[11] DME. National Energy Efficiency Strategy of the Republic of South Africa. Pretoria: Department of Minerals and Energy; 2009.
[12] DME. Energy Efficiency Accord. Pretoria: Department of Minerals and Energy; 2005.
[13] DTI. The 2010/11 - 2012/13 Industrial Policy Action Plan (IPAP II). Pretoria: Department of Trade and Industry; 2010.
[14] DEA. National Climate Change Response Green Paper 2010. Pretoria: Department of Environmental Affairs; 2010.
[15] DME. Draft Integrated Resources Plan for Electricity for South Africa 2010-2030. Pretoria: Department of Minerals and Energy; 2010.
[16] DEAT. Long Term Mitigation Scenarios Strategic Options for South Africa. Pretoria: Government Printer.; 2007.
[17] The Presidency. President Jacob Zuma to attend High Level Segment of the COP-15 scheduled for 18 December 2009 in Copenhagen. http://www.thepresidency.gov.za/show.asp?include=president/pr/2009/pr12081345.htm&ID=1933&type=pr (accessed 9 December 2009)
[18] NBI. Energy Efficiency Accord signatories 2010. Johannesburg: National Business Initiative; 2010.
[19] DoE. Project Design Document. Department of Energy 2010 http://www.energy.gov.za/files/esources/kyoto/kyoto_frame.html (20 December 2010)
[20] UNFCCC. CDM registered projects by region. UNFCCC 2010: URL: http://cdm.unfccc.int/Statistics/Registration/RegisteredProjByRegionPicChart.html (accessed 20 December 2010)
[21] DME, NBI. Assessment study of the Energy Efficiency Accord. Pretoria/Johannesburg: Department of Minerals and Energy and National Business Initiative; 2008.

Influence of the methods of estimation of the wind velocity distribution parameters on the accuracy of wind energy calculations

T. Niedoba[1] & B. Soliński[2]
[1]*Faculty of Mining and Geoengineering,*
University of Science and Technology in Krakow, Poland
[2]*Faculty of Management,*
University of Science and Technology in Krakow, Poland

Abstract

Many parameters influence wind energy resources, as well as the amount of energy that can be produced by wind power plants. They are connected with wind and climatic conditions, the location of the power plant, as well as those associated with the power plant itself.

The basic parameter of wind energy is the velocity of flowing air, whose variation is very high. As a result, permanent changes in power and wind energy amounts occur, which was shown in the paper on the basis of selected locations in Poland and Netherlands.

The methodology of "Wind Atlas", elaborated by RISO in 1989 and being currently the world standard, is used to evaluate wind energy resources. In this methodology, the Weibull distribution function is applied to describe the distribution of wind velocity, whose parameters are calculated on the basis of empiric data, with the moments method as the method of approximation. This is, however, a possible source of errors, which are then included in the produced amount of energy. It is crucial, then, to apply the proper approximation methods and to know the possible occurring errors.

The measurement of the wind velocity should be very precise, because even a small error like 5% in evaluation of wind velocity produces about a 16% miscalculation in the forecasting of the total amount of energy produced.

Three methods of Weibull distribution function approximation for wind velocity were presented in the paper: the least square method, the maximum likelihood method and the moments method. The given results were verified and

WIT Transactions on Ecology and the Environment, Vol 144, © 2011 WIT Press
www.witpress.com, ISSN 1743-3541 (on-line)
doi:10.2495/ECO110351

the impact of the applied approximation methods on the preciseness of the produced wind velocity distribution function, as well its statistical parameters and energetic resources, was evaluated.

Keywords: wind velocity, Weibull distribution function, approximation methods, wind energy.

1 Introduction

Many parameters have an influence on the quantity of wind energy resources and the amount of achievable energy obtained by wind power plants. They are connected with both wind, climatic and landscape conditions, as well the parameters of the wind power plant itself.

The basic parameter of wind energy is air flow velocity and its influence directly on wind energy resources. The characteristic feature of air flow is continuous changeability – noticeable in a whole spectrum of a time period – which is connected with continuous wind power and energy changes. The methodology of "Wind Atlas" is being applied to evaluate energy resources, which was established by RISO in 1989 [1–4] and is currently the world standard. This methodology assumes the Weibull distribution function in purpose of wind velocity occurring distribution function in certain areas description and the parameters of this distribution function are given on basis of empirical data, applying the moment method as the approximation method. However, this approximation is a source of numerous errors, which are then transformed into given energy amounts. Thus, for the purpose of more precise determination of energy resources it is imperative to apply correct approximation methods and to be aware of the possible occurring errors.

2 Dependence of quantity of created energy on wind velocity

The significant dependence of wind energy on air flow velocity is the reason that the wind velocity measurement should be more precise, because even a very small error, e.g. of 5% in wind velocity evaluation, can result in a 16% error in the forecasted energy evaluation. This is caused by the fact that the amount of energy is growing proportionally to the cube of wind velocity (1), e.g. if wind velocity grows twice then the energy produced grows eight times. The basics of wind energy calculation were widely described in [2, 5–8].

$$E = \frac{1}{2}\rho v^3 t \cdot \ (\mathrm{Wh/m^2}) \tag{1}$$

where: ρ - air density ($\mathrm{kg/m^3}$); v – wind velocity (m/s); t – time.

It is possible to notice that on figure 1, presenting the wind power in dependence on its velocity for air density ρ= 1,25 $\mathrm{kg/m^3}$ (line A). For velocity of 6 m/s the wind achieves the power of 135 $\mathrm{W/m^2}$, and for the velocity twice higher, of 12 m/s, the given power is eight times higher, which would achieve 1080 $\mathrm{W/m^2}$. Assuming that the wind velocity is stable in time equal to t (e.g. t=1

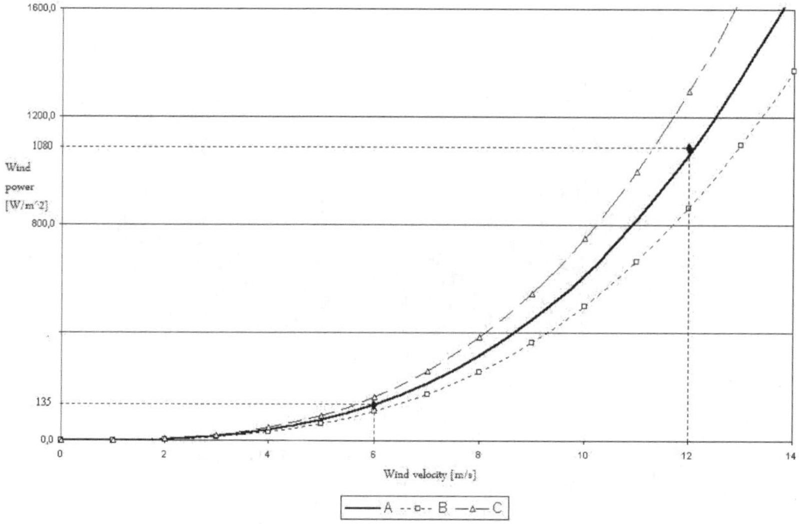

Figure 1: Theoretical wind power in dependence on air velocity and air density. Source: [5].

hour), then the produced energy will be equal to, respectively, 0,135 kWh/m^2 and 1,08 kWh/m^2. From this relationship it is clearly visible that the locations characterized by higher wind velocities are more profitable.

3 Wind velocity distribution function

Wind velocity is not a stable value and is changing continuously. Generally, every moment the wind velocity varies. This changeability is a characteristic

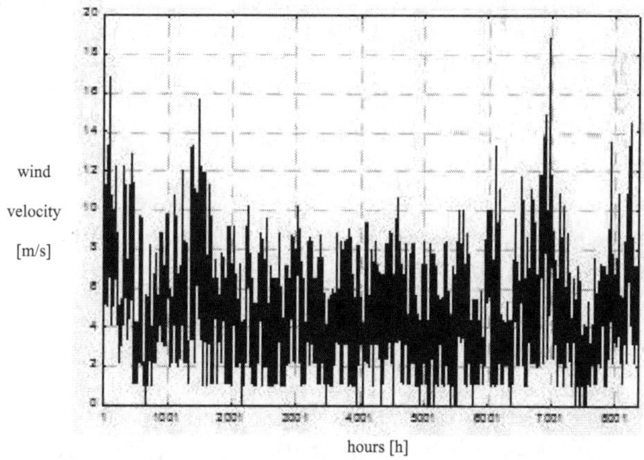

Figure 2: Hourly changeability of wind velocity during one year in the area of Rymanow. Source: personal elaboration.

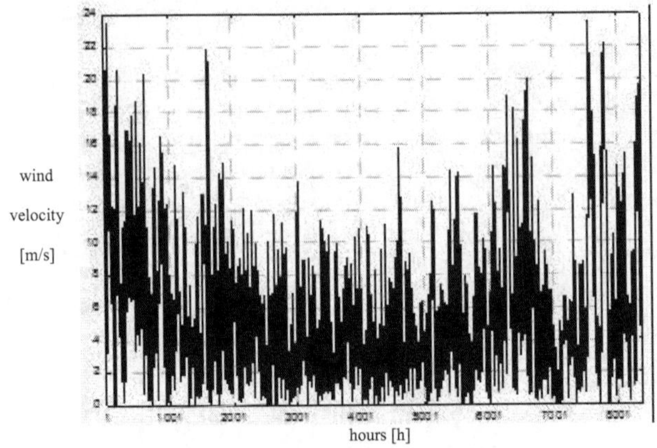

Figure 3: Hourly changeability of wind velocity during one year in the area of Lelystad. Source: personal elaboration.

feature of flowing air, which is also very clear when averaging data. For hour averaging – the standard being applied to determine wind energy resources (currently also 10 minutes averaging is being applied) – this changeability was shown on figures 2 and 3 for selected two locations in Poland and Netherlands. On reviewing this data it is possible to notice the continuous fluctuation of mean hour wind velocity.

Variation of wind velocity influences on amount of obtained energy, so it is very crucial to collect precise data by high quality anemometers and data loggers what is the subject of many scientific works [2, 5, 7, 9–11].

As a result of taking into consideration such a big variation in wind velocity during the following time units, it is advisable to apply the probabilistic distribution function to evaluate wind energy resources. The distribution function being applied in wind energy is a two-parametric Weibull distribution function, which is described by a scale parameter A and shape parameter k. The formulae (2) of Weibull density function, in which $f(v)$ is frequency of occurring wind velocity during certain period of time, is in the following form [5–7, 11,12]:

$$f(v) = \frac{k}{A}\left(\frac{v}{A}\right)^{k-1} \exp\left[-\left(\frac{v}{A}\right)^{k}\right] \tag{2}$$

and the distribution function of random variable v of Weibull distribution function is determined by the equation (3).

$$F(v) = 1 - \exp\left[-\left(\frac{v}{A}\right)^{k}\right] \tag{3}$$

Parameters A and k influence the distribution of a random variable, as well as on the form of the density curve, and determine the characteristics of the wind velocity and thus its frequency distribution. It is crucial then to operate by the correct parameters given by approximation. Otherwise, each inaccuracy will be magnified to an even bigger inaccuracy of energy evaluation (according to formulae 1).

4 Approximation methods

As a result of approximating the unknown distribution function of the researched random variable, it is advisable to apply one of several available approximation techniques. The most frequent methods being applied are: the least square method, the maximum likelihood method and the moments method.

4.1 Least squared method (LSM)

Let assume that (X, Y) are random variables and set $\{(x_1,y_1)\ldots,(x_n,y_n)\}$ is a random sample. If between variables X and Y functional dependence occurs $Y=\Phi(X;\Theta_1,\ldots,\Theta_k)$, where $\Theta_1,\ldots,\Theta_k$ are unknown parameters, then we may assume point $(\hat{\Theta}_1,\hat{\Theta}_2,\ldots,\hat{\Theta}_k)$ as the evaluation of these parameters, which minimize the function.

$$L(\Theta_1,\Theta_2,\ldots,\Theta_k)=\sum_{i=1}^{n}(y_i-\Phi(x_i;\Theta_1,\ldots,\Theta_k))^2\,. \quad (4)$$

If function Φ is a differential in relation to variables $\Theta_1,\ldots,\Theta_k$ then to determine point ($\hat{\Theta}_1,\ldots,\hat{\Theta}_k$) the following set of equations has to be solved:

$$\frac{\partial L}{\partial\Theta_j}(\hat{\Theta}_1,\ldots,\hat{\Theta}_k)=0 \quad j=1,\ldots,k \quad (5)$$

If $Y=\Phi(X,\Theta_1,\Theta_2)$ then, if it is possible, the new variables $\tilde{x}$ and $\tilde{y}$ are being introduced to create the dependence between new variables as linear, so $\tilde{y}=a\tilde{x}+b$, where coefficients $a=a(\Theta_1,\Theta_2)$, $b=b(\Theta_1,\Theta_2)$ are being determined as linear regression coefficients and then the parameters Θ_1 and Θ_2 are determined by reverse transformation [13].

4.2 Maximum likelihood method (MLM)

Let assume that X is a random variable and that the density function of this variable $f(x,\Theta_1,\ldots,\Theta_k)$ is known, where $\Theta_1, \Theta_2,\ldots, \Theta_k$ are unknown parameters. Let $(x_1,\ldots,x_n)$ be n-dimensional random sample for variable X.
The likelihood function is the function given by equation (6).

$$\begin{aligned}L(\Theta_1,\ldots,\Theta_k)&=f(x_1;\Theta_1,\ldots,\Theta_k)\cdot f(x_2;\Theta_1,\ldots,\Theta_k)\cdot\ldots\cdot\\&\cdot f(x_k;\Theta_1,\ldots,\Theta_k)\end{aligned} \quad (6)$$

In purpose of determination of unknown parameters values $\Theta_1, \Theta_2,\ldots, \Theta_k$ the point ($\hat{\Theta}_1,\ldots,\hat{\Theta}_k$) is selected, for which function $\ln L(\Theta_1,\ldots,\Theta_k)$ achieves the maximum. It means that in point ($\hat{\Theta}_1,\ldots,\hat{\Theta}_k$) the set of equation (7) is satisfied.

www.witpress.com, ISSN 1743-3541 (on-line)

$$\frac{\partial f}{\partial \Theta_j}\left(\hat{\Theta}_1,\ldots,\hat{\Theta}_k\right)=0 \quad for \quad j=1,\ldots,k \tag{7}$$

By solving this set of equations, the estimators of the unknown distribution function parameters are given by maximum likelihood method [13].

4.3 Moments method (MM)

If $(x_1,\ldots,x_n)$ is a big random sample then random variable X is variable, which distribution function, which depends on parameters $(\Theta_1,\ldots,\Theta_n)$. The moments method consists in assuming the solution of set of equations, which is given by equating individual moments from sample with theoretical moments, as the evaluation of parameters $\Theta_1,\ldots,\Theta_n$, so:

$$\frac{1}{n}\sum_{i=1}^{n} x_i^2 = m_l\left(\Theta_1,\ldots,\Theta_k\right) \text{ for } l=1,\ldots,k. \tag{8}$$

where $m_l\left(\Theta_1,\ldots,\Theta_k\right)=\int_{-\infty}^{+\infty} x^l f\left(x;\Theta_1,\ldots,\Theta_k\right)dx$, and $f\left(x;\Theta_1,\ldots,\Theta_k\right)$ is density of random variable X density function. The moments method may be applied where we have the analytic form of moments $m_l\left(\Theta_1,\ldots,\Theta_k\right)$ [14].

5 Approximation of the wind velocity distribution function

The methods mentioned above were applied to an approximation done on the basis of empirical data originating from Rymanow (Poland) – data concerning a period in 1998, collected by the Laboratory of Wind Energy Monitoring AGH (www.wiatr.krakow.pl) and data from Lelystad (Netherlands) – data relating to a period in the year 1998, originating from researches KNMI (table 1). This data was selected partially, because it lacked some measurements, in way ensuring to have pairs of data for every hour both from location of Rymanow, as well from location of Lelystad. In this way, the 2 sets containing 8363 measurements each were given, which in turn gave the coefficient of data availability equal to 96%.

Table 1: Characteristics of empirical data concerning wind velocity.

Data	Height of measurement [m n.p.g]	Averaging time	Number	Mean value [m/s]	Standard deviation [m/s]	Minimum [m/s]	Maximum [m/s]
Rymanow 1998 POLAND	20	1 h	8363	5,591	4,081	0,0	23,6
Lelystad 1998 NETHERLANDS	10	1 h	8363	4,850	2,545	0,0	18,9

Source: personal elaboration.

As a result of the application of the approximation methods mentioned above, the Weibull distribution function parameters were given, which are presented in table 2. Furthermore, the appropriateness of given function was evaluated by residual deviation given by formulae (9).

$$s_r = \sqrt{\frac{\sum_{i=1}^{k}\left[F(v_i) - \hat{F}(v_i)\right]^2}{n-2}} \tag{9}$$

where: $F(v_i)$ – theoretical distribution function; $\hat{F}(v_i)$ – empirical distribution function, n- set number.

Table 2: Results of the wind velocity distribution function approximation.

Data	Parameters/statistics	MLM	LSM	MM
Rymanow 1998 Poland	A	6,153	6,119	6,00
	k	1,414	1,368	1,33
	Residual deviation s_r [%]	0,204	0,199	0,202
	Mean velocity [m/s]	5,599	5,598	5,517
	Standard deviation [m/s]	4,015	4,140	4,189
Lelystad 1998 Netherlands	A	5,476	5,186	5,50
	k	1,999	1,744	1,93
	Residual deviation s_r [%]	0,217	0,249	0,245
	Mean velocity [m/s]	4,853	4,619	4,878
	Standard deviation [m/s]	2,538	2,733	2,633

On analyzing the data, it is possible to say that for wind conditions in the researched locations, all the applied methods gave relatively low values of rest deviations, from 0,199 to 0,249. It proves wellness of the approximation of researched random variable (wind velocity) distribution function. For the considered approximation techniques in the area of Rymanow, every value of the rest deviations were relatively low and similar. The smallest error was obtained for LSM, which equated to 0,199% and the highest for MLM (0,204%). However, it is possible to observe the difference in given Weibull distribution function parameters evaluations. These differences were about 2,5% for A parameter and 6,3% for k parameter, which certainly would be transformed into a differential of wind power density and consequently into given energy amounts. Similarly, for the location of Lelystad, the rest deviations are also low, but the smallest error was obtained for MLM (0,217%) and the highest for LSM (0,249%). Analysis of the given Weibull distribution function parameters in this case indicates relatively significant differences, equal to 5,7% for A parameter and 13,2% for k parameter (in relation to MM, being applied in the "Wind atlas" methodology in program WasP).

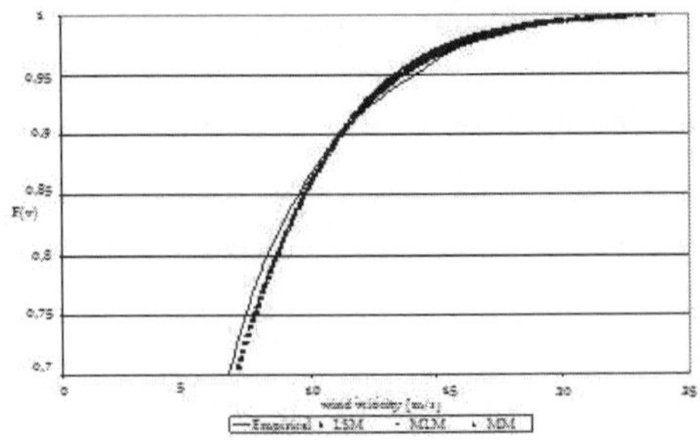

Figure 4: Distribution function curves for wind velocity, given by various approximation methods for the location of Rymanow.

6 The influence of approximation methods on the quantity of wind energy production

Wind energy resources may be determined by the application of power density factor (W/m^2), which is given by the equation (10) [2,5-7,10-11].

$$P = \frac{1}{2}\rho \int_0^\infty v^3 \left\{ \frac{k}{A}\left(\frac{v}{A}\right)^{k-1} \exp\left[-\left(\frac{v}{A}\right)^k \right] \right\} dv \tag{10}$$

where: k – shape parameter, A – scale parameter.

Table 3: Amounts of power density given on the basis of parameters given by various approximation methods.

Data from location	Parameter	MLM	LSM	MM
Rymanow	Wind power density [W/m^2]	311	324	322
Lelystad	Wind power density [W/m^2]	134	134	141

Source: personal elaboration

On the basis of Weibull distribution function parameters, the wind power density amounts were calculated from equation (10), which are presented in table 3. It is noticeable that these values differ between themselves. It proves that the method of Weibull distribution function parameters estimation has an influence on the quantity of evaluated wind energy resources. The analysis of occurring differences between wind energy parameters and suitable statistics is presented in table 4.

Table 4: Comparison of deviations from real values for mean value, standard deviation and mean wind power density.

Data	Parameters	Empirical data	MLM		LSM		MM	
			value	Deviation from value [%]	value	Deviation from value [%]	value	Deviation from value [%]
Rymanow 1998 POLAND	Mean wind velocity	5,591	5,599	0,14	5,598	0,13	5,517	-1,34
	Variance	16,655	16,120	-3,31	17,140	2,83	17,548	5,09
	Mean power density	330	311	-6,11	324	-1,85	322	-2,48
Lelystad 1998 NETHERLANDS	Mean wind velocity	4,850	4,853	0,06	4,619	-5,0	4,878	0,57
	Variance	6,472	6,441	-0,47	7,469	13,35	6,933	6,65
	Mean power density	138	134	-2,99	134	-2,99	141	2,13

The results of this analysis identify the high differential of given parameters by LSM in relation to values given from empirical data. It is mainly visible for the location of Lelystad (5% for mean value and above 13% for variance), which may prove the questionable usefulness of this method in analyses concerning wind velocity, and particularly in certain cases (for the location of Rymanow the deviations are acceptable). The rest of the errors for MLM and MM are quite similar, but it is possible to notice some advantage of MLM, for which these deviations (concerning mean value and standard deviation) for both locations are smaller than for MM.

7 Conclusions

On the basis of the conducted research and analyses, it is possible to formulate the following conclusions:

1. Weibull distribution function describes correctly the inconsistency of wind velocity, which is proven by very low values of rest deviation.
2. The approximation methods being applied result in slightly different suitability to the empirical data, although the level of rest deviation is similar.
3. The relatively big differential of Weibull distribution function parameters evaluation given by the application of various approximation methods is visible.
4. The evaluations of statistics (mean value and standard deviation) for described techniques also result in a relatively big differential.
5. The amount of wind power density given by the application of researched approximation methods differ between themselves, which proves that the selected method of Weibull parameters estimation influences the values of evaluated wind energy resources.

6. Relatively high deviations for Weibull distribution function parameters evaluation with application of LSM (location Lelystad) may identify an inherent risk in this method application during analyses concerning wind energy.
7. Deviations for MM and MLM are relatively low and similar, but a certain advantage in MLM is visible, for which these deviations for both locations (concerning mean value and standard deviation) are lower than for MM.

The authors are going to continue and broaden their research, based on a wider array of wind conditions (new locations) and on non-classical methods of distribution function parameters estimation.

Acknowledgement

The paper is the effect of statutory work no. *11.11.100.276.*

References

[1] Troen I., Petersen E.L., *European Wind Atlas*, Riso National Laboratory, Roskilde, Denmark, 1989.

[2] Petersen E.L., Mortensen N.G., Landberg L., Hojstrup J., Frank H., *Wind Power Meteorology*. Riso National Laboratory, Roskilde, Denmark 1997.

[3] Petersen E.L., Troen I., Frandsen S., Hedegaard K., *Windatlas for Denmark. A rational method for wind energy siting.* Risoe-R-428, Riso National Laboratory, Roskilde, Denmark 1981.

[4] Bowen A.J., Mortensen N.G., *Exploring the limits of WAsP program.* European Union Wind Energy Conference, 20-24 May, Goteborg, Sweden, 1996.

[5] Soliński B., *Wpływ parametrów energetycznych wiatru i elektrownii wiatrowej na wielkość wytworzonej energii elektrycznej.* Proceedings of Waste Management School, 15-18 czerwca 2004, Kraków-Rytro, 2004.

[6] Hau E., *Wind turbines: fundamentals, technologies, application and economics.* Springer Berlin 2000.

[7] Burton T., Sharpe D., Jenkins N., Bossanyi E., *Wind Energy Handbook.* John Wiley & Sons Ltd., Chichester 2001.

[8] Soliński I. Soliński B.: *Energetyka wiatrowa w Polsce*. Wyd. IGSMiE PAN, Kraków,. Polityka Energetyczna, vol.7, iss. 1, 2004.

[9] Albers A., Klug H.: *High Quality Wind Speed Measurements for Site Assessment*, DEWI Magazine, vol. 15, August, 1999.

[10] Harrison R., Hau E., Snel H., *Large wind turbines – design and economics.* John Willey & Sons, Chichester 2000.

[11] Soliński I., Ostrowski J., Soliński B., *Energia wiatru. Komputerowy system monitoringu*. Wydawnictwa AGH, Kraków 2010.

[12] Parzen E.: *On estimation of probability density function and mode,* Ann. Math. Statist., vol. 33, 1962.

[13] Gajek L., Kałuszka M.: *Wnioskowanie statystyczne*, Warszawa, WNT, 2000.
[14] Koronacki J., Mielniczuk J.: *Statystyka dla studentów kierunków technicznych i przyrodniczych*, Wyd. Nauk.-Techn., Warszawa, 2001.

Cultural evolution, energy use and human development

C. Viniegra
LawSchool, Universidad Panamericana, Mexico

Abstract

This analysis makes reference to life's biological laws and the biophysical boundaries to which all living beings are constrained in order to propose a theoretical model that links restriction and capacity to evolutionary strategies. The latter is related to cultural evolution, which identifies human culture as a phenomenon that stays within life's general evolution. After the model is presented, a statistical analysis is used for validation. Despite great diversity of cultural and historical backgrounds in present societies, data from economic, energy and technological features of more than 100 countries show striking regularities related to the way that physical and social resources are applied to the process of economic and human development, and show consistency with the proposed strategies of the model (Status Quo, Survival, Transformation and Growth). Among the most striking features discovered by this analysis, it is worth noticing that, apparently, there is a minimal threshold of energy expenditure and Research and Development (R&D) investment required to achieve a high level of social development. Furthermore, in order to achieve a positive and strong cooperative effect between energy and R&D, worldwide evolution of such indexes seem to indicate that economic and social evolution are constrained to similar scaling laws than those discovered for biological species.

Keywords: cultural evolution, economics, scaling laws, negentropy, energy use, life's laws.

1 Introduction

While the modern era has been a time of advances of every kind, human beings still face the persistence of ancient problems such as curable diseases, social

www.witpress.com, ISSN 1743-3541 (on-line)
doi:10.2495/ECO110361

marginalization and wars. At the same time, humankind faces new challenges because of the deterioration of the environment, changes in climate conditions of the planet, exhaustion of key resources such as farming fields and clean water, along with the growth of social pathologies such as addictions and new violence phenomena that range from the family to the work environment. When thinking about old and new issues, decades of strategies aimed to solve them have accumulated and results are the opposite of what was expected and the provision of ideas seems insufficient to explain or solve even one important issue at its root.

Due to the recurrence of crisis of all types and to the probable perspective of disasters, such as a global climate change that can create, among other things, hunger and massive migration of people, it has become necessary to review our understanding of how human culture develops and relates to general evolution and use of the environment.

2 Life's laws and the human being

Life is the process in which a finite amount of matter self-organizes, metabolizes (uses energy from the environment), grows, reproduces and evolves [1].

2.1 Self-organization, metabolism and growth

According to the second law of thermodynamics, life generates higher levels of order [2] by importing negative entropy for itself, also known as negentropy or syntropy, while exporting a larger amount of entropy to the environment. In the aforementioned thermodynamic exchange, life increases order in a process called data exchange [3]. In this way, energy obtained from the environment is used through metabolism to organize molecules in coherent sequences of information, known as genetic information. The byproduct of this process is heat release or the exchange of entropy to the environment.

Generation of higher levels of order or information exchange require energy use and dissipation of heat. In this process, the cells of all living organisms are geared to increase the successful occupation of an ecological niche with a minimum of energy expenditure and seem to be bounded by the laws of mass and energy transfer at a microscopic scale since all of them have very large surface-to-volume ratios in order to minimize local concentration and temperature gradients. Also, large multicellular organisms need vascular systems in order to distribute nutrients and oxygen in an efficient way. This has been reflected in a universal mathematical rule called Kleiber's Allometric Law of Scaling [4].

2.2 Reproduction

Living beings deteriorate over time and, with deterioration, life eventually disappears. Nonetheless, the phenomenon of life, through the genetic mechanisms of reproduction, inherits an updated copy of its genetic information to a new being whose entropy is low enabling the continuity of the process of life. When defining the guidelines by which life operates, the first one is that in spite of being circumscribed to the universal laws of physics, **life uses everything within its reach to be continuous through time**.

In the reproduction process of living beings, two different strategies can be observed, explained under the "r/K Selection Theory" [5].

Living beings oriented towards an "r" reproduction strategy exist in unstable environments and non-saturated ecological niches where high speed of reproduction is crucial for survival. The characteristic features on this strategy are low energy investment to the descendants and high fecundity, low body volume, early maturity, short life expectancy and wide dispersal of the population. Rodents are a mammal example. Scientific literature points "r" selected species as opportunists.

Living beings oriented towards a "K" reproduction strategy exist in stable environments and niches close to their maximum load capacity; their fecundity rate is low, energy investment from parents to descendants is high, maturity is slow and life expectancy is long. Big mammals are the main examples of species under "K" selection. Scientific literature points "K" selection species as balance species.

2.3 Evolution

The physical margins where life is possible seem small, but there is enough room within them to generate adaptation needs [6]. These are covered in the genetic traits that persist or get discarded through the reproduction process. Regardless of the fact that the most popular interpretation on Darwin's proposition, "survival of the fittest", is survival of the biggest, strongest, and fiercest, a better interpretation on Darwin's work is that fitness should be understood as the selection of genetic features that guarantee, in the long run, the survival of the best adapted [7].

As an example, it can be said that in similar environments, discounting energy availability, living beings with high metabolic rates will be better adapted when energy is abundant and, given that abundance, it may be necessary to compete in a numerous population. On the contrary, in an environment with low energy availability, living beings with low metabolic rates and greater willingness to collaborate will be more successful [8].

2.4 Restriction and capacity

Life is limited by nature's laws, but given the opportunity, it widens its capacities to accumulate negentropy and responds to environmental changes using the evolution mechanism. In this light, one may conclude that there are two working axes that determine the success or failure of the changes and their persistence through time.

2.4.1 Restriction – the environment, the external

The list of environmental variables that affect living beings is long, but the most bounding resource is energy. For example for autotrophic organisms (mainly photosynthetic) the main constraint is solar energy transformed in biomass. For heterotrophic organisms, living out of autotrophic organisms, added to the previous limitation, the conversion efficiency is roughly 10% for each trophic level under consideration (herbivores, carnivores and detrivores). The fact, that life occurs in a limited space, with limited energy of solar input, makes obvious that life is bounded by spatial and energy constraints.

2.4.2 Capacity – what is of each, the intrinsic

With life's existence, in spite of the higher transfer rate of entropy towards the universe, a greater order is produced in its operating range. In the long term and as a second directive of life, **life works to widen its possibilities of information exchange or negentropy accumulation**. At the limit, life will achieve as much order as the available energy and entropy level of the environment will allow.

The dependence on the environment and its changes produce exchanges among the time permanence directive versus the capacity to generate negentropy (negentropic capacity) directive. The latter prevents that the speed in which life spreads through the environment to be constant since changes in the environment and restriction levels define the necessity to modulate the evolution process in order to guarantee permanence in time.

The acceleration in capacity growth requires larger amounts of energy, either due to a population increase or to an increase in the complexity of the living beings that are part of the process. However, in spite of the irrevocable calling of life to generate as much order as possible, the speed to do it will vary in time as a function of the acceleration and deceleration of the capacity to generate negentropy.

2.5 Metabolic model

This paper has described the features of living beings, biological and physical laws that limit them and two axes of action, restriction and capacity. In that light, and using the environment as the external and capacity as the internal or what is intrinsic to life, it is possible to propose an analysis of the gathered elements in a model generated by the metabolic strategies produced by intersecting restriction and capacity (see Figure 1).

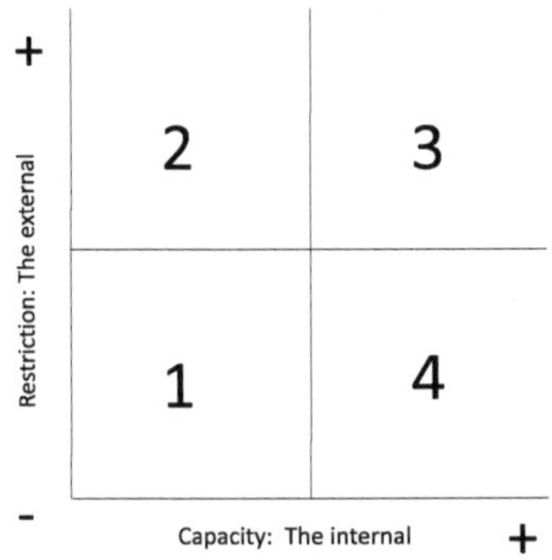

Figure 1: Schematic representation of the model.

Quadrant 1. Low restriction/Low capacity: Status Quo. On these coordinates, available resources supplied by the environment surpass the metabolic needs of the population. The environment is unstable either by reproductive competition and/or environmental variations. Additionally, there is no change in the capacity to generate negentropy. The reproductive rate is maximized but given the low negentropic capacity of the population, the resources' consumption rate is lower than the environment's replacement rate.

Quadrant 2. High restriction/Low capacity: Survival. On this quadrant, the energy availability is compromised and there are no chances for expansion due to the impossibility to increase negentropic capacity. In metabolic terms, all energetic non-essential pathways disappear and energy use rate is minimized. If the environment's replacement rate is overrun, there is a reduction of population or death.

Quadrant 3. High restriction/High capacity: Transformation. The shortage of resources is compensated by an accelerated growth in negentropic capacity to take on new opportunities. There are potential but not present threats; every new function is enabled to seize a new opportunity within the environment.

Quadrant 4. Low restriction/High capacity: Growth. High availability of resources allows the colonization of the environment. It is possible to increase the negentropic capacity and the metabolic rate at the same time. Competition is produced by the population's increase and metabolisms that consume more energy are favoured.

2.6 Hemispherical analysis of the model

By dividing the plane into four sections with two lines one vertical and another horizontal, the resulting hemispheres will be: vertical cut, left and right hemispheres; horizontal cut, top and bottom hemispheres.

Vertical

- Left. Low evolution and capacity. Environmental conditions overcome life's long term evolutionary dynamics, options are limited. Negentropy generation is limited.
- Right. High evolution and capacity. The long term dynamic for capacity growth, works unbounded in any environmental conditions. Negentropy's generation grows, as well as energy use and the exchanged entropy with the environment.

Horizontal

- Top. There is little waste of energy, cooperation prevails over competition. The challenge imposed by the environment is more important than rivalry, thereby metabolism is optimized. In addition, efficiency levels used to incorporate energy to life process are increased. On the reproductive side, selection is oriented towards "K"; there is low variation in environmental factors and low rivalry level.

- Bottom. Energy use is maximized; competition prevails over cooperation; reproductive competition is more important than the environment's challenge. Efficiency is not important, but final metabolic capacity is. Reproductive selection is oriented towards "r"; there is instability due to environmental factors and/or because a high degree of rivalry.

2.7 Ecological niches

To enhance the analysis it is also important to make reference to different types of living beings and the ecological niches they hold within the general model while explaining Figure 2.

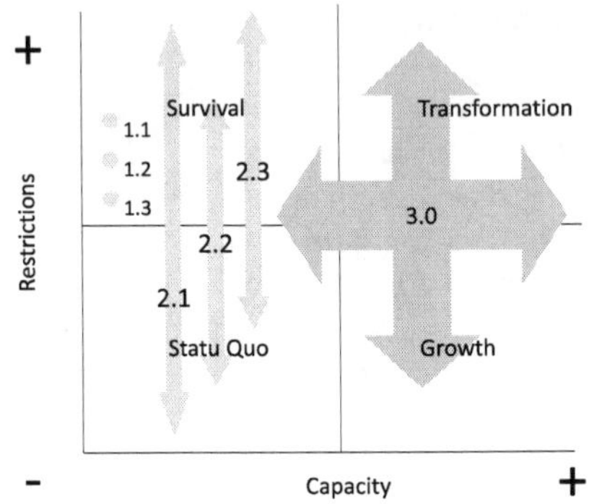

Figure 2: Ecological niches.

Autotroph. They populate the different restriction levels by differentiation of species, there is no locomotion or behaviour; in Figure 2 a species is a point, the whole forms a negentropic capacity line (photosynthesis). They evolve through physical features or "hardware". (Points: 1.1, 1.2, 1.3 in Figure 2.)

Non human Heterotrophs. They acquire the capacity so that each species populates a vertical line and, through locomotion, they generate a new adaptation strategy: migration, which means capacity to run away from danger, attack, pursue pleasure (instinct); behaviour emerges and is always closely related to "hardware" configuration. Species manage to settle on new capacity spaces by differentiating their energy strategy: herbivorous, carnivorous, and omnivorous. They also evolve through "hardware". Locomotion accelerates evolutionary competition. (Lines 2.1, 2.2, 2.3 in Figure 2.)

Human Heterotrophs. They abandon evolution by "hardware" (physical features) and represent a new evolutionary strategy for life; evolution through "software". They are capable of choosing any quadrant, be it by migration, aversion to pain, attraction to pleasure and by cultural development. They have the ability to choose their behaviour, which adds extra capacity mechanisms and information to the ones granted by genetic inheritance. They manage to change the pattern of resource usage in a quick way through a new mechanism, cultural evolution [9] that is inherited to their descendants through knowledge transmission (cultural reproduction) which means the accumulation of negentropy in culture. (Arrows 3.0 in Figure 2.)

2.8 Cultural evolution

In biological terms, the human being is close to the extreme of "K" selection within the reproduction strategy. This means that it improves his descendants' survival possibilities by investing in them a great amount of energy. "K" selection also means there is a long maturity period for individuals and a longer life span.

Unlike some species under "K" selection, such as elephants and whales, the human being does not have great body volume or exceptional physical capacities, but it consumes huge amounts of energy. As stated before, the main difference in man's survival strategy lies within the reduction of physical and genetic competition with other species in order to evolve quickly through a new strategy already defined as cultural evolution.

Exactly like genetic evolution, cultural evolution generates negentropy through the use of energy exchanged for higher levels of information expressed in knowledge, art and technology. Furthermore, just as it happens on genetic evolution, new levels of information (knowledge) generate new possibilities to access energy sources and for new ways to make use of the environment (technology). The latter is done in a way similar to what happened when energy strategies like the herbivorous, the carnivorous, and the omnivorous appeared. In philosophy, a rise in the understanding of the universe is defined as an increase in the consciousness level or approximation to the truth. In biology, the difference between conscious (man) and unconscious beings (the others) is also pointed out.

Even though the theme of consciousness is one of the most debated points in religious theology and secular philosophy, both approaches point at the need to answer the main ontological and cosmological questions: What exists? What is life for? Is the universe infinite? Does time have a beginning and an end? What is the essence of matter? It may be presumed that whoever manages to obtain an answer to any of these questions would be on a higher level of consciousness that those who do not.

Moreover, making definitions based on features that separate unconscious animals from conscious man has become complicated since ethology reports more complex behaviours in animals than the ones assumed before (i.e. the chimpanzee's use of tools, cooperation actions, learning and problem solving within various species), while neurosciences acquire better understanding of the human brain and report higher intervention of the subconscious in people's actions and motives than ever imagined [10].

Therefore, it is possible to say that the gradient going from the inert to the subconscious and to the conscious, is related to the non living, life and life's level of organization according to the capacity to generate negentropy. This positively correlates energy use and the level of consciousness, where the effective use of energy (energy efficiency), is more important than total use of it.

2.9 Energy and human development

Figure 3 shows the relationship for 126 countries, between the Human Development Index (HDI) and per capita energy consumption (ToE = Tons of oil Equivalent) and the best adjusted curve seems to be a logarithmic model. It is worth noticing that all countries with HDI value higher than 0.9 use more than 2ToE per capita. Also, all countries with per capita consumption lower than 0.7 ToE have an average HDI of 0.6. The corresponding empirical equation can be written as follows:

$$HDI = k\ln(ToE/ToE_0) + B \tag{1}$$

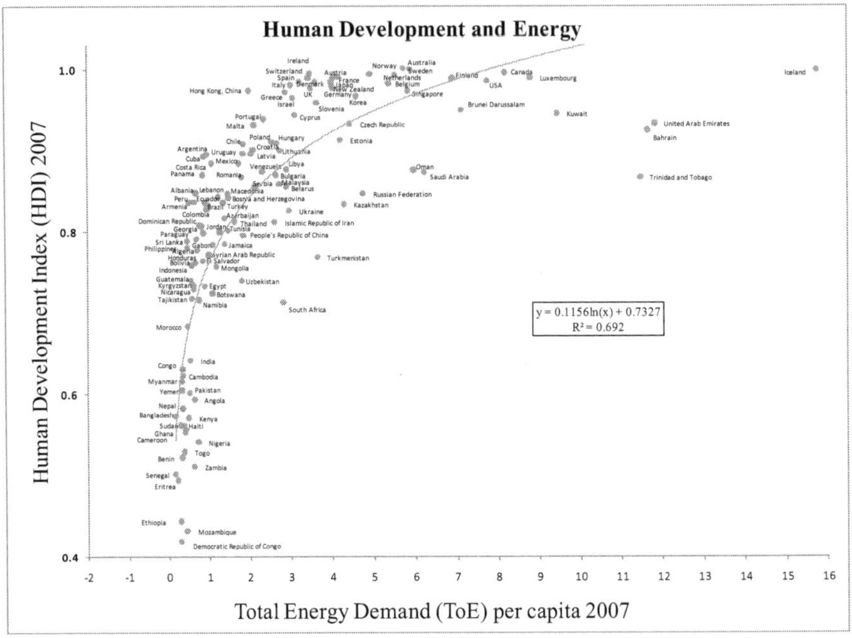

Figure 3: Human development and energy demand. Source: HDI, United Nations Development Programme (www.undp.org), total energy demand, International Energy Agency (www.iea.org). ToE is a standardized unit equal to the amount of energy obtained through the combustion of a ton of crude oil. It is equal to 41.868 Giga Joules and to 11.63 Mega Watts/hour.

Expression (1) has a threshold value when $ToE = ToE_0 = 1$, and, $HDI = B = 0.737$, beyond which, the values of ToE increase very fast in order to approach the highest level of human welfare with per capita values of ToE higher than 3. The value of the empirical constant, $k = 0.1156$, seems to be a global efficiency factor, since the correlation shown in Figure 3 was taken from a sample of 126 countries (more than 90% of human population) and would be related to an intrinsic property of present human societies.

This is a very important consideration in terms of planning the future because it implies that the improvement of the quality of life for nearly 5 billion people, living with HDI lower than 0.7, will require a very large increase of global energy expenditure. Otherwise, the social and political tensions across the world will increase breaking down the present level of international agreements. For example, terrorism and widespread violence will be difficult to eradicate in a world where most of the people experience a very low HDI environment.

2.10 Energy and technology development

We can state that total energy use of a society E_T has three terms

$$E_T = E_M + E_G + E_{RD} \quad (2)$$

where E_M is the energy required for maintenance, E_G is the energy required for growth and E_{RD} is the energy to be invested in R&D (capacity growth). Here it is not intended to estimate each fraction of the total energy budget. Instead it is assumed that there should be a power law between E_{RD} and E_T and we also assume that energy expenditure is a reflection of the technological capacity of each society, measured as a normalized innovation index, I_N, defined as the number of patents registered in the US Patent Office each year and assigned to nationals of each given country and normalized per million people. This index has two interesting properties; it involves the largest patent registrar in the world where international companies compete for the largest share in the US market, and also, it only takes into account the patents where the assignee (the owner of the patent rights) is from a given country. This way, the innovation index is strongly related to innovation supported by companies or individuals that wish to risk money in patent registration expenditures and have residence in a given country. The power law is of the following form

$$E_T = E_{T0}(I_N/I_{N0})^n \quad (3)$$

Figure 4 shows the log–log correlation implied in the power law between ToE and I_N; it should be noted that from the previous list of 126 countries, only 74 have patents assigned to individuals or companies resident in those countries. It is worth noticing that the variation coefficient is only $R^2 = 0.51$, suggesting that the total energy expenditures related to technological development require a more careful partitioning. But it shows a general trend that may help to establish worldwide country typology in the way that energy expenditures are related to innovation. This preliminary model of energy expenditure follows the empirical equation

$$ToE = 2.1394 I_N^{0.2108} \quad (4)$$

In Figure 4 it can also be seen that the variable defined to evaluate capacity (patents/million inhabitants) increases through various magnitude orders that range from less than 0.1 patents a year per million inhabitants, up to more than 200. This way, developing countries are clearly separated from developed ones. It is worth noting that there is not a single case of a developing country with low energy consumption and high patents' registration rate.

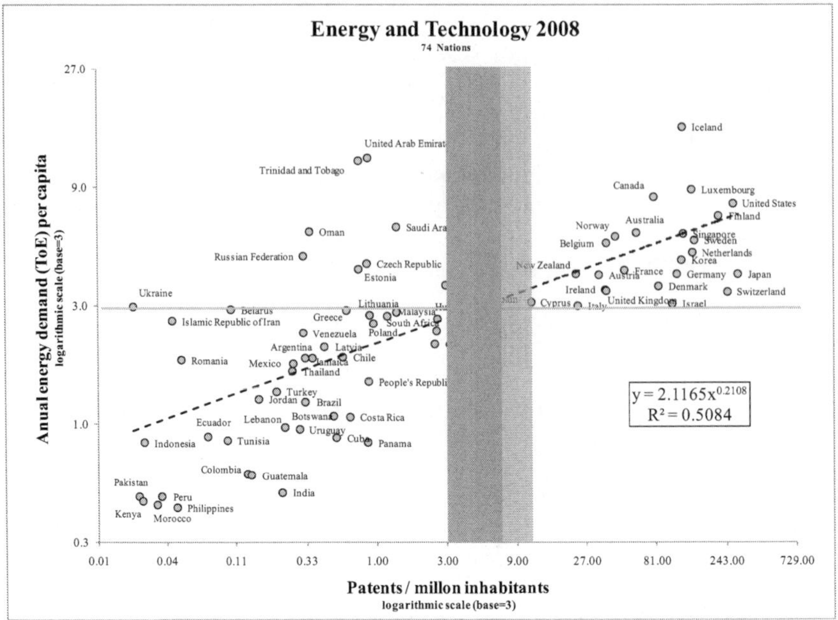

Figure 4: Energy and technology. Source: energy International Energy Agency (www.iea.org) and patents from the U.S. Patent Office (www.uspto.gov).

Again this seems to indicate the existence of threshold around $I_N = 3$ patents per million people, as a requisite to reach an energy expenditure, per capita, of 3 ToE. In Figure 4 there is a shadowed region that covers from 3 to 10 patents a year per million inhabitants and seems to be the transition zone between development scales and contains countries under change process, from underdevelopment to development.

The slope of the empirical power equation in Figure 4 may be used as an indication of the comparative efficiency reciprocal interaction between energy expenditures and technological development. There are some interesting cases to discuss. For example, the United Arab Emirates, Trinidad and Tobago, several oil exporter countries and some ex Soviet countries, have a value of I_N lower than 0.33 patents per million although they have energy expenditures higher than 3 ToE per capita. However, they cannot be considered members of the highly developed nations club since their HDI is lower than 0.9.

The most suggestive aspects in Figure 4 are the empty spaces at the right bottom and left top quadrants. Empty spaces indicate possible hindrances or forbidden zones, top and bottom, of a scaling rule between energy use and development. In fact when all points (2,115) from a data base with a data set from 1978 to 2007 were plotted (Figure 5) and analyzed with a potential model, the result was a regression equation equal to $y = 1.63x^{0.26}$, with a variation coefficient $R^2 = 0.59$, where "y" is equal to primary energy supply in ToE units and "x" is equal to patents registered in year per million inhabitants.

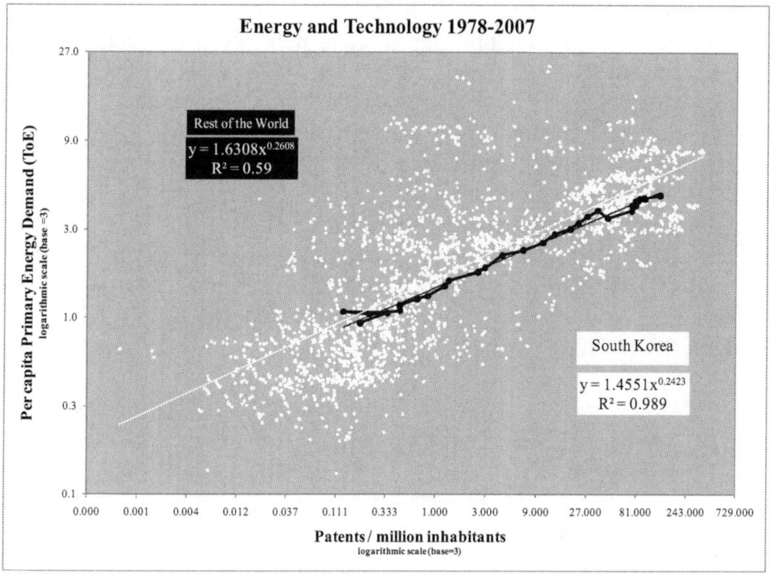

Figure 5: Evolution of energy and technology from 1978 to 2007. South Korean performance is indicated in black with a correlation function $y = 1.4551x^{0.2423}$ ($R^2 = 0.989$). Source: energy International Energy Agency (www.iea.org) and patents from the U.S Patent Office (www.uspto.gov).

After making a diachronic (related to time) analysis with an online tool named "Motion Chart" from "Google Docs", it was oberved that that the nation with best performance, and the only one that advances several orders of magnitude in both axes is South Korea. A nation that frequently is referred in specialized publications as an example of success given the fast achievement in economic and social development. For this reason, a specific analysis was made only with South Korea's data, also presented in Figure 5, here the set of couples of values (ToE, I_N) were plotted vs. time and the regression analysis showed the following empirical equation

$$ToE = (1.45)I_N^{0.24}, \ (R^2 = 0.98), \qquad (4)$$

This is a striking result suggesting that this equation is indeed a scale-up equation between energy and technology and supports the notion that the power law of Figures 4 and 5, is the "developtmental pathway" for social and economic improvement. Here, the scaling coefficient is approximately "¼".

Figure 6 shows the diacronic analysis of, ToE vs. I_N, for Costa Rica, Mexico, South Korea and Saudi Arabia. The only growing up curve belongs to South Korea. The other three cases oscillated as if they were caught in an under development trap. Curiously enough, the main numerical difference of those plots is not the slope (the power ¼) but the scale-up reference value ToE_0. Apparently, this is related to a given choice of resource allocation. A high level of (ToE_0> 1.45) would imply to much attention paid to energy expenditure with low level of technology development, as seems to be the case of Saudi Arabia. A low level of ToE_0 would imply that there is a weak coupling between technological development and energy expenditures. And there is a third case, Mexico; here the trajectory is apparently superimposed to the ascending trajectory of South Korea but in fact, it has oscillated in approximately the same place of the phase plot during the last 30 years. This also may be related to the lack of long term commitment in order to establish a working link between technological development and energy expenditures. It is worth noticing that Mexican R&D expenditures have been maintained at a level lower than 0.4% of GDP during the last three decades and the gap between scientific publication and technological innovation is well known.

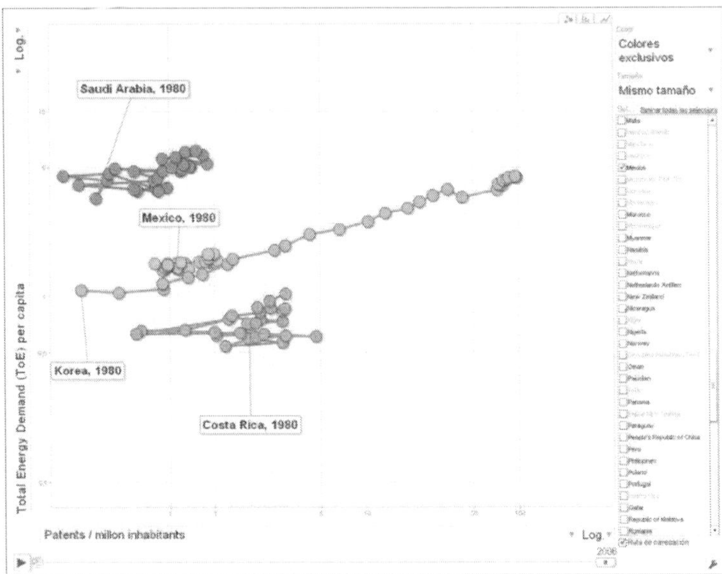

Figure 6: Diachronic analysis (selected countries). Diachronic analysis was made with "motion chart" from Google® docs (docs.google.com).

It is also worth noticing that the concepts "economy of scale", and "marginal analysis" used in economics, come from the same principle as the biologic scaling rules: large systems are more efficient, but the costs they generate, including externalities – at some stage – grow faster than the benefits they generate, to the point where obtained additional benefit is lower than the cost increase. An example of probable boundaries and efficiency differentials is observed in the position held by Switzerland and the United States (Table 1). It is necessary to highlight that in a closed system like the earth, the non inclusion of externalities to the overall development cost accumulates systemic risks in the form of a reduction of the biospheres capacity to regenerate resources, which means the consecution of the life expansion directive without balance with the permanence in time directive.

Table 1: Energy efficiency of Switzerland and the USA.

Concept	Switzerland	United States of America
GDP per capita (USD)	**64,011**	**46,716**
Energy demand (ToE) per capita	**3.42**	**7.74**
Patents a year / million inhabitants	**246.07**	**267.61**
Human development index	**.960**	**.956**

3 Conclusions

- All living beings are found in any of four possible metabolic strategies due to the exchanges generated within the relationship between restriction and capacity.
- When seen through the lens of energy use and exchange of information, human culture abides to the fundamental laws of life and can be considered to remain within life's general evolution. Given that human societies follow similar development pathways as the ones proposed on the model and when measured by their use of resources and by their technological output.
- The self determination capacity humans have to design their own existence relates to the capacity to move to any of the four metabolic strategies. Thus, the cultural diversity of human kind can be understood, as the cultural evolution of human groups which have selected different survival strategies.

References

[1] Nielsen, P. E., A New Molecule of Life. *Scientific American*, December 2008.

[2] Rubí, M. J., Does Nature Break the Second Law of Thermodynamics? *Scientific American*, October 2008.

[3] Brillouin, L., Negentropy. *Journal of Applied Physics*, pp. 1152-1163, 1953.

[4] Kleiber, M., Body Size and Metabolic Rate. *Physiological Reviews*, 27, pp. 511-541, 1947.

[5] Pianka, E.R., On r and K selection. *American Naturalist*, 104, pp. 592-597, 1970.

[6] Darwin, C., *The Origin of Species*, John Murray: London, 1859.

[7] Shermer, M., Darwin Misunderstood. *Scientific American*, p. 34, February 2009.

[8] Kropotkin, P., *Mutual Aid: A Factor of Evolution,* Forgotten Books: London, (Conclusion), 1902.

[9] Teilhard de Chardin, P. *The Phenomenon of Man*, Evergreen Review, Kindle p. (Kindle location 429), 1955 (Kindle 2009)

[10] Zaltman, G., *Marketing Metaphoria.* Harvard Business Press: Boston, pp. 47-61, 2008.

Section 7
Soil contamination

Remediation of arsenic contaminated soil using soil washing with an acidic and reducing solution

J. G. Kim, J. H. Lee, Y. C. Cho & S. H. Park
Korea Institute of Geoscience and Mineral Resources, Korea

Abstract

Soil contamination with arsenic (As) is a serious environmental concern due to its highly toxicity and carcinogenic property. Iron oxides are known to be the most important factor for controlling As concentration in soil pore water. They adsorb large amount of As and coprecipitate with As. We tried to develop a new washing method of As contaminated soil by the dissolution of iron oxide and desorption of As employing pH and redox potential adjustment. An As contaminated soil sample was collected at a rice paddy field near a copper smelter, Korea. The As concentration of the soil sample was determined with an aqua-regia extraction method. Washing solutions were prepared to be 0.001–0.1N HCl and 0–3% Na-dithionite. The soil and washing solution were mixed at 1:4 ratio and the mixtures were reacted for 15, 30, 45 and 60 minutes. After the reaction, the soil and washing solution were separated with a centrifugation. The pH, Eh and concentrations of As, Fe and Mn of the washing solution were determined and the As concentration of the washed soil sample was also determined. The separated washing solution was treated with H_2O_2, $CaCl_2 \cdot 2H_2O$ and cationic organic polymer to remove As. The soil washing with the 0.01N HCl_3% Na-dithionite solution for 30 minutes was the most effective in the As removal from the soil reducing As concentration from 43 mg/kg to 18 mg/kg. The As concentration of the washing solution was reduce from 19.2 mg/L to 0.76 mg/L after the waste water treatment.
Keywords: soil washing, arsenic, pH, Eh.

WIT Transactions on Ecology and the Environment, Vol 144, © 2011 WIT Press
www.witpress.com, ISSN 1743-3541 (on-line)
doi:10.2495/ECO110371

1 Introduction

The solubility, mobility and toxicity of As in soils are depending on its oxidation state [1]. Arsenite (As^{3+}) and arsenate (As^{5+}) are the most common species of arsenic under surface environment [2]. The concentration of As in soil pore-water is mainly governed by adsorption and precipitation. The adsorption is controlled by iron oxide and aluminum oxide in acidic soil and by carbonate in calcareous soil [3]. Iron (Fe)-, manganese (Mn) and calcium (Ca)-arsenic minerals are the most common As bearing mineral in oxidizing soil and As sulfides are the most common mineral in reducing soil. We developed a soil washing method for the As contaminated soil using the dissolution of As adsorbents and As containing minerals with the pH and Eh adjustment of washing solution.

2 Material and method

Arsenic contaminated soil was collected from a paddy field near a copper smelter operated for 50 years. The collected soil was air-dried and gently ground with a rubber stopper to pass 2 mm sieve. The texture, mineralogy, ignition loss, pH, electrical conductivity, contents of water soluble cation and anion, and contents of Fe and Mn oxide of the soil sample were determined. The As content of the soil was determined with an aqua-regia extraction method and the As was fractionated into water soluble, Al-oxide, Fe-oxide, reducible, oxidizable and residual forms. The soil washing solution containing 0.001–0.1 N HCl and 0–3% Na-dithionite and the soil reacted at 1soil and 4washing solution for 15–60 minutes. After the reaction and the separation of soil and washing solution, the contents of As in the soil and the washing solution were determined. The separated solution was treated to remove As and SO_4^{2-} with H_2O_2, $CaCl_2{\cdot}2H_2O$ and cationic polymer. After the treatment, the concentrations of As and SO_4^{2-} of the washing solution were determined.

3 Results and discussion

The physical, mineralogical and chemical properties of the soil are shown in table 1.

Table 1: Physical, mineralogical and chemical properties of the soil. Q: quartz, F: feldspar, C: chlorite, M: mica, K: kaolinite.

Texture	Min.	pH	EC	I.L.	Water soluble cation				Water soluble anion			DCB extrac.	
					Ca	Mg	Na	K	Cl	SO4		Fe	Mn
					mg/kg								
Silt loam	Q, F, C, M, K	6.13	0.78 (mS/cm)	5.7 (%)	116	64.6	499	61.9	187.5	169.2		2351	281.1

The As content of the soil determined with the aqua-regia extraction method was 43.0 mg kg^{-1} and Fe-oxide, reducible and residual forms were the dominant As fractions (table 2).

Table 2: As fractions in the soil.

Water soluble	Al-oxide	Fe-oxide fo	Reducible	Oxidizable	Residual
-----	-----	----- % (mg.kg) -----	-----	-----	-----
2 (0.78)	11.4 (4.51)	34.5 (13.69)	23.5 (9.30)	2.4 (1.10)	26.2 (10.30)

Amount of extracted As during the soil washing increased with decreasing Eh (fig. 1) and increasing amount of extracted Fe (fig. 2).

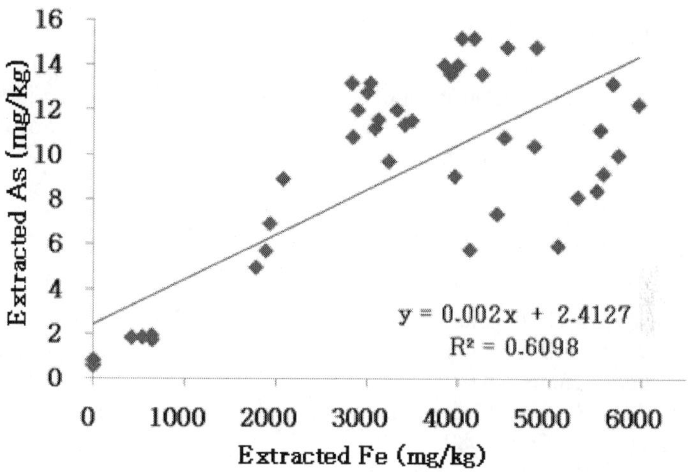

Figure 1: Relationship between amount of extracted As and Fe.

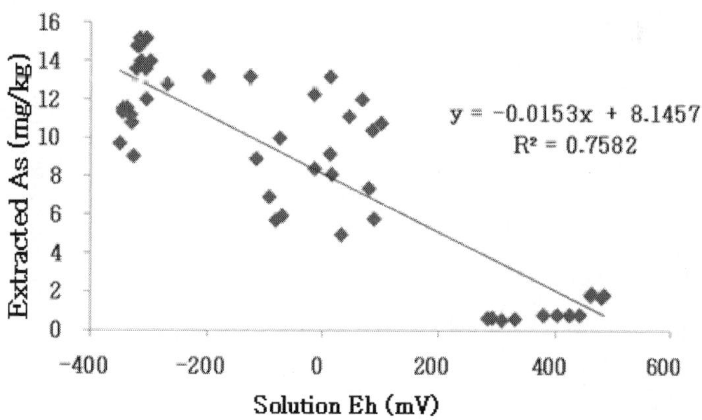

Figure 2: Relationship of washing solution Eh and amount of extracted As.

The optimum condition for As extraction from the soil was 0.01 N HCl_3% Na-dithionite washing solution and 30 minutes of reaction. The As concentration of the soil was reduced from 43.0 mg kg^{-1} to 18.3 mg kg^{-1} after the soil washing with the optimum condition. The As concentration of the washing solution reduced from 5.0 mg l^{-1} to 0.7 mg L_1 and SO_4^{2-} concentration from 10,400 mg l^{-1} to 2,440 mg l^{-1} with waste water treatment. The XRD and SEM data of the precipitate after the waste water treatment confirmed the presence of Fe-oxide and gypsum. The soil characteristics and chemical properties of the waste water indicate that As was extracted from the soil by dissolution of Fe-oxide with decreasing Eh and pH. The removal of As and SO_4^{2-} from waste water was due to the precipitation of Fe-oxide and gypsum by the treatment.

4 Conclusion

Extraction As from the soil during soil washing was mainly due to the increased solubility of Fe-oxide with the reduction of pH and Eh. The removal of As and SO_4^{2-} from the waste water was mainly due to formation of Fe-oxide and gypsum during the waste water treatment.

References

[1] Masscheleyn, P.H., Delaune, R.D. & Patrick, W.H., Effect of redox potential and pH on arsenic speciation and solubility in a contaminated soil. Environ. Sci. Technol., **25**, pp. 1414-1419. 1991.

[2] Cullen, W.R. & Reimer, K.J., Arsenic speciation in the environment. Chem. Rev., **89**, pp. 719-764, 1989.

[3] Muhammad, S., Arsenic chemistry in soils: a review of thermodynamic predictions and field observations. Water, Air, and Soil Contamination, **93**, pp. 117-136. 1997.

Measurement of concentrations of the natural radio-activities and Cs-137 in soil samples in the Al-Baha region of Saudi Arabia

A. A. Al-Zahrany & K. S. Al-Mogabes
Institute of Atomic Energy Research,
King Abdulaziz City for Science and Technology, Saudi Arabia

Abstract

Measurement of concentrations of naturally occurring radio nuclides together with Cs-137 deposited in soil have been carried out in the Al-Baha region of the Kingdom of Saudi Arabia. Soil samples collected from about fifty locations have been analyzed using high resolution hyper pure germanium spectrometers. Concentrations of the gamma emitting radio nuclides from naturally occurring uranium-radium and thorium series, Potassium-40, and Cesium-137 have been determined for the selected locations. Gamma-ray dose rates have been calculated from the measured concentrations of natural radioactivity in soil.
Keywords: natural radioactivity, Cs-137 fallout, NORM, radioactivity level in soil, Saudi Arabia.

1 Introduction

Naturally occurring radioactive material (NORM), arising from the decay of uranium-radium and thorium series and potassium-40, has been found in the earth's crust and soil, the underground water and even in living tissues and organs of any living organism. This presence has been recognized since the early 1930's. However it received minimal attention until the last few decades, when the role of terrestrial radiation as the main contributor to the collective effective dose of the world's populations has been recognized UNSCEAR [1–3]. Moreover, measurement of concentrations of some man-made radio nuclides such as Cs-137 and others in soil is too important to evaluate the contribution of

WIT Transactions on Ecology and the Environment, Vol 144, © 2011 WIT Press
www.witpress.com, ISSN 1743-3541 (on-line)
doi:10.2495/ECO110381

those man-made radio nuclides to the population's effective dose as well as to evaluate the magnitude of the radioactive fallout in the region.

Measurement of the naturally occurring radiation and radioactivity in Saudi Arabia started in the early 1980's [4–9]. However an intensive and systemic study of the environmental radiation levels and concentrations of natural and man-made radio nuclides in soil, water and other segments of the environment has been started by the Institute of the Atomic Energy Research (IAER) of KACST since 1992. The present work is a part of this systematic study. It is devoted to the study of the concentrations of naturally occurring radio nuclides and Cs-137 in soil, and to the assessment of the gross gamma dose rates in selected locations in Al-Baha region of the Kingdom of Saudi Arabia.

Al-Baha region is situated in the South-Western part of the Kingdom of Saudi Arabia and located between latitudes of 41°E to 42°E and longitudes of 19°N to 20°N. Al-Baha region is situated on a chain of Sarawat Mountains at an elevation of more than 2500 meters above sea level. It consists of six cities which are Al-Baha, Baljurashy, Al-Mandaq, Al-Makhwah, Qelwah, and Al-Qara. Al-Baha city represents the centre of the region. The surface area of Al-Baha region is approximately 36000 square kilometres and divided into 31 administrative centres. Moreover, it is one of the most densely populated provinces in the Kingdom of Saudi Arabia with a population of 411,888 [11].

2 Materials and methods

2.1 Sampling and sample preparation

In this research, fifty soil samples were collected from different locations throughout the Al-Baha region. The coordinates of sampling locations were determined by Global Positioning System (GPS) and the map of Al-Baha region is shown in Figure 1. The samples were randomly selected from undisturbed sites situated in open areas far away from buildings and any other constructions, and the distance between the neighbouring sites was about 5-15 km. The soil sample representing a given site was collected from three different points forming an equilateral triangle of 10m length at a depth ranging from 0 to 30 cm below the surface and then the collected soil from the different points with equal portions was mixed together.

The mixed soil samples were dried in open air for enough time and then dried an oven at 100 °C for about 24 hours to achieve a constant dry weight. After that, the dried soil samples were ground to obtain homogenous powder and sieved through a mesh size 2mm diameter. Finally, a one liter volume of each soil sample was weighed and transferred to one litre Marinelli beakers [9, 10]. After sealing the Marinelli beaker with a standard volume soil sample they were stored for one month to reach the secular equilibrium between the radium and thorium content of the soil samples and their daughters [9, 10].

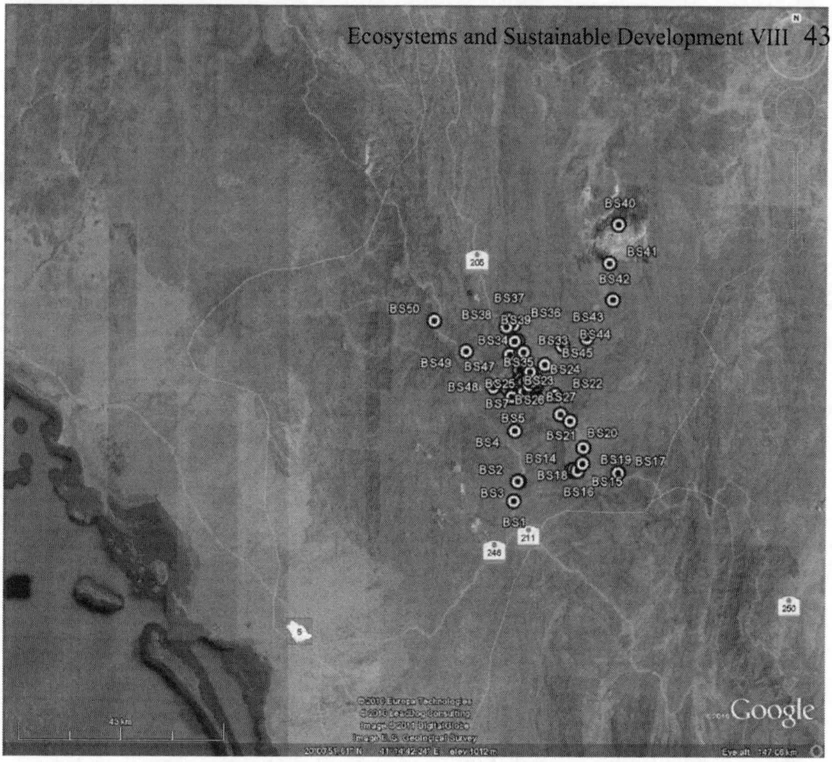

Figure 1: Map of Al-Baha region showing the locations of the collected samples (map produced by using Google Earth software).

2.2 Counting system and calibration

The samples were analyzed spectroscopically and gamma-ray spectra emitted from soil samples were measured using different high resolution hyper-pure germanium spectrometers with energy resolution better than 1.95 keV for the 1332 keV gamma-ray line of Co-60, and with relative efficiencies ranging from 20 to 40%. Energy and efficiency calibration of the spectrometers in the energy range from 60 to 3000 keV have been carried out in the same geometrical configuration used for measuring soil samples using a set of Ba-133, Eu-152, Eu-154, Am-241 and Ra-226 standard sources. All these standard sources were distributed homogeneously in one litre Marinelli beaker.

Gamma-ray spectra emitted from the soil samples were collected in 8000 channels of the PC for a time period of 24 hours to obtain satisfactory statistical errors. The different full-energy peaks of measured spectra were identified and analyzed using Gamma Vision and Genie 2000 analysis software's.

3 Results and discussion

Concentrations for U-238, Th-232, Ra-226, K-40, and Cs-137 of fifty sites, obtained from the analysis of the studied samples are given in table 1, in Bq/Kg

Table 1: Concentrations of radionuclide in soil samples and exposure rate in different sites of Al-Baha Region.

Sample No.	^{238}U (Bq/Kg)	^{232}Th (Bq/Kg)	^{226}Ra (Bq/Kg)	^{40}K (Bq/Kg)	^{137}Cs (Bq/Kg)	Exp. Rate Calc.(μR/hr)	*Longitude North*	*Latitude& East*
1	10.6	10.33	10.63	327.29	0.47	2.85	41 26 306	19 48 108
2	5.68	6.75	6.94	250.89	0.22	2.02	41 26 730	19 50 245
3	9.95	11.6	12.32	408.27	5.02	3.40	41 26 897	19 50 245
4	13.14	11.8	13.05	311.16	2.24	3.03	41 26 346	19 55 737
5	53.83	41.97	36.41	493.62	0.28	7.41	41 25 972	19 59 389
6	7.79	7.06	8.05	386.23	0.25	2.72	41 25 470	20 00 362
7	9.57	8.3	7.62	132.29	0.09	1.65	41 27 390	19 59 919
8	4.74	4.36	3.95	129.91	2.03	1.13	41 27 700	19 59 815
9	2.95	3.65	3.79	134.08	0.53	1.09	41 28 418	20 00 217
10	6.58	5.8	4.9	84.61	0.01	1.09	41 28 701	19 59 510
11	2.76	1.94	2.02	64.92	0.91	0.55	41 28 535	19 59 854
12	4.78	2.72	3.79	120.51	1.08	0.96	41 28 649	20 00 504
13	5.54	4.32	4.62	168.23	5.48	1.34	41 28 836	20 00 722
14	7.21	7.18	5.77	285.87	0.24	2.14	41 33 062	19 51 455
15	10.54	11.58	9.76	319.66	4.87	2.84	41 33 385	19 51 530
16	7.98	12.14	9.46	352	6.96	3.01	41 33 751	19 51 416
17	8.91	14.1	12.2	273.61	0.19	2.97	41 38 387	19 51 208
18	19.67	4.61	4.01	138.41	0.5	1.19	41 34 313	19 52 191
19	2.94	4.46	3.38	189.11	0.07	1.37	41 34 364	19 53 907
20	14.37	16.72	9.07	487.51	0.32	3.90	41 32 804	19 56 793
21	2.96	1.81	2.57	149.44	0.42	0.95	41 31 678	19 57 520
22	7.66	6.67	7.68	310.8	4.38	2.33	41 31 045	19 59 576
23	4.78	5.69	5.52	231.48	3.65	1.77	41 28 975	20 01 129
24	8.78	7.47	8.49	125.37	LLD	1.61	41 28 414	20 01 075
25	14.27	15.91	12.78	322.03	3.61	3.35	41 28 083	20 02 064
26	16.89	14.11	17.06	379.86	LLD	3.75	41 27 905	20 01 355
27	7.61	9.93	10.06	304.24	1.39	2.68	41 27 940	20 00 647
28	11.33	6.36	9.84	267.94	5.28	2.26	41 27 068	20 00 709
29	6.39	6.34	7.35	418.95	3.97	2.77	41 26 727	20 00 833
30	10.94	9.77	10.65	326.31	14.9	2.80	41 26 738	20 00 990
31	9.89	10.87	10.9	368.86	0.81	3.09	41 26 950	20 01 259
32	10.24	8.95	9.86	335	3.56	2.74	41 26 859	20 02 310
33	5.34	4.64	4.84	324.7	5.35	2.07	41 27 012	20 04 148
34	10.23	10.3	9.44	511.69	0.04	3.59	41 27 325	20 04 150
35	6.3	9.41	7.69	439.87	12.39	3.10	41 26 605	20 05 356
36	2.26	2.98	3.11	281.65	0.7	1.66	41 26 149	20 07 094
37	6.32	3.45	5.27	205.17	0.04	1.49	41 25 496	20 07 654
38	2.74	1.35	2.43	279.82	0.34	1.49	41 25 334	20 06 871
39	11.31	10.6	10.93	421.34	9.48	3.30	41 26 266	20 05 319
40	8.73	11.12	7.37	412.71	0.35	3.08	41 38 513	20 17 863
41	18.1	8.44	18.81	141.35	0.11	2.40	41 37 441	20 13 707
42	9.3	12.39	10.5	592.43	0.66	4.16	41 37 826	20 09 762
43	14.47	14.07	15.02	395.25	3.54	3.69	41 34 755	20 05 646
44	6.57	7.14	6.93	185.24	0.64	1.76	41 31 865	20 04 795
45	13.01	12.21	15.27	324.3	8.13	3.26	41 29 799	20 02 818
46	14.03	11.77	13.42	428.37	9.54	3.57	41 27 383	20 00 970
47	2.47	3.3	3.56	248.91	1.67	1.56	41 25 732	20 03 880
48	3.76	3.12	3.87	298.81	0.46	1.79	41 23 815	20 00 431
49	17.78	16.87	16.18	468.09	3.19	4.28	41 20 656	20 04 267
50	20.76	13.42	16.08	366.31	7.96	3.58	41 16 917	20 07 543

for dry soil. The uranium-238 was calculated from the 63.29 keV photo-peak which was completely isolated from any other photo-peaks in the spectra, while the Ra-226 was calculated from, 295.21, 351.92 and 609.31 keV photo-peaks arising from its daughters in equilibrium. The 269.41, 338.40 and 911.07 keV photo-peaks were used to calculate the concentration of Th-232 in equilibrium with its daughters. The 661.66 and 1460.75 keV gamma-ray lines were used to calculate the concentrations of Cs-137 and K-40 respectively. For Ra-226 and Th-232 the given concentrations are the average ones for the three most intensive photo-peaks. The relative errors of the measured concentrations lie between 5% for the higher experimental concentrations and increases to about 20% for lower concentrations.

The minimum and maximum concentrations for U-238, Th-232, Ra-226, K-40, and Cs-137 together with the mean values over the whole region are given in table 2.The U-238 concentrations ranged from 2.26 Bq/kg to 53.93 Bq/kg with average value of 9.89Bq/kg, The Th-232 concentrations ranged from 1.35 Bq/kg to 41.97 Bq/kg with average value of 9.04Bq/kg, The Ra-226 concentrations ranged from 2.02 Bq/kg to 36.41 Bq/kg with average value of 9.10Bq/kg, The K-40 concentrations ranged from 64.92 Bq/kg to 592.43 Bq/kg with average value of 298.49Bq/kg, and The Cs-137 concentrations ranged from 0.01 Bq/kg to 14.90 Bq/kg with average value of 2.88Bq/kg. For the radio nuclides U-238, Th-232, Ra-226, and K-40 the highest concentration were found in soil sample number 5, this may due to the geological formation of the location.

Table 2: The range and average concentrations of radio nuclides and exposure rate in soil samples of Al-Baha region.

Sample No.	^{238}U (Bq/Kg)	^{232}Th (Bq/Kg)	^{226}Ra (Bq/Kg)	^{40}K (Bq/Kg)	^{137}C (Bq/Kg)	Exp. Rate Calc. (μR/hr)
Avg.	9.89	9.04	9.10	298.49	2.88	2.53
Min.	2.26	1.35	2.02	64.92	0.01	0.51
Max.	53.83	41.97	36.41	592.43	14.9	7.85

Table 3 shows comparison of concentration ranges of U-238, Th-232, Ra-226, K-40 and Cs-137 in the studied soil samples with the national and the international studies. It is seen from this table that the mean concentrations for all naturally occurring radio nuclides averaged over in the whole region are comparatively approximately similar to the national level [8, 12, 13]. Moreover, the mean concentrations for all naturally occurring radio nuclides averaged over in the whole region were found less than the international level for soil in United States, China, and Japan [2].

The gamma ray effective dose for each location, calculated from measured concentrations of Th-232, Ra-226 and K-40 and the conversion factors, that are used to convert concentrations to dose rate, given in reference[15] and corrected to give the effective dose rate. Using the mean values of the calculated dose rates (2.53 μR/hr) the average annual effective dose for Al-Baha region was evaluated to be 0.21mSv/year, which is comparatively less than the annual average dose (worldwide) from all natural sources of exposure, which is equal to 2.4 mSv/year [2, 3].

Table 3: Comparison of average concentrations of radio nuclides in soil samples of Al-Baha region to the national and international studies [2, 8, 12, 13].

Region	^{238}U (Bq/Kg)	^{232}Th	^{226}Ra	^{40}K	^{137}Cs
This work	9.89	9.04	9.10	298.49	2.88
Riyadh Region	12.91	17.79	17.91	200.9	2.18
Eastern Region	8.82	3.08	9.79	136.44	0.66
Tabouk Region	15.49	21.49	15.08	232.9	1.10
United State	35	35	40	370	-
China	33	41	32	440	-
Japan	29	28	33	310	-

4 Conclusions

The level of natural radioactivity and man-made radionuclide such as Cs-137 in soil samples from Al-Baha region has been measured using gamma-ray spectroscopy. The mean concentrations for all naturally occurring radio nuclides averaged over in the whole region were found less than the international level for soil in United States, China, and Japan. Moreover, the average annual effective dose for Al-Baha region was found comparatively less than the world average.

References

[1] United Nation Scientific Committee on the effect of Atomic Radiation. UNSCEAR 1993, Report to the general assembly, New York, United Nations, 1993.

[2] United Nation Scientific Committee on the Effects of Atomic Radiation (UNSEAR). Source, effects and risks of ionizing radiation, New York: United Nations, 2000.

[3] United Nation Scientific Committee on the Effects of Atomic Radiation (UNSEAR). Source, effects and risks of ionizing radiation, New York: United Nations, 2008.

[4] Martin, P. Estimate of Gamma-ray dose equivalent rates due to natural radioactivity sources in Saudi Arabia. *Proc. of Int. Sym. Appl. Technol. Ionizing Radiation, Vol.3-1647, 1982.*

[5] Owain, R. S., Environmental radiation monitoring., M.Sc. Thesis, College of Engineering, King Saud University, Riyadh, 1984.

[6] Al-Hussan, K. A. and Wafa, N., Environmental radiation background level in Riyadh city, *Radiation Protection Dosimetry, Vol.40, No.1, 1992.*

[7] Al-Haj, A. N., Measurement of natural radiation background level of Riyadh City, M.Sc. Thesis, College of Engineering, King Saud University, Riyadh, 1987.

[8] Al- Area, A. M. Al-Dinah, S. K. Al-Marsha, A. I. Al-Yamani, A.Y., Al-Zhrany, A. A., Bagazi, A. O., Farouk, M. A. and Shabana, E. I.,

Measurements of natural and man-made radio nuclides in the soil in Riyadh region. *Proceedings of ASIA Congress on Radiation Protection, Beijing, 212-215 (1993).*

[9] Al-Zahrany, A. A., Comparative study of radioactivity in ground water and soil in one region of the K.S.A, M.Sc. Thesis, College of Engineering, King Saud University, Riyadh, 1994.

[10] International Atomic Energy Agency, IAEA. Measurement of radio nuclides in food and the environment. Vienna, Austria: Technical Reports Series no. 295. 1989.

[11] Central Department of Statistics and Information in Kingdom of Saudi Arabia. http://www.cdsi.gov.sa/english/index.php

[12] Al-Zahrany, A. A., Al-Mogabes, K. S., and Farouk, M. A. Measurement of concentrations of the natural radio-activities and Cs-137 in the soil in the Eastern Region of Saudi Arabia. *Proceedings of 3rd Arabian Conf. for Peaceful Uses of Atomic Energy, Damascus, Syria, 9-13 Dec 1996.*

[13] Al-Kheliewi, A. S., Al-Mogabes, K. S., and Al-Zahrany, A. A. Measurement of natural and man-made radio nuclides in soil of Tabouk region. *Proceedings of 3rd Arabian Conf. for Peaceful Uses of Atomic Energy, Damascus, Syria, 9-13 Dec 1996.*

[14] IAEA-TECDOC-566, The Use of Gamma-Ray Data to Define the Natural Radiation Environment, 1990.

[15] International Atomic Energy Agency, IAEA, The use of gamma-ray data to define the natural radiation environment. Vienna, Austria: Technical Report Series No.566, IAEA, 1990.

Section 8
Waste management

Carbon dioxide balance and cost analysis for different scenarios of solid waste management

D. Panepinto & G. Genon
DITAG Department, Politecnico di Torino, Italy

Abstract

The final destination of municipal solid wastes requires the realization and localization on the territory of important industrial plants (separation plants, WtE systems, final landfills, and also other innovative solutions, like digesters or gasifiers), together with the important aspects of cost; careful consideration is also required about the effects of externality, chiefly concerning carbon balance, with reference to GHG production. Taking into account the necessity to define the choice for the rational individuation of the most correct waste management system, in the present work we evaluated a specific Italian case relative to the waste management for a city located in Piedmont (North Italy): it is important to consider that the amount of wastes that are produced in this city corresponds to about 540.000 t/y. For the correct management of this load we considered different scenarios, by introducing thermal structures (direct incineration and gasification plants) and other complementary systems (mechanical separation plants, anaerobic digesters). The alternative scenarios have been carefully studied and compared from the environmental and economic point of view in order to obtain suitable choice criteria. The results of the specific case analysis clearly indicate the environmental and economic convenience of the solution based on the direct incineration.

Keywords: incineration, gasification, anaerobic digestion, carbon dioxide balance, cost analysis, social costs.

1 Introduction

The final destination of municipal solid wastes requires the realization of important industrial plants (separation plants, thermal systems, final landfills, and eventually, biological reactors); for a correct choice, the aspect of costs must

WIT Transactions on Ecology and the Environment, Vol 144, © 2011 WIT Press
www.witpress.com, ISSN 1743-3541 (on-line)
doi:10.2495/ECO110391

be carefully considered, but also the effects of externality must be taken into account, and in particular carbon balance must be evaluated in order to define the specific contribution to GHG production.

As regards the environmental impact it is possible to observe that the BAT adoption has led to a substantial reduction of many pollutants emission, but against this the impact arising from carbon dioxide release must be considered, and with this purpose the best energetic exploitation of the system must be obtained.

In the general structure of integrated waste treatment and disposal a strong improvement of selective collection, aimed at reuse and production of secondary materials, can today be observed. Downstream from this opportunity, there is however a very important flow of undifferentiated waste, to which also the residuals from the material valorization must be added; for this flux it is required to individuate suitable prospects for disposal. From this flow, that in any case represents a high percentage of the total waste production, the fraction useful for reuse (metal, glass, the organic compounds) is effectively removed and the remaining part is basically formed from a combustible fraction, a minor wet organic fraction and a residual substantially inert mineral fraction. On the basis of this composition, and by considering the interest for the thermoelectric co-generative production from non-fossil sources, it is convenient to propose a perspective of energy recovery of this fraction, with a double aim: to find a correct destination for the disposal of this stream, and at the same time to contribute to production of energy in a compatible way.

A first choice that is today largely discussed concerns the energetic valorization of the undifferentiated flow (through direct combustion in incinerators or through treatment in plants using innovative technology, chiefly gasification plants), or, instead, the proposal of mechanical separation systems, downstream the selective collection, systems that appear essentially aimed at separation of three fractions, a fuel fraction (known as RDF, which can be designed for specific plants or introduced into the market of the fuels), a wet fraction (for which, with the possibility of biological stabilization in anaerobic digestion plants, also prospects of indirect energy valorization could be individuated), and finally a mineral fraction, for which a final landfill destination seems to be the only practical option possibility.

Between the two alternative above mentioned schemes, the first is certainly the most diffuse today, both in Italy and in the most technologically equipped part of the Europe, for simplicity reason, economic convenience, liberation from the problems of secondary flows except the necessity to dispose of the obtained ashes. As concerns the second scheme, the attention is today directed to it with always more important development hypothesis, on account of the fact that this scheme is considered more environmentally compatible, less impacting on the area in terms of necessary infrastructures, and more oriented in the direction of development of synergies with other plants that are already present on the territory.

In the present work we have carefully evaluated these different technologies for waste disposal and at the same time for the valorization of the wastes; in fact on the basis of the necessity to define a strategy of rational choice for the correct management of the wastes, the aim of the work is to individuate in a general way an evaluation methodology useful to define the correct solution for the management of the wastes produced in the examined area.

2 Methodology

As indicated in the introduction, the aim of this study is the comparison and the consequent best choice among scenarios for residual waste management, downstream from selective collection of the fractions that can be reused.

The first scenario that in this perspective is considered is that of combustion of the residues in conventional moving-grate system; this technology is today widely spread at national level and even at world level.

In opposition to this solution, great interest has been attributed to the perspective of waste gasification, chiefly in the scheme that is followed by energy use, but also potentially with the consideration about production of chemicals. In this sense it is surely convenient to mention the perspective of mechanical separation of undifferentiated waste into three main flows: a dry flow destined to a energetic valorization, a wet flow of organics destined for an anaerobic digestion with production of biogas and finally a stream with a predominant mineral composition, substantially inert, to be disposed of in a landfill. For a correct individuation of the more appropriated waste management system the two above reported technologies must be evaluated.

A correct approach to these aims must include the following steps:

- definition of an overall flow chart, which identifies the need for different equipment or plant systems that are needed for each scenario; this quantitative scheme must consider the appropriate mass balances for the individual components of the input wastes, and their intermediate and final destination;
- identification, by literature data, from notice of other plants or eventually from pilot tests, of the expected performance for the various equipment, in particular as regards the emission flows and the capacity for power generation (thermal and electric energy);
- evaluation of the effects of energy immission in the area where the supposed plant will be realized, and connected possibility to replace existing energetic generation capacities and corresponding pollutants flows;
- determination of the building costs for the all equipment that is introduced in each scenario, of the operative costs corresponding to the various items of consumption, of the economic returns from the sale of energy or of any materials;
- evaluation of the external costs, as they can be defined on the basis of emission flows and their cost of externality; these costs can be found, for example, in the ExternE project , or they can be determined on the basis of

quality scenarios for the receiving environments and consequent evaluation of the effects of this quality on the external costs;

- From the point of view of the definition of external costs from the emissions of the disposal equipment, all the emissions to air, surface and underground water, soil should be conceptually considered; there is, in this context, a precise and well-proven method of approach that presents a particular reference to systems for waste disposal. In this paper, with the aim to arrive to a first comparison (orders of magnitude), but especially taking into account the results that are reported in previous literature studies , where the relative importance of different forms of impact is defined, it was considered sufficient to take into account in the evaluation of the externality costs the only external cost of CO_2 emission, that can be correctly estimated by its contribution to the GHG and to the consequent climate change;
- As a further consideration about the discussion on the acceptability of this important simplification, another aspect should be taken into account, in addition to the fact that this form of impact is the more significant, and so the used approximation is certainly minimized; the carbon dioxide impact can be exactly calculated on the basis of mass and energy balance, as opposed to other forms of impact, for which different specific technological structures could lead to large variations in the emission levels. In this sense the results that will be obtained at the end of the elaboration can be considered as significant and acceptable.

- Definition of the final social cost (industrial plus external) are obtained for the disposal by also taking into account the contemporary strategy of energy production.

The development of a calculation procedure so defined seems to propose a general scheme, but the specific results are obviously a consequence of different local problems (pre-existing scenarios for power generation, environmental sensitivity of the interested area, quantitative dimension of the problem to be solved, ability to handle significant technological innovations); hence it must be stressed the fact that it is impossible to establish an optimal technology choice in a general sense, but this outcome should result only from a specific analysis for a particular case. It seems impossible to explore all the different possibilities that are present in the areas where this study could be addressed, so it was decided in this paper to illustrate only a specific particular case by considering with this simplification to clarify the modalities of implementation of the presented methodology; from this reconstruction of scenario some more general considerations should be extracted.

3 Objective of the study

The different above illustrated technology schemes have been carefully evaluated in a specific Italian case study: it consists in the correct individuation of waste management modality for final destination of wastes that are produced in the city of Piedmont, in the north of Italy. The amount of the wastes produced in this city corresponds to about 540.000 t/y (constituted by 470.000 t/y of MSW and 70.000 t/y of sovvalli). In this city the realization of an incineration plant with a capacity of about 420.000 t/a was already decided upon, but as a consequence of a higher estimated production a new waste management overall solution is required (of course the incineration plant already under construction will be part of any strategy).

For the correct individuation of the waste management we studied three different scenarios:

- scenario A: it corresponds to the thermal treatment with direct combustion in a large incineration plant of the whole amount of the wastes produced in the city;
- scenario B: it is similar to scenario 1, but with the adoption, together to an incineration plant, also of a gasification plant, by sharing the total waste production between the two thermal plants;
- scenario C: firstly the separation of the whole amount of the wastes produced in the city is considered, and from this operation the direct combustion in an incineration plant of the obtained dry fraction and the management in a dry anaerobic digestion plant of the wet fraction are introduced.

On account of the observed impossibility of potentiality considerations to send the RDF that could be obtained to a cement plant, or to other existing plants, this opportunity has not been considered.

The three introduced scenarios have been evaluated by means of the following tools:

- analysis of the operating systems that must be present in the three different scenarios, and corresponding definition of the required plants and the landfill necessity;
- carbon dioxide balance; it has been performed by using mass and energy balance, with individuation of two emission factors (expressed as tCO_2/t waste and tCO_2/MWh);
- cost analysis;
- definition of the social (industrial plus external) costs for the analyzed scenarios, with application of the above-mentioned tools.

3.1 Fluxes characterization

As concerns the product composition of waste in input to the various systems that are present in the different scenarios, there are some differences. In scenarios A and B we assumed to treat a MSW with the product composition that is reported in table 1. In the scenario C, as we mentioned, the MSW (composition as table 1) is initially sent to a process of mechanical separation. From this treatment two flows (dry and wet respectively), are obtained, with a product composition that has been obtained by numerical calculations on the basis of procedure of literature (J.R. Burton et al. [1]); the results are shown in table 2.

In table 3 the operative parameters are reported that were used in the different elaborations. For all scenarios we considered a cogenerative disposition.

Table 1: MSW product composition.

PRODUCT COMPOSITION (scenarios A e B)	
Product fraction	%
Paper	25
Paperboard	7
Textile	5
Wood	1
Plastic	19
Glass and inert	9
Metal	3
Domestic organic	22
Green materials	2
Fine materials	5

Table 2: Treated waste product composition.

PRODUCT COMPOSITION (scenario C)		
Product fraction	Wet fraction (in input at digester)	Dry fraction (in input at incineration plant)
Paper	11%	32%
Paperboard	-	9%
Textile	-	7%
Wood	-	1%
Plastic	8%	25%
Glass and inert	-	13%
Metal	-	4%
Domestic organic	81%	2%
Fine materials	-	7%

4 Elaborations and results

4.1 Analysis of the systems

In the following table we can see the operating plants that are present in the three different scenarios and the potentiality of them. We can also see the landfill necessity for each of the considered scenarios (the value has been obtained with elaboration of the data reported in table 3).

Table 3: Parameters used in the elaborations.

Parameters	Value
MSW LHV [MJ/kg], scenarios A and B	11,5
Dry waste fraction (treated waste) LHV [MJ/kg], scenario C	13,7
Plant availability [h/a], scenarios A, B and C	7.800
Combustion energetic yield ($\eta_{plant\ inc}$), scenarios A, B and C	90% (ATO-R [2])
Gasification energetic yield ($\eta_{plant\ gas}$), scenario B	80% (ATO-R [2])
Electric energy generation efficiency, scenarios A and B ($\eta_{el\ t}$) [%], scenarios A, B and C	19% (TRM [3])
Thermal energy generation efficiency, scenarios A and B ($\eta_{th\ t}$) [%] scenarios A, B and C	51% (TRM [3])
Electric energy generation efficiency ($\eta_{el\ b}$) [KWh/Nm^3_{biogas}], scenario C	2,5 (A. Canovai [4])
Thermal energy gen. efficiency ($\eta_{th\ b}$) [KWh/Nm^3_{biogas}], scenario C	1,9 (A. Canovai [4])
Avoided CO_2 , electric recover – EF_{el} [g CO_2/kWh], scenarios A, B and C	675 (GME [5])
Avoided CO_2, thermal recover – EF_{th} [gCO_2/kWh], scenarios A, B and C	210 (GME [5] modification by Burton [1])
Biogas production (η_{biogas}) [% on organic in input at the digester], scenario C	20 (De Baere and Boelens [6])
Digestate production [% on organic in input at the digester], scenario C	80 (De Baere and Boelens [6])
Bottom ash production [% on waste in input at thermal plant], scenarios A, B and C	22
Fly ash production [%on waste in input at thermal plant], scenarios A, B and C	4

4.2 Carbon dioxide balance

As we have seen, in account of the fact that the carbon dioxide is one of the main GHG, a carbon dioxide balance study is required for the impact evaluation on climate change of the different proposed scenarios.

In order to obtain a carbon dioxide balance we propose the following considerations:

- for the scenario A and B, where only thermal treatment of the waste is introduced, we considered that all the carbon that is contained in the entering waste (C_{total}) leads to carbon dioxide; so the following equation can be used:

$$CO_{2\ produced}\ [t] = C_{total} * PM\ CO_2/PM\ C * \text{Total waste} \quad (1)$$

- for scenario C, we considered the carbon dioxide coming from separation, the carbon dioxide coming from incineration and obviously, also in this case, the carbon contained in the input waste and transformed in carbon dioxide; we considered also the carbon remaining in the digestate. So for this scenario it is valid the following equation:

$$CO_{2\ produced}\,[t] = CO_{2\ separation} + CO_{2\ incineration} + CO_{2\ biogas} + (C_{digestate} - C_{soil\ remain}) * 44/12 \qquad (2)$$

with:

$CO_{2\ separation} = 67.5$ kg $CO_2/\ t_{waste}$ (Blengini and Genon [7] modified with elaboration of table 3]

$C_{soil\ remain} = 10\%\ C_{digestate}$ (Boldrin et al. [8]; Moller et al. [9])

- for the three scenarios we took into account the carbon dioxide coming from substituted energetic plants.

$$CO_{2\ avoided} = Power_{\ th,\ el} * availability\ (h/y) * EF_{\ th,el} \qquad (3)$$

- for the energy production we utilize the following equation:

$$Energy_{\ thermal\ plant}\ [MWh] = LHV * tot_{\ waste} * \eta_{plant\ inc,\ gas} * \eta_{el,th,\ I} \qquad (4)$$

$$Energy_{\ anaerobic\ digestion}\ [MWh] = tot_{\ waste} * \eta_{biogas} * \eta_{th,\ el,\ b} \qquad (5)$$

Table 4: Analysis of the systems.

	Scenario A	Scenario B	Scenario C
Waste treated [t/y]	540.000	540.000	540.000
Incineration plant[t/y]	540.000*	420.000**	408.400***
Gasification plant[t/y]	-	120.000	-
Separation plant[t/y]	-	-	540.000
Anaerobic digestion plant [t/y]	-	-	131.600
Landfill (bottom ash) [t/y]	118.800	118.800	89.848
Landfill (fly ash) [t/y]	21.600	21.600	16.336
Landfill (digestate) [t/y]	-	-	105.280
Total landfill [t/y]	140.400	140.400	202.776

* 540.000 t/y = 470.000 t/y (MSW) + 70.000 t/y (sovvalli)
** 420.000 t/y = 350.000 t/y (MSW) + 70.000 t/y (sovvalli)
*** 408.400 t/y = 338.400 t/y (dry fraction from separation) + 70.000 t/y (sovvalli)

So, on the basis of the above reported considerations and on the basis of the data that are reported in tables 1, 2 and 3, in table 5 we can see the results of the carbon dioxide balance.

By analyzing the results of table 5 we can see the environmental advantage for the scenario A that introduces the thermal treatment of the total amount of the produced waste with direct combustion in incineration plant.

Table 5: Results of the carbon dioxide balance.

	Scenario A	Scenario B	Scenario C
Treated wastes [t/y]	540.000	540.000	540.000
Produced energy [MWh]	1.087.619	1.062.779	1.092.159
Produced CO_2 [t/y]	841.812	841.812	866.408
t CO_2 / t wastes	1.56	1.56	1.60
t CO_2 / MWh	0.77	0.79	0.79
Avoided CO_2 [t/y]	365.673	357.321	375.504
t CO_2 / t wastes*	0.88	0.90	0.91
t CO_2 / MWh*	0.44	0.46	0.45
*at net of the avoided CO_2			

In this context it is important to highlight that, also by considering only the fossil fraction of carbon dioxide, the same result is obtained; in fact it must be considered that the definition of the share of fossil and biogenic carbon dioxide depends only on the waste composition in input to the various scenarios, and not on the management systems that are adopted.

4.3 Cost analysis

After these results it is important to consider also the results of the cost analysis (table 6).

Table 6: Cost analysis.

	Scenario A	Scenario B	Scenario C
Capital cost [€]	250.000.000	300.000.000	305.000.000
Amortization cost [€/t]	46.3	55.6	56.5
Operative cost [€/t]	76	90	112
Energy revenue [€/t]	102	97	97
External cost, CO_2 [€/t]	16.7	17	17.2

The capital cost and the operative cost are defined on the basis of the market costs relative to the different plants that are present in each scenario, by taking also into account a scale effect.

The amortization cost is calculated with the following equation:

$$\text{Amortization cost } [€/t] = (\text{Capital cost} * 10\%) / \text{waste total} \tag{6}$$

with:

10% = amortization percentage [sources: capital Italian market].

The energy revenue is calculated on the basis of the market price relative to the electric and thermal energy produced and of also of the prizes for energy obtained from renewable sources (in Italy green certificates). The external cost is defined on the basis of the CO_2 external cost (Rapporto ExternE [10]).

Also, in this case, scenario A leads to the best costs.

4.4 Definition of the social cost

After we individuated and examined the environmental and economic aspects of the scenarios. We have performed a global comparison of them by using the tool of the social cost.

The social cost is the result of the addition (in our study with ratio 1:1) of the environmental aspects that have been defined by means of the calculated balances (table 5) and transformed in an external cost, and of the economic aspects, as they have been defined from a specific analysis (table 6). We obtain, in this way, a final overall index of global evaluation of the analyzed aspects; by this index it is possible to arrive to an organic and global comparison of the different proposed scenarios.

On the basis of the above reported discussion the social cost is defined by the following equation:

$$\text{Social cost } [€/t] = \text{industrial cost} + \text{external cost} \tag{7}$$

The industrial cost is composed by the following parameters: “amortization cost”, “operative cost” and “energy revenue”.

By analyzing the results that are reported in the table it shows that the best scenario is the scenario A, where the thermal treatment in an incineration plant of the total amount of waste produced in the studied city is used.

5 Conclusions

On the basis of the reported elaborations it is possible to arrive at the following specific considerations, with a value that is referred to the considered local situation:

Quantitative considerations. From a quantitative point of view, a fundamental problem that is observed is the residual necessity of landfill. By analyzing the results of table 4 we can observe that the bottom ash landfill necessity, in the scenario C, is lower in comparison with the values for the only thermal systems, but, in the same scenario C, the necessity of a big volume for the management of digestate arises. In fact the only real destination for the digestate, in the analyzed context, on the base of the national legislation, is its landfill disposal. Plus, the possibility to send it to thermal treatment must in fact be excluded: in consideration of the very large flow rate it would require the construction of a new facility, and also it is very unlikely a place in an existing conventional grate system for biomass thermal utilisation could be found. In addition to these reasons, the thermal destination must be excluded also on account of thermotechnical considerations: the digestate flow present has a significant water content and consequently it would significantly worsen the conditions of combustion. It is also impossible to suggest its use for agricultural purposes as fertilizer or a soil improver, on account of existing strict quality requirements imposed by Italian law (Legislative Decree 217/2006).

Environmental considerations. From the point of view of the environmental impact we distinguish between local and global impact. From the local point of view the main criticality is represented by the incineration emissions: for their limitation it is necessary to realize a structure that has been conceived on the basis of the BAT; in this case its suitability, and consequently avoidance of local environmental impact, can be considered as acquired. On this basis the use of incineration for MSW or for the treatment only of a separated dry fraction is not considered to lead to appreciable differences. In the scenario of the MBT plant there are two additional elements must be taken into account as the origin of impact: the separation plant and the digester; from these plants, chiefly, the impact from the separator must be evaluated on account of odour emissions, while for the digester we can establish that with a convenient and efficient plant structure the impact can be eliminated. From the point of view of the bottom ashes there are no differences among the various scenarios, while from the point of view of the management of the digestate its landfill disposal, without further stabilization operations, could lead to local impacts due to volatilization phenomena. Finally, from the point of view of global impact, this must be considered chiefly with reference to carbon dioxide emissions. From the previously discussed calculations (table 5), it can be observed an advantage for the direct total thermal treatment.

Economic considerations. From the point of view of the costs, we observe (table 6) that the investment cost is lower for the scenario A that uses the direct combustion in an incineration plant. The scenario C has an investment cost that can be compared with that of the scenario B. Also from the point of view of the operative cost the scenario A results are the more convenient. For the scenario C the adoption before the incineration plant of a another step of pre-treatment necessarily entails an important increase in the operative costs, in comparison with the two thermal scenarios. Finally, from the point of view of the energetic revenue, this very important and meaningful aspect is higher in the scenario A. The external costs of the CO_2 for the three scenarios obviously reflect the trends of the CO_2 emissions at net of the avoided CO_2 avoided.

For the above discussed thematic considerations a large confirmation can be obtained by analyzing the elaborations results of the social cost: in fact it is able to put together the environment and the economic aspects (table 7). By analyzing this index, with the purpose of a general overview of the analyzed scenarios, it is clear that the scenario is more favorable for option A.

On the basis of the above reported discussion, from the general point of view, the scientific relevance of the conducted work consists of the definition of a logical scheme for comparison that can be adapted to different management and disposal requirements.

In fact, as has been shown, the created comparison methodological scheme, valid and tested for the specific analyzed case, can be considered a useful

Table 7: Social cost.

	Scenario A	Scenario B	Scenario C
Amortization cost [€/t]	46.3	55.6	56.5
Operative cost [€/t]	76	90	112
Energy revenue [€/t]	102	97	97
Industrial cost [€/t]	**20.3**	**48.6**	**71.5**
External cost, CO_2 [€/t]	16.7	17	17.2
Social cost [€/t]	**37**	**68.6**	**88.7**

approach for the definition of the best solution for specific territorial necessity of waste management.

References

[1] J. R Burton et al., "Waste sorting and RDF – Production in Europe", Elsevier Applied Science Publishers, 1985;

[2] ATO-R – Verifica della fattibilità di un impianto di trattamento termico dei rifiuti a tecnologia innovativa nella provincia di Torino, Ingegneria Ambientale, Quaderni, 51 (2010);

[3] TRM – Impianto di termovalorizzazione dei rifiuti della provincia di Torino – Italia. "Bilanci termici e di massa", 2006;

[4] A. Canovai, Consorzio Italiano Compostatori, "La proposta operativa: modello tecnico-economico impiego sistema UMICA", workshop, Salerno 18 March 2009;

[5] GME – Gestore Mercati Energetici. Website: <http://www.mercatoelettrico.org/It/Default.aspx> accessed 12 July 2010;

[6] L. De Baere and J. Boelens (OWS) "Rest or mixed waste sorting – digestion – separation for the recovery of recyclables and energy", website: <http://www.ows.be>, accessed 12 July 2010;

[7] Blengini GA., Genon G. (Politecnico di Torino),"LCA del sistema integrato dei RSU nella Provincia di Torino", Provincia di Torino, July 2008;

[8] Boldrin A. et al., "Composting and compost utilization accounting of greenhouse gases and global warming contributions", Waste Management and Research, 2009:27, 800-812;

[9] Moller J., et al., "Anaerobic digestion and digestate use: accounting of greenhouse gases and global warming contribution", Waste Management and Research, 2009:27, 813-824;

[10] Valore di mercato (Rapporto Externeè), website: <http://www.externe.info>; accessed 12 July 2010.

Reducing municipal waste through promoting integrated sustainable waste management (ISWM) practices in Surabaya city, Indonesia

D. G. J. Premakumara, M. Abe & T. Maeda
Institute for Global Environmental Strategies, Japan

Abstract

Solid waste management is considered to be one of the most serious environmental issues confronting urban areas in developing countries. In keeping with global trends, many cities are focused on integrated approach based on the 3Rs (reduce, reuse and recycle) and trying to reduce the amount of waste generation at source rather than later at the end-of-pipe. However, these efforts are still very limited with a mix of results and relatively few efforts have been made to regulate organic materials that usually comprise over 50% of the total waste generation in the cities. Despite the fact, Surabaya city, the second largest in Indonesia with a population of three million people has shown to have succeeded in reducing its municipal waste generation by more than 20% over the last four years, introducing waste separation and reduction activities at household, community-managed waste collection and polluter pay principle, promoting composting practice by setting up composting centers and distributing thousands of compost baskets to residents, and organising community cleanup campaigns and waste recycling activities with local NGOs, private sector (formal and informal) and the media. Surabaya's achievement exemplifies how a city can reduce a large amount of waste in a short period of time spending limited budget by establishing Integrated Sustainable Waste Management System (ISWM), which seeks institutional, social, environmental, political, technical and financial aspects while emphasising the critical role of involving various stakeholders, covering waste prevention and resource recovery, including interactions with other urban systems and promoting and integrating different habitat scales from household, neighbourhood to city.

Keywords: integrated sustainable waste management, composting, recycling, participation, informal sector, community-based organisations.

WIT Transactions on Ecology and the Environment, Vol 144, © 2011 WIT Press
www.witpress.com, ISSN 1743-3541 (on-line)
doi:10.2495/ECO110401

1 Introduction

Solid waste management is considered to be one of the most serious environmental issues confronting urban areas in developing countries (UN-HABITAT [1]). Rapid urbanisation, increasing industrialisation, rising incomes and more sophisticated form of consumerism are leading to an increase in the amount of solid waste (Casanova [2]), which is in turn creating enormous challenges for municipal authorities in providing effective waste collection services to its residents. The *Global Waste Management Market Assessment* (2007) indicates that municipal solid waste generation worldwide in 2006 at two billion tonnes with a 37% increase forecast by 2011 Key Note Publication Ltd [3]. The daily waste generation in urban areas of Asia was about 0.7 million tonnes in 1999 and it is expected to reach about 1.8 million tonnes per day by 2025 (World Bank [4]). It is common for many municipalities in developing countries to spend 20-50% of their available budget on solid waste management, even though 30-60% of all the urban solid wastes remain uncollected and less than 50% of the population is served (UNEP [5]). In addition, many landfills in developing countries are poorly designed and uncontrolled, which creates serious environmental impacts (Zurbrugg *et al.* [6]). Further, existing landfills have almost reached their maximum capacity and finding new sites that offer environmentally sound and economically viable options is becoming increasingly difficult.

To address these issues, many cities in developing countries focus on establishing an integrated solid waste management system based on the 3Rs and trying to reduce the amount of waste generation at source rather than later at the end-of-pipe. However, these efforts are still very limited with a mix of results and relatively few efforts have been made to regulate organic materials that usually comprise over 50% of the total waste generation in the cities (Dulac [7]). The legitimate question arises how cities in developing countries can be deled with municipal solid waste in an efficient, effective and sustainable manner.

Despite the fact, Surabaya city, the second largest in Indonesia with a population of three million has shown succeed in reducing its municipal waste generation by more than 20% over the last four years, introducing waste separation and reduction activities at household, community-managed waste collection services and polluter pay principle, promoting composting practice by setting up composting centres and distributing thousands of compost baskets to its residents, and organizing community cleanup campaign and waste recycling activities with local NGOs, private sector (formal and informal) and the media.

This paper therefore aims to examine the key factors in the success of the Surabaya's solid waste management system by elaborating the Integrated Sustainable Waste Management (ISWM) framework. It is structured as follows: Section 2 gives a general concept of the ISWM and its relevance as an analytical framework. Then, section 3 analyses the municipal solid waste management in Surabaya city and discusses the city's new waste management system in establishing material recycling society, promoting compost utilisation. Section 4 analyses in details the key factors supported for the successful implementation of

Surabaya's new waste management system based on the ISWM framework, taking into account the stakeholder participation, waste prevention and resource recovery through interaction of other urban systems and integration of different habitat scales, and integration of different aspects of sustainability rather than overemphasis on technical aspects alone. Finally, section 5 concludes by drawing lessons learned from the evidence and discussions. The paper provides both general information and case study analysis based on the information gathered through literature review, field observations and discussions with the relevant stakeholders.

2 The concept of integrated sustainable waste management (ISWM)

2.1 Moving towards sustainable solutions for municipal solid waste management

Municipal solid waste is commonly defined as household waste or domestic waste and wastes of a similar nature generated by commercial, industrial, public institutions and from public spaces, such as streets, parks and gardens [5]. In the past, municipal solid waste management was seen largely as a technical problem with engineering solutions [1]. It was primarily included waste collection, land disposal and incineration. However, this thinking has changed during the 1980s and 1990s, when it became apparent that land filling and incineration have significant environmental impacts, as well as not enough to deal with the large volumes of solid waste generated in the cities.

Due to these issues and the emergence of the sustainable development concept resulted to foster the idea of sustainable municipal solid waste management. Effective management of municipal solid waste is essential to achieve the sustainable development, as mentioned in Agenda 21: "environmentally sound management of wastes was among the environmental issues of major concern in maintaining the quality of the Earth's environment and especially in achieving environmentally sound and sustainable development in all countries" (Commission on Sustainable Development [8]).

Gertsakis and Lewis [9] suggest that true sustainability will require significant increases in the efficiency of resource use recovering more valuable products from the municipal waste stream. It became apparent that the practice of the 3Rs is the basic requirement for sustainability in municipal waste management. According to Smith and Scott [10], the waste management hierarchy is a protocol to maximise the recovery options and to minimise disposal through open dumping, limiting negative impact on the environment and natural resources as much as possible is the basis for achieving sustainable waste management.

The sustainable municipal waste management, however, cannot be separated from good governance (Gilbert *et al.* [11]). It has been shown that municipal authorities could not successfully collect and remove waste without the active cooperation from the service users [1]. Further, cities came to know that the 3Rs technology depend on institutional, governance and policy frameworks, which are highly varied, and complex, and directly related to local conditions. To address these issues, the concept of ISWM was emerged and it identifies the importance of integrating all stakeholders involved in municipal waste management, the elements of the municipal waste management system and aspects of the local context that should be taken into account for achieving sustainable municipal waste management in developing countries (van de Klundert [12]).

2.2 Integrated sustainable waste management (ISWM) framework

The ISWM, as shown in Figure 1, is a framework that was first developed by WASTE, a Dutch non-governmental organisation (NGO), and its partners in developing countries in the 1980s, and further developed by the Collaborative Working Group on Solid Waste Management in Low and Middle Income Countries (CWG) in the mid 1990s (Anschutz *et al.* [13]). Since then, it has become popular as an approach to reach better, more sustainable solutions to municipal waste management in developing countries (Anschutz *et al.* [14]).

Figure 1: The ISWM framework.

3 The municipal solid waste management system in Surabaya city

Surabaya, the second largest city in Indonesia located in eastern Java, serves as an important commercial and industrial capital of the province. The city of three million people generated over 1,500 metric tonnes of municipal waste a day in the following sectors: residential (68%), markets (16%), commercial/industrial (11%), streets and open spaces (5%) (Medina [16]). Further, the composition of municipal waste generated in Surabaya shows that more than 50% is organic and ideal for composting [16]. The municipal solid waste management of the city consist of three sub-components, including waste collection, transportation and final disposal.

3.1 Before the project implementation: collection and disposal oriented system

3.1.1 Waste collection

In residential areas, people use different types of storage bins, such as plastic or metal bins and the fixed brick-boxes in front of the house for their household waste storage. People are simply put their garbage into the bins and collectors pick it up, usually the next day. Since the introduction of the community primary collection (Copricol) law in 1980, the collection of waste from residential areas is managed by the Rukun Warga (RW), a neighbourhood association, organises the waste collection scheme in the area collecting fees from households, hiring waste collectors, providing pushcarts, and paying salaries to waste collectors for their services in collecting and transporting waste from household units to transfer stations. Collectors pick up waste from the storage bins using bamboo baskets and take it to a transfer station in handcarts.

Collection of waste from commercial, institutional establishments and from street sweeping is carried out by the Cleansing and Landscaping Department of the municipal authority. The city's waste collection coverage is about 70% in 2005, the remaining 30% is left in the streets, ditches, and open spaces, which blocks the drainage system, contaminates water resources, and results in increased insect and rodent populations (Silas [17]). This situation is at its worst during the rainy season, particularly in low-lying areas where most of the urban poor reside.

3.1.2 Transfer station

A transfer station is used to temporally dispose the collected waste from residential areas, before it will transport to the landfill site aiming to reduce the hauling distances for collection trucks, thus lowering transportation costs. Some of the transfer stations are relatively modern, provided with attendants and equipped with mechanical transfers and waste compaction, whereas others are simply a large steel container, concrete bin, and/or open space.

3.1.3 Transportation

The collected waste is then transported to the landfill site at Benowo, about 22 miles west, in an area of 40 acres (16 ha). The Cleansing and Landscaping Department of the Municipality is in-charge of transporting the waste accumulated at transfer depots to the final disposal site using municipal containers and trucks. Due to heavy traffic, it takes a long time to reach them, which limits each truck to a maximum of two trips to the disposal site per day. Lacking pollution controls, the landfill site is an open dump and produces smell and odour to environment. Sometimes, it is on fire, particularly in dry seasons and smoke disturb to the people and the environment. It was also estimated that the landfill site has a remaining lifespan of five years, and is facing difficulties in finding a new site due to a scarcity of public land.

3.1.4 Recycling activities

Surabaya city does not have a municipal recycling programme, but scavengers have been recovering materials from waste for recycling and reuse for decades. According to some studies, scavengers play an important role in the waste management system and the economy of Surabaya, recovering an estimated 30% of waste generated by the city's residents, creating income opportunities for urban poor and a supply of inexpensive raw materials to industry [16, 17] these studies further finds that scavenging can also create problems. For searching reusable and recyclable materials, scavengers spread the waste and sometimes scatter it, making collection more difficult. Direct contact with waste also impacts scavenger health.

3.1.5 Financial aspects

The residents pay two separate user fees for waste collection services. One is to the RW for collecting waste from households, and the other is to the city office for transportation and disposal. The monthly fee for the collection of waste from households is decided collectively by the community members and is usually paid together with other community fees. The monthly free for transportation and disposal is calculated by depend on the living condition of the residential area and is paid together with the monthly water supply or electricity bill. Annual user fees collected by the city amount to only half of the 6% city budget allocated for waste management.

3.2 Alternative municipal waste management: implementation of resource recovery system

As a practical solution for the growing issue, a new municipal solid waste management system was developed under a city to city environmental cooperation between Surabaya and Kitakyushu city, Japan in 2004.

3.2.1 Designing a pilot project in Kampong Rungkut Lor

The Kitakyushu International Techno-Cooperative Association (KITA) in Kitakyushu and Pusdakota, a local NGO in Surabaya worked together in designing a pilot project in Kampong Rungkut Lor, a low-income neighbourhood located adjacent to the largest industrial area in the city. It was a typical low-income neighbourhood with about 200 families and the issue of solid waste has often ignited arguments.

In the new system, people were educated to separate organic materials at source, before putting the waste out for collection. The organic waste is then collected separately by Pusdakota twice a week and transported to the nearby composting center. At the center, it is processed using the Takakura Composting Method (TCM), named after a composting expert, Koji Takakura from JPEC, a consulting firm in Kitakyushu, was developed based on the traditional windrow composting method. The composting centre gradually produced a good quality of compost in large quantities from separately collected organic waste in two or three weeks and sold to the nearby farmers, private vendors and residents.

Though, programme faced typical constraints in the beginning, such as a lack of community participation and community willingness to separate waste at source, as a result of regular education and public campaigns, as well as the actual visible benefits observed on the ground, the mindset of community members changed, which led to increased participation, discouraged out-side dumping, and resulted in greener and cleaner community. In responding to the community requests, TCM was further modified for use at the household level.

Some households involved making compost from their kitchen waste using a simple compost basket and lessons provided by Pusdakota and used their product for fertilising household gardens. Pusdakota also started to function as a material recycling facility (MRF) by collecting inorganic waste, including recyclable, separately, thus encouraging further waste segregation at source and linking informal recycling business into the municipal waste management system. The pilot experience shows that if it works well, a new waste management system can reduce at least half of waste that needs to be land filled.

3.2.2 Moving towards city-wide programme

Surabaya city realised the benefits of the pilot project and took leadership to scale-up the new waste management system by adopting the same composting method in partnership with other local NGOs and a women's group called Pendidikan Kesehahteraan Keluarga (PKK). A system of environmental carders (one carder for every ten households) was established throughout the city to share information on the new waste management system, assist new families in starting household composting, and educate them about the benefits of keeping the environment clean and green. Further, they involved in following-up with monitoring of household composting and troubleshooting by helping families with their composting activities. In 2009, there were about 26,700 environmental carders, covering all 8,800 neighbourhoods in the city.

The households who have general knowledge of the functions of using the compost bins were given free bins and to date, the city has already distributed about 19,000 composting bins, purchasing from Pusdakota (one basket costs about US$ 10) aiming to support for their voluntary activities. In addition to household composting, the city supports for composting projects at the community level. Currently, 16 community-based composting centres have been functioned in the city to process waste collected from neighbourhoods, markets, streets, and parks. These composting centres have produced about 300 metric tons of compost per month, which are used by the city for parks and street greening programmes.

Since 2005, Surabaya city has been organised a different types of neighbourhood competitions, including Cleanest District Award, Green and Clean Competition, and Free from Waste Competition in partnership with several media partners in the city (Java Pos, JTV (local TV), Suara Surabaya (local radio), Radar Surabaya (local newspaper) to motivate and strengthen community participation in the city's new waste management system, and encourage them to improve the neighbourhood environment. About 15 small-and-medium scale recycling businesses have been promoted with the Unilever

Peduli Foundation (UPF), an NGO affiliated with Unilever Indonesia to support the activities of informal waste pickers and their networks. Further, city issued a local regulation No. 1/2006 on community-based solid waste management and incorporated this new strategy in the preparation of the midterm development plans of 2006-2010.

As a result, the waste collection ratio was increased by 90% and daily average waste generation, which used to be more than 1,500 tonnes a day in and before 2005, has decreased 1,300 tonnes in 2007 and 1,150 tonnes in 2008 as shown in figure 2. The city has perceptively become cleaner and greener, as acknowledged by many residents and with its achievements honoured with some national and international awards, including the Adipura Award (Clean City Award) from the government of Indonesia for three consecutive years since 2006 and the Energy Global Award from Austria in 2005.

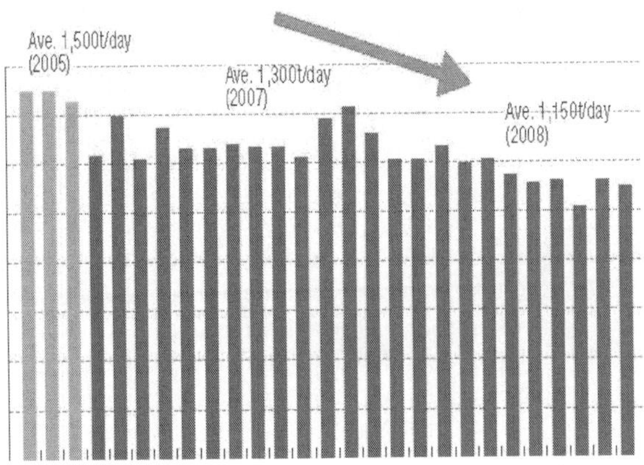

Figure 2: Waste reduction by 20%.

4 The key factors for the successful implementation of the municipal waste management system in Surabaya city

This section discusses the key factors for the successful implementation of the municipal solid waste management system in Surabaya city based on the ISWM framework elaborating its three key elements (system elements, stakeholders and aspects) that need to be addressed for municipal waste management system to work well and to work sustainably over time.

4.1 Integration of waste management system elements

Surabaya's municipal waste management system has been proved effective in reducing a significant amount of waste generation in a short period of time by establishing waste separation at source, neighbourhood-managed waste

Table 1: Municipal waste management system at habitat scales.

Habitat scale	Collection and disposal system (before the project implementation)	Resource recovery system (new waste management system)
Household level	Storage at source	Separation at source Household composting Recycling Storage at source
Neighbourhood level	Primary collection	Primary collection Sorting and pre-treatment Recycling Composting
City level	Secondary collection Final disposal	Secondary collection Sorting and pre-treatment Recycling Composting Final disposal

collection, recycling and composting at different habitat scales, and integrating them as much as possible at the city level as shown in table 1.

The hierarchy advocated in many industrialised countries with high standards of living (giving high-priority for waste reduction), though, Surabaya's experience reveals that exploring possibilities of engaging various stakeholders in recycling organic waste through composting would alone significantly improve the situation and alleviate the problems of municipal waste management. The reason is that organics are usually the largest component of municipal solid waste in developing countries. The greatest reduction of waste collection and disposal can be achieved by diverting organic waste component from the waste stream. Composting organic waste, either at household level or on a community level, can prove to be a worthwhile effort. A typical challenge to establish a market and demand for the compost products can be overcome by integrating with other urban systems, such as urban agriculture and public parks.

Like many other cities in Indonesia, recycling and recovery activities in Surabaya is carried out by the informal sector on all levels of the municipal waste management system. Such work is done in a very labour-incentive way and for income earning. This gave an opportunity for the city to integrate the existing recycling systems in order to maximise the use of the waste as a resource, to protect and develop people's livelihoods and to reduce the costs to the city of managing the municipal waste, rather than establishing new recycling streams.

4.2 Integration of different stakeholder groups

Stakeholders are all groups and individuals who have a stake, an interest and play different roles in the municipal waste management system [15]. Traditionally, solid waste management is a municipal responsibility and provided all the elements of the waste management system. However, the experience of Surabaya city has shown that cooperation and coordination between different stakeholders is essential to develop an effective and sustainable municipal management system as shown in figure 3.

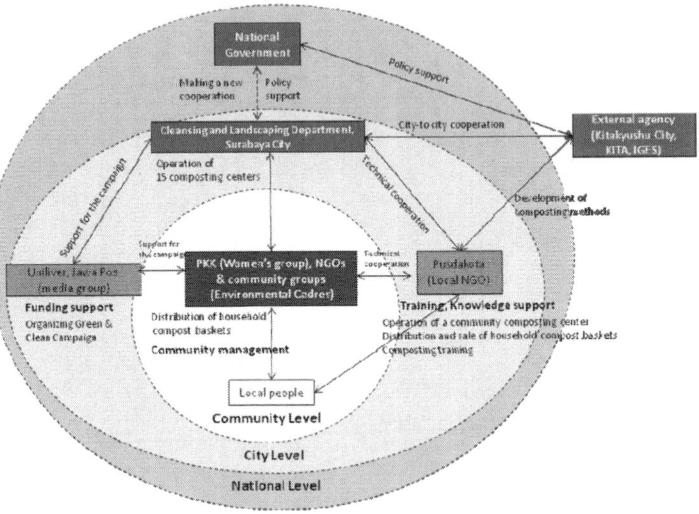

Figure 3: Stakeholder partnership in Surabaya city.

A municipal solid waste system consists of three main groups of stakeholders: the providers (who actually offer the service), the users (who are the clients) and the other external agents. Users or waste generators are key stakeholders in municipal waste management and their participation is essential for the successful implementation of municipal solid waste management programmes. They can play a range of roles, such as, waste prevention, separation at source, placing waste outside for collection, reuse, making composting using organic materials, participating in design of a waste service, participating in clean-ups, paying for waste management services, and monitoring and supervising the operation of services. Community involvement leads to more responsible behaviour, increased environmental awareness, and a higher willingness to pay among users of a waste management system. It can also lead to empowerment of groups of stakeholders that have had limited access to decision-making power and resources and build social capital within the community.

The role of community-based organisations (CBOs), NGOs, research institutes, and universities is essential in supporting communities and the informal sector, including organising, providing training, advocacy, and support them with research, technical and financial assistance. National and local governments can establish enabling environment for local communities and the informal sector by changing legislation and recognising them as candidates for service contracts. To involve all these groups and to address the constraints they face requires a change in the attitudes of governments and politicians and giving some incentives encouraging community participation.

4.3 Key aspects of ISWM

The Surabaya's experience show that technologies for municipal solid waste management need to be simple, low-cost, easy-to-follow methods, adopt to the physical environment, topography and other physical requirements, preferably locally manufactured and based on indigenous technology, geared towards efficiency and optimum utilisation of equipment. However, sustainable solid waste management is definitely not only a technical challenge. Understanding and taking into account the environmental impacts, financial and economic benefits, social and cultural aspects, and the institutional, political and legal framework, is essential for planning and implementation of a sustainable solid waste management system.

5 Conclusion

The key conclusion from the Surabaya case study is that a successful solution for the municipal solid waste management needs to consider all of the three aspects of ISWM. The success of the ISWM system based on the 3Rs depends on the partnerships among all stakeholders, including waste generators, private sector (formal and informal), national and local governments, and international agencies. A clear vision, political commitment and change in attitudes as well as technical, financial, legal, and institutional capacity enabled these partnerships and maximised its impact.

References

[1] UN-HABITAT, *Solid Waste Management in the World's Cities,* Earthscan Ltd: London and Washington, D.C., p. V, 2010.

[2] Casanova, L., Modernizing solid waste in the Asian Tigers (Key Sheet 1). *Solid Waste Management in the World's Cities,* ed. UN-HABITAT, Earthscan Ltd: London and Washington, D.C., pp. XXVI-XXVII, 2010.

[3] Key Note Publications Ltd, *Global Waste Management Market Assessment 2007,* Key Note Publication Ltd: Hampton, 2007.

[4] World Bank, *What a Waste: Solid Waste Management in Asia,* the World Bank: Washington, D.C., p. 1, 1999. www.worldbank.org/urban/solid_wm /erm/CWG%20folder/uwp1.pdf

[5] UNEP, *Developing Integrated Solid Waste Management Plan: Training Manual (Volume 4: ISWM Plan)*, UNEP: Osaka and Shiga, p. 2, 2009.

[6] Zurbrugg, C., Drescher, S., Rytz, I., Maqsood Sinha, A. H. Md., Enayetullah, I., Decentralised composting in Bangladesh, a win-win situation for all stakeholders, *Resource Conservation and Recycling,* 43, pp. 281-292, 2005.

[7] Dulac, N., *The Organic Waste Flow in Integrated Sustainable Waste Management: Tools for Decision-Makers-Experiences from the Urban Waste Expertise Programme (1995-2001),* WASTE: Gouda, 2001.

[8] Commission on Sustainable Development, *Report of the United Nations Conference on Environment and Development, Rio de Janeiro, 3-14 June 1992, vol.1, Resolutions adopted by the conference*, United Nations Publication, 1992

[9] Gertsakis, J., Lewis, H., *Sustainability and the Waste Management Hierarchy – A Discussion Paper,* EcoRecycle: Victoria, 2004. www.cfd.mit.edu.au/content/download/189/1390/file/Sustainability%20and %20the%20Waste%20Hierarchy.pdf.

[10] Smith, P. G., Scott, J. S., *Dictionary of Water and Waste Management,* Elsevier: Butterworth and Heinemann, 2005.

[11] Gilbert, R., Stevenson, D., Girardet, H., Stren, R., *Maling Cities Work: The Role of Local Authorities in the Urban Development,* Earthscan Ltd: London, 1996.

[12] van de Klundert, A., *The Sustainability of Alliances between Stakeholders in Waste Management: Using the Concept of Integrated Sustainable Waste Management – Working Paper for UWEP/CWG*, 2000. www.gdrc.org/uem/waste/ISWM.pdf

[13] Anschutz, J., Ijgosse, J., Scheinberg, A., *Putting ISWM to Practice,* WASTE: Gouda, 2004.

[14] Anschutz, J., Keita, M., Rosario, A., Lapid, D., Rudin, V., *UWEP City Case Studies,* WASTE: Gouda, 2005

[15] van de Klundert, A., Anschutz, J., *Integrated Sustainable Waste Management – the Concept*, WASTE: Gouda, 2001.

[16] Medina, M., Community-based organics diversion, *BioCycle,* pp. 57-59, 2010.

[17] Silas, J., *Waste Management Policies in Surabaya: An Integrated Approach,* Kitakyushu Initiative: Kitakyushu, 2002. www.kitakyushu.iges.or.jp/successful_practices/solid_waste_management. html.

[18] Wilson, D. C., *Presentation on collection, sorting and recycling,* Regional Conference on Partnership in Municipal Solid Waste Management, ERM: UK, 2000

Section 9
Ecotoxicity

Community patterns of benthic macroinvertebrates in streams in relation to temperature variation using the Self-Organizing Map

S.-H. Park[1], D.-H. Kim[1], W.-S. Cho[1], M. Bae[2], Y.-S. Park[2], Y.-E. Na[3] & T.-S. Chon[1]
[1]Department of Biological Sciences, Pusan National University, Korea
[2]Department of Biology, Kyung Hee University, Korea
[3]Institute of Agricultural Science and Technology, Rural Development Administration, Korea

Abstract

Benthic macroinvertebrates were collected from 2,000 sample sites in the major river basins in South Korea from 1997 to 2002. In total 5 phyla, 9 class, 23 order, 111 family and 727 species of benthic macroinvertebrate were recorded during the survey period. Communities from the unpolluted sites (BMWP ≥ 45) were selected and were trained by the Self-Organizing Map (SOM). The benthic macroinvertebrates were accordingly clustered based on topography of river basins in a larger scale, and subsequently according to temperature variation within the river basins in a smaller scale. The scope of species distribution was also illustrated through visualization of the SOM. Some selected species appeared to be locally adaptable while other species were widely distributed across the river basins. Patterning communities based on the SOM was efficient in monitoring species distribution and adaptability of the species in response to temperature variation.

Keywords: benthic macroinvertebrate, community, stream, temperature variation, self-organizing map.

1 Introduction

A trend in climate change is currently undeniable and ubiquitous [1], and the effects on global flora and fauna are significant in a wide range [2] resulting in

WIT Transactions on Ecology and the Environment, Vol 144, © 2011 WIT Press
www.witpress.com, ISSN 1743-3541 (on-line)
doi:10.2495/ECO110411

potential responses such as species extinct, *in situ* adaptation, and migration to tolerable areas [3]. There are also examples of species that may have succumbed to the pressures of contemporary climate change [4]. In correspondence with climate change, large-scale distributional changes in species have been recorded in a numerous account [e.g., 5].

Spatial distribution of benthic macroinvertebrates communities is closely associated with a variety of natural (e.g., temperature, pH) and anthropogenic (e.g., organic matter) factors [6]. Temperature is one of the key factors in determining distribution and abundance of benthic macroinvertebrates [7]. Although there are reports regarding longitudinal and seasonal variations [8], spatial distribution of benthic macroinvertebrates specifically related to the temperature gradient has not been studied extensively.

A difficulty for checking community response to environmental stress lies on complexity of data, since benthic macroinvertebrate consist of multi-variables (i.e., diverse taxa), varying in a non-linear fashion. The self-organizing map (SOM) is an efficient tool for mining non-linear data and has been extensively used for patterning communities since 1990s [9, 10]. The SOM efficiently classified benthic macroinvertebrate communities in polluted streams and illustrated community response to anthropogenic disturbances pertaining to the sample sites [7, 9]. In this study, we elaborated the SOM clustering and visualization in the large scale community data to address the distribution pattern of benthic macroinvertebrates in relation to topographic and temperature variation in the national scale.

2 Methodology

2.1 Ecological data

Benthic macroinvertebrates communities were sampled in relatively clean to intermediately polluted areas in South Korea. The samples were collected from the 2,000 sites in the major river basins in Korea (fig. 1) from 1997 to 2002 by using a Surber sampler (30 cm x 30cm, 500μm).

Two longest rivers in South Korea, Nakdong and Han Rivers, were further divided into 2 and 3 regions, respectively, by considering topographic characteristics. The densities of collected individuals were converted to the unit area ($1m^2$). The details for sampling benthic macroinvertebrates have been conducted according to the National Natural Environment Monitoring Protocol in Korea [11]. Community composition and other ecological information on the collected data could be referred to Park *et al.* [12].

Overall 5 phyla, 9 class, 23 order, 111 family and 727 species of benthic macroinvertebrates were recorded during the survey period. Shannon diversity index and the revised Biological Monitoring Working Party (BMWP) [13] were measured to present biological water quality. In order for the polluted communities not to be mixed with the unpolluted communities for analysis, the 584 sites with BMWP values equal to or over 45 were selected for training with the SOM. The rare species collected less than 10 sites during the survey period were not included in the input data. Table 1 presents the number of sample sites, information on location, and biological indices for the major river basins used for analysis.

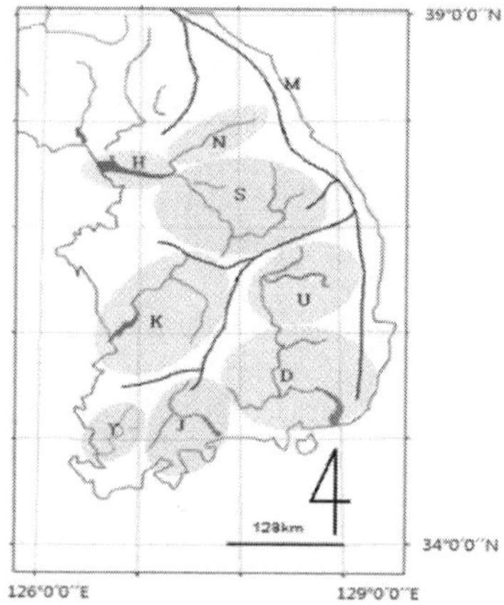

Figure 1: The major river basins for collecting benthic macroinvertebrates in South Korea (H: Han River, N: Bukhan River (North trajectory in Han River), S: Namhan River (South trajectory in Han River), K: Geum River, J: Seomjin River, Y: Yeongsan River, U: North region in Nakdong River, D: South region in Nakdong River), M: Mountainous range.

Table 1: The sample sites, location of the river basins, and diversity and BMWP indices for benthic macroinvertebrate communities.

Rivers	Abbreviation	Sites (No.)	Latitude(°) (Range)	Longitude(°) (Range)	BMWP (Range)	Shannon Diversity (Mean ± S.D.*)
Bukhan	N	79	37.5-38.5	127.5-129.5	48.4-132.6	2.9±0.3
Namhan	S	81	37.5-36.8	127.5-129.5	47.9-121	3.1±0.3
Han	H	49	36.8-38.5	126-127.5	51.3-150.8	2.8±0.3
Geum	K	69	35.8-36.8	126-128	47.2-138.6	2.9±0.3
Upstream of Nakdong	U	101	35.8-36.8	128-129.5	46.6-133.6	3.0±0.3
Downstream of Nakdong	D	42	34.5-35.8	128-129.5	58-147	2.9±0.3
Seomjin	S	136	34.5-35.8	127-128	46-134.2	2.8±0.3
Yeongsan	Y	33	34.5-35.8	126-127	46.-117	2.7±0.3

* Standard deviation

In order to check the longitudinal variation within the river systems in smaller scale, the data were rearranged after training the data for the whole river basins in South Korea. The data for south and north regions in the Nakdong River basin (138 species from 139 sites) were combined for training. Similarly the communities in the Namhan (S) and Han River basins (H) (167 species from 129 sites), and communities in the Bukhan (N) and Han River basins (H) (144 species from 127 sites) were separately combined for analysis. Beside the

Namhan and Bukhan River basins in mountainous area, only a few number of sample sites with high altitude were surveyed: four sites (over 600m) in the Nakdong River basin and one site (over 460m) in the Han River basin. The samples with high altitude in the Nakdong and Han River basins were not included in the data for analysis. For training the data within the same river basin, the species collected less than 5 sites were additionally excluded for analysis. In total 5 phyla, 9 class, 22 order, 107 family and 590 species were included in the input data for training.

2.2 Data analysis

We utilized the SOM for patterning macroinvertebrate communities [12, 14]. The output layer consists of *L* x *M* computation nodes in the SOM. With each neuron being represented as *j*, the output layer is arranged in two dimensions for convenience of visual understanding. Suppose a community data containing *S* species, and the density of species, *i*, is expressed as a vector x_i. The vector, x_i is considered to be an input layer to the SOM. In the network each node, *j*, is supposed to be connected to each node, *i*, of the input layer. The connectivity is represented as the weights, w_{ij} (t), adaptively changing in each iteration of calculation, *t*. Initially, the weight is randomly assigned in small values. Each neuron of the network computes the summed distance between the weights and the distance d_j (t) at output node *j* network is calculated as shown below:

$$d_j(t) = \sum_{i=0}^{S-1} \left(x_i - w_{ij}(t)\right)^2 \tag{1}$$

The input values with greatly different numerical values in densities are avoided for training. The data were transformed by natural logarithm in order to emphasize the differences in low densities. Subsequently the transformed data were proportionally normalized between 0 and 1 in the range of the maximum and minimum density for each species collected during the survey period.

The neuron responding maximally to a give input vector is chosen to be the winning neuron, the weight vector of which has the shortest distance to the input vector. The winning neuron and possibly its neighboring neurons are allowed to learn by changing the connecting weights in the manner to further reduce the distance between the weight and the input vector as shown below:

$$w_{ij}(t+1) = w_{ij}(t) + a(t)\left(x_i - w_{ij}(t)\right)Z_j \tag{2}$$

where Z_j is assigned to 1 for the winning (and its neighboring) neuron(s) while it is assigned 0 for the rest neurons, and *a(t)* denotes the fractional increment of the correction. Detailed algorithm could be referred to Zurada [15], Chon *et al.* [16]. After training, the Ward's linkage method [17] was applied to the weights of the SOM for clustering the patterned nodes. Training and clustering were carried out under the MATLAB environment [18].

3 Results

3.1 Overall pattern

The sample sites were clustered accordingly after training by the SOM (figs. 2(a) and 2(b)). A majority of communities were separately clustered according to the river basins while other communities were observed together in different clusters. The samples from the river basins K, U, S and N were separately grouped in clusters 4, 8, 1 and 2, respectively. In addition, cluster 6 accommodated the samples from both south (D) and North (U) regions in the Nakdong River. The sample sites belonging to these clusters showed unique patterns with community compositions. The communities collected from J, H and Y, however, appeared together in a mixed manner in clusters 3, 5 and 7 (fig. 2(a)). Detailed comparison of communities in different clusters is beyond the scope of this study and will be reported elsewhere.

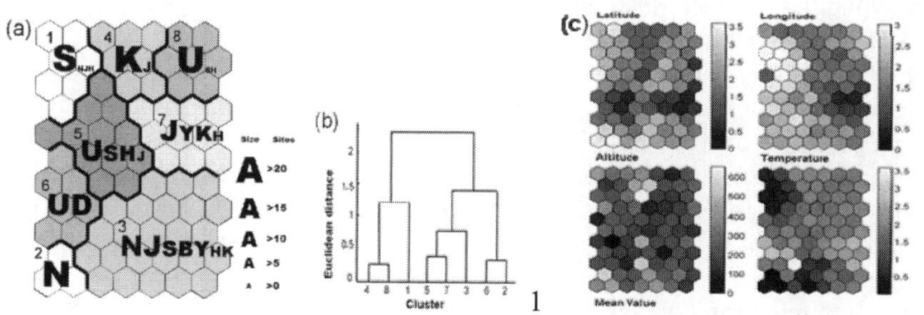

Figure 2: Community clustering in the 8 river basins by the SOM. (a) Clusters of the sample sites on the SOM. The size of the symbols indicates the number of sample size grouped within the clusters. (b) Dendrogram based on the Ward's linkage method. (c) Latitude, longitude, altitude and temperature corresponding to the SOM shown in (a). The numbers in the vertical bar indicate the real values minus the minimum values (latitude: 34.5°, longitude: 126°, altitude: 0m, temperature: 10.8°).

The patterned nodes (fig. 2(a)) were further compared with latitude, longitude, altitude and temperature at the corresponding positions on the SOM (fig. 2(c)). Clustering of communities was not clearly matched to the profile of latitude in the large scale. This indicated that overall community compositions of benthic macroinvertebrates in South Korea are not in accordance with the south-north longitudinal variation. In the profile of longitude, however, the gradient was observable: the values of longitude were higher in the left area, while the values were lower in the right area in the figure showing longitude (fig. 2(c)). This is understandable that topography in the Korean peninsula is divided into two regions in the east and west across the mountainous range longitudinally passing from north to south on the right side of the peninsula (fig. 1).

Consequently higher, colder mountainous areas are located in the east, while lower, warmer areas are placed in the west.

Altitude, however, did not clearly match community clustering (fig. 2(c)), while temperature was broadly in accordance with the profile of the longitude (fig. 2(c)). Temperature was generally higher in the area corresponding to the area of lower longitude (i.e., west area), while lower in the area matching to the area of higher longitude (i.e., east area) (fig. 2(c)). At the top and bottom left areas in fig. 2(c), values of temperature were especially lower, corresponding to the Bukhan (10.8°C±0.7) and Namhan (11.3°C±0.6) River basins, respectively.

3.2 Patterning within the river basins

Considering that the sample sites were clustered mainly according to the river basins in the large scale as shown in fig. 2(a). The two largest rivers in Korea, Nakdong and Han, were analyzed along the longitudinal axes. Fig. 3 shows clustering of the sample sites from the Nakdong River basin by the SOM. The communities were grouped along with the longitudinal axis (figs. 3(a), 3(b) and 3(d)). Cluster 1 mainly covered the sample sites from U (north region in the Nakdong River), while cluster 2 showed the middle range of the Nakdong River (fig. 3(c)), and cluster 3 corresponded to D (south region in the Nakdong River). The gradient of the latitude was associated with the vertical gradient in community clustering (figs. 3(a) and 3(d)). The profiles of longitude and altitude shown in fig. 3(d), however, did not match community clustering (fig. 3(a)). Temperature, in contrast, was in accordance with the gradient of the latitude and community clustering (fig. 3). The higher range in temperature was observed in the south region in the Nakdong area, while the lower range in temperature was shown in the north region in the Nakdong area (figs. 3(c) and 3(d)).

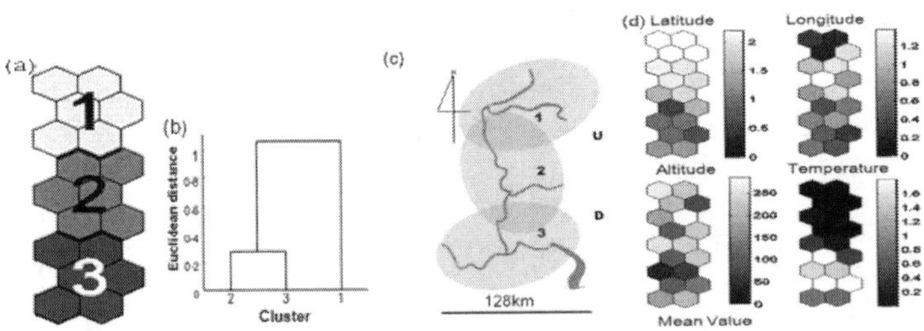

Figure 3: Community clustering in the Nakdong River basin by the SOM. (a) Clusters of the sample sites on the SOM. (b) Dendrogram based on the Ward's linkage method. (c) The Nakdong River basin matching to clusters. (d) Latitude, longitude, altitude and temperature corresponding to the SOM shown in (a). The numbers in the vertical bar indicate the real values minus the minimum values (latitude: 34.51°, longitude: 128°, altitude: 0m, temperature: 12.6°).

Since two main trajectories are formed from the Han River (fig. 1), communities from the Bukhan and Namhan River basins were separately patterned. The main stream of Han River (H) was added to each trajectory. The clusters accordingly revealed the longitudinal differences in the Namhan and Han River basins (figs. 4(a) and 4(b)). The samples were divided into 3 clusters (fig. 4(a)). The cluster 1 presented the upstream region in the Namhan River (S), while cluster 2 matched the junction between the Namhan and Han Rivers. Cluster 3 showed communities from the Han River (H). Similar to the case of the Nakdong River basins (fig. 3(c)), communities were accordingly patterned along with the longitudinal axis of the river basin (figs. 4(a) and 4(c)). The gradient of the latitude was observed from upstream to downstream in the Namhan River basin (fig. 4(d)). Longitude also showed the gradient from left to right region in the Namhan River basin. This type of concurrent changes in both latitude and longitude are understandable by considering that the Namhan River runs diagonally toward south-east from the river mouth (fig. 1). Combining latitude and longitude together, diagonal changes in community patterns from south-east to north-west were demonstrated in the Namhan River basin (figs. 1, 4(c) and 4(d)). Consequently temperature matched to both latitude and longitude.

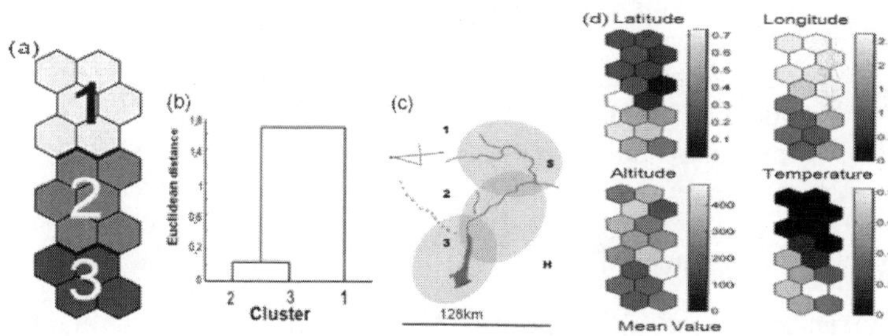

Figure 4: Community clustering in the Han and Namhan River basins by the SOM. (a) Clusters of the sample sites on the SOM. (b) Dendrogram based on the Ward's linkage method. (c) The Han and Namhan River basins matching to clusters. (d) Latitude, longitude, altitude and temperature corresponding to the SOM shown in (a). The numbers in the vertical bar indicate the real values minus the minimum values (latitude: 36.8°, longitude: 126°, altitude: 0m, temperature: 11.3°).

The communities in the Bukhan River basin were also divided into 3 clusters based on the SOM (figs. 5(a) and 5(b)). However, the community pattern was not formed in accordance with the longitudinal axis.

Cluster 2 presenting the upstream region in the Bukhan River (N) occupied the bottom position in community clustering, while cluster 1 corresponding to the junction of two rivers was placed in the top position. Cluster 3 matching the Han River basin (H) was located in the middle position in community clustering (fig. 5(a)). Latitude and longitude (fig. 5(d)) neither showed a clear matching to community clustering (fig. 5(a)) in this case. Altitude, however, was in accordance with community clustering. The high level in altitude (fig. 5(d)) corresponded to the top and bottom clusters shown in fig. 5(a). Profiles of temperature were in accordance with variation in altitude (fig. 5(d)). Since the Bukhan River is located in the mountainous area, and a numerous sample sites were selected in high altitude as stated above, the correspondence between altitude and community clustering was demonstrated in this river basin.

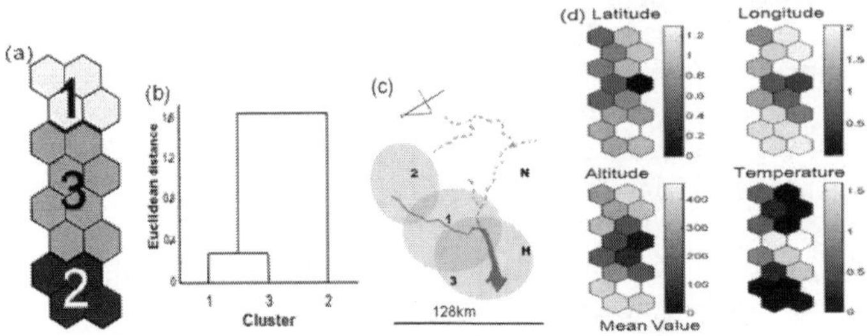

Figure 5: Community clustering in the Han and Bukhan River basins by the SOM. (a) Clusters of the sample sites on the SOM. (b) Dendrogram based on the Ward's linkage method. (c) The Han and Bukhan River basins matching to clusters. (d) Latitude, longitude, altitude and temperature corresponding to the SOM shown in (a). The numbers in the vertical bar indicate the real values minus the minimum values (latitude: 36.8°, longitude: 126°, altitude: 0m, temperature: 10.8°).

3.3 Distribution of species

The abundance profiles of the collected species were also accordingly displayed on the corresponding nodes in the SOM through visualization (figs 6–8). The species were broadly divided into broad and narrow groups in distribution. The species occurring in the narrow area were further divided into two types at the high and low temperature zones. Fig. 6(a) and 6(b) show the species distribution in narrow ranges in the Nakdong River basin for low and high temperature, respectively. Bottom region (fig. 6(a)) indicated lower latitude with high temperature, while upstream (fig. 6(b)) presented the opposite condition.

In the narrow range species such as *Ephemera strigata* and *Gammarus* sp were observed in the downstream zone (i.e., low temperature) (fig. 6(a)), while species such as *Hydropsyche* KUa,b and *Cincticostella castanea* were limitedly observed in the upstream zone (i.e., high temperature) (fig. 6(b)). Species including *Sieboldius albardae* were collected in a broad zone (fig. 6(c)).

Similar to the case of the Nakdong River basin, the scope of species distribution in the Namhan River basin showed the narrow (figs. 7(a) and 7(b)) and broad (fig. 7(c)) ranges. The collected species were divided into two narrow ranges matching the Han (bottom zone: fig. 7(a)) and Namhan (upper zone: fig. 7(b)) River basins. In the narrow range species such as *Laccotrephes japonensis* and *Ischnura asiatica* were observed at the bottom zone (i.e., higher temperature) (fig. 7(a)). Species including *Zaitzevia nitida* and *Iron maculates* were narrowly presented in the upper zone in the Namhan River basin (fig. 7(b)). Species such as *Ephemera separigata* was collected in a broad range (fig. 7(c)).

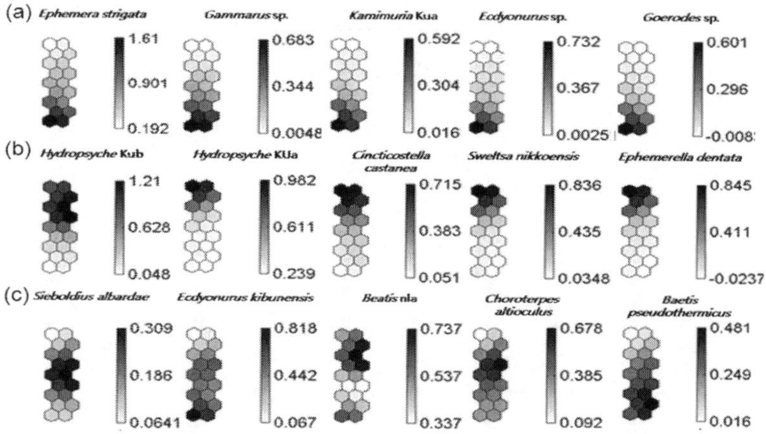

Figure 6: Species distribution in the Nakdong River basin. Gray levels in vertical bars indicate density in no. of individuals per m^2. (a) Species narrowly occurring in high temperature zone, (b) Species narrowly occurring in low temperature zone, and (c) Species occurring in a broad zone.

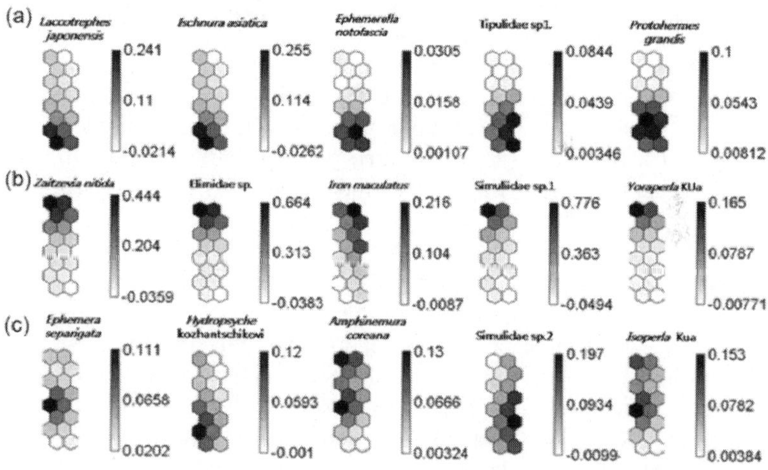

Figure 7: Species distribution in the Han and Namhan River basins. Gray levels in vertical bars indicate density in no. of individuals per m^2. (a) Species narrowly occurring in high temperature zone, (b) Species narrowly occurring in low temperature zone and, (c) Species occurring in a broad zone.

The scope of species distribution in the Bukhan River basin also showed the narrow (figs. 8(a) and 8(b)) and broad (fig. 8(c)) ranges in distribution. The spatial distribution of the collected species in narrow ranges were divided into the Bukhan (bottom part: fig. 8(a)) and Han (middle part: fig. 8(b)) River basins. The species occurring in the Bukan River basin (fig. 8(a)) matched the area of high altitude in the SOM visualization (fig. 5(d)), while the species grouped in the Han River basin (fig. 8(b)) corresponded to low altitude in the SOM (fig. 5(d)). In the narrow range species such as *Ecdyonurus bajkovae* and *Drunella cryptomeria* were observed in the Bukhan River basin (fig. 8a), while species such as *Cheumatopsyche brevilineata* were collected in the Han River basin (fig. 8(b)). Species including *Ephemera orientalis* and *Ecdyonurus levis* were observed in a broad area (fig. 8(c)).

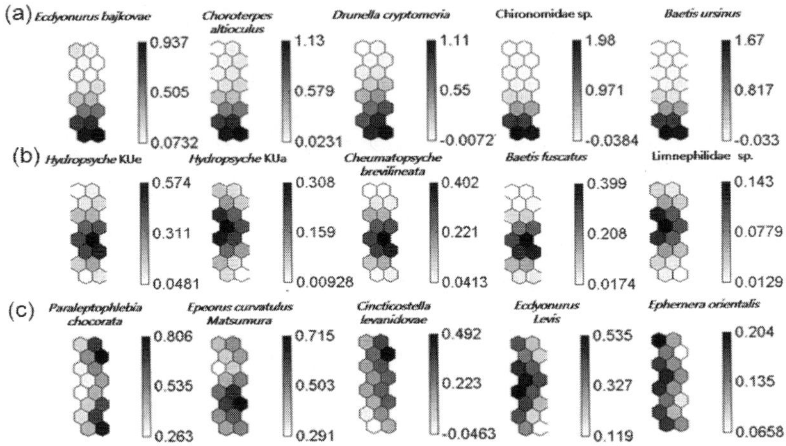

Figure 8: Species distribution in the Han and Bukhan River basins. Gray levels in vertical bars indicate density in no. of individuals per m^2. (a) Species narrowly occurring in high altitude zone, (b) Species narrowly occurring high altitude zone, and (c) Species occurring in a broad zone.

4 Discussion and conclusion

In this study, we demonstrated that benthic macroinvertebrates communities trained by the SOM efficiently reflected effects of topographic difference and temperature variation (figs. 2–5). The spatial difference was accordingly illustrated by the SOM. Temperature variation was further identifiable with latitude (fig. 3), longitude and latitude (fig. 4), and altitude (fig. 5) in different river systems.

Through visualization the longitudinal differences in community compositions was demonstrated within the river basin systems. The narrow and broad ranges regarding temperature variation were efficiently elucidated in figs. 6, 7 and 8. The spatial ranges in relation to temperature variation may be useful for estimating the level of thermal tolerance. Considering that temperature may continuously rise according to climate change, species occurring at the narrow range in the warm areas may disperse upward. For instance, *Ephemera strigata* was limitedly observed in D would be the candidate species for upward distribution along with increase in temperature. In contrast the species presented at high latitude in low temperature, such as *Baetis ursinus* in N may extinct in South Korea along with the rise in temperature in the long period. Considering there are other numerous biological and abiological factors in determining position of target species, however, it is too early to conclude that these species would either flourish or extinct in South Korea. At least these results, however, indicate that the species stated above deserve attention for their abundance in relation to the long-term temperature increase.

In the Bukhan River basin, altitude was in accordance with the community patterns, while latitude and longitude did not clearly match community clustering (fig. 5). The Bukhan River basin is located near to the mountainous area as stated above. The effect of altitude difference on community patterns is consequently understandable. The results supported that temperature variation is not only dependent upon latitude and longitude but also dependent upon altitude. In the future the interrelationships among latitude, longitude and altitude could be studied in revealing temperature effects on communities in an integrative manner.

Seasonality may be another source of variation, especially in a temperature zone including Korea. In this study, however, the survey was fixed in the season covering late spring and summer. Seasonal differences could be further checked in the long term data. As is well known, pollution also plays an important role in determining community compositions [9]. The unpolluted sample sites were selected to reveal the natural effect of temperature variation by selecting the samples with the BMWP values over 45 in this study. In the future the effect of temperature on polluted communities needs to be further studied.

In conclusion, the SOM was efficient in patterning and visualizing community compositions and accordingly clustered benthic macroinvertebrates corresponding to the topography in a larger scale and to temperature variation in a smaller scale. The species distribution in relation to temperature variation was also identifiable with the SOM. Revealing the scope of spatial distribution would be important in characterizing community compositions and species adaptability to temperature variation. The SOM could serve as an efficient ecological map to illustrate community distribution in response to environmental stressors.

Acknowledgement

This study was carried out with the support of "Cooperative Research Program for Agricultural Science & Technology Development (project No.20080454000)", RDA, Republic of Korea.

References

[1] Karl, T. R. and Trenberth K. E., Modern global climate change, *Science*, 302, pp. 1719-1723, 2003.

[2] Root, T. L., Price, J. T., Hall, K. R., Schneider, S. H., Rosenzweig, C. and Pounds, J. A., Fingerprints of global warming on wild animals and plants, *Nature*, 421, pp. 57-60, 2003.

[3] Coope, G. R., *Insect faunas in ice age environments: why so little extinction? In: Extinction Rates* (eds Lawton, J. H., May, R. M.), pp. 55-74, Oxford University Press, New York, 1995.

[4] Pounds, J. A., Bustamante, M. R., Coloma, L. A., Consuegra, J. A., Fogden, M. P. L., Foster, P. N., Marca, E. L., Masters, K. L., Merino-Viteri, A., Puschendorf, R., Ron, S. R., Sánchez, G. A., Still, C, J. and Young, B. E., Widespread amphibian extinctions from epidemic disease driven by global warming, *Nature*, 439, pp. 161-167, 2006.

[5] Hickling, R., Roy, D. B., Hill, J. K., Fox, R. and Thomas, C. D., The distributions of a wide range of taxonomic groups are expanding polewards, *Global Change Biology*, 12, pp. 1-6, 2006.

[6] Cao, Y., Bark, A. W. and Williams, W. P., Analysing benthic macroinvertebrate community changes along a pollution gradient: A framework for the development of biotic indices, *Water Research*, 31, pp. 884-892, 1997.

[7] David, D. L., Jerry, L. W. and Daniel, G. P., The effect of trout farm effluent on the taxa richness of benthic, *Aquaculture*, 147, pp. 37-55, 1996.

[8] Leunda, P. M., Oscoza j., Mirandaa, R. and Ariñoa, A. H., Longitudinal and seasonal variation of the benthic macroinvertebrate community and biotic indices in an undisturbed Pyrenean river. *Ecological Indicators*, 9, pp. 52-63, 2009.

[9] Song, M. Y., Hwang H. J., Kwak, I. S., Ji C. W., Oh, Y. N., Youn, B. J. and T. S. Chon, Self-organization mapping of benthic macroinvertebrate communities implemented to community assessment and water quality evaluation, *Ecological modeling*, 203, pp. 18-22, 2007.

[10] Park, Y. S., Verdonschot, P., Chon, T. S. and Lek, S., Patterning and predicting aquatic macroinvertebrate diversities using artificial neural network, *Water Research*, 37, pp. 1749-1758, 2003.

[11] Ministry of Environment, The national natural environment monitoring protocol, Ministry of Environment, Seoul, 1997.

[12] Park, Y. S., Song, M. Y., Park, Y. C., Oh, K. H., Cho, E. and Chon, T. S., Community patterns of benthic macroinvertebrates collected on the national scale in Korea, *Ecological Modeling*, 203, pp. 26-33, 2007.

[13] Walley, W. J. and Hawkes, H. A., A computer-based development of the Biological Monitoring Working Party score system incorporating abundance rating, biotope type and indicator value, *Water Research*, 31(2), pp. 201-210, 1997.

[14] Kohonen, T., Self-organized formation of topologically correct feature maps, *Biological Cybernetics*, 43, pp. 59-69, 1982.

[15] Zurada, J., *Introduction to artificial neural systems*, ISBN, 1992.
[16] Chon, T. S., Park, Y. S., Moon, K. H. and Cha, E. Y., Patterning communities by using an artificial neural network, *Ecological Modelling*, 90, pp. 69-78, 1996.
[17] Ward, J. H., Hierarchical grouping to optimized an objective function, *Journal of the American Statistical Association*, 58, pp. 236–244, 1963.
[18] The Mathworks Inc., Matlab version 6.1, Massachusetts, 2001.

Engineered ecosystem development for Agro-Process Intensification

G. Akay[1,2,3] & S. Fleming[1]
[1]*Process Intensification and Miniaturization (PIM) Centre, School of Chemical Engineering and Advanced Materials, Newcastle University, Newcastle upon Tyne, UK*
[2]*The Institute of StemCell Biology and Regenerative Medicine, Newcastle University, Newcastle upon Tyne, UK*
[3]*Institute for Nanoscale Science and Technology, Newcastle University, Newcastle upon Tyne, UK*

Abstract

The principles of Process Intensification and Miniaturization technology, originally developed for chemicals processing, biotechnology and tissue engineering were applied to achieve plant growth and crop yield enhancement, which can be described as AgroProcess Intensification (A-PI). The basic principle of A-PI is the enhancement of multiple interactions between plant roots, water, nutrients, bacteria using micro-(bio)reactors as synthetic rhizosphere in soil (SRS) which is a highly porous nano-structured hydrophilic ionic macro-porous polymer. If soil fertility is not limited by water, nutrients or bacteria then these soil additives do not have any function, although they can still be used as root delivery system for plant protection. In this study, we used nutrient and bacterium depleted soil with or without water stress to grow soybean in order to demonstrate the effect of the SRS in enhancing biomass growth under stress. Through an extensive scanning electron microscopy studies, it is shown that AgroProcess Intensification is achieved through the association of the plant roots with SRS which retains both soil water and nutrients and transfer them to the plant while protecting the nitrogen fixing bacterium. Implications of this method are discussed in terms of engineering of ecosystem to grow crops and biomass in substandard soil under water and fertilizer stress.

Keywords: AgroProcess Intensification, agriculture, fertilizer stress, nitrogen fixation, soil additives, symbiosis, synthetic rhizosphere, water stress.

WIT Transactions on Ecology and the Environment, Vol 144,
www.witpress.com, ISSN 1743-3541 (on-line)
doi:10.2495/ECO110421

1 Introduction

The objective of this study is to investigate the use of highly hydrophilic, ionic, elastic nano-structured macro-porous polymers prepared by high internal phase emulsion polymerisation and subsequently modified/functionalised and utilised as soil additives in order to enhance the growth of plants under fertilizer and water stress. The basic polymers are generically known as PolyHIPE Polymers (PHPs) which are modified and used as Synthetic rhizosphere (SRS) media. If used on a large scale in nutrient deficient soil under water stress, these materials can achieve:

1. Water regulation/conservation; 2. Fertilizer slow release/conservation; 3. Delivery of plant growth regulators; 4. Biological control of pathogens; 5. Reforestation; 6. Root infection for nitrogen fixation; and 7. Mixed bacterial inoculants for enhanced crop yield/nitrogen fixation.

The biodegradability of these materials can be controlled. If left within the soil for several years, they will become part of the ecosystem. Although their concentration in soil is low, due to their very low density and very high water adsorption capacity, they substantially enhance the soil quality while supporting the beneficial bacteria which are included initially in them. The use of bioactive SRS-media as soil additives is particularly important in nitrogen fixing plants.

1.1 Processing in ecosystem

Nature's processing in photosynthesis through plants involves two domains; plant leaves and plant roots. Although the synthesis involving leaves and roots can also be deemed highly efficient, the accessibility to conversion energy (solar energy) by the leaves and to nutrients, water and catalysts (soil microorganisms and plant exudates) are on the other hand highly inefficient since the mass transfer area density TAD (surface area per unit volume) in both cases is very small and in the case of plant root system, to some extent, it is also random. This randomness emanates from the fact that soil is the carrier medium for water, nutrients and microorganisms. Although there are inherent mechanisms to promote root interactions with nutrients/water/microorganisms, these mechanisms are not efficient due antagonistic interactions between water and nutrients / microorganisms; as uncontrolled water supply results in the depletion of nutrients and microorganisms from the rhizosphere where the many-body interactions take place. Nevertheless, water is also essential for nutrient and bacterium delivery to plant roots. But this process is far from efficient. One of the outcomes of the antagonistic effects is the washing of fertilizers from the soil; it is estimated that 50-70% of the fertilizers are not utilised by the plant [1]. Furthermore, due to the presence of other microorganisms, even when single useful bacterium cultures are reintroduced into soil at very high population levels, it is immediately subjected to competitive and antagonistic effects from the indigenous microflora which results in population decline [2]. Finally, water holding capacity of soil is not large and water is distributed over a wide volume in soil.

1.2 Need for engineered ecosystem

Technology led increases in crop yield achieved during 1970s and 1980s has not been replicated since then and the rate of enhancement has been falling [3]. Increasing demand for food as well as energy crops poses huge challenges for the sustainability of terrestrial ecosystems. Current trends in agriculture add to these concerns. They include, diminishing returns of fertilizer application, water stress and rapid approach to climate-adjusted genetic yield potential ceiling which indicates that there is a lack of large exploitable 'yield gap' in agriculture [4].

In order to combat these challenges, several strategies are currently under development [5]. We have recently shown [6–8] how plant growth and crop yield enhancement can be achieved through simultaneous water regulation, fertilizer release and symbiosis; thus addressing most of the concerns cited above. Here, targeting of water, biological nitrogen fixation and delivery of fertilizer (including nitrogen) is achieved by associating plant roots with a synthetic symbiotic media. This technique differs fundamentally from other biomass/crop enhancement techniques using soil conditioners or slow release fertilizers or inter-cropping or biological nitrogen fixation because, in the present technique, the symbiotic-media is associated with the plant roots. Hence the delivery of the nutrients, water and bacterial activity, including root infection (leading to nodule formation for biological nitrogen fixation by bacteria in legumes) are regulated by the synthetic rhizosphere which is populated by root hairs.

1.3 A model for AgroProcess Intensification

In order to achieve selective root/water/nutrient/microorganism many-body interactions without any antagonistic effects while protecting useful soil bacteria, conserving water and nutrients, we use a symbiotic-media as soil additives. These materials act as micro-bioreactors for the efficient delivery of water and nutrients to the plant roots and preserve the activity of bacteria by creating a synthetic rhizosphere, essentially modifying the ecosystem. However, in order to ensure that the plant roots selectively seek and associate with SRS, several restrictions are placed on the type of material used as SRS; they include high water adsorption rate and capacity, elasticity, pore size, nano-porosity of the pore walls, ability to act as an efficient support for bacteria so as to utilise the 'confinement phenomenon' to achieve BioProcess Intensification [9, 10]. AgroProcess Intensification through the application of process intensification and miniaturization principles is only possible if the root growth selectively targets the symbiotic media and root hairs and primary roots extensively form a network through it. The observed extensive root association clearly indicates that the roots are directed towards the source of water and nutrients. It is this property of the plant roots that allows the extensive root and symbiotic-media association and makes the process selective rather than random.

The many body interactions within SRS results in process intensification through the same mechanism as that observed in process intensification and miniaturization [11] in which the mass transfer is enhanced by the reduction in diffusion path, accessibility of catalytic sites and enhancement of TAD. In addition, 'side reactions' are inhibited, catalyst concentration/activity is maintained through the protection of useful bacteria and high concentration of root exudates is present in SRS. Roots growing out of the SRS are already infected with bacteria and needs no further nodulation.

2 Experimental

2.1 Preparation of synthetic rhizosphere media polymers

All chemicals needed for the preperation of PolyHIPE Polymers (PHP) were purchased from Sigma-Aldrich. They include styrene, divinyl benzene, sorbitan monooleate (Span 80), potassium persulphate and concentrated sulphuric acid (98%). The polymer was prepared using the following method [12] which is conducted in three stages; emulsification, polymerisation and functionalization (in this case, sulphonation was used to render the polymer hydrophilic and elastic). The aqueous phase composition in the emulsification stage consisted of 5 wt% concentrated sulphuric acid, 94 wt% deionised water and 1 wt% potassium persulphate. The oil phase was made up of 76 wt% styrene, 14 wt% sorbitan monooleate (Span 80; surfactant) and 10 wt% divinyl benzene (DVB, used as cross-linking agent). The two phases were mixed in a 12 cm diameter mixing vessel equipped with a stirrer with two flat paddles set at right angles close to the bottom of the vessel. The total polymer volume of one batch was 250 ml of which 90% (225ml) was aqueous phase and 10% (25 ml) is oil phase. The oil phase was added to the mixer which was set at 300 rpm. The aqueous phase was added with a peristaltic pump at the rate of 45 ml/minute (i.e. dosing time of 5 minutes), followed by a mixing time of 1 minute. The emulsion was then drained from the mixer into 5 x 50 ml plastic tubes, capped, inverted and placed in an oven at 60°C for 8 hours for polymerisation to take place. The polymer was then cut into 4 mm discs then dried overnight in an oven at 60°C.

Sulphonation of PHP was achieved by soaking the discs in 98% sulphuric acid for 2 hours, then microwaving in a conventional 1 kW kitchen microwave oven for 30 seconds followed by turning over the samples and repeating the irradiation for another 30 seconds. Total irradiation time was 150 seconds [13]. The sulphonated discs were washed with deionised water for 30 minutes twice, followed by 60 minutes soaking in 2.5N ammonium hydroxide. Excess ammonium hydroxide was then removed by neutralising with acetic acid and washing with distilled water, bringing the pH down to 6.2 ± 0.8. These polymers measured ca. $5 \times 5 \times 5\ mm^3$.

2.2 Scanning Electron Microscopy (SEM)

Dry samples of polymer were prepared for examination with SEM by gold coating using a Polaron E1500 Sputter Coater. Polymers with roots growing through are first treated to fix the roots using 2% glutaraldehyde/phosphate buffer solution, then dehydrated in progressively more concentrated ethanol – 10%, 25%, 50%, 75%, 100%, 10 minutes per concentration. Samples were then critical-point dried with liquid CO_2, followed by gold coating the same as the dry polymer samples [12, 13]. Samples were then examined using a Cambridge S240 scanning electron microscope.

2.3 Soil preparation

The growth medium used was made up of 75% Wilkinsons (UK supplier) soil and 25% horticultural sand. The soil mixture was washed to remove some of the nutrients producing a product with lower fertility so that any effect the polymer was having on plant growth would be evident earlier. The soil was washed by adding 10 litre of water to 2.5 kg of dry soil. The washing was monitored by measuring the conductivity of the water over the washing period. The conductivity of tap water was 315 μS/cm. After 24 hours, with occasional stirring, the conductivity increased to 775 μS/cm (average of 3 samples). The water was drained and further 10 litres of water added. After a further 24 hours, the conductivity averaged 506 μS/cm. The conductivity due to soil electrolytes was therefore reduced from 460 μS/cm to 191 μS/cm.

Soil was then dried in trays and sterilised in an autoclave in 500g aliquots in autoclavable plastic bags. The nitrogen, phosphorus and potassium content of the soil were measured, in triplicate, before and after washing.

2.4 Planting of soybeans

Soybeans (*Glycine* max, variety *Pan*) from South Africa were used in these experiments. Plants were grown in the soil/sand mixture in a greenhouse in 13cm diameter pots. PolyHIPE Polymer was added to the soil at 0.5% by weight [6–8]. These pots hold 500g of soil; therefore 2.5g of PHP was added to each pot. Polymer was only added to the soil below the level of the seed.

Four treatments were used in this experiment: (1) Soybeans under normal watering; (2) Soybeans under reduced watering; (3) Soybeans with PHP under normal watering and (4) Soybeans with PHP under reduced watering. Normal watered plants were watered twice a week with 100ml water. Reduced watered plants were watered twice a week with 50ml water. Each treatment was replicated 6 times. Plants were harvested after 15 weeks and the length, fresh weight and dry weight of the shoots and the fresh weight and dry weight of the roots were recorded.

3 Results and discussion

3.1 Soil characteristics

The results of the soil analysis before and after washing are shown in Table 1. After washing, nitrogen content was reduced by 36.4%, phosphorus by 32.2% and potassium by 46.5%. There was no bacterium in the soil due to sterilization.

Table 1: Soil nutrient content before and after washing.

Nutrient type	Before washing	After washing
Nitrogen	0.22wt%	0.14wt%
Phosphorus	8.66 μg/ml	5.87 μg/ml
Potassium	203 μg/ml	108.5 μg/ml

3.2 Biomass growth

Growth of biomass in terms of soybean height, shoot and root weight (fresh or dry) under normal and reduced (droughted) watering are shown in Figures 1–3. As seen in these figures, the addition of SRS-polymer significantly increases the shoot length and weight of both shoots and roots compared to the control plants.

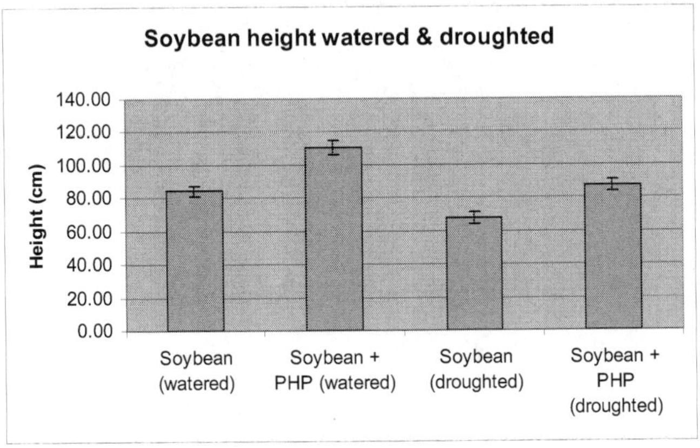

Figure 1: Average soybean height under normal and reduced (droughted) watering conditions.

Figure 1 shows that the addition of PHP has had a significant effect on the shoot length both of normally watered plants and droughted plants. Watered plants with SRS-polymer were 30.8% longer than control plants. Although droughted plants were 20.2% shorter than the normally watered control plants, the addition of SRS-polymer increased the length of droughted plants by 29.7% compared to the droughted plants with no SRS-polymer.

Figure 2 shows that compared to the control group (no SRS-polymer) the following enhancements in biomass growth were obtained when SRS-polymer was used: Normally watered shoot fresh weight increased by 84.2%. Droughted shoot fresh weight increased by 71.8%. Normally watered root fresh weight increased by 87.5% compared to control. Droughted root weight with SRS-polymer increased by 26.9%.

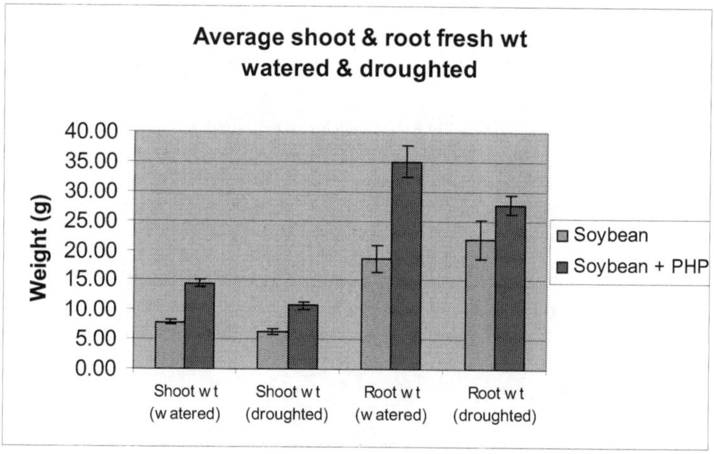

Figure 2: Average soybean shoot and root fresh weight under normal and reduced (droughted) watering conditions.

Figure 3 shows that compared to the control group (no SRS-polymer), the following enhancements in biomass growth were obtained when SRS-polymer was used: Normally watered shoot dry weight increased by 111.9%. Droughted shoot dry weight increased by 106.4%. Watered root dry weight increased by 50.8%. Droughted root dry weight increased by 4.4% compared to control.

The trends in the enhancements in biomass yield both in fresh and more significantly in dry weight are similar and consistent with our earlier results [6–8]. It is interesting to note that under drought conditions, although dry shoot weight increases by 106.4%, the increase in dry root weight is only 4.4%, indicating that the available roots are more efficient in the presence of SRS-polymer in sustaining the growth of biomass compared with the control (no SRS-polymer).

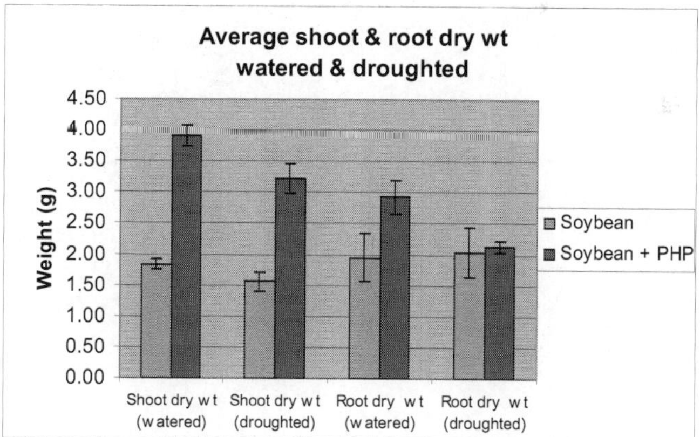

Figure 3: Average soybean shoot and root dry weight under normal and reduced (droughted) watering conditions.

3.3 Root association with SRS-polymer

When rigid sulphonated PHP is used as soil additives, the observed enhancement in biomass yield is not as significant as with elastic PHP which also has large pore and interconnect hole size. Root association with SRS-polymers is especially important when biologically active SRS-polymers are used for root infection by nitrogen fixing bacteria [6–8].

Figure 4 shows the soybean root association with SRS-polymer. The internal structure of the SRS-polymer is shown in Figures 5 and 6. Figure 5 illustrates the sulphonated PHP micro-structure used in the experiments while Figure 6 illustrates the root association.

Figure 4: Roots of soybean and SRS-polymer after a growth of 15 weeks.

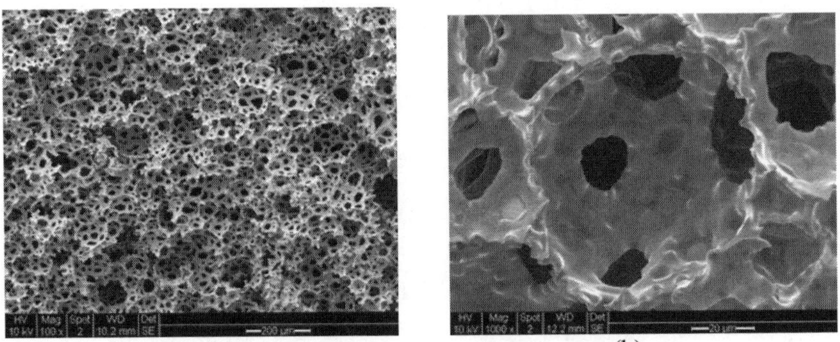

(a) (b)

Figure 5: Micro-structure of sulphonated PolyHIPE Polymer before it is used as SRS-media. (a) Magnification 100x (Scale bar 200 μm); (b) Magnification 1000x (Scale bar: 20 μm).

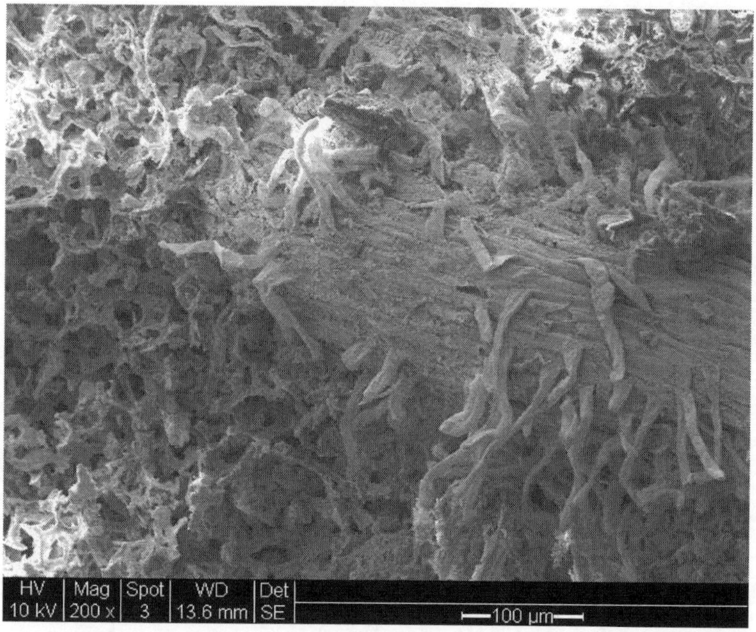

Figure 6: Fracture surface of SRS-media polymer showing soybean root growing through it with root hairs penetrating into the SRS-media.

It is clear from Figures 5 and 6 that although the pore and interconnect size of the SRS-media polymer is small compared with the size of soybean root, root penetration takes place and root hairs penetrate into the media. Under these conditions water and fertilizer transport to the root system is enhanced since diffusion path associated with this transport is now microns, rather than centimetres. As neither SRS-media polymers or soil contain any bacteria, the plant growth is solely dependent on the soil nitrogen and we have not observed any root exudates presence in the SRS-media. Root exudates are only present when SRS-media has supported bacteria.

4 Conclusions

Sulphonated elastic PolyHIPE Polymers were used as Synthetic Rhizosphere media in order to enhance biomass yield in soybean under fertiliser and water deficiency in the absence of any nitrogen fixation from air. Fertiliser levels falls during plant growth as a result of wash-out and plant metabolism. In our previous studies (for example grass growth), the effect of the SRS-media started showing after second or third harvest under normal watering. As nutrients were already depleted, the beneficial effects of the SRS-media became clear immediately even with normal watering. Under water stress, the observed effect is similar to those observed in our previous studies [6–8]. We have also shown that, under water stress, although there is substantial enhancement in biomass yield, increase in root weight in the presence of polymer is not significant in the presence of SRS-media, indicating that the available root mass is sufficient to sustain enhanced biomass growth.

Acknowledgement

We are grateful to the Engineering and Physical Sciences Research Council (EPSRC) for supporting this research and providing PhD studentship for Steve Fleming.

References

[1] Abraham, J., and Pillai, V. N. R., Membrane-encapsulated controlled-release urea fertilizers based on acrylamide copolymers, J. Appl. Polymer Sci., 60, 2347-2351 (1996).

[2] Bashan, Y., Inoculants of plant growth-promoting bacteria for use in agriculture. Biotechnol. Adv., 16, 729-770 (1998).

[3] Conway, G. and Toenniessen. G., Feeding the world in the twenty-first century. Nature, C55-C58 (1999).

[4] Tilma, D., Cassman, K. G., Matson, P. A., Naylor, R. and Polasky, S., Agricultural sustainability and intensive production practices. Nature 418, 671-677 (2002).

[5] Ashraf, M., Ozturk, M., and Athar, H. R., Salinity and water stress: Improving crop efficiency Springer, Heidelberg (2009).

[6] Akay, G., Burke, R.D., Synthetic symbiosis system as soil additives to deliver active ingredients through plant roots for enhanced plant and crop yield. International Patent Publication Number WO/2010/040996 (2010).

[7] Burke, D. R., Akay, G. and Bilsborrow, P., Development of novel polymeric materials for AgroProcess Intensification. J. Appl. Polymer Sci. 118, 3292-3299 (2010).

[8] Akay, G. and Burke, R. D., Synthetic symbiosis system as soil additives to deliver active ingredients through plant roots for enhanced plant and crop yield. Plant and Soil (submitted for publication).

[9] Akay, G., Erhan, E., Keskinler, B., Bioprocess intensification in flow through micro-reactors with immobilized bacteria, Bioengineering Biotechnology, 90, 180-190 (2005).

[10] Akay, G., Microporous polymers. International Patent Publication Number WO 2004/005355, (2004).

[11] Akay, G., Process intensification and miniaturization. p. 183-199, In: S. Lee, ed. Encyclopedia of chemicals processing, Vol. 1. Taylor and Francis, New York, (2006).

[12] Akay, G., Bokhari, M. A., Byron, V. J. and Dogru, M., Development of nano-structured materials and their application in bioprocess-chemical process intensification and tissue engineering. In: Chemical Engineering

Trends and Developments, Ed: M.A. Galan and E. M. Del Valle, Wiley, London. pp. 171-196 (2005).

[13] Akay, G., Noor, Z. Z., Calkan, O. F., Ndlovu, T. M., and Burke, D. B., Process for preparing a functionalised polyHIPE polymer, US Patent 07820729 (2010).

Ecotoxicity of anionic surfactants AKYPO®

E. Jurado, M. Fernández-Serrano, J. Núñez-Olea,
M. Lechuga & F. Ríos
*Department of Chemical Engineering, Faculty of Sciences,
University of Granada, Spain*

Abstract

In this paper, the toxicity values of ether carboxylic derivatives surfactants with commercial name AKYPO®, and the anionic surfactant linear alkylbenzene sulfonate (LAS), have been determined by applying the 24-h immobilization test with *Daphnia magna* (freshwater crustacea), the LumiStox® 300 test which employs the luminescent bacterium *Photobacterium phosphoreum* of the strain *Vibrio fischeri,* and the 72-h algal growth-inhibition test with the microalgae *Selenastrum capricornutum*, using culture-growth inhibition as the effect criterion. Three AKYPO®, with different alkyl chain and degree of ethoxylation, and LAS have been tested. For all tests, the results indicated that *Vibrio fischeri* is more sensitive to toxic effects from AKYPO® and LAS than *Daphnia magna* or the microalgae. The results demonstrate that the toxicity is lower for the AKYPO® with the shortest alkyl chain. The anionic surfactant LAS presents intermediate toxicity values.

Keywords: anionic surfactants, ecotoxicity, AKYPO®, Daphnia magna, microalgae, Vibrio fischeri.

1 Introduction

The enormous worldwide use of surfactants in detergent and cosmetic formulations, which are generally dumped into water systems, requires them to be as inocuous as possible for the environment: low toxicity and easily biodegraded ones. The aspect of environmental impact of chemicals is mainly governed by their ecotoxicity which is relatively high in the case of surfactants because of their surface activity and the resulting action against biological membranes [1]. This increasing worry impels the development of new

www.witpress.com, ISSN 1743-3541 (on-line)
doi:10.2495/ECO110431

surfactants. The amphoteric character of anionic surfactants facilitate their accumulation in living organisms. Anionic surfactants mainly show eye and skin irritation potentials. Because of the high number of surfactants in contact with humans, many in vitro methods have been developed for the prediction of the eye irritation potential of surfactants [2].

The anionic surfactants tested in this research, AKYPO® series from Kao Corporation S.A., improve the foam quality of the detergent reducing irritation level, therefore they are used as co-surfactants in detergents which have to be in contact with the skin. AKYPO® are ether carboxylic derivatives and are commercialised in concentrated acid form. Figure 1 shows the chemical structure of AKYPO®.

$$\mathbf{R\text{-}O(CH_2\text{-}CH_2O)_n\text{-}CH_2\text{-}COO^{-}X}$$

Figure 1: Chemical structure of AKYPO®.

The objective of this study is to compare the ecotoxicity of the anionic tensioactive linear alkylbenzene sulfonate (LAS) with these novel surfactants.

Many types of bioassays are available to establish the toxicity levels of compounds for aquatic organisms, but many of these tests are also time-consuming and not routinely applicable. Moreover, the use of higher organisms as test species may also be ethically undesirable. There is a need to replace acute toxicity tests on fish with more effective assays. Although several bioassays using microorganisms have been described, most of the bacterial screening tests have been based on luminiscence measurements, because in this way they are rapid, reproducible, and simple to use, they cause no ethical problems, and they are cost-effective [3]. The characteristics of speed, reliability, and normalization of the toxicity results by bioassays with luminescent bacteria make them ideal for gathering data on toxicity, which can be compared and statistically studied for establishing correlations between toxicity as well as the chemical structure and/or different properties of the compounds assayed. Assays using luminescent bacteria are gaining wide acceptance for the quick and simple determination of the toxicity of chemical compounds in surface water and wastewater as well as in extracts of solid matrices. This explains the fact that, together with the *Daphnia* assay, these are listed as an approved bioassay to characterize toxic and dangerous wastes. *Daphnia magna* has proved to be a sensitive and simple laboratory model for predictive toxicity studies [4].

In a general way, the use of toxicity data has been extended to invertebrates and fish, considering them more important and representative than the toxicity assays with primary producers, as the former are more sensitive to toxins. However, some studies [5] have shown that in some cases, plants are far more sensitive, for example in assays with metals, industrial effluents, pesticides, and cationic surfactants.

For continued advancement in the search for relationships between toxicity and structural parameters in the field of surfactants, in the present work the ecotoxicity assay with luminescent bacteria, *Daphnia magna,* and microalgae is applied to different ether carboxylic derivatives surfactants. Surfactant toxicity has been measured by the LumiStox® assay according to the UNE-EN ISO 11348-2 guideline [6], using luminescent bacteria of the strain *Vibrio fischeri* as test microorganisms and by applying the 24-h inmobilization test with *Daphnia magna* according to the UNE-EN ISO 6341 guideline [7]. Also, the 72-h algal growth-inhibition test was performed with the microalga *Selenastrum capricornutum.*

The results indicate that *Vibrio fischeri* is more sensitive to toxic effects originating from AKYPO® and LAS than is *Daphnia magna* and microalgae.

2 Materials and methods

2.1 Surfactants

The surfactants used in this study are the commercial ether carboxylic derivatives AKYPO® LF2, AKYPO® RLM-25 and AKYPO® RLM-100 supplied by Kao Corporation S.A. (Tokyo, Japan). Table 1 shows the degree of ethoxylation (n) and the alkyl chain length (R) of the surfactants. The surfactant LAS is also supplied by Kao Corporation S.A. (Tokyo, Japan). The rest of the reagents used were PA quality and supplied by Panreac.

Table 1: Description of the surfactants employed in the tests.

Commercial Name	Structure
AKYPO® LF 2	R: 8 n: 8
AKYPO® RLM-25	R: 12-14 n: 3
AKYPO® RLM-100	R: 12-14 n: 4.5

2.2 Acute toxicity tests

Three toxicity tests were undertaken: the LumiStox® 300 test which employs the luminescent bacterium *Photobacterium phosphoreum*, the 24-h immobilization test with *Daphnia magna* (freshwater crustacea), and the 72-h algal growth inhibition test with *Selenastrum capricornutum.*

In the first one, measurements were taken with the measuring system LumiStox® 300, which consists of an instrument for measuring bioluminescence and an incubation unit according to the UNE-EN ISO 11348-2 guideline [6]. The toxicity measurement is based on the luminous intensity of the marine bacteria of the strain *Vibrio fisheri* NRRL-B-11177 after a certain exposure time to a toxic substance. The luminescent bacteria, dehydrated and frozen at −18ºC, were reactivated with the suspension supplied by Dr. Lange. The assay conditions were pH 7.0, ClNa concentration of 2%, all the measurements duplicated for incubation times of 15 and 30 min. When necessary, the sample was filtered prior to the assay.

The toxicity value was measured as EC_{50} or EC_{20}, which are, respectively, the surfactant concentrations that inhibit 50 and 20% after 15 and 30 min of exposure.

Acute toxicity tests with *Daphnia magna* were performed in Standard Reference Water (SRW) according to the UNE-EN ISO 6341 guideline [7]. The tests were performed in 100 mL polystyrene vessels, with 50 mL of SRW in each one. 20 neonates (<24 h) were transferred to vessels containing different concentrations of the test chemical, and the vessels were closed with a polyethylene cap. The neonates were separated from adults every day. There was no feeding and no aeration during the tests and the tests were run at 20±1ºC. Immobilization was determined visually after 24 h. For each surfactant, controls and at least five concentrations were used for the determination of the mobility inhibition of 50% of *Daphnia* population (IC_{50}).

The 72-h algal growth-inhibition test with the microalga *Selenastrum capricornutum* was administered according to the OECD 201 guideline [8]. The procedure consists of filling culture vials with appropriate volumes of nutrient medium and solutions of the surfactant being tested. At the beginning of the test, an inoculum of algae was added to the vials to be tested and to the vials of control, and were kept under stable and predetermined incubation conditions. Inocula were cultivated at 25±1ºC and constant uniform illumination (8000 lux). After 24, 48, and 72 h the algal density was determined to establish whether growth had been inhibited or stimulated with respect to control. Cell density was estimated by the optical density of the culture at 670nm.

For all the tests, the surfactant concentration and one control were performed in triplicate for each organism tested.

3 Results and discussion

The toxicity was determined for different AKYPO®. The initial concentrations of the surfactant were between 100 and 500 mg/L, depending on the surfactant assayed. For LumiStox® system, the initial values of luminous intensity measured were corrected by a factor that takes into account the natural decrease in luminous intensity, even in the absence of the toxic sample:

$$fk = \frac{I_t(0)}{I_0(0)} \quad (1)$$

with $I_0(0)$ and $I_t(0)$ being the readings of luminous intensity in the well containing concentration 0 at time 0 and t.

The percentage of inhibition (inhibitory effect) was calculated by the expression:

$$H_t = \frac{(I_{0t}(c) - I_t(c))}{I_{0t}(c)} 100 \tag{2}$$

where

$$I_{0t}(c) = \overline{fk}\, I_0(c) \tag{3}$$

with $\overline{fk}$ being the average correction factor of the control samples, $I_0(c)$ and $I_t(c)$ being readings of light intensity in the well containing concentration c at time 0 and t.

The Gamma function, the ratio between the light intensity lost by the bacterial solution and that remaining after exposure to the toxic sample, can be determined by the equation:

$$\Gamma_t = \frac{H_t}{100 - H_t} = \frac{\overline{fk} \cdot I_0(c) - I_t(c)}{I_t(c)} \tag{4}$$

From the results, a linear relationship can be deduced between the function Γ and the concentration of the surfactant used, in the following form:

$$\log(c) = b \cdot \log(\Gamma) + \log(a) \tag{5}$$

Figure 2 provides an example of the linearization for AKYPO® LF2. The values of EC_{20} and EC_{50}, expressed as mg/L, are the concentrations of surfactant

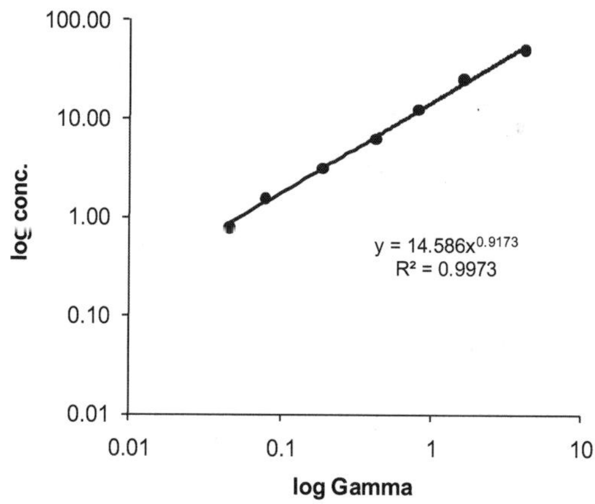

Figure 2: Linear relationship between the function Γ and concentration according to Eq. (5).

that inhibit 20 and 50%, and are calculated, giving Γ values of 0.25 and 1, respectively. Table 2 shows the results for the different surfactants, in decreasing order of toxicity, for incubation times of 15 and 30 min.

Table 2: Acute toxicity data for AKYPO® and LAS for the tests with *Vibrio fischeri* (values of EC_{50} and EC_{20} in mg/L).

Surfactant	**EC_{20} (15 min)**	**EC_{50} (15 min)**	**EC_{20} (30 min)**	**EC_{50} (30 min)**
AKYPO® RLM-25	0.42	3.58	1.90	4.74
AKYPO® RLM-100	4.51	14.18	4.39	15.19
LAS	9.41	27.58	8.29	26.50
AKYPO® LF2	22.93	134.59	36.44	181.65

IC_{50} values for the tests with *Daphnia magna* were calculated using linear-regression analysis after conversion of dose-response curves by logarithmic transformation of the concentrations (Figure 3). Table 3 shows (in decreasing order of toxicity) the IC_{50} values for the tests with *Daphnia magna,* for the different surfactants assayed.

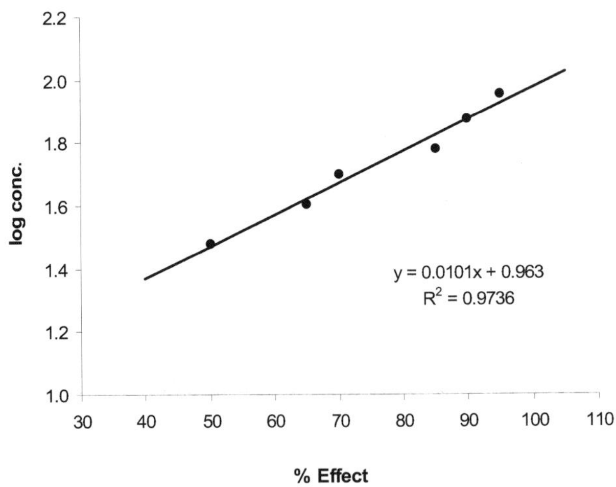

Figure 3: Linear relationship between the % effect and concentration for the tests with *Daphnia magna.*

Table 3: Acute toxicity data for AKYPO® and LAS for the tests with *Daphnia magna* (values of IC_{50} in mg/L).

Surfactant	IC_{50}
AKYPO® RLM-25	3.478
LAS	10.69
AKYPO® RLM-100	18.74
AKYPO® LF2	120.95

EC_{50} values for the tests with microalgae were calculated using linear regression analysis based on the dose-response curves (Figure 4). Table 4 shows (in decreasing order of toxicity) the EC_{50} values for the tests with microalgae, for the different surfactants tested.

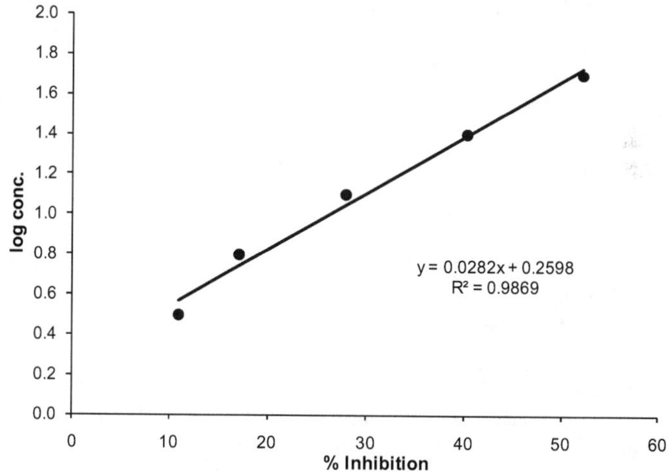

Figure 4: Linear relationship between the % inhibition and concentration for the tests with microalgae.

Table 4: Acute toxicity data for AKYPO® and LAS for the tests with microalgae (values of EC_{50} in mg/L).

Surfactant	EC_{50}
AKYPO® RLM-25	7.08
AKYPO® RLM-100	26.01
AKYPO® LF2	76.26
LAS	151.07

The results presented in Tables 2, 3, and 4, and in the graphs in Figure 5 show that *Vibrio fischeri* was more sensitive to toxic effects from AKYPO® and LAS than was *Daphnia magna* and microalgae, according to the results shown by García et al. [9] for non ionic surfactants alkylglucosides.

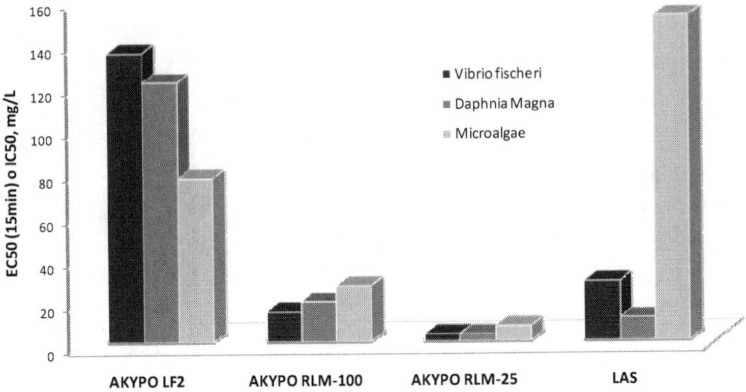

Figure 5: Determination of AKYPO® and LAS toxicity using *Daphnia magna, Vibrio fischeri*, and microalgae.

The results show that the toxicity is lower for the AKYPO® with the shortest alkyl chain. The degree of ethoxylation (n) has the reverse effect: the bigger degree of ethoxylation the smaller toxicity. The relation between LAS toxicity with AKYPO® toxicity depends on the test assayed. Using the bioluminescent bacteria *Vibrio fischeri* only the AKYPO® LF2 results less toxic than the LAS. With *Daphnia magna* AKYPO® LF2 and AKYPO® RLM100 are less toxic than the surfactant LAS. However, according to microalgae test, LAS presents the smallest toxicity. In fact, as may be seen in Figure 5, LAS is the surfactant with the more dispersal toxicity values.

Guilhermino et al. [10] indicate 0.22 mg/L as border value for *Daphnia magna* test. So, in view of the results (Table 3), all the surfactants tested present acceptable toxicity.

4 Conclusions

Toxicity values of three ether carboxylic derivatives anionic surfactants with commercial name AKYPO® and the well known anionic surfactant LAS have been determined using three methods: 24-h immobilization test with *Daphnia magna*, the LumiStox®300 test with the luminescent bacterium strain *Vibrio fischeri* and the 72-h algal growth-inhibition test with the microalgae *Selenastrum capricornutum*. The results obtained show that *Vibrio fischeri* is more sensitive to toxic effects from AKYPO® and LAS than *Daphnia magna* and microalgae. The influence of AKYPO® structure is in the sense that the toxicity is lower for the AKYPO with the shortest alkyl chain, and the bigger degree of ethoxylation the smaller toxicity. The relation between LAS toxicity with AKYPO® toxicity depends on the test assayed because LAS is the surfactant with the more dispersal toxicity values. However, according with Guilhermino et al. [10] and the results obtained for *Daphnia magna* test, all the surfactants tested present acceptable toxicity.

References

[1] Steber J., Guhl W., Steker N. and Schröder F. (1995) Alkyl poliglycosides-ecological evaluation of a new generation of nonionic surfactants. *Tenside Surfactants Detergents*, 32 (6), 515-521.

[2] Cserháti T., Forgács E., and Oros G. (2002) Biological activity and environmental impact of anionic surfactants. *Environment International,* 28, 337– 348.

[3] Farré M., García M., Tirapu L., Ginebreda A. and Barceló D. (2001). Wastewater toxicity screening of non-ionic surfactants by Toxalert® and Microtox® bioluminiscence inhibition assays. *Analytica Chimica Acta*, 427 (2), 181-189.

[4] Sandbacka M., Christianson I. and Isomaa B. (2000). The acute toxicity of surfactants on fish cells, *Daphnia magna* and fish - A comparative study. *Toxicology in Vitro*, 14, 61-68.

[5] Rand G. (1995). Aquatic Toxicology: Effects, Environmental fate and Risk Assessment. Gary M. Rand (Ed). 1083 p.

[6] UNE EN ISO 11348-2 (NORMA UNE EN ISO 11348-2). Determinación del efecto inhibidor de muestras de agua sobre la luminiscencia de *Vibrio fischeri* (Ensayo de bacterias luminiscentes).

[7] UNE-EN ISO 6341 (NORMA UNE EN ISO 6341). Determinación de la inhibición de la movilidad de Daphnia magna Straus (Cladocera, Crustacea) (Ensayo de toxicidad aguda).

[8] OECD Guideline for Testing of Chemicals. (1984). Alga, growth inhibition test.

[9] García M., Ribosa I., Campos E. and Sánchez-leal J. (1997). Ecological properties of alkylglucosides. *Chemosphere*, 35 (3), 545-556.

[10] Guilhermino L., Diamantino T., Silva M.C., and Soares A. M. V. M. (2000) Acute Toxicity Test with *Daphnia magna:* An Alternative to Mammals in the Prescreening of Chemical Toxicity? *Ecotoxicology and Environmental Safety Environmental Research* Section B, 46, 357-362.

Author Index

Sustainable Chemistry

Edited by: **G. RENIERS**, *University of Antwerp, Belgium and* **C.A. BREBBIA**, *Wessex Institute of Technology, UK*

As the demands of society increase in their quest for safer, better and more convenient products, the chemical industry is faced on the one hand with the possible depletion of natural resources and on the other with a large share of responsibility for the damage that may be caused to the environment and the population by their products.

The main area of chemical activity consists of the conversion of various base metals into substances and materials with new chemical and physical properties. A substantial amount of research is being carried out regarding this conversion process, which now must incorporate sustainability requirements. In addition, other associated activities, such as the chemical supply chain, performance measurements and different types of management tasks, must meet sustainability standards.

Achieving sustainable chemistry based on clean processing routes, efficient use of resources, renewable materials, adequate management systems and other activities is not only essential for the future of a competitive chemical industry but also for a sustainable and healthy society and environment. Key to sustainable chemistry are innovative and cleaner technologies and the development of appropriate business models, performance measurements, and better integrated management.

This book contains papers on the latest academic and industrial research in the field presented at the first international conference convened on the sustainable chemistry.

WIT Transactions on Ecology and the Environment, Vol 154
ISBN: 978-1-84564-558-8 eISBN: 978-1-84564-559-5
Forthcoming 2011 / apx 250pp / apx £95.00